Multidisciplinary Approaches to the Study of Stone Age Weaponry

Vertebrate Paleobiology and Paleoanthropology Series

Edited by

Eric Delson
Vertebrate Paleontology, American Museum of Natural History
New York, NY 10024, USA
delson@amnh.org

Eric J. Sargis
Anthropology, Yale University
New Haven, CT 06520, USA
eric.sargis@yale.edu

Focal topics for volumes in the series will include systematic paleontology of all vertebrates (from agnathans to humans), phylogeny reconstruction, functional morphology, Paleolithic archaeology, taphonomy, geochronology, historical biogeography, and biostratigraphy. Other fields (e.g., paleoclimatology, paleoecology, ancient DNA, total organismal community structure) may be considered if the volume theme emphasizes paleobiology (or archaeology). Fields such as modeling of physical processes, genetic methodology, nonvertebrates or neontology are out of our scope.

Volumes in the series may either be monographic treatments (including unpublished but fully revised dissertations) or edited collections, especially those focusing on problem-oriented issues, with multidisciplinary coverage where possible.

More information about this series at www.springer.com/series/6978

Multidisciplinary Approaches to the Study of Stone Age Weaponry

Edited by

Radu Iovita

MONREPOS Archaeological Research Centre, Museum for Human Behavioural Evolution, Römisch-Germanisches Zentralmuseum (RGZM), Leibniz-Forschungsinstitut für Archäologie, Neuwied, Germany

Katsuhiro Sano

The University Museum, The University of Tokyo, Tokyo, Japan

Springer

Editors
Radu Iovita
MONREPOS Archaeological Research Centre
Museum for Human Behavioural Evolution,
Römisch-Germanisches Zentralmuseum (RGZM),
Leibniz-Forschungsinstitut für Archäologie
Neuwied
Germany

Katsuhiro Sano
The University Museum
The University of Tokyo
Tokyo
Japan

ISSN 1877-9077 ISSN 1877-9085 (electronic)
Vertebrate Paleobiology and Paleoanthropology Series
ISBN 978-94-017-7601-1 ISBN 978-94-017-7602-8 (eBook)
DOI 10.1007/978-94-017-7602-8

Library of Congress Control Number: 2016934928

Cover Illustration: Photo: Marie Sjøvold

Printed on acid-free paper

This Springer imprint is published by Springer Nature
The registered company is Springer Science+Business Media B.V. Dordrecht

Foreword

Each new succeeding generation of researchers draws upon the accumulative knowledge and creative thinking of previous generations but it sometimes takes the concerted efforts of a group of talented researchers working together to completely reinvigorate a subject by introducing fresh approaches to ideas and redefining new directions in research. This impressive volume "Multidisciplinary approaches to the study of Stone Age weaponry" does precisely that and by bringing together a very wide field of specialists with complimentary interests and novel ideas, it presents a significant contribution to the advancement of this subject.

Much of the work presented in this volume is totally new and it is a great credit to the energy and efficiency of the editors, Radu Iovita and Katsuhiro Sano, that they have managed to gather together so many papers containing the ideas of leading specialists (one hesitates to use the term "cutting-edge") and to marshal together the results in so comprehensive a manner. The book is divided into four sections ("experimental applications", "archaeological applications", "measures of weapon performance" and "measures of weapon curation"), each of which is self-contained and can be read on its own, but equally and as a whole, reflects the multidisciplinarity and the impressive inter-weaving of new and updated approaches on the subject of Stone Age weaponry.

Many of the guiding principles underpinning today's ideas on Stone Age weaponry lie deeply rooted in the past scholarship of nineteenth century antiquarian archaeologists such as Sir John Evans. As a pioneer in these studies he experimented in the manufacture of artefacts out of stone, bone and horn (antler) including stone javelin and arrow heads. However, it is worth recalling that he too was aided in this by the observations of a number of co-researchers including Dr. Ferdinand Keller (who wrote about the size and use of arrowheads in the Swiss lake villages), Worthington G. Smith (who undertook some of the earliest refitting studies at Caddington), and F.C.J. Spurrell (who published on refitting artefacts from Stoneham's Pit, Crayford and also carried out early knapping experiments). Although many of these seminal studies were concerned with processes of manufacture and use, there does not seem to have been much time devoted to experimenting with ancient forms of weaponry. Evans himself fully acknowledged the importance of studying flint weapons, but rather took for granted that the size differences reflected different forms of projectile use without investigating their effectiveness much further. In the first edition of "Ancient Stone Implements, Weapons and Ornaments of Great Britain", published in 1872, he wrote "The variation in size probably arises from some of them having tipped spears to be held in the hand for close encounters, while others may have been attached to lighter shafts, and formed javelins to be thrown at objects at some distance; and the majority of the smaller kind were, beyond doubt, the heads of arrows discharged from bows".

Further background on the subsequent history and the development of studies into Stone Age weaponry is eloquently summarized by the editors in the preface. But as a minor aside and on a personal note, I would like to add a few words about my own introduction to the subject. When Christopher Bergman and I set up our projectile experiments in November 1981 (results presented in the following year's Prehistoric Society Spring Conference in London and published in 1982), our aim was to test the efficiency of microlithic points as arrowheads. There were few other similar experiments reported at that time but we drew inspiration from studies like John Whitthoft's 1968 paper on damaged Eskimo stone arrowpoints and Semenov's equally important book on prehistoric technology, which appeared in 1964. Christopher was also a member of a highly active group of flintknappers and researchers who were then engaged in similar experiments using bone and flint projectiles, based at the Institute of Archaeology in London. In retrospect, our experiment involving a dead roe deer dangling by its hind legs from a tree in a garden in North Oxford, where I was then living, now seems rather bizarre and unsophisticated. Indeed, the spectacle drew looks of horror and consternation from some of the co-habitants of my college accommodation and caused a brief sensation locally. Despite the wholly unnatural positioning of the animal, we tried to be as authentic as possible in other aspects of the experiment in using replicas of microliths and copying their hafting positions from waterlogged Mesolithic examples of arrowshafts with slotted flint points. The deer was shot repeatedly using a bow of known draw strength (40 lb at 26 inches) and itself a facsimile of the famous Holmegaard bow. Besides the well-known impact fractures that were published in 1982, the results taught us a series of valuable lessons and it is sometimes the unintended consequences of experiments that are most instructive but infrequently reported. At a common-sense level, the experiment revealed that although fletching of the arrows helps stabilize their flight, it made little difference to accuracy over relatively short distances and it could be argued that such additions to arrows were simply unnecessary if the prey were being hunted at close quarters, as is the practice among the Kalahari San for example. A more interesting observation was that many of the points passed cleanly through the animal without incurring obvious damage to the tip and it became clear to us that this must also have happened regularly in the past. In such instances, provided the shafts were undamaged, the arrows could simply be reused and it made us realize that unbroken archaeological specimens may have penetrated the soft tissue but not have struck bone. Our results therefore suggested a more holistic approach would be rewarded by including further research on hafting methodology, use-wear and residue analysis and it is gratifying to know that subsequent work along these lines continues to be followed up in this excellent volume.

So to all existing and future generations of experimenters and researchers, I would like to warmly commend this volume written by some of the finest practitioners amongst the current generation of scholars and who have demonstrated that the subject of Stone Age weaponry is still highly relevant and very much to the fore in studies of our human past.

Nick R.N.E. Barton
Institute of Archaeology
University of Oxford
Oxford, OX1 2PG, UK

References

Barton, R. N. E., & Bergman, C. A. (1982). Hunters at Hengistbury: some evidence from experimental archaeology. *World Archaeology, 14*, 237–248.

Bergman, C. A., & Newcomer, M. H. (1983). Flint arrowhead breakage: Examples from Ksar Akil, Lebanon. *Journal of Field Archaeology, 10*, 238–243.

Evans, J. (1872). *The ancient stone implements, weapons and ornaments of Great Britain*. London: Longmans, Green & Co.

Keller, F. (1866). *The lake dwellings of Switzerland and other parts of Europe* (1st ed.). Translated and arranged by John Edward Lee. London: Longmans, Green and Co.

Semenov, S. A. (1964). *Prehistoric technology: An experimental study of the oldest tools and artefacts from traces of manufacture and wear*. London: Cory, Adams, & Mackey.

Smith, W. G. (1894). *Man the primeval savage. His haunts and relics from the hill-tops of Bedfordshire to Blackwall*. London: Edward Stanford.

Spurrell, F. C. J. (1880). On the implements and chips from the floor of a Palaeolithic workshop. *Archaeological Journal, 37*, 294–299.

Spurrell, F. C. J. (1884). On some Palaeolithic knapping tools and modes of using them. *Journal of the Anthropological Institute, 13,* 109–118.

Whitthoft, J. (1968). Flint arrowpoints from the Eskimo of northwestern Alaska. *Expedition, 10*, 30–37.

Preface

Twenty Years Since Knecht: New Answers, New Questions and New Approaches to the Study of Stone Age Armatures

Although it goes beyond a simple "proceedings volume", this book is in part the result of a workshop with the same name organized by us in September 2011, at the University of Mainz, Germany. Both of us had previously dealt with the question of identifying weapons in the archaeological record using different methodologies (Sano 2009; Iovita 2011), yet we had both also begun new experimental projects aimed at solving some of what we had felt were methodological insufficiencies in existing protocols (Iovita et al. 2013, 2016; Sano and Oba 2015; Sano et al. 2016). It was in the context of the work for these projects that we began to realize how many researchers around the world were simultaneously working on weapon technologies and trying to find what role these might have played in shaping the course of human evolution. More importantly, we quickly found that many of these researchers were working independently, starting from different premises, but also having different background questions in mind, and representing different scientific traditions and schools. In short, weapons were all of a sudden globally fashionable, and that was and remains a good thing. Yet not everything about a new scientific trend is positive. We immediately realized that the sharp increase in interest carried with it the potential for duplication on the one hand, and for competition on the other. While competition and debate are healthy elements of any scientific enterprise, history has shown that that too much competition at the beginning of a scientific "trend" can lead to a stifling of creativity and an acrimonious atmosphere, a combination that could ultimately hinder real progress. It was precisely with the goal of avoiding these problems that Multidisciplinary Approaches to the Study of Stone Age Weapons—the workshop (Fig. 1), and, later, the book, were conceived.

The decision to elect Heidi Knecht's (1997) seminal work, *Projectile Technology*, as a model for our book was a conscious one. As we outline below, Knecht's volume represented the culmination of a decade and a half of intensive research (the 1980s and early 1990s) when several theoretical, methodological and, not least of all, empirical-archaeological lines of work came together to form a synthesis of the meaning of [projectile] weapons in human evolution. More than fifteen years later, we believe we are now moving towards another synthesis, albeit with very different drivers and actors, and it is this Zeitgeist that we hope to have captured.

In contrast to Knecht's volume, the research presented in this book is all archaeological, or at least archaeologically oriented. However, its multidisciplinarity stems from the multidisciplinarity of archaeology itself, which, as a historical science, draws from knowledge accumulated in other disciplines, including physics and materials science, cognitive and behavioural science, biology, and cultural anthropology. Much like Knecht's book, this volume unites archaeological perspectives from a variety of time periods and from all five continents with a large assortment of new analytical and experimental studies.

Fig. 1 Participants at the 2011 conference on Multidisciplinary Approaches to the Study of Stone Age Weapons. The names of speakers are indicated by the numbers in the silhouettes below

Because of the aforementioned explosion in research that has taken place in the last ten years and the diversity of the approaches, we have decided to structure this volume in five parts, whose unequal sizes we believe accurately reflect the current distribution of work in the field: *recognizing ancient Stone Age weapons* (with new experimental approaches and the application of their results to archaeological material in two separate parts), the *evaluation of the performance and efficacy of different weapon systems*, the *maintenance and curation* of Stone Age armatures, and, finally, some of the *behavioural and cognitive implications* of producing and utilizing different types of weapons.

In conclusion, we still believe that discussion and exchange at regular intervals is the key towards a healthy and productive debate which ultimately pushes breakthroughs, and therefore, we wanted to gather many different points of view in Mainz, and, now, in this volume. We are also hoping that, rather than being an authoritative work on the study of weapons in the Stone Age, our book will instead be the starting point for a longer conversation about the role of this unique human technology in the evolutionary history of our species.

We thank the participants in the Stone Age weaponry workshop for their fruitful discussion. Special thanks are due to all the contributors to this volume for their informative papers and patience. We are grateful to all the reviewers for their time and effort. Finally, we thank Eric Delson and Eric Sargis, the Editors of the Vertebrate Paleobiology and Paleoanthropology Series, for their support throughout the publication process.

Radu Iovita
Katsuhiro Sano

References

Iovita, R. (2011). Shape variation in Aterian tanged tools and the origins of projectile technology: a morphometric perspective on stone tool function. PLoS ONE, 6, e29029. doi: 10.1371/journal.pone.0029029

Iovita, R., Schönekeß, H., Gaudzinski-Windheuser, S., & Jäger, F. (2013). Projectile impact fractures and launching mechanisms: results of a controlled ballistic experiment using replica Levallois points. *Journal of Archaeological Science*, *48*, 73–83.

Iovita, R., Schönekeß, H., Gaudzinski-Windheuser, S., & Jäger, F. (2016). Identifying weapon delivery systems using macrofracture analysis and fracture propagation velocity: a controlled experiment. In R. Iovita & K. Sano (Eds.), *Multidisciplinary approaches to the study of Stone Age weaponry* (pp. 13–27). Dordrecht: Springer.

Knecht, H. (1997). *Projectile technology*. New York: Plenum Press.

Sano, K. (2009). Hunting evidence from stone artefacts from the Magdalenian cave site Bois Laiterie, Belgium: a fracture analysis. *Quartär*, *56*, 67–86.

Sano, K., Denda, Y., & Oba, M. (2016). Experiments in fracture patterns and impact velocity with replica hunting weapons from Japan. In R. Iovita & K. Sano (Eds.), *Multidisciplinary approaches to the study of Stone Age weaponry* (pp. 29–46). Dordrecht: Springer.

Sano, K., & Oba, M. (2015). Projectile experimentation for identifying hunting methods with replicas of Upper Palaeolithic weaponry from Japan. In J. Marreiros, N. Bicho & J. Gibaja Bao (Eds.), *International conference on use-wear analysis, Faro, Portugal. Use-wear 2012* (pp. 474–486). Newcastle upon Tyne: Cambridge Scholars Publishing.

Contents

Contributors

Note * indicates preferred address for correspondence

Nuno Bicho
ICArEHB. FCHS, Universidade do Algarve, Faro, Portugal

Marcelo Cardillo
Consejo Nacional de Investigaciones Científicas y Técnicas, Instituto Multidisciplinario de Historia y Ciencias Humanas, Universidad de Buenos Aires, Ciudad Autónoma de Buenos Aires, Argentina

João Cascalheira
ICArEHB. FCHS, Universidade do Algarve, Faro, Portugal

Pierre Cattelain
Cedarc/Musée du Malgré-Tout, GDRE Préhistos, CReA-Patrimoine, Université Libre de Bruxelles, ULB – Site de Treignes, Treignes, Belgium

Judith Charlin
Consejo Nacional de Investigaciones Científicas y Técnicas, Instituto Multidisciplinario de Historia y Ciencias Humanas, Universidad de Buenos Aires, Ciudad Autónoma de Buenos Aires, Argentina

Steven E. Churchill
*Department of Evolutionary Anthropology, Duke University, Durham, NC, USA
Evolutionary Studies Institute, University of the Witwatersrand, Johannesburg, South Africa

C. Clarkson
School of Social Science, The University of Queensland, Brisbane, QLD, Australia

Yoshitaka Denda
Miyagi Prefectural Board of Education, Sendai, Japan

Rob Dinnis
School of History, Classics and Archaeology, University of Edinburgh, Edinburgh, UK

Richard Fullagar
Centre for Archaeological Science, School of Earth and Environmental Sciences, University of Wollongong, Wollongong, NSW, Australia

Sabine Gaudzinski-Windheuser
*MONREPOS Archaeological Research Centre and Museum for Human Behavioural Evolution, Römisch-Germanisches Zentralmuseum (RGZM), Leibniz-Forschungsinstitut für Archäologie, Neuwied, Germany
Institut für Altertumswissenschaften, Arbeitsbereich Vor- und Frühgeschichtliche Archäologie, Johannes-Gutenberg-University of Mainz, Mainz, Germany

Juan Gibaja
Consejo Superior de Investigaciónes Cientificias, Institución Milá i Fontanales, Barcelona, Spain

Wallace Karl Hutchings
Department of Sociology and Anthropology, Thompson Rivers University, Kamloops, BC, Canada

Radu Iovita
MONREPOS Archaeological Research Centre and Museum for Human Behavioural Evolution, Römisch-Germanisches Zentralmuseum (RGZM), Leibniz-Forschungsinstitut für Archäologie, Neuwied, Germany

Frank Jäger (Deceased)
Physikalisch-Technische Bundesanstalt (PTB), Braunschweig, Germany

Michelle C. Langley
*Institute of Archaeology, University of Oxford, Oxford, UK
Archaeology and Natural History, School of Culture History and Language, College of Asia and the Pacific, Australian National University, Acton, ACT, Australia

Marlize Lombard
Department of Anthropology and Development Studies, University of Johannesburg, Johannesburg, South Africa

João Marreiros
ICArEHB. FCHS, Universidade do Algarve, Faro, Portugal

Annemieke Milks
Institute of Archaeology, University College London, London, UK

Masayoshi Oba
Yamagata Prefectural Center for Archaeological Research, Kaminoyama, Yamagata, Japan

Telmo Pereira
ICArEHB. FCHS, Universidade do Algarve, Faro, Portugal

Jean-Marc Pétillon
TRACES, CNRS, Université Toulouse 2, Toulouse, France

Hugues Plisson
University of Bordeaux, CNRS, MCC, INRAP, PACEA, Pessac, France

Matthew Pope
Institute of Archaeology, University College London, London, UK

Joseba Rios-Garaizar
Traceology Group, Consorcio CENIEH (Centro Nacional de Investigación Sobre La Evolución Humana), Burgos, Spain

Veerle Rots
Chercheur Qualifié du FNRS, TraceoLab/Préhistoire, University of Liège, Liège, Belgium

Paul E. Salem
Department of Evolutionary Anthropology, Duke University, Durham, NC, USA

Katsuhiro Sano
The University Museum, The University of Tokyo, Tokyo, Japan

Holger Schönekeß
Physikalisch-Technische Bundesanstalt (PTB), Braunschweig, Germany

Benjamin J. Schoville
Department of Archaeology, University of Cape Town, Rondebosch, South Africa
School of Human Evolution and Social Change, Institute of Human Origins, Arizona State University, Tempe, AZ, USA

Michael J. Shott
Department of Anthropology and Classical Studies, University of Akron, Akron, OH, USA

Lyn Wadley
Evolutionary Studies Institute, Centre of Excellence, Palaeosciences, University of the Witwatersrand, Johannesburg, South Africa

Mina Weinstein-Evron
Zinman Institute of Archaeology, University of Haifa, Haifa, Israel

John C. Whittaker
Department of Anthropology, Grinnell College, Grinnell, IA, USA

Jayne Wilkins
Department of Archaeology, University of Cape Town, Rondebosch, South Africa
School of Human Evolution and Social Change, Institute of Human Origins, Arizona State University, Tempe, AZ, USA

Alla Yaroshevich
Israel Antiquities Authority, Jerusalem, Israel

Yossi Zaidner
Zinman Institute of Archaeology, University of Haifa, Haifa, Israel

Part I

Recognizing Weapons: Experimental Approaches

Chapter 1
When Is a Point a Projectile? Morphology, Impact Fractures, Scientific Rigor, and the Limits of Inference

Wallace Karl Hutchings

Abstract Archaeologists have long sought a reliable means to identify whether certain pointed stone artifacts represent weapon armatures, and more specifically, whether specific types of pointed artifacts are associated with specific weapon technologies. These attempts have generally relied on ethnographic data; morphological, and more recently, morphometric, criteria; experimentation; use wear analyses; residue analyses; and combinations thereof. This paper is concerned with the reliability of established methods of identification of the stone arming tips of ancient weaponry, and in particular established means of differentiating weapon delivery technologies. The author presents a critical review of major attempts to isolate criteria intended to identify such artifacts and technologies; identifies deficiencies in the methodologies and criteria employed to date; and concludes that due to underlying subjective methods and a lack of comprehensive experimentation, current methods for identifying weapon armatures and delivery technologies lack sufficient scientific rigor.

Keywords Diagnostic impact fracture • Morphology • Projectile points • Residues • Tip cross-sectional area • Use-wear

Prehistoric spears, javelins, spearthrower darts, and arrows can be readily recognizable items when recovered in their entirety. The length of a weapon shaft, its overall size and weight, its balance, and the presence or absence of a notched versus a dimpled nock, for example, are often indicators of such an implement's function. Unfortunately, the hafts of these weapons were constructed from organic materials, and apart from rare instances of unusual preservation, are not often preserved in archaeological contexts; the archaeologist generally recovers only the lithic artifact, commonly referred to as a "point", that once served as the armature component of the weapon.

In common parlance we tend to use the terms "point" and "projectile point" rather ambiguously. The former commonly implies the latter, while the latter is commonly, though incorrectly, used in reference to the armature of a spear, or other similar weapon, which is a thrusting weapon *per se*, rather than a projectile weapon (the term javelin is used herein to refer to a spear-like weapon that is thrown). Of course, archaeologists seldom recover direct evidence of the weapon delivery technology employed by ancient hunters. As a result, they face some important challenges: (1) how to determine whether an individual lithic artifact actually functioned as a weapon armature, and (2) how to recognize the delivery technology associated with that armature. We need to know both before we can be sure whether a specific pointed artifact is indeed a projectile point.

The first challenge is commonly tacked via generalization and analogy; repeated classes of artifacts are observed in repeated associations. Considered as an aggregate, when similar forms of lithic artifacts are recovered repeatedly in similar contexts; pointed lithic bifaces embedded in animal bone in kill sites for example; they might reliably be classified as components of weaponry, though the specific form of the weapon may still be unknown. The second challenge is more formidable. Of particular significance to the discussion that follows, is the simple fact that lacking direct evidence of specific weapon technologies, archaeologists have from necessity attempted to identify secondary criteria that demonstrate associations with specific weaponry.

This paper presents a critical review of contributions that have led to the popular acceptance of certain mainstream criteria as diagnostic indicators of weapon identification and use, and identifies deficiencies in those criteria. The purpose is to argue for more rigorous experimental methods, and to

W.K. Hutchings (✉)
Department of Sociology and Anthropology, Thompson Rivers University, 900 McGill Road, Kamloops, BC V2C 0C8, Canada
e-mail: khutchings@tru.ca

Radu Iovita and Katsuhiro Sano (eds.), *Multidisciplinary Approaches to the Study of Stone Age Weaponry*, Vertebrate Paleobiology and Paleoanthropology, DOI 10.1007/978-94-017-7602-8_1

draw attention to the need to recognize the limits of inferences that reasonably can be drawn from our work. The author emphasizes that an important aspect of science is its self-correcting nature; errors and approximations are an inherent and necessary part of the scientific process. The critical review that follows is likewise offered as a necessary part of this process, and is not intended to be a critique of the individuals whose contributions are discussed; indeed, without such contributions there can be no tradition of archaeological science.

The author also acknowledges a bias in the discussion that follows; in particular, a strong bias towards North American perspectives and associated point forms, due to the underlying fact that this discussion grew out of an initial concern for the objective identification of North American Paleoindian weaponry. Thus the focus tends to be on generalized North American artifact forms that might reasonably have functioned as various hafted weapon armatures, and on bifacial "points" rather than microliths.

Identifying Weapons and Delivery Technologies

Studies Based on Morphology, and Morphological Types

Numerous researchers (e.g., Evans 1957; Forbis 1962; Wyckoff 1964; Corliss 1972; Thomas 1978; Shott 1997; see also Shea 2006) have sought to identify prehistoric weapons and delivery technologies through examination of the one surviving component of these systems: their lithic armatures. Such research has commonly involved investigation of neck, shoulder, or stem widths of points; or various measures reflecting overall point size or shape. The underlying assumptions pertaining to these analyses are that:

1. a relatively thin, triangular, leaf-shaped, or lanceolate, pointed artifact was probably a "point" (which is *understood* to be a weapon armature);
2. neck, shoulder, and stem widths of points reflect the diameter of weapon shafts or foreshafts ("hafts") of weaponry;
3. spear and javelin hafts are large, dart hafts smaller, and arrow hafts smaller still; and
4. spear and javelin points are big and heavy; arrow-points are small and light; and spearthrower dart point sizes and weights lie somewhere in between.

These assumptions have long-standing historical precedent in the North American archaeological literature. For example,

> These [points] will be discussed in two categories: (1) small, thin, light, finely chipped specimens believed to have served on arrows; and (2) larger, thicker, heavier and more crudely chipped specimens we believe were used on darts thrown with atlatls. That such a distinction actually existed over vast areas of America is no longer denied by many archaeologists (Baker and Kidder 1937: 51).

Despite the simplicity of such assumptions, an influential study by Fenenga (1953) of 884 points from the American Midwest, Southwest, and California, suggested that there may be some basis for these distinctions. Fenenga (1953) demonstrated that a frequency plot of either point neck widths, or overall weights, produced a bi-modal distribution suggesting mutually exclusive point groupings. Even though no data were presented to establish the actual sizes and weights of prehistoric weapon shafts themselves, the bi-modal distribution was interpreted as reflecting the morphological differences between spearthrower and bow projectiles.

The issue was later addressed by Thomas (1978) who employed a sample of 132 hafted arrow points and 10 hafted spearthrower dart points drawn from ethnographic collections, as well as archaeological specimens recovered from Pueblo Bonito (New Mexico), to determine the relationship between point size and the diameter of the actual foreshaft it was attached to. Thomas (1978) noted a correlation between arrow foreshaft diameter and arrow point neck width, but was unable to document a similar relationship between spearthrower dart foreshafts and their respective points. Despite this, the data suggested that arrow foreshafts were significantly smaller than spearthrower dart foreshafts, and arrowheads themselves were significantly smaller than dart tips. Furthermore, a discriminant analysis based on considerations of length, width, thickness, and neck width of the points correctly classified approximately 86% of the study sample (Thomas 1978: 471). Thomas's approach provided no mechanism for dealing with unnotched points.

Shott's (1997) reassessment of Thomas's data utilizes a significantly enlarged sample of hafted dart points, and considers shoulder width as an alternative to neck width, since he found the latter variable to be inadequate:

> A neck width threshold of 9 mm correctly classifies 38 of 39 dart points, but misclassifies as darts 82 of 132 arrow points (62.1 percent). A threshold value of 8.5 mm produces identical results for darts but misclassifies 89 arrow points (67.4 percent). Even a threshold of 10.4 mm, one standard deviation lower than Chatters et al.'s (1995:757) mean for inferred dart points, misclassifies 57 arrows (43.2 percent) (Shott 1997: 98).

Employing shoulder width criteria and a larger sample, Shott was better able to classify dart points, however, Shott's (1997: 99) *overall* ability to distinguish dart and arrow points, at 85% success, is essentially equivalent to that of Thomas's at 86% (1978: 471).

The relevant measure in these approaches was selected because of its relation to shaft or foreshaft diameter. The latter is the more important variable, however, since within reasonable limits of variation it is more closely related to weapon performance, and is, therefore, indicative of the weapon system. Even a casual perusal of archery equipment, for example, whether ancient or modern, will convince the reader that much more variation exists in point dimensions than in shaft or foreshaft dimensions for a given weapon kit. So while points are much more common in the archaeological record, the data recovered directly from the study of shafts and foreshafts, rather than inferred from point metrics, are expected to be a better reflection of weapon system design considerations.

Published metric data and scale photographs are readily available for hundreds of dart foreshafts recovered from dry cave sites throughout the American Great Basin and Southwest (e.g., Kidder and Guernsey 1919; Guernsey and Kidder 1921; Loud and Harrington 1929; Guernsey 1931; Harrington 1933; Woodward 1937; Heizer 1938; Fenenga and Heizer 1941; Cosgrove 1947; Jennings 1957; Smith 1963; Smith et al. 1963; Taylor 1966; Dalley and Petersen 1970; Berry 1976; Dalley 1976; Janetski 1980; Hattori 1982; Tuohy 1982; Pendleton 1985; Salls 1986). This literature indicates that while most recovered dart foreshafts are approximately 0.8–1.1 cm in diameter, many are less than 0.6 cm in diameter (Hutchings 1997). In comparison, the mean diameter of arrow foreshafts from Thomas's ethnographic sample (n = 118) is 0.7 cm, while the mean diameter of arrow foreshafts from his archaeological sample (n = 14) is 0.9 cm (Thomas 1978: Tables 1 and 2). In 1981, a 0.6 cm diameter dart foreshaft with a hafted stone point, along with five other dart foreshafts ranging from 0.4 to 0.6 cm, was recovered from NC Cave, Lincoln County, Nevada (Tuohy 1982). In reference to Thomas's (1978) study, these finds, as well as 56 other dart foreshafts from cave sites in the vicinity of Lake Winnemucca, Nevada, prompted Tuohy (1982: 97) to comment:

> I am not convinced that enough data have been marshalled [sic] to segregate arrow foreshafts from dart foreshafts on the basis of size or variability in dimensions such as length, width, weight, or shaft diameters, and the new data from "NC" Cave and the Winnemucca Lake foreshafts from a cache support this contention.

Studies such as those of Fenenga (1953), Thomas (1978), and the archaeological specimens from the American Great Basin and Southwest referred to above rely on a number of normative assumptions. First, they assume that the point samples are representative of one specific technology, the one they are found associated with (e.g., arrow *or* dart, etc.), and are not transferable between coexisting technologies. Second, they assume that there is no meaningful variation *within* a single technology, that point and shaft dimensions did not vary to adapt to application (e.g., larger projectiles for larger game). Third, they assume that the study samples are representative of that technology through time and space. Furthermore, in differentiating points based on metric attributes, particularly attributes of size, these studies can be impacted by both subtle and dramatic instances of repair and resharpening. While some researchers (e.g., Shott 1997) have attempted to compensate for this, it is a much more complex concern in this specific regard than has generally been recognized. In particular, it may be difficult to determine how often an artifact has been recycled, and whether it was recycled within or between technologies. As an example that explores the implications of each, a hypothetical point that was created for use as a spear armature may conceivably be recycled into an arrow point. If the recycling is noticeable, the resulting point may be treated separately by the analyst who may be inclined to decide that the morphology had been adversely affected by the recycling event, therefore excluding its metric data from the aggregate. It is possible, however, that the recycling may have resulted in an arrow point of ideal morphology (i.e., just because it was recycled, does not necessarily mean that the end product was not exactly what was desired for the new end use; and, in addition, as an "arrow point" the piece had *never* been resharpened). Certainly, the fact that this hypothetical point was eventually hafted as an arrow armature tells us that it was considered an acceptable point for that technology, so treating it as a resharpened point may skew our research. Had a second hypothetical arrow point been *manufactured* with identical proportions it is likely that the metric data derived from it would not be considered comparable with the first. Of course, data derived from shaft and foreshaft diameters avoid such issues altogether, and benefit from being more closely related to the phenomena we are interested in (i.e., the propulsion technology rather than just the points).

In choosing to study examples of hafted arrow points, Thomas's (1978) sample was unavoidably recent by way of preservation bias. This was an inevitable consequence of the research parameters, and it may have biased the sample by assuming a priori that small, late period, and ethnographic arrow points and shafts are representative of bow technology throughout time. In contrast, we must accept that the absence of point types known to be associated with arrows does not constitute evidence for the absence of the bow; if we choose to rely on characteristics of size and suitability we must keep in mind that many early lithic points are of a size and weight suitable for use with the bow, even if not ideal, and we have no empirical proof that hafting methods are discrete indicators of delivery technologies. In fact, Browne (1940: 211) noted that even North American Folsom Paleoindian points would make highly efficient arrow points.

More recently, tip cross-sectional area (TCSA) and tip cross-sectional perimeter (TCSP) have been proposed as criteria to be used in combination with other data (e.g., use-wear, context) to distinguish weapon armatures (Hughes 1998; Shea et al. 2001; Shea 2006; Sisk and Shea 2009). Reminiscent of older morphological studies, morphometric criteria are derived from both ethnographic and experimental studies of relatively recent weaponry, and consider characteristics deemed to optimize mechanical and aerodynamic efficiency. TCSA and TCSP identify as a possible projectile or spear point, any suitable pointed object that falls within the range of variables known from ethnography and experimentation to be acceptable for use as a weapon tip, substituting area and perimeter measures to identify delivery technologies. As such TCSA/TCSP may be considered suitable for identifying additional objects with similar mechanical and aerodynamic characteristics as those known from ethnography and experimentation, even though they do not offer any empirical evidence in and of themselves that objects so classified were actually employed as weapon armatures. Perhaps more to the point, when applied to assemblages distant in time or space from those on which the measures were developed, these criteria imply that the same values of mechanical and aerodynamic efficiency were of primary concern to the people who produced those distant assemblages. Stated another way, they constitute a tautologous argument that assumes a priori that we already know the range of objects that constitute an acceptable weapon technology. While there can be no doubt that underlying mechanical and aerodynamic principles do not change, there is no reason to expect that mechanical and aerodynamic considerations or priorities were the same, and of equal value, to all people in all places and times. At the very least, we might reasonably expect periods of ancient weaponry development and experimentation to produce variability beyond the TCSA/TCSP ranges expected for more recent, well-developed technologies. For these reasons, TCSA and TCSP cannot be considered valid indicators of projectile function within assemblages where the ranges of pertinent variables have not *already* been established.

Further complicating the issue of acceptable morphology, Ahler (1971) found evidence suggesting that artifacts that might otherwise be readily labeled as bifacial "projectile" points were not always used primarily as projectile armatures, but were often used as knives and multi-purpose tools. When faced with the tautology of existing morphometric criteria, as well as evidence that readily recognizable "projectile points" were at times not used to arm projectiles, one is forced to question the usefulness of morphological and morphometric methods of identification.

Studies Based on Microwear, Residues, and Impact Fractures

Microwear analyses have been proven effective in differentiating modes of contact between stone tools, location and orientation of use contact, hafting, and even materials against which tools were used (e.g., Semenov 1964; Tringham et al. 1974; Keeley 1980; Tomenchuk 1985; Kay 1996; Dockall 1997; Rots 2003, 2004). Unfortunately, since the direction of contact for spears, javelins, darts, and arrows can be identical, and the use of each weapon type might be expected on identical contact materials, microwear analyses have not demonstrated an ability to identify specific weapon technologies *per se* independent of relational analogues (i.e., independent of ethnographic or direct historic analogies). Hafting traces are also not necessarily diagnostic of a specific weapon technology since spears, javelins, darts, and arrows can exhibit common patterns of hafting wear (although the area of contact may be noticeably larger and more intense for a spear point than an arrow point).

Organic residues may tell us what materials a lithic artifact has been in contact with (Hardy and Raff 1997; Hardy and Kay 1999; Hardy et al. 2001), perhaps also identifying the area of hafting. Three flakes bearing a tar mastic, and recovered from a Mid-Pleistocene bone-bearing deposit, are considered by Mazza et al. (2006: 1317) to constitute evidence for hurled weapons, despite a lack of any wear traces or other corroborating evidence apart from their association with bone (see also Hardy et al. 2001; Boëda et al. 1998). Of course there may be many conceivable reasons to haft pointed lithic artifacts, but ultimately, evidence of hafting, even if associated with longitudinal wear traces, is not indicative of any specific weapon technology.

So-called "diagnostic impact-fractures" (DIFs) have been touted by many analysts (e.g., Witthoff 1968; Frison 1974; Ahler and McMillan 1976; Frison et al. 1976; Odell 1977; Frison 1978; Roper 1979; Barton and Bergman 1982; Bergman and Newcomer 1983; Fischer et al. 1984; Odell 1988; Shea 1988; Woods 1988; Holdaway 1989; Dockall 1997) to be indicative of weapon impact. For example, Bergman and Newcomer (1983: 241–243) describe three types of DIFs identified during their projectile experiments; the burin-like fracture, the flute-like fracture, and the bending fracture. A forth type of DIF, the bending fracture-initiated spin-off, was identified by Fischer et al. (1984). Bergman and Newcomer (1983) employed DIFs to suggest that certain Upper Paleolithic artifacts may constitute projectile armatures. Likewise, Fischer et al. (1984)

employed DIFs to suggest that certain Mesolithic and Neolithic artifacts may constitute projectile armatures.

While most archaeologists have restricted such analyses to formal points, Odell (1988), has labeled large numbers of modified and unmodified flakes from a site in the Lower Illinois Valley (USA) as projectile armatures. Relying primarily on the identification of DIFs, he suggests: (1) that diagnostic projectile impact fractures may often be observed on simple retouched flakes, as well as unretouched waste flakes and detritus; (2) that the practice of employing suitable waste flakes as functional projectile points may be widespread; and that (3) this phenomenon will have repercussions on studies of technology and foraging efficiency. Odell's (1988) waste flake analysis is based on previous comparative studies of impact-related breakage patterns, most notably Odell and Cowan (1986), which used an extensive series of shooting experiments employing replicated bifacial points and unmodified flakes as armatures on both javelins (the authors use the term "spears") and arrows. Various other types of use-wear such as edge-rounding, surface polish, and linear striations were also used to identify projectile impact-related damage. Odell's (1988: 344–345) tabulated data are unclear with respect to the percentage of the study sample represented by waste flakes and detritus, versus morphological projectile points. He does, however, state that only 3% of the functional projectile points from the Smiling Dan site sample are found among "… modified type collection objects" (presumably, morphological projectile points) (Odell 1988: 346), suggesting that 97% of the site's projectile points were not, for lack of precise terminology, traditional points.

This author finds little reason to doubt the possibility that prehistoric peoples made greater use of materials usually classified as debitage and detritus than has been popularly recognized, yet there are several problems inherent in the use of impact breakage patterns and wear traces as evidence of projectile use.

These problems arise due to both the general morphology and functional nature of lithic projectiles:

1. lithic projectiles generally exhibit little use-wear or haft related polish (Kay 1996) prior to catastrophic failure; and
2. impact fractures are generally location- and orientation-specific forms of damage that can be caused as much through thrusting, or even dropping (Hutchings 1991, 2011), as from projectile impact.

In fact, it is possible to produce flakes and blades during simple core reduction which unintentionally exhibit sympathetic or repercussive fractures that often appear similar to projectile point impact fractures; an issue also noted by Fischer et al. (1984: 24). For example, the thin distal and lateral margins of flakes and blades can be damaged when they strike the ground after removal from a core, or from being dropped into a pile for subsequent use by the flintknapper. Such damage would constitute an impact fracture per se, but not an impact fracture caused by use as any type of weapon armature. Given a site with a relatively large population of waste flakes, blades, and other debitage, a significant number of pieces might be expected to exhibit so-called DIFs, but even though these fractures were caused by an impact, they are certainly not diagnostic of any weapon use.

A Study of Impact Fractures Among Debitage

An investigation of modern (replicative) flintknapping debris intended to explore the incidence of "impact fractures" on discarded flint debitage (Hutchings 1991: Appendix F), demonstrated that 72.4% of a sample of 246 pieces of flint chipping debris were suitable, with respect to overall morphology and weight, for hafting as practical arrowheads. Of these, 15 pieces (6.1% of the original sample) were found to exhibit damage suggestive by location, distribution, and morphology, of projectile use according to the macroscopic criteria of Odell and Cowan (1986), as well as those of Odell (1988) and others (e.g., Ahler 1971; Roper 1979; Barton and Bergman 1982; Bergman and Newcomer 1983; Fischer et al. 1984). In fact, three of the haftable pieces which exhibited DIFs also exhibited simple side-notches that could facilitate hafting; one of these three exhibited simple, uniform, bilateral side-notches.

> The results obtained by this simple study demonstrate a high probability of observing projectile impact-like breakage patterns among discarded waste flakes and other debitage and detritus. Over 6% of the sample produced erroneous "use-wear". The overall morphology of these pieces, and the current definition of what constitutes a projectile point, would suggest not only that they came in contact with some target material, but that they were shot or thrown at the target material as projectile points (Hutchings 1991: Appendix F, emphasis in original).

These results have been duplicated by Pargeter (2011) who found diagnostic impact fractures on 1.8% of an assemblage of experimental knapped debris, and as much as 2.4% of a trampled experimental debitage assemblage. As a result, Pargeter suggests that erroneous diagnostic impact fractures can be expected on approximately 3% of a lithic assemblage. Pargeter (2011: 2885) also noted the occasional formation of smooth, semi-circular notches on the trampled debris (see also Lombard and Pargeter 2008). Villa et al. (2009b: 449) also note that fractures like those associated with weapon impacts can result from processes of manufacture and trampling. Likewise, Sano (2009) found that relatively low frequencies of erroneous DIF types can be expected on knapped, retouched, and trampled assemblages.

Discussion

At the heart of shortcomings in the methodologies discussed are two critical issues; the confounding factor of equifinality, and the extent to which we can make reasonable inferences based on the parameters of our replicative experiments. The methodologies discussed have been derived from analogues and tested by replicative experimentation, but experimental hypothesis testing in and of itself does not guarantee robust results. Hypotheses can take several different forms, each of which may be scientifically valid, but each of which may be best suited to differing circumstances. In the study of complex phenomena, where numerous and confounding variables may produce instances of equifinality, direct testing of hypotheses may result in a lack of robusticity since testing often ceases once a finite number of positive results are generated. For example, given a phenomenon (P_1), we may hypothesize that P_1 is the result of a suspected behavior (b_1), so we conduct an experiment to determine whether b_1 results in the production of P_1. If it does (i.e., we observe a positive result), we have verified our hypothesis, but we have not demonstrated that other behaviors ($b_{\ldots x}$) could not also produce P_1. If we do not continue to test alternative hypotheses, we have succeeded only in accommodating the data by demonstrating a correlation.

> [The] accommodation process suffers from a lack of empirical sufficiency…. Data are, in fact, used in model construction (the models are fitted to the data observed), and only those dimensions of the data *supporting* model construction are considered… however, the resulting model cannot be tested because relevant data have already been used in construction.
>
> The accommodation approach to explanation has the appeal of common sense but is different from the usual scientific process of interpretation. A scientific approach fits data to models through *falsification* procedures, thereby assessing the utility of interpretations for explaining observations. The fitting of data to models requires that *all* causes for observed patterning be considered and compared to model implications. In other words, relations among phenomena are predicted from theory and compared to the actual, empirically measured relationships defined for the data. Data are not simply interpreted in terms of the model [Rigaud and Simek 1987: 48, emphasis in original].

In the study of complex phenomena, it may be more suitable to test the consequent of the hypothesis. This form of hypothesis testing generally takes the form of a predictive if:then statement, and relies on the concept of coherence. For example, if the hypothesis is correct, then we predict that we will also observe other specific phenomena ($P_{\ldots x}$). In this form of testing, confidence in the hypothesis increases as more and more instances of coherence are observed (i.e., more $P_{\ldots x}$ are successfully predicted).

Arguably, in the study of complex phenomena hypothesis testing by falsification may produce the most robust results. In this form of testing, the hypothesis is considered valid provided it cannot be falsified; naturally, a successfully falsified hypothesis must be abandoned and an alternative sought.

The falsification process has demonstrated that DIFs are not, individually, or in small assemblages, diagnostic of weapon impact as they have been shown to be produced in low frequencies by knapping, retouch, and trampling activities. Granted, it has been adequately documented, both experimentally and at kill sites, that at an assemblage level, significant frequencies of DIFs (present on approximately 40% or more of a pointed tool assemblage) are indicative of weapon impacts (e.g., Frison 1974; Fischer et al. 1984; Bratlund 1996; Villa et al. 2009a, b). At present, specific weapon delivery technologies cannot, however, be reliably differentiated via DIFs.

The validity of morphological analyses (including morphometrics) rests on the rigor of an appropriate analogue, but can be seriously confounded by significant morphological similarities and overlaps between technologies. The robusticity of morphological analyses must be considered increasingly suspect with increased spatial and temporal separation from our analogue. Considered either individually or together, the presence of wear traces and residues on lithic artifacts can potentially indicate the area of contact, direction of motion, and the contact material. As such, they offer the greatest potential relative to the approaches discussed herein for the recognition of weaponry. In the absence of relational analogues, however, they are likewise, in and of themselves, incapable of *differentiating* weapon technologies.

Does all of this mean that we cannot trust our ability to identify *any* stone weapon armatures? Of course not. While the *individual* weapon identification methodologies critiqued above have failed, in the author's opinion, to demonstrate adequate scientific rigor, when combined as *multiple lines of evidence* they benefit from the principle of coherence, and so are best employed in concert to provide identifications with varying levels of robusticity. The confounding factor of equifinality, however, still renders the multiple lines of evidence approach incapable of differentiating specific weapon delivery technologies. Of course the repeated associations of certain pointed artifact types and morphological characteristics, not only with kill sites and animal remains, but with good ethnographic analogues, are sufficiently reliable in most instances that we can be comfortable identifying *classes* of pointed artifacts as weapon armatures.

Where we start to encounter problems is when our assemblages are very small, or we wish to know whether a *specific* artifact actually served as a weapon armature; more tenuous still is our ability to associate the actual use of a specific artifact with a *specific weapon technology* (cf. Callow 1986; Villa and Lenoir 2006; Mussi and Villa 2008). These issues represent serious challenges and impact our basic ability to accurately reconstruct the lives of ancient peoples since weapon technologies affect not only subsistence focus and success, but also basic settlement patterning, resource scheduling, and myriad other issues that make up a culture. In fact, archaeologists intuitively recognize the inability of these methods to generate convincing results when important issues, such as those related to higher order concepts, are at stake. The question of Neanderthal use of ranged weaponry, for example, is one such issue. The same weapon identification methodologies discussed above have been employed to suggest that Middle Paleolithic Levallois and Mousterian points represent hafted weapon armatures rather than tools used for unspecialized tasks such as cutting, scraping, or other (and multiple) purposes (Shea 1988, 1990, 1991, 1993, 1995a, b, 1997, 2003a, 2006, 2009; Solecki 1992; Shea et al. 2001; Sisk and Shea 2009). This suggestion carries with it the connotation that Neanderthals possessed more sophisticated cognitive capacities than often credited, since they were capable of complex behaviour (Shea 2003a, b; O'Connell 2006; see also McBrearty and Brooks 2000; cf. Shea 2011). My concern here is not to argue whether these Middle Paleolithic artifacts are, or are not, actual projectile points or even weapons, but rather, whether from a scientific point-of-view, the proffered evidence is logical and supportable, or whether it is instead attempting to reach beyond the limits of reasonable and supportable inference.

Due to the significance of the cognitive implications, the validity of the Neanderthal weaponry data has been met with apprehension and even skepticism (see Bordes 1961; Holdaway 1989; Anderson-Gerfaud 1990; Holdaway 1990; Debénath and Dibble 1994; Plisson and Béyries 1998; Kuhn and Stiner 2001). When one considers the suggested antiquity of projectile weaponry (e.g., Thieme 1997; cf. Shea 2006) it seems tempting to compare our Paleolithic cousins to recent hunter-gatherers, using ranged weapons to safely and efficiently harvest game; but in this instance ranged weapon use has yet to be demonstrated empirically. Even considering the evidence from Umm el-Tlel, Syria, of a Levallois point embedded in the cervical vertebra of a wild ass (Boëda et al. 1999), in the absence of a haft element, or some other supportive evidence, one can only speculate whether the point was thrust or thrown. In fact, we cannot even be sure that the piece was thrust as a spear armature, since it is entirely *possible* that it was thrust as a simple form of hafted dagger, perhaps to deliver a *coup de grâce*; a possibility also recognized by Sisk and Shea (2009: 2044), and indeed by Boëda et al. (1999) when they conclude that the piece was *minimally* "hafted onto the distal extremity of a shaft", and that "the use of Levallois points as projectile weapons is only one of several functional possibilities" (Boëda et al. 1999: 401). Other examples of lithic artifacts embedded within animal bone are discussed by Villa et al. (2009a: 856–857), but are notably more recent. Even taking preservation issues into account, the fact that such associations are not more common suggests that at least some caution is indicated with respect to the issue of Neanderthal ranged weaponry. Considering also the increasing interest in the role of weapon technologies in human dispersal (e.g., McBrearty and Brooks 2000; Brooks et al. 2006; Shea 2006; Villa and Lenoir 2006; Churchill and Rhodes 2009; Shea and Sisk 2010), it seems a most propitious time to exercise caution and evaluate methodological robusticity.

Of course, it should not be the case that we require significant issues and implications before properly assessing the validity of a given methodology. The simple experimental results presented above, and replicated by others (Sano 2009; Pargeter 2011), illustrate obvious problems inherent in employing impact breakage patterns as evidence of projectile function at the level of the individual artifact. Due to their fragility, we should actually *expect* narrow, fragile tips of pointed artifacts to exhibit fracture damage, the majority of which may be directed parallel to the long axis of the artifact. These fractures can exhibit near-identical morphologies, and may have been caused by some form of "impact", but there are many causes of impact, apart from, and in addition to, weapon use. From this perspective, so-called "diagnostic impact-fractures" are little more than a series of tip fractures found on both projectile points and other pointed lithic implements and debitage, that are not *diagnostic* of anything other than breakage, and the term "impact fracture", as it pertains to a visual means of identifying weapon armatures, is rendered meaningless.

Does this mean that we must discard these studies? Not necessarily, as it is entirely possible that there is something useful here, but we do have to be aware of how these results were derived in order to assess their applicability and scientific validity. For example, the breakage patterns associated with thrusting- and projectile-weapons use, when observed on point *types* already *known* to generally have been used as weapon armatures, may reasonably be employed to construct functional hypotheses – such hypotheses are reasonably derived by invoking the Direct Historical Approach. One cannot, however, with respect to artifacts of *unknown* function, employ the correlation between projectiles and breakage patterns as evidence of use as a projectile; we are all aware that correlation is not causation.

Conclusion

Since hafted spear points may conceivably be used on similar contact materials, and with similar directions of use, as javelins, darts, and arrows, we can expect some similarities in the microwear, hafting traces, residues, TCSA/TCSP, and impact fractures exhibited by each. The end result is that archaeological identifications of weapon technologies that rely on visual recognition of morphology, or any morphology-based metrics; including DIFs, either alone, or in connection with microwear traces; residues; or TCSA/TCSP; are incapable of independently and reliably identifying specific weapon technologies. The simple fact that pointed artifacts *known* to never have been used as weapons would be identified as "projectile points" serves to falsify the hypothesis that commonly employed existing criteria reliably indentify weaponry. The author concedes, however, that by employing multiple lines of evidence, we can increase the robusticity of weapon identification.

In research where reasonable relational analogs exist to support the identification of specific projectile technologies, and specific artifacts as projectiles (e.g., by employing the Direct Historic Approach), this is much less of an issue. Unfortunately, where studies have been undertaken that rely on our ability to identify weapon technologies in the archaeological record *in the absence of good, relational analogs*; those that ultimately rely on the common methodologies discussed herein, rather than on direct evidence related to propulsion technology (i.e., associated with notched, or dimpled shafts; or other direct evidence of bow, spearthrower, or javelin technologies) have been built on a tenuous foundation.

As anthropologists, we are driven by our curiosity and our desire to find answers to questions regarding the human past, but as scientists we must be careful to adhere to the precepts of good science so that we can be confident in our results; only then can we proceed to build upon existing research. Replicative experimental archaeology can, and has, contributed significantly to our understanding of weapon technologies, but we must take care that our methodologies are rigorous, and that our inferences are supported logically and empirically. To accomplish this, replication studies must successfully eliminate confounding factors and alternative explanations if they are to avoid instances equifinality.

References

Ahler, S. (1971). *Projectile point form and function at Rodgers shelter, Missouri*. Missouri archaeological society research series 8. Columbia: University of Missouri.

Ahler, S. A., & McMillan, R. B. (1976). Material culture at Rodgers shelter: A reflection of past human activities. In W. R. Wood & R. B. McMillan (Eds.), *Prehistoric man and his environments: A case study in the Ozark highland* (pp. 163–199). New York: Academic Press.

Anderson-Gerfaud, P. C. (1990). Aspects of behaviour in the Middle Palaeolithic: Functional analysis of stone tools from southwest France. In P. A. Mellars (Ed.), *The emergence of modern humans* (pp. 389–418). Edinburgh: Edinburgh University Press.

Baker, W. E., & Kidder, A. V. (1937). A spear thrower from Oklahoma. *American Antiquity, 3*, 51–52.

Barton, R. N. E., & Bergman, C. A. (1982). Hunters at Hengistbury: Some evidence from experimental archaeology. *World Archaeology, 14*, 237–248.

Bergman, C. A., & Newcomer, M. H. (1983). Flint arrowhead breakage: Examples from Ksar Akil, Lebanon. *Journal of Field Archaeology, 10*, 238–243.

Berry, M. S. (1976). Remnant cave. In G. F. Dalley (Ed.), *Swallow shelter and associated sites*. University of Utah anthropological papers 96 (pp. 115–127). Salt Lake City: University of Utah Press.

Boëda, E., Connan, J., & Muhesen, S. (1998). Bitumen as hafting material on Middle Paleolithic artifacts from the El Kown Basin, Syria. In T. Akazawa, K. Aoki & O. Bar-Yosef (Eds.), *Neandertals and modern humans in Western Asia* (pp. 181–204). New York: Plenum.

Boëda, E., Geneste, J. M., Griggo, C., Mercier, N., Muhesen, S., Reyss, J., et al. (1999). A Levallois point embedded in the vertebra of a wild ass (*Equus africanus*): Hafting, projectiles and Mousterian hunting weapons. *Antiquity, 73*, 394–402.

Bordes, F. (1961). *Typologie du paléolithique ancien et moyen*. Bordeaux: Delmas.

Bratlund, B. (1996). Hunting strategies in the late glacial of northern Europe: A survey of the faunal evidence. *Journal of World Prehistory, 10*, 1–48.

Brooks, A., Yellen, J., Nevell, L., & Hartman, G. (2006). Projectile technologies of the African MSA: Implications for modern human origins. In E. Hovers & S. Kuhn (Eds.), *Transitions before the transition* (pp. 233–256). New York: Plenum/Kluwer.

Browne, J. (1940). Projectile points. *American Antiquity, 5*, 209–213.

Callow, P. (1986). The flint tools. In P. Callow & J. Cornford, (Eds.), *La Cotte de St. Brelade* (pp. 251–314). Norwich (UK): Geo Books.

Churchill, S., & Rhodes, J. (2009). The evolution of the human capacity for "killing at a distance": The human fossil evidence for the evolution of projectile weaponry. In J.-J. Hublin & M. P. Richards (Eds.), *The evolution of hominin diets: Integrating approaches to the study of palaeolithic subsistence* (pp. 201–210). Dordrecht: Springer.

Corliss, D. W. (1972). *Neck width of projectile points: An index of culture continuity and change* (p. 29). Pocatello, Idaho: Occasional Papers of the Idaho State University Museum.

Cosgrove, C. B. (1947). *Caves of the Upper Gila and Hueco areas in New Mexico and Texas* (Vol. 24, No. 2). Cambridge: Papers of the Peabody museum of American archaeology and ethnology.

Dalley, G. F. (1976). *Swallow shelter and associated sites*. University of Utah anthropological papers 96. Salt Lake City: University of Utah Press.

Dalley, G. F., & Petersen, K. L. (1970). Additional artifacts from Hogup Cave. Appendix X. In M. C. Aikens (Ed.), *Hogup cave*. University of Utah anthropological papers 93 (pp. 283–286). Salt Lake City: University of Utah Press.

Debénath, A., & Dibble, H. L. (1994). *Handbook of Paleolithic typology, Vol. 1: Lower and Middle Paleolithic of Europe*. Philadelphia: University of Pennsylvania Press.

Dockall, J. E. (1997). Wear traces and projectile impact: A review of the experimental and archaeological evidence. *Journal of Field Archaeology, 24*, 321–331.

Evans, O. F. (1957). Probable uses of stone projectile points. *American Antiquity, 23*, 83–84.

Fenenga, F. (1953). The weights of chipped stone points: A clue to their functions. *Southwestern Journal of Anthropology, 9*, 309–323.
Fenenga, F., & Heizer, R. F. (1941). The origins and authenticity of an atlatl and atlatl dart from Lassen County, California. *American Antiquity, 7*, 134–141.
Fischer, A., Hansen, P. V., & Rasmussen, P. (1984). Macro and micro wear traces on lithic projectile points: Experimental results and prehistoric examples. *Journal of Danish Archaeology, 3*, 19–46.
Forbis, R. G. (1962). The old woman's buffalo jump, Alberta. *Contributions to anthropology* (pp. 56–123). Ottawa: Department of Northern Affairs and National Resources, National Museum of Canada, Bulletin 180.
Frison, G. C. (1974). *The Casper Site*. New York: Academic Press.
Frison, G. C. (1978). *Prehistoric hunters of the high plains*. New York: Academic Press.
Frison, G. C., Wilson, M., & Wilson, D. J. (1976). Fossil bison and artifacts from an early altithermal period arroyo trap in Wyoming. *American Antiquity, 41*, 28–57.
Guernsey, S. J. (1931). Explorations in Northeastern Arizona: Report on the archaeological fieldwork of 1920–1923. *Papers of the Peabody Museum of American Archaeology and Ethnology* (Vol. 12, No. 1). Cambridge: Harvard University.
Guernsey, S. J., & Kidder, A. V. (1921). Basket-maker caves of Northeastern Arizona: Report on the explorations 1916–17. *Papers of the Peabody Museum of American Archaeology and Ethnology* (Vol. 8, No. 2). Cambridge: Harvard University.
Hardy, B., & Kay, M. (1999). Stone tool function at Starosele: Combining residue and use-wear analysis. In V. P. Chabai & K. Monigal (Eds.), *The Paleolithic of Crimea II: The Paleolithic of Western Crimea* (Vol. 2, pp. 197–210). Liége, Belgium: Etudes et Récherches de l'Université de Liége, ERAUL 87.
Hardy, B., Kay, M., Marks, A. E., & Monigal, K. (2001). Stone tool function at the Paleolithic sites of Starosele and Buran Kaya III, Crimea: Behavioral implications. *Proceedings of the National Academy of Sciences, USA, 98*, 10972–10977.
Hardy, B., & Raff, R. A. (1997). Recovery of mammalian DNA from Middle Paleolithic stone tools. *Journal of Archaeological Science, 24*, 601–611.
Harrington, M. R. (1933). *Gypsum Cave, Nevada*. Los Angeles: Southwest Museum Papers 8.
Hattori, E. M. (1982). *The archaeology of Falcon Hill, Winnemucca Lake, Washoe County, Nevada*. Carson City: Nevada State Museum Anthropological Papers 18.
Heizer, R. F. (1938). A complete atlatl dart from Pershing county, Nevada. *New Mexico Anthropologist, 11*, 69–71.
Holdaway, S. (1989). Were there hafted projectile points in the Mousterian? *Journal of Field Archaeology, 16*, 79–86.
Holdaway, S. (1990). Mousterian projectile points – Reply to Shea. *Journal of Field Archaeology, 17*, 114–115.
Hughes, S. S. (1998). Getting to the point: Evolutionary change in prehistoric weaponry. *Journal of Archaeological Method and Theory, 5*, 345–408.
Hutchings, W. K. (1991). *The Nachcharini composite projectile: Design theory and the study of hunting systems technology at Mugharet En-Nachcharini in the Post-Natufian Levant. M.A. thesis*, University of Toronto.
Hutchings, W. K. (1997). The Paleoindian fluted point: Dart or spear armature? The identification of Paleoindian delivery technology through the analysis of lithic fracture velocity. PhD dissertation, Simon Fraser University.
Hutchings, W. K. (2011). Measuring use-related fracture velocity in lithic armatures to identify spears, javelins, darts, and arrows. *Journal of Archaeological Science, 38*, 1737–1746.
Janetski, J. C. (1980). Wood and reed artifacts. In J. D. Jennings (Ed.), *Cowboy cave* (pp. 75–95). Salt Lake City: University of Utah Press, University of Utah Anthropological Papers 104.
Jennings, J. D. (1957). *Danger cave*. Salt Lake City: University of Utah Press, University of Utah Anthropological Papers 27.
Kay, M. (1996). Microwear analysis of some Clovis and experimental chipped stone tools. In G. H. Odell (Ed.), *Stone tools: Theoretical insights into human prehistory* (pp. 315–344). New York: Plenum Press.
Keeley, L. H. (1980). *Experimental determination of stone tool uses*. Chicago: University of Chicago Press.
Kidder, A. V., & Guernsey, S. J. (1919). *Archaeological explorations in Northeastern Arizona*. Washington, D.C.: Smithsonian Institution Bureau of American Ethnology, Bulletin 65.
Kuhn, S. L., & Stiner, M. C. (2001). The antiquity of hunter-gatherers. In C. Panter-Brick, R. H. Layton, & P. Rowley-Conwy (Eds.), *Hunter-gatherers: An interdisciplinary perspective* (pp. 99–142). Cambridge, UK: Cambridge University Press.
Lombard, M., & Pargeter, J. (2008). Hunting with Howiesons Poort segments: Pilot experimental study and the functional interpretation of archaeological tools. *Journal of Archaeological Science, 35*, 2523–2531.
Loud, L. L., & Harrington, M. R. (1929). *Lovelock Cave* (Vol. 25). USA: University of California Publications in American Archaeology and Ethnology.
Mazza, P., Martini, F., Sala, B., Magi, M., Colombini, M., Giachi, G., et al. (2006). A new Palaeolithic discovery: Tar-hafted stone tools in a European Mid-Pleistocene bone-bearing bed. *Journal of Archaeological Science, 33*, 1310–1318.
McBrearty, S., & Brooks, A. (2000). The revolution that wasn't: A new interpretation of the origin of modern human behavior. *Journal of Human Evolution, 39*, 453–563.
Mussi, M., & Villa, P. (2008). Single carcass of *Mammuthus primigenius* with lithic artifacts in the Upper Pleistocene of northern Italy. *Journal of Archaeological Science, 35*, 2606–2613.
O'Connell, J. F. (2006). How did modern humans displace neanderthals? Insights from hunter-gatherer ethnography and archaeology. In N. J. Conard (Ed.), *When neanderthals and modern humans met* (pp. 43–64). Tübingen: Kerns Verlag.
Odell, G. H. (1977). The application of micro-wear analysis to the lithic component of an entire prehistoric settlement: Methods, problems, and functional reconstructions. PhD dissertation, Harvard University.
Odell, G. H. (1988). Addressing prehistoric hunting practices through stone tool analysis. *American Anthropologist, 90*, 335–356.
Odell, G. H., & Cowan, F. (1986). Experiments with spears and arrows on animal targets. *Journal of Field Archaeology, 13*, 95–212.
Pargeter, J. (2011). Assessing the macrofracture method for identifying Stone Age hunting weaponry. *Journal of Archaeological Science, 38*, 2882–2888.
Pendleton, L. S. A. (1985). Material culture: Artifacts of wood. In D. H. Thomas (Ed.), *The archaeology of Hidden Cave, Nevada*. Anthropological Papers of the American Museum of Natural History (Vol. 61, no. 1, pp. 251–261). New York: American Museum of Natural History.
Plisson, H., & Béyries, S. (1998). Pointes ou outils triangulaires? Données fonctionnelles dans le moustérien levantin. *Paléorient, 24*, 5–24.
Rigaud, J., & Simek, J. F. (1987). "Arms too short to box with God": Problems and prospects for Paleolithic prehistory in Dordogne, France. In O. Soffer (Ed.), *The Pleistocene Old World: Regional perspectives* (pp. 47–61). New York: Plenum Press.
Roper, D. C. (1979). Breakage patterns of central Illinois woodland projectile points. *Plains Anthropologist, 24*, 113–121.

Rots, V. (2003). Towards an understanding of hafting: The macro- and microscopic evidence. *Antiquity, 77*, 805–815.
Rots, V. (2004). Prehensile wear on flint tools. *Lithic Technology, 29*, 7–32.
Salls, R. A. (1986). The La Brea atlatl foreshafts: Inferences for the Millingstone horizon. *Pacific Coast Archaeological Society Quarterly, 22*, 21–30.
Sano, K. (2009). Hunting evidence from stone artefacts from the Magdalenian cave site Bois Laiterie, Belgium: A fracture analysis. *Quatär, 56*, 67–86.
Semenov, S. A. (1964). *Prehistoric technology*. Bath: Adams and Dart.
Shott, M. J. (1997). Stones and shafts redux: The metric discrimination of chipped-stone dart and arrow points. *American Antiquity, 62*, 86–101.
Shea, J. J. (1988). Spear points from the Middle Paleolithic of the Levant. *Journal of Field Archaeology, 15*, 441–450.
Shea, J. J. (1990). A further note on Mousterian spear points. *Journal of Field Archaeology, 17*, 111–114.
Shea, J. J. (1991). The behavioral significance of Levantine Mousterian industrial variability. PhD dissertation, Harvard University.
Shea, J. J. (1993). Lithic use-wear evidence for hunting by Neandertals and early modern humans from the Levantine Mousterian. In G. L. Peterkin, H. M. Bricker & P. Mellars (Eds.), *Hunting and animal exploitation in the Later Palaeolithic and Mesolithic of Eurasia* (pp. 189–197). American Anthropological Association Archaeological Paper 4.
Shea, J. J. (1995a). Behavioral factors affecting the production of Levallois points in the Levantine Mousterian. In H. L. Dibble & O. Bar-Yosef (Eds.), *The definition and interpretation of Levallois technology* (pp. 279–292). Madison: Prehistory Press.
Shea, J. J. (1995b). Lithic microwear analysis of Tor Faraj rockshelter. In D. O. Henry (Ed.), *Prehistoric ecology and evolution: Insights from Southern Jordan* (pp. 85–97). New York: Plenum Press.
Shea, J. J. (1997). Middle Paleolithic spear point technology. In H. Knecht (Ed.), *Projectile technology* (pp. 79–106). New York: Plenum Press.
Shea, J. J. (2003a). Neandertals, competition, and the origin of modern human behavior in the Levant. *Evolutionary Anthropology, 12*, 173–187.
Shea, J. J. (2003b). The Middle Paleolithic of the East Mediterranean Levant. *Journal of World Prehistory, 17*, 313–394.
Shea, J. J. (2006). The origins of lithic projectile point technology: Evidence from Africa, the Levant, and Europe. *Journal of Archaeological Science, 33*, 823–846.
Shea, J. J. (2009). The impact of projectile weaponry on late Pleistocene hominin evolution. In J.-J. Hublin & M. P. Richards (Eds.), *The evolution of hominid diets: Integrating approaches to the study of Paleolithic subsistence* (pp. 189–199). Dordrecht: Springer.
Shea, J. J. (2011). *Homo sapiens* is as *Homo sapiens* was: "Behavioral modernity" in Paleolithic archaeology. *Current Anthropology, 52*, 1–35.
Shea, J. J., Davis, Z., & Brown, K. (2001). Experimental tests of middle Paleolithic spear points using a calibrated crossbow. *Journal of Archaeological Science, 28*, 807–816.
Shea, J. J., & Sisk, M. L. (2010). Complex projectile technology and *Homo sapiens* dispersal into western Eurasia. *Paleoanthropology, 2010*, 100–122.
Sisk, M. L., & Shea, J. J. (2009). Experimental use and quantitative performance analysis of triangular flakes (Levallois points) used as arrowheads. *Journal of Archaeological Science, 36*, 2039–2047.
Smith, G. A. (Ed.). (1963). *Archaeological survey of the Mojave river area and adjacent regions*. San Bernardino: San Bernardino County Museum Association.
Smith, G. A., Schuiling, W. C., Ritner, L. M., Sayles, J., & Jillson, P. (1963). The archaeology of Newberry Cave, San Bernardino County, Newberry, California. In G. A. Smith (Ed.), *Archaeological survey of the Mojave river area and adjacent regions* (pp. 138–197). San Bernardino: San Bernardino County Museum Association.
Solecki, R. L. (1992). More on hafted projectile points in the Mousterian. *Journal of Field Archaeology, 19*, 207–212.
Taylor, W. W. (1966). Archaic cultures adjacent to the northeastern frontiers of Mesoamerica. In G. F. Ekholm & G. R. Willey (Eds.), *Archaeological frontiers and external connections* (pp. 59–94). Washington: Handbook of Middle American Indians 4.
Thieme, H. (1997). Lower Paleolithic hunting spears from Germany. *Nature, 385*, 807–810.
Thomas, D. H. (1978). Arrowheads and atlatl darts: How the stones got the shaft. *American Antiquity, 43*, 461–472.
Tomenchuk, J. (1985). The development of a wholly parametric use-wear methodology and its application to two selected samples of Epipaleolithic chipped stone tools from Hayonim Cave, Israel. PhD dissertation, University of Toronto.
Tringham, R., Cooper, G., Odell, G., Voytek, B., & Whitman, A. (1974). Experimentation in the formation of edge damage: A new approach to lithic analysis. *Journal of Field Archaeology, 1*, 171–196.
Tuohy, D. R. (1982). Another Great Basin atlatl with dart foreshafts and other artifacts: Implications and ramifications. *Journal of California and Great Basin Anthropology, 4*, 80–106.
Villa, J., & Lenoir, M. (2006). Hunting weapons of the Middle Stone Age and the Middle Palaeolithic: Spear points from Sibudu, Rose Cottage and Bouheben. *South African Humanities, 18*, 89–122.
Villa, J., Boscato, P., Ranaldo, F., & Ronchitelli, A. (2009a). Stone tools for the hunt: Points with impact scars from a Middle Paleolithic site in southern Italy. *Journal of Archaeological Science, 36*, 850–859.
Villa, J., Soressi, M., Henshilwood, C., & Mourre, V. (2009b). The still bay point of Blombos cave (South Africa). *Journal of Archaeological Science, 36*, 441–460.
Witthoff, J. (1968). Flint arrowpoints from the Eskimo of northwestern Alaska. *Expedition, 10*, 30–37.
Woods, J. C. (1988). Projectile point fracture patterns and inferences about tool function. *Idaho Archaeologist, 11*, 3–8.
Woodward, A. (1937). Atlatl dart foreshafts from the La Brea pits. *Bulletin of the Southern California Academy of Sciences, 36*, 41–60.
Wyckoff, D. G. (1964). The cultural sequence of the Packard Site, Mayes County, Oklahoma. *Oklahoma River Basin Survey Project: Archaeological Site Report, No. 2*. Norman: University of Oklahoma Research Institute.

Chapter 2
Identifying Weapon Delivery Systems Using Macrofracture Analysis and Fracture Propagation Velocity: A Controlled Experiment

Radu Iovita, Holger Schönekeß, Sabine Gaudzinski-Windheuser, and Frank Jäger

Abstract In the last few decades, zooarchaeological studies have demonstrated beyond doubt that the hunting abilities of hominins were quite formidable from quite early on. Unfortunately, direct evidence for the use of weapons in hunting is quite rare and depends heavily on the preservation of organic elements. In particular, in the absence of such evidence, it is notoriously difficult to pinpoint the first appearance of complex, mechanically-assisted projectiles (such as darts and arrows) in the archaeological record. In this chapter, we present data from a controlled ballistic experiment with the aim of establishing patterns in the formation of impact fractures that would allow for the discrimination of thrusting spears, (hand-thrown) javelins, and spearthrower darts and arrows. By controlling for the weapon tip shape, weight, and raw material, impact angle (IA), as well as target composition, we are able to focus on the key elements that separate the different launching systems: velocity and kinetic energy output. The results show that fracture scar length is proportional to kinetic energy at impact, but only if the impact is perpendicular, as acute IAs reduce the energy requirements for the production of large, typical impact fractures. We also confirm previous results of Hutchings (JAS 38:1737–1746, 2011) regarding the relationship between precursory loading rate and fracture propagation speed, documenting a weak linear relationship between the two in our sample. We conclude by discussing the implications of this study for identifying different weapon armatures in the archaeological record.

Keywords Controlled experiments • Weapon delivery mechanisms • Spears • Javelins • Spearthrower darts • Wallner lines • Fracture velocity

Weapon Delivery Systems in an Evolutionary Perspective

Weapons take up a very important role in current portrayals of technological evolution and especially in the discussions about the comparative fitness levels of archaic v. anatomically modern humans (e.g., McBrearty and Brooks 2000). To a great degree, this is because weapons are perhaps the last unchallenged remnant of the image of early hominins as "Man the Hunter" (Lee and DeVore 1968). Nearly all such portrayals focus on the struggle of prehistoric hominins with large, dangerous animals, for which the natural physical disadvantages of the former must have been overcome by the use of increasingly advanced technology. In the absence of such technology, direct close combat scenarios have been inferred, and to a large extent, also documented in the form of trauma patterns on the human skeletal remains. This is especially the case for Neandertals, whose patterns of trauma have been found to match those of rodeo riders (Berger and Trinkaus 1995), although these results no longer seem as strong as they once were (see Trinkaus 2012).

But despite the comical picture of the more ancient hominins battling enormous beasts and getting hurled up in the air by them, recent research in zooarchaeology has shown that the hunting abilities of quite ancient hominins must have been quite formidable (e.g., Farizy et al. 1994; Gaudzinski 1995; Marean and Kim 1998; Gaudzinski and

R. Iovita (✉) · S. Gaudzinski-Windheuser
MONREPOS Archaeological Research Centre and Museum for Human Behavioural Evolution, Römisch-Germanisches Zentralmuseum (RGZM), Leibniz-Forschungsinstitut für Archäologie, 56567 Neuwied, Germany
e-mail: iovita@rgzm.de

S. Gaudzinski-Windheuser
e-mail: gaudzinski@rgzm.de

S. Gaudzinski-Windheuser
Altertumswissenschaften, Arbeitsbereich Vor- Und Frühgeschichte, Johannes-Gutenberg-University of Mainz, Schönborner Hof, Schillerplatz 11, 55116 Mainz, Germany

H. Schönekeß · F. Jäger (*Deceased*)
Physikalisch-Technische Bundesanstalt (PTB), Braunschweig, Germany
e-mail: schoenekess@ptb.de

Radu Iovita and Katsuhiro Sano (eds.), *Multidisciplinary Approaches to the Study of Stone Age Weaponry*, Vertebrate Paleobiology and Paleoanthropology, DOI 10.1007/978-94-017-7602-8_2

Roebroeks 2000; Gaudzinski-Windheuser 2005). Many sites across Europe demonstrate the mass-hunting of bovids, cervids, and equids in the form of monospecific faunal assemblages with good evidence for cut marks (Gaudzinski and Roebroeks 2000; Gaudzinski 2005). Moreover, hominins did not only systematically hunt large and dangerous herd herbivores, but also solitary animals such as rhinoceros (Bratlund 1999). Evidence from several sites in Europe also strongly suggests that Neandertals regularly hunted cave bears, possibly during hibernation (Kindler 2012). Thus, if Neandertals (and possibly earlier *Homo heidelbergensis*) were accomplished hunters, the pressing questions become: what kind of weapons did early hominins use? To what extent can weapon types be correlated with technological and biological evolution?

Almost ubiquitous in the literature is the model of unilinear evolution from thrusting spears to javelins (hand-thrown spears), to mechanically-assisted weaponry, namely spearthrower-propelled darts and bows and arrows (see Fig. 2.1). The latter two are nowadays referred to as "complex projectiles" (Shea and Sisk 2010; Sisk and Shea 2011) and are considered, beyond their improved functionality in terms of impact energy and killing capacity, to also imply complex cognitive processes, both in their production and use (Lombard and Haidle 2012; see also Lombard and Wadley 2016).

We owe this unilinear model in large part to chronological milestones established by extraordinary discoveries of well-preserved organic remains, which allow us to reconstruct, in those specific cases, the whole technology. For instance, the discovery of 300–400 kyr old sharpened wooden implements at Schöningen in Germany [universally interpreted as spears or javelins (Thieme 1997; Rieder 2003)] in association with more than 20 horses seems to provide a baseline for the technology available to the earliest hominins, both in terms of age and quality of the evidence. At the other end, the oldest preserved arrow shafts from Stellmoor, also in Germany (Rust 1943), are only about 10 ka old (Fischer and Tauber 1986; Weber et al. 2011), whereas the oldest spearthrower, from the Solutrean site of Combe Saunière in France, is probably only a few thousand years older, at ca. 18–20 ka (Cattelain 1989; Roque et al. 2001). Between these milestones, there is only indirect evidence of weapon delivery systems, mostly in the form of traces left on stone points, which may point to their use as

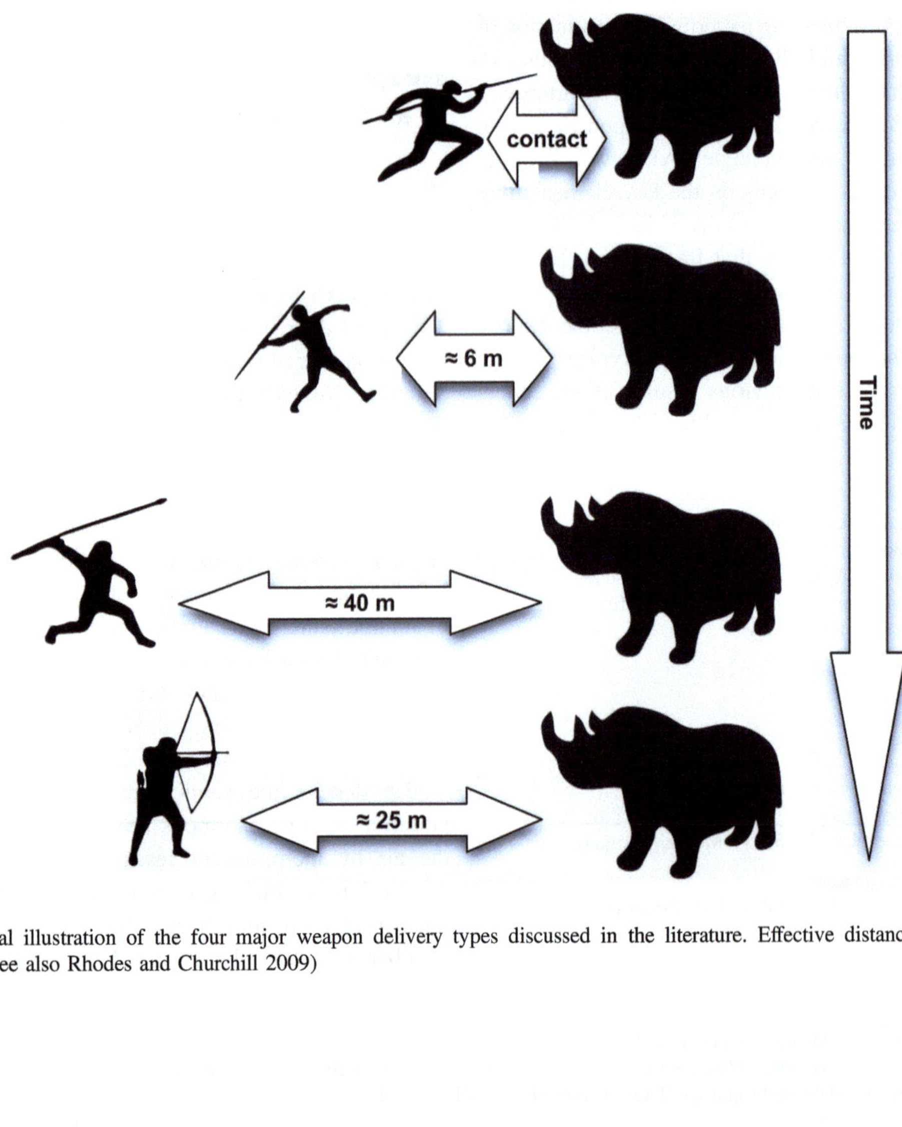

Fig. 2.1 Graphical illustration of the four major weapon delivery types discussed in the literature. Effective distances are from Churchill (Churchill 1993; see also Rhodes and Churchill 2009)

weapon tips. In particular, it seems that we are dependent upon lithic evidence to establish the first appearance of mechanically-aided projectile weaponry, since there are many lithic industries between 300 ka and 20 ka which feature implements that are similar to components of flying projectiles, but no preserved organic parts that could confirm their place within the technology.

The Contribution of Lithic Use Wear Analysis

Especially for the reasons outlined above, the study of impact-related lithic use wear has proven of critical importance. Systematic investigations of macroscopic damage, also known as "diagnostic impact fractures" (DIFs, Lombard 2005) have a long history, with a particularly active period in the 1980s (e.g., Witthoft 1968; Ahler and McMillan 1976; Frison et al. 1976; Barton and Bergman 1982; Huckell 1982; Bergman and Newcomer 1983; Fischer et al. 1984; Odell and Cowan 1986; Geneste and Plisson 1989; Plisson and Beyries 1998). The last decade has seen a renewed interest in experimentation (e.g., Shea et al. 2001; Lombard et al. 2004; Pargeter 2007, 2011; Lombard and Pargeter 2008; Sisk and Shea 2009; Schoville 2010), as well as searching the archaeological record for evidence of different weapon types (Sano 2009; Villa et al. 2009a, b; Lombard 2011; Rots et al. 2011; e.g., Brown et al. 2012; Lazuén 2012; Wilkins et al. 2012). Including the indirect evidence, the picture of unilinear evolution of weapon technologies appears to contain a few surprises. The most recent data from Africa (Wilkins et al. 2012) hints at the appearance of hafted, stone-tipped spears or javelins by about 500 ka, presenting evidence similar to that offered in the 1980s to infer the same function for Levallois points from Neandertal sites (e.g., Shea 1988, but see Plisson and Beyries 1998). Similarly, several recent papers have made arguments in favor of the appearance of bow and arrow technology by 60 (Lombard and Phillipson 2010; Lombard 2011) or even 70 ka (Brown et al. 2012) in South Africa, much before its appearance in Europe, where it is pre-dated by the spearthrower.

However, demonstrating the existence of complex projectiles beyond any doubt requires an extra step beyond demonstrating the existence of weapons per se. In addition to evidence for violent impact, a convincing argument must be made that a launching mechanism, either a spearthrower or a bow, would have been used. Because the latter are almost never preserved, most studies that have tried to identify the launching mechanism so far (e.g., Lombard and Phillipson 2010) have combined the use-wear evidence with metric attributes, which are known to be capable of discriminating between arrow, (spearthrower) dart, and spear tips on a purely metric basis (Shott 1997; Shea 2006). Tip cross sectional area (TCSA) and tip cross sectional perimeter (TSCP), ballistic parameters relevant to a tip's ability to penetrate the target (but see Newman and Moore 2013; Clarkson 2016) have thus been used in conjunction with use wear data to conclude that the studied weapons would have been small and light, and would most likely have been launched with the aid of a propelling device (Sisk and Shea 2011).

Although the combination of multiple lines of evidence may be more or less plausible depending on the case, the only evidence which can establish a weapon tip as having been launched by a mechanical device is evidence of a pattern consistent exclusively with speeds or impact energies beyond what can be achieved with hand delivery (or even through accidents, see also Hutchings 2016). Ideally, we should establish quantifiable, objective criteria for distinguishing between types of weaponry based upon direct consequences of their physical properties. The best approach to understand these sorts of causal relationships is through experiments performed under controlled conditions, where the effect of each relevant variable can be evaluated separately.

Controlled v. Replicative Experiments

In order to infer the existence of any past human behavior from archaeological remains, it is necessary to identify which aspects of that behavior are intrinsically connected with those remains. Because both behaviors and their effects on artifacts are very complex, a simplifying model must be sought in order to focus on the relevant variables and lessen the labor costs of reconstruction. If the behavior is "hunting with penetrative weapons" and the remains are "lithics with use wear", it is necessary to identify which aspects of the former are intrinsically (and causally) related to the latter. Such a project requires a variety of types of experimental work, which are equally important and interdependent. First, as many possible types of hunting must be replicated and the full spectrum of associated lithic use wear catalogued. This first step has been well covered in the past 20 years of experimental research. Second, all identified wear patterns that could be replicated through post-depositional damage or accidents should be eliminated from the catalogue. This type of experiment has come to the foreground in the last few years, especially with respect to the effect of trampling on edge damage (Sano 2009; Schoville 2010; Pargeter 2011). Finally, the types of wear that are expected from theory should be evaluated in controlled and repeatable conditions, in order to focus in on specific physical variables of interest, such as output velocity or kinetic energy.

This last step is crucial to being able to separate wear that is directly caused by violent impact from wear that is caused by ancillary factors. When an experimenter attempts to investigate one variable (e.g., launching system), but in the course of the experiment changes one of the other variables (for instance, tip shape, which is difficult to control through flintknapping), many other variables may be changed as well (such as edge angle, point angle, thickness, etc.), thereby influencing the type and size of the effect (in this case, size, shape, etc. of the damage). It is tempting then to think that the changes in the result are caused by the target variable (in this case, launching system), but the influence of the other unaccounted factors cannot be ruled out, unless they are held constant. In this study, we present two series of experiments aimed at separating thrust spears from flying projectiles, carried out in the most controlled fashion possible, while simultaneously preserving the essential physical properties of the target attributes. More specifically, we control for tip and shaft shape, raw material and weight, while measuring impact velocity and angle.

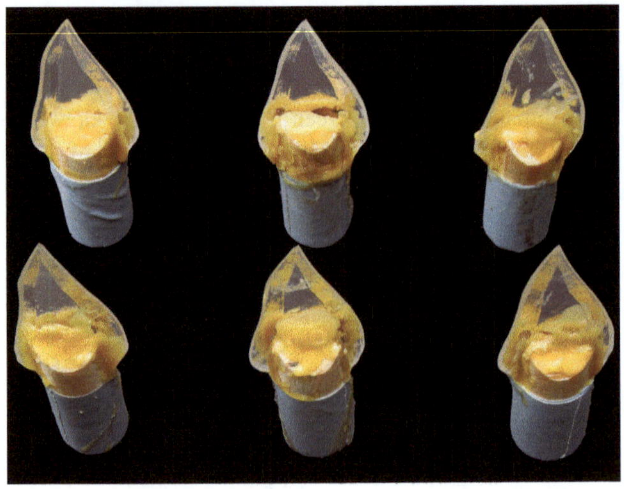

Fig. 2.2 Six hafted copies of the Jabrud Levallois point, including the foreshafts

Materials and Methods

In order to examine the fundamental differences between the damage sustained by points in thrusting v. projectile use, we employed two different series of controlled laboratory experiments. Both series used the same weapon tips and the same target, the differences between them being reduced to the relevant physical quantities related to the load rate and kinetic energies applied to the target at impact.

General Setup

For both experiments we used nearly identical soda-lime glass copies of a Levallois point from Jabrud, Syria (Rust collections, University of Cologne, original dimensions (mm): 64.5 (length), 36.5 (max. width), 6 (max. thickness) (see Fig. 2.2). The copies were cast in soda-lime glass, a standard material science experimental substance with known brittle fracture properties, using a waffle-iron-like metal form. Glass is a good substitute for other siliceous rocks, including obsidian, that might have been used by prehistoric people to make projectile points, mainly because it breaks in the same way. Naturally, due to differences in the formation processes of flints or silcretes, these rocks will probably exhibit some differences from glass, whereas obsidian, which is volcanic glass, will be most similar. However, carrying out the tests with any one of these rock types would produce results that are different from the others, and glass presents the advantage of being easily shaped through heating. The shape of the cast points is identical visually, but the points needed to be filed on the ventral side in order to avoid having jagged irregularities, due to the accumulation of glass droplets when the form was closed. Filing the points adds to the convexity of their ventral side, departing from the typical Levallois point shape. In order to check the variability of the glass-casted copies, the same simple measurements were collected on a random sample of 53 points, and the following coefficients of variation (CVs) were obtained: 0.009 (length), 0.026 (max. width), and 0.041 (max. thickness), and 0.035 (weight). The CV values are extremely small, suggesting that the points can be treated as essentially the same, at least with respect to their ballistic properties.

Both series of experiments likewise used the same target, which was fastened onto a 22 kg steel box (see Fig. 2.3), featuring an inner compartment which pivots around an axle at the bottom, allowing the control of the AI in 15-degree intervals, from 90° to 30°. A further slot for a 20° angle was never used in the experiments, because the spear frequently hit the upper part of the tilted steel frame by accident. The inner compartment is further divided in two sections, both filled with 20% ballistic gelatin, and separated by interchangeable plates of bone-like polyurethane. The synthetic bone plates, manufactured by SynBone AG, are 6 mm thick, and are specially designed to be used in ballistic testing. They mimic the structure of mammalian bone (including cortical and trabecular bone) and are covered in a thin layer of rubber, which is similar to the periosteum. Originally held in place by a steel slot, the bone plate was eventually sandwiched between a 2.5 cm thick slab of gelatin on the impact side and an 8.5 cm thick block on the other side. This was done so as to mimic the position of a bone within

Fig. 2.3 View from above of the target with the spear having exited at the back. Note the four layers, from left to right: Leather, gelatin, polyurethane "bone" plate, gelatin

muscle and other soft tissue more realistically, allowing for elastic absorption of the impact. Finally, ≈2 mm-thick scraps of cow leather were pressed against the outer gelatin, completing the target.

Likewise, in both experiments, the points were slot-hafted in machined pine wood foreshafts, which were then screwed onto metal spear shafts. We used natural beeswax as an adhesive. Beeswax is known from ethnographic as well as Paleolithic archaeological contexts (d'Errico et al. 2012) and presents the advantage of a quick and strong bind. However, it is equally easy to remove the points by simply heating the foreshaft, allowing for a time-efficient firing of more specimens.

Projectile Experiment

The first experiment (described in greater detail in Iovita et al. 2014) focused on isolating the two major causes of damage production, namely projectile impact velocity and angle (IA) from other situationally variable factors that can influence these patterns in a real-life hunting episode. In this series, the foreshafts containing the hafted points (n = 234) were attached to an aluminium tube (total weight of the spear ≈266 g) and propelled by an air-gun into the target from close range (free flight 24 cm). The launching mechanism does not allow for the spinning of the spears, so that the target is always hit identically. This does differ from a variety of hand-thrown motions, but was chosen for reducing complexity in the number of variables. The air gun chamber is capable of pressures between 1.25 and 15 bar, resulting in projectile speeds (for the 266 g spear) between ca. 7 and 30 m/s. The velocities were measured by a transient recorder connected to a set of light curtains (see Fig. 2.4).

The largest final velocity produced by our air gun (≈30 m/s, 110 km/h) corresponds to some atlatl-delivered darts (Raymond 1986; Stodiek 1993; Hughes 1998; Baugh 2003). Stodiek's high-speed camera (1993:194) recorded an approximately 70 km/h (≈20 m/s) final velocity at entry into the target at 25 m distance – but with a spear that weighed only 90 g. This corresponds to ≈19 J of kinetic energy, much lower than our highest values of over 100 J. Hutchings and Brüchert (1997) obtained much higher velocities than any other study, and for their dart (of similar weight, 273.4 g),

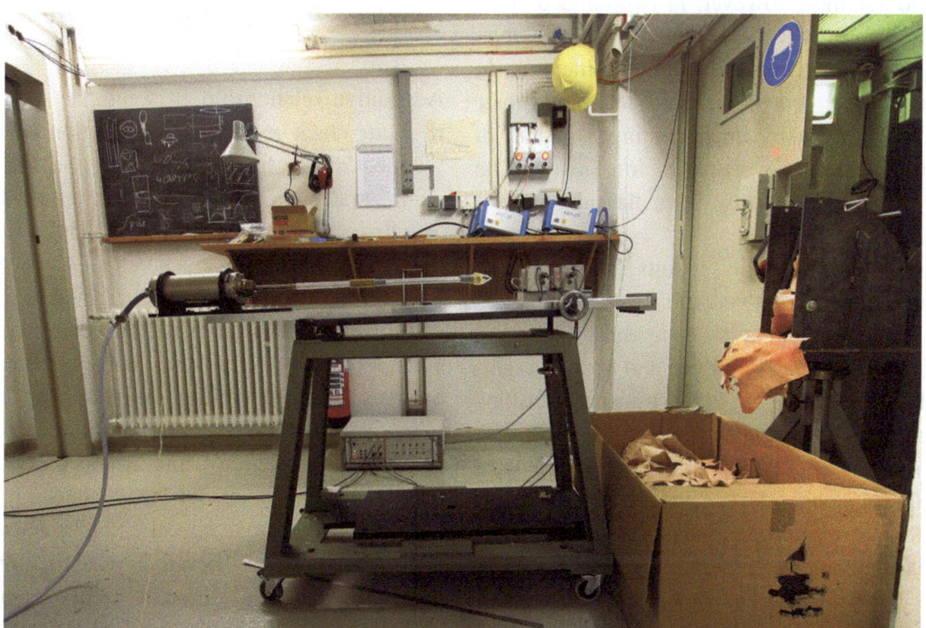

Fig. 2.4 Air gun for the projectile experiments. A detailed description, including all measurements can be found in Iovita et al. (2014)

they report initial velocities in the range of 34.9–64 m/s, with an average value of 42.5 m/s. Using their estimate of velocity loss of approximately 10% over the effective distance of 15 m to the target, and taking into account that our experiment measured essentially impact velocities, our highest values overlap with the lower range obtained by that study. Based on previous published results, we consider our range to be relatively comprehensive in terms of velocities and kinetic energies at impact, ranging between hand-thrown and atlatl-delivered flying projectiles (Fig. 2.4).

Thrusting Experiment

Since thrusting can deliver similarly high energies to those obtained by high-speed launching projectiles, but at considerably smaller speeds, the type of applied load is quite different. This has been previously mentioned in the literature (e.g., Shea et al. 2001; Hutchings 2011). With the exception of one experiment aimed at investigating the effect of thrusting and throwing loads on the bone morphology of the human shoulder (Schmitt et al. 2003), most experiments aimed at replicating thrusting did not measure the speed of the thrusting motion and did not calculate the associated kinetic energy of the spear. Shea et al. (2001:810) cite measured results out of the forensic literature on knife stabbing (e.g., Chadwick et al. 1999; Horsfall et al. 1999), putting the speed of the spear itself between 1 and 1.5 m/s, and the resulting energies at about 28–63 J. The speeds used in our experiment are in a similar range, from 1.1 to 2.7 m/s. However, in two-handed spear thrusting, at least a portion of the attacker's own weight is added to the spear in an attempt to increase momentum. In our second experimental series, we tried to approximate this in a controlled fashion, by implementing a system of weights that would guarantee a sustained force exerted on the target both during and shortly after the impact (see Fig. 2.5 below).

In our setup, the same Levallois points (n = 43) were hafted to their foreshafts and screwed into a metal bar, which was suspended from the ceiling at both ends, making a pendulum with 2 fixed points. 10, 20, 30, and 60 kg weights were successively added to the bar in a symmetrical fashion and the spear was launched from a variety of distances (30, 50, and 70 cm from the target). The impact velocities were measured using the same light curtain and transient recorder system used for the projectile experiments, making use of an attached metal bar hanging from the underside of the main spear shaft. The impact angle (IA) was always 90° in this experimental series. In case the spear bounced back from the target, it was caught by one of the experimenters in order to avoid double hits.

Macrofracture Analysis

Following the launch, each point was removed from the haft and the fracture type was recorded, along with the scar length and the missing length and placed in a labelled plastic bag. The fractures were identified and photographed with the aid of a Leica M420 macroscope, with magnifications up to 40x. The fracture typology followed those used by the major experiments in the 1980s, e.g., Barton and Bergman (1982), Fischer et al. (1984), and Odell and Cowan (1986), also reviewed more recently by Dockall (1997), and Sano (2009; Iovita et al. 2014). We use six major categories of damage:

1. flute-like (subsumed under longitudinal)
2. burin-like (subsumed under longitudinal)
3. transverse/snap
4. spin-off (secondary fracture)
5. tip crushing
6. microscopic (incipient or very small fractures)

In each case, only the largest fracture was used for the classification, although the presence of multiple fractures was noted. Because it was not always possible to refit the missing flakes back on to the tip, multiple fractures were only noted where they were believed to have hit the plate several times, which was corroborated with skipping marks on the plate. Finally, for each fracture, the type of initiation and termination was recorded according to the Ho Ho Committee (Cotterell et al. 1979) definitions.

Fracture Propagation Velocity

A positive relationship between the loading rate and the resulting crack or fracture propagation velocity was implied by the Griffith fracture concept (Griffith 1921) and later demonstrated experimentally (Richter 1974; Kerkhof 1975). The first attempts in archaeology to infer impact loading rate from measurements of crack velocity were made in the 1980s by Tomenchuk (1985), who first tried to use this method to distinguish between pressure and percussion flakes. He calculated crack velocity from visible secondary fracture characteristics, such as Wallner Lines (Wallner 1939) and fracture wings (*Bruchschwingen*) according to the methodology prescribed by Kerkhof and Müller-Beck (1969), Kerkhof (1975). This methodology was extended by Hutchings, who used it to demonstrate pressure flaking for some Clovis flutes (Hutchings 1999), and, more recently, for quantifying the difference between different weapon launching systems (Hutchings 2011; Sahle et al. 2013). These publications contain very detailed accounts of the procedure for identifying and calculating instantaneous fracture propagation

Fig. 2.5 Thrusting pendulum, in this photo loaded with six 10 kg weights

velocity, and so a review of these protocols will not be repeated here. In practice, this velocity can be calculated trigonometrically, by measuring several angles related to the intersection of a Wallner Line and the direction of the crack front or another Wallner Line. The geometry of the intersection is determined by the ratio of the transverse wave speed (c_2) and the crack front velocity (or fracture propagation velocity, $\dot{c}$), and since the former is a constant for each material, the velocity of the crack front at the point of intersection is a function of the intersection's geometry.

The pieces were taken out of their bags by a student and re-labeled with different codes, so as to avoid confirmation bias. The correspondence to the original experiment numbers was noted on a separate piece of paper to be checked at the end of the measurement series. One of us (RI) then photographed the pieces through an adapter to the microscope and performed the measurements on them using ImageJ (Fiji). Whenever possible (i.e., whenever multiple Wallner lines were visible), multiple measurements along the fracture surface were taken between 20 and 50% of the fracture length, and the maximum measurement was repeated three times. Unfortunately, this was only possible in a few cases, the majority of fractures exhibiting only one feature suitable for measurement. The measurements were repeated three times using the same point (recorded on a saved copy of the original image) at one week intervals, so as to avoid repetition bias.

Results

Macrofracture Patterns

Essentially, the results from the thrusting experiments can be differentiated from the ones of the projectile experiments by a much higher proportion of tips that were actually damaged. As can be seen in Fig. 2.6, the proportions of longitudinal macrofractures in the two samples (61:147 and 23:19, for projectiles and thrust spears respectively) are markedly different and the difference is statistically significant (Chi-squared = 9.03, df = 1, $p < 0.01$). This makes sense from a ballistic point of view. Given a similar amount of kinetic energy, the slower impact will be less likely to break the target, and the point will be more likely to suffer damage itself.

This can also be seen in the comparison between the proportions of the different fracture types in the thrust spears and the three sets of 10 identical shots at low (≈9.5 m/s), middle (≈15.5 m/s) and high (≈30 m/s) speeds (see Fig. 2.7, see also Iovita et al. 2014).

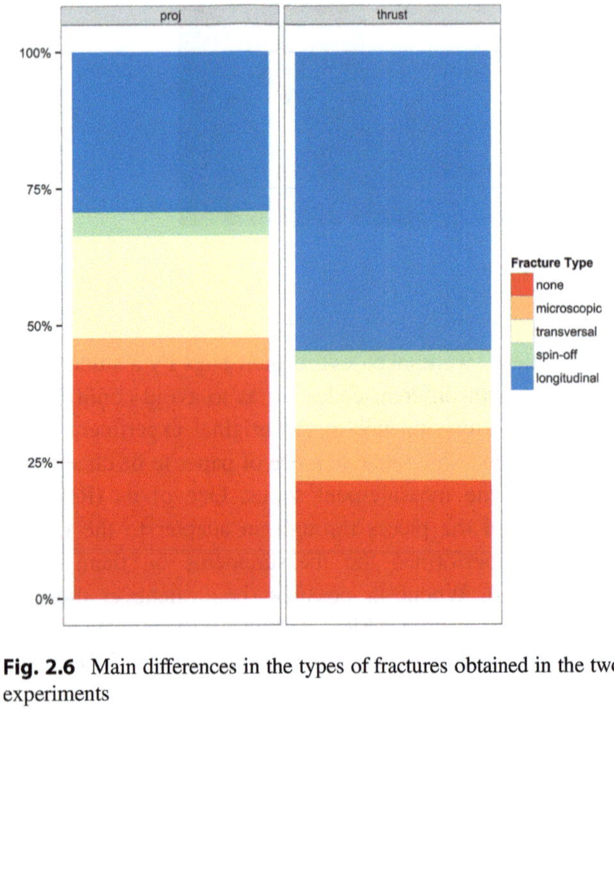

Fig. 2.6 Main differences in the types of fractures obtained in the two experiments

In terms of the scar lengths, there are no significant differences between thrusting and projected points of similar energies (see Fig. 2.8). The projectiles in the fastest quarter of the speed range obtained significantly larger damage scars than any of the thrusting spears, confirming the expectation that kinetic energy determines the size of the flake(s) removed at impact.

However, when the data are analyzed in raw form, there appears to be no relationship between kinetic energy and scar length, unless the impact angle (IA) is kept constant at 90°. In the latter case, the relationship is linear and the regression is significant ($t = 3.13$, $p < 0.01$ for projectiles, $t = 2.07$ and $p < 0.05$ for thrust spears) but weak for both projectiles and thrust spears ($R^2 = 0.4$ for projectiles, but only 0.1 for thrust spears; see also Fig. 2.9).

Fracture Propagation Velocity Results

Although secondary fracture features such as Wallner Lines and fracture wings are difficult to see on reflecting materials such as glass, it was nevertheless possible, through the use of a light diffuser and by focusing the objective to slightly below the surface (where Wallner Lines actually form), it was possible to obtain photographs on which measurements were possible (see Fig. 2.10) on a total of 48 pieces ($n_{thrust} = 15$, $n_{proj} = 33$). Since all the specimens presented here are made from the same soda-lime glass, and in order to allow an easier comparison with other raw materials from other experiments, the values reported are in terms of the ratio between the fracture propagation velocity ($\dot{c}$) and the transverse wave velocity (c_2), rather than in m/s.

Hutchings (2011, Table 3, see also Fig. 8) proposes the use of three loading rate ranges, corresponding roughly to $\dot{c}/c_2$ ratios between 0 and 0.10 (quasi-static), 0.10 and 0.38 (rapid), and above 0.38 (dynamic). In Hutchings's experiments, values of up to 0.58 of the transverse wave speed were calculated for both dart and arrow armature impacts.

In our experiments, we were largely able to confirm Hutchings's results, as can be seen from Fig. 2.11. The relationship between weapon velocity and fracture velocity expressed as the $\dot{c}/c_2$ ratio is linear and the regression of weapon velocity on the $\dot{c}/c_2$ ratio is significant, although the relationship is noisy ($R^2 = 0.34$, $p < 0.01$).

If we restrict the analyses to the range of kinetic energies common to both flying projectiles and thrust spears (in this experiment 6.4–58.2 J), there is no relationship between fracture velocity and kinetic energy, further confirming that the method is not sensitive to the amount of energy, but rather the precursory loading rate. We were unfortunately not able to statistically test the effect of angles on the $\dot{c}/c_2$ values obtained, because of too small samples of readable

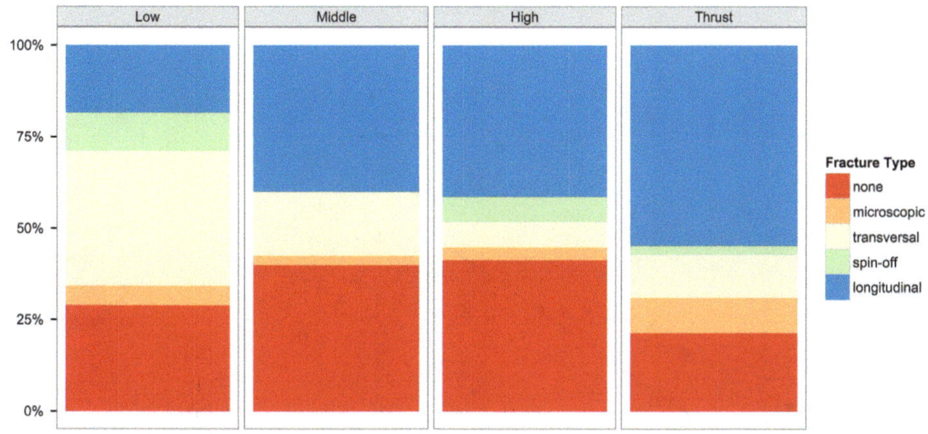

Fig. 2.7 Comparison of the types of fractures by controlled speed interval at IA = 90° (see text for the velocities for each trial)

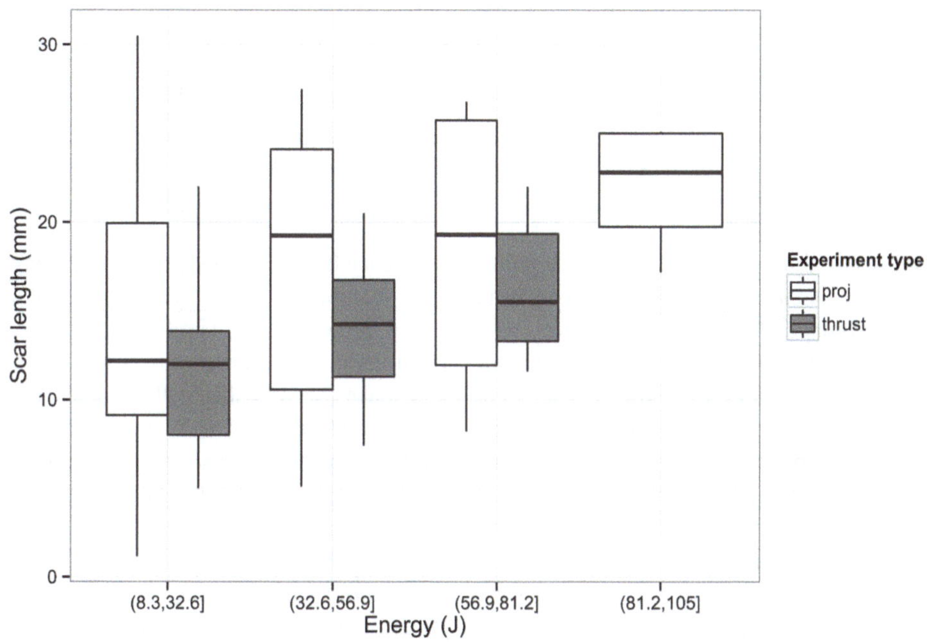

Fig. 2.8 Graph of scar length against the four equal intervals of kinetic energy. None of the thrust spears produced energies in the highest interval

secondary fracture characteristics per angle category. However, as recent results show (Iovita et al. 2014), acute impact angles significantly improve the chance of a fracture developing at any given speed and also influence the size of the resulting fracture, so the effect of the impact angle should be further investigated in the future.

Despite the many commonalities, there are also some differences from Hutchings's results that we wish to discuss. First, although our range for our thrusting spears, 0.07–0.22, is almost identical to that of Hutchings (2011, Table 3, 0.02–0.21), our median was 0.15, well within his "rapid" range. In that study, 84% of the fractures on spears remained in the "quasi-static" range with values not exceeding 0.10. It is possible that we used slightly higher velocities for our thrust spears, but it is difficult to compare our results directly with those of hand-thrust spears. Second, we did not obtain even a single fracture speed beyond 0.38, that is, within the "dynamic range". The obvious explanation for this is that the speeds we used are within the lower part of the range of darts and arrows used by Hutchings (2011) and derived from his earlier performance experiments with spearthrowers (Hutchings and Brüchert 1997). In the latter experiments, initial velocities in average of ≈42 m/s for a dart of similar weight were recorded, estimating ≈38 m/s at target entry. Due to the limitations of our machine, we could not accelerate our spear to more than 30 m/s. However, given that the bone plates would almost

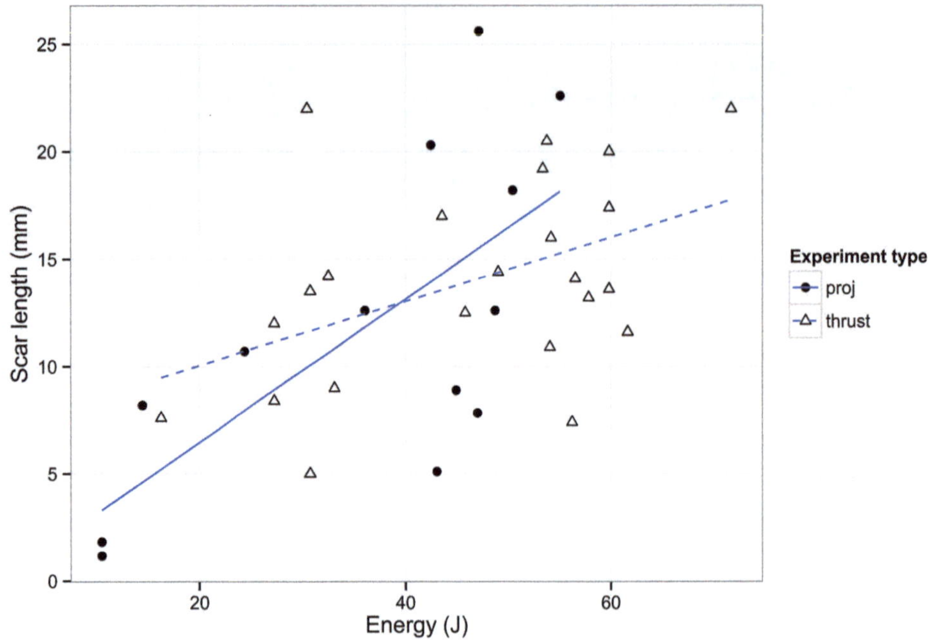

Fig. 2.9 Plot of kinetic energy against scar length, showing a weak, but significant linear relationship for both thrust and flying weapons (IA = 90°). Only points with longitudinal macrofractures were counted, as it is difficult to measure other types of breaks accurately

always break upon impact at this velocity, leaving the points undamaged, we did not pursue the objective of increasing launching speed any further. All other experiments with spearthrowers recorded launching speeds well below those of Hutchings and Brüchert (1997) and it is likely that different constructions would have resulted in different weapon launching power. For example, it is a well-known fact that many of the bows of modern hunter-gatherers are quite small and weak, such as those of the San bushmen, which often cannot draw more than 8–10 kg (Bartram 1997). Therefore, it is probable that the velocities obtained by Hutchings and Brüchert (1997), and hence, the fracture velocities from the later experiments (Hutchings 2011) represent an upper limit for what one might encounter archaeologically in the early development stages of complex projectile technology.

Discussion and Conclusions

As the present experiments show, simple macrofracture analysis does not help to distinguish thrusting from projectile impacts. If even controlled experiments where points are shot only once in standardized targets do not provide sufficiently nuanced patterns for the detection of different projectile weapon speeds, it is difficult to imagine that one could succeed after the reintroduction of all the ancillary factors likely to occur in realistic hunting scenarios. As we have shown quantitatively and with a large sample, the main difference between thrust spears and flying projectiles seems to be that thrust spears suffer more damage to the points than flying projectiles of similar kinetic energies. This result confirms previous observations (e.g., Lombard et al. 2004) and makes sense ballistically, since spear thrusting is very slow, even compared with simple projectiles. For comparable kinetic energies, at slow speeds and high momentum, the weapon tip penetrates the bone plate more frequently, but without shattering it. The result is often point failure through shearing after continuing to be pushed forward. In contrast, with high-speed impacts, the bone plate often shatters instantly, leaving the point intact.

In an archaeological setting, however, this difference is unlikely to survive, due to the likely repeated use, as well as curation and recycling of undamaged points, which would make it difficult to distinguish weapon technologies based on frequency of damage. However, one possible implication of the higher number of damaged points for thrusting spears is an expectation of a higher number of bones in faunal assemblages exhibiting embedded spear tips. However, the only type of damage on bone that is possibly diagnostic for projectile use is the perforated scapula (Smith 2003). Unfortunately, such evidence is extremely rare, and becomes common in periods where complex projectiles are already known (e.g., Bratlund 1991, see also Gaudzinski-Windheuser 2016). Other studies (Rots et al. 2011) have identified gestures that may be associated with spear thrusting, such as a twisting motion, aiming either to cause more internal damage, or to facilitate pulling the spear out. This would result in the co-occurrence

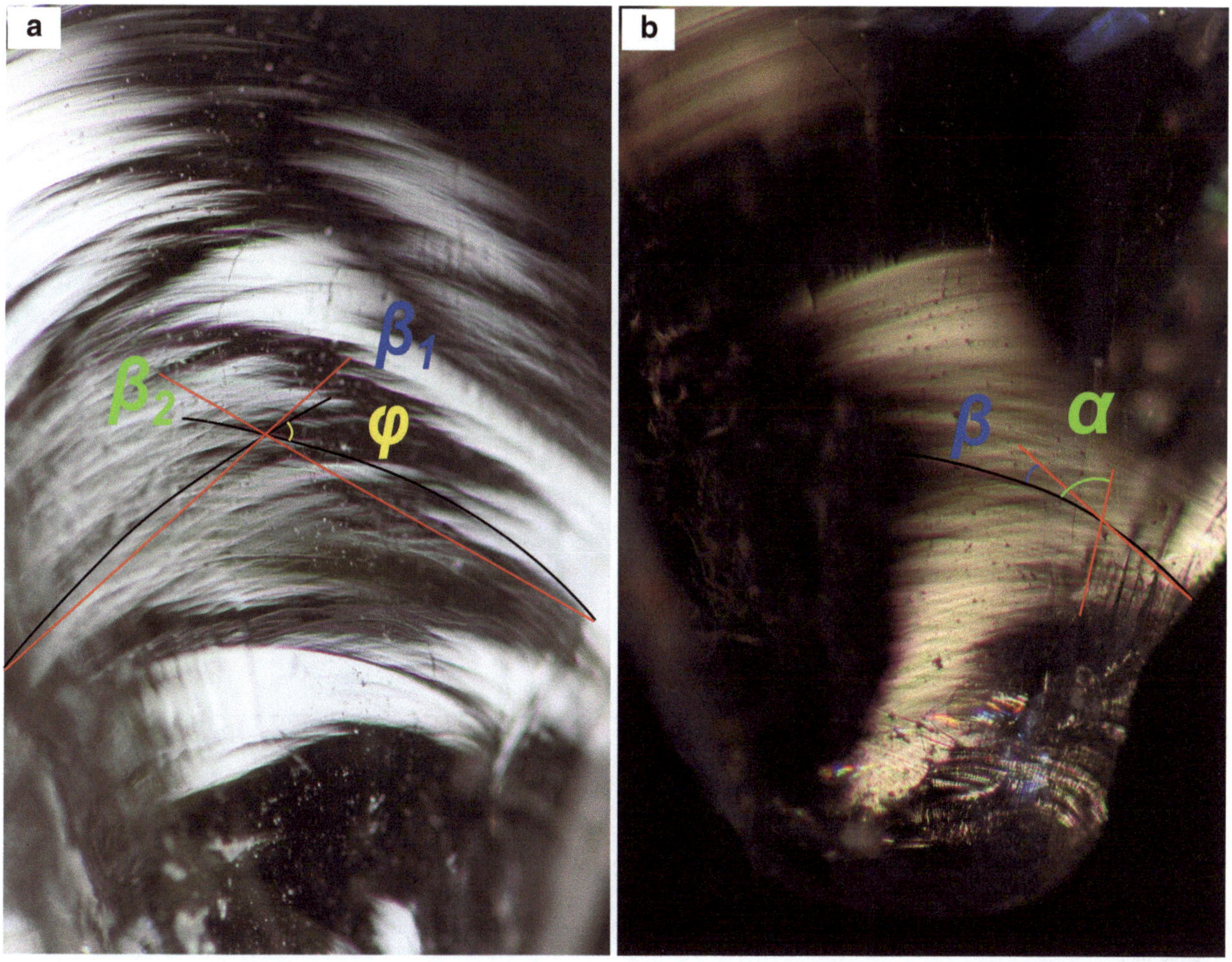

Fig. 2.10 Two fractures exhibiting clear crossing sets of Wallner-Lines. **a** v = 21.5 m/s, $\frac{c'}{c_2} = \frac{sin(\varphi)}{\sqrt{cos^2(\beta_1) + cos^2(\beta_2) + 2cos(\beta_1)cos(\beta_2)cos(\varphi)}} = 0.34$; and **b** $v = 13.6\,m/s$, $\frac{c'}{c_2} = \frac{cos(\alpha)}{cos(\beta)} = 0.27$

of an impact and a torsion fracture, but this has not been so far researched in a systematic fashion, and unfortunately, twisting motions in and of themselves could be the result of other tasks than spear thrusting.

The so far best method for identifying weapon technology based on use wear is through the calculation of fracture propagation velocities using one of Kerkhof's (1975) protocols, also outlined in Hutchings (2011). We were able to confirm that a linear relationship between projectile speed and fracture speed does exist and is independent of the kinetic energy output. However, the relationship is relatively weak ($R^2 = 0.34$), with a large variation of fracture velocities observed for any particular projectile speed, something also mentioned by Hutchings (2011). This means that, given a sufficiently large sample representing only one, or two very different weapon launching technologies (one slow, one very fast), one could determine roughly in which range of speed the respective weapons might have been used. However, this picture is complicated by the fact that the rapid range, where the majority of the fracture speeds in this experiment fell, contains many fractures that have nothing to do with weapon use, such as flintknapping and dropping accidents, and possible other intentional tasks that involve impacts.

Moreover, projectile speeds of up to 30 m/s, common to many complex projectiles, could also exhibit fracture speeds that only span the rapid range, meaning that this method is only useful for the identification of extremely fast complex projectiles. While it has been demonstrated that atlatls can

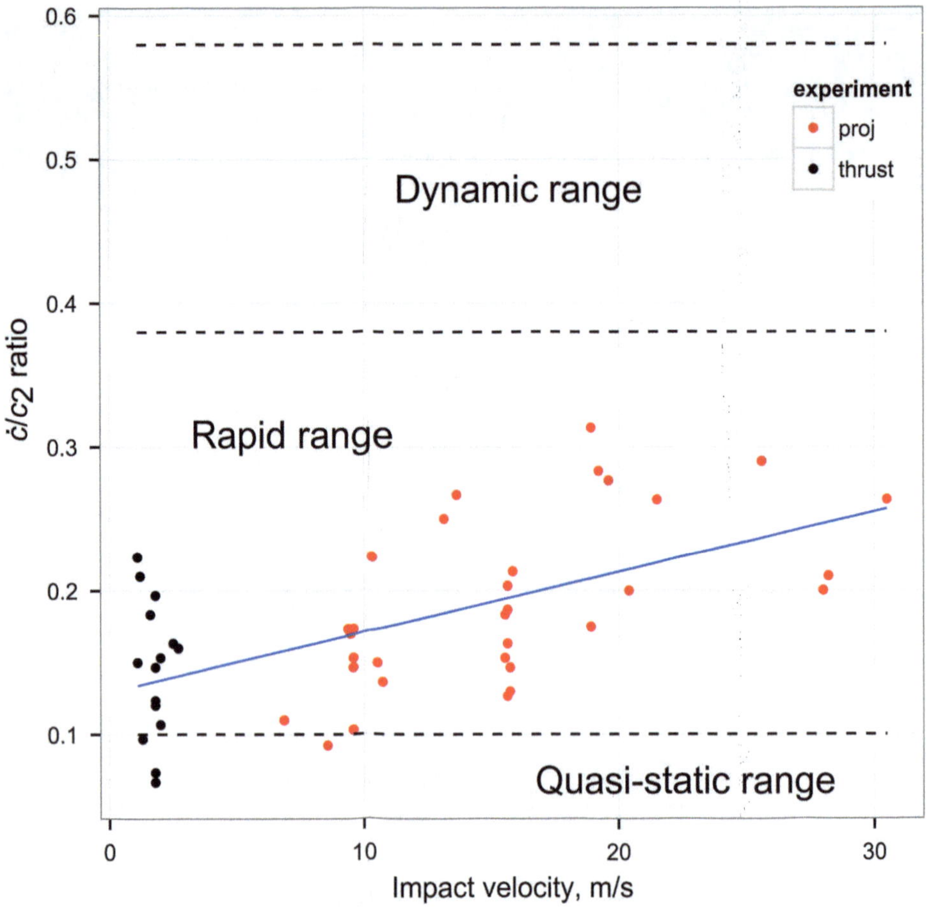

Fig. 2.11 Graph of weapon impact velocity against fracture propagation velocity, expressed as the $\dot{c}/c_2$ ratio. Note the lack of points in the dynamic range

accelerate darts up to 64 m/s (Hutchings and Brüchert 1997), more studies of different constructions are needed to assess how typical such speeds might have been in the past.

Similarly, fracture propagation velocities are useful for identifying thrusting spears, if very low fracture speeds ($\dot{c}/c_2$ ratio below 0.10) are observed. However, if fractures only spanning the rapid range are observed in an assemblage, it will be impossible to distinguish between any of the three major types of launching system, unless the fractures are associated with other attributes (such as standardized shape) that can be used as independent clues.

In conclusion, we must remember that hunter-gatherers often use a mixture of weapons, depending on the type of game and hunting situation (e.g., Churchill 1993). Tool recycling and site function are expected to play an important role in determining the frequency of usable data for answering questions about the type of launching system, especially if weapon tips are made of a single piece (as is the case for many tanged arrowheads or bifacial points), being more likely to change function following incurring too much damage. In any case where a complex projectile technology is suspected, a thorough study of the fracture velocities should be carried out. Such a study may reveal a few specimens that can be placed in the dynamic range, demonstrating beyond doubt that this is the case; however, a negative result does not rule such a technology out and for such cases only a suite of converging lines of evidence can provide a convincing argument in favor of complex projectile technology.

Acknowledgements This work is part of the research strategy of the MONREPOS Archaeological Research Centre and Museum for Human Behavioural Evolution. The authors wish to thank Jörg Herbst, Heinz Hertel, Maik Röhr, and Hans-Georg Staats (PTB) for help with carrying out the experiments and speed measurements. Thanks to Jürgen Richter (University of Cologne) for the authorization to make a mold of the original specimen, to Leslie Pluntke (RGZM Restoration Laboratory) for the production of the mold, and to Susanne Greiff (Archaeometry Center, RGZM) for advice with the process. Many thanks to Walter

Mehlem (Praehistoric Archery, Kruft, Germany), who designed and constructed the target and to Wolfgang Mehr (and the glass-casters of Meka Glas GmbH, Kaufbeuren, Germany), who produced the glass copies. Thanks to Yvonne Völlmecke (University of Mainz), who sorted the fractured points and prepared them for the microscopic analysis and fracture propagation speed calculations. Funding for this work was provided by the "Neandertal Projectile Technology" (NeProTec) grant from the German Research Foundation (DFG). Finally, the authors thank the anonymous reviewers, whose comments improved the paper.

References

Ahler, S. A., & McMillan, R. B. (1976). Material culture at rodgers shelter: A reflection of past human activities. In W. R. Wood & R. B. McMillan (Eds.), *Prehistoric man and his environments: A case study in the Ozark highland* (pp. 163–199). New York: Academic Press.

Barton, R. N. E., & Bergman, C. A. (1982). Hunters at Hengistbury: Some evidence from experimental archaeology. *World Archaeology, 14*, 237–248.

Bartram, L. E. (1997). A comparison of Kua (Botswana) and Hadza (Tanzania) bow and arrow hunting. In H. Knecht (Ed.), *Projectile technology* (pp. 321–343). New York: Plenum Press.

Baugh, R. A. (2003). Dynamics of spear throwing. *American Journal of Physics, 71*, 345.

Berger, T. D., & Trinkaus, E. (1995). Patterns of trauma among the Neandertals. *Journal of Archaeological Science, 22*, 841–852.

Bergman, C. A., & Newcomer, M. H. (1983). Flint arrowhead breakage: Examples from Ksar Akil, Lebanon. *Journal of Field Archaeology, 10*, 238–243.

Bratlund, B. (1991). A study of hunting lesions containing flint fragments on reindeer bones at Stellmoor, Schleswig-Holstein, Germany. *CBA Research Report, 77*, 193–207.

Bratlund, B. (1999). Taubach revisited. *Jahrbuch des Römisch-Germanischen Zentralmuseums Mainz, 46*, 61–174.

Brown, K. S., Marean, C. W., Jacobs, Z., Schoville, B. J., Oestmo, S., Fisher, E. C., et al. (2012). An early and enduring advanced technology originating 71,000 years ago in South Africa. *Nature, 491*, 590–593.

Cattelain, P. (1989). Un crochet de propulseur solutréen de la grotte de Combe-Saunière 1 (Dordogne). *Bulletin de la Société Préhistorique Française, 86*(87), 213–216.

Chadwick, E., Nicol, A., Lane, J., & Gray, T. (1999). Biomechanics of knife stab attacks. *Forensic Science International, 105*, 35–44.

Churchill, S. E. (1993). Weapon technology, prey size selection, and hunting methods in modern hunter-gatherers: Implications for hunting in the Palaeolithic and Mesolithic. *Archeological Papers of the American Anthropological Association, 4*, 11–24.

Clarkson, C. (2016). Testing archaeological approaches to determining past projectile delivery systems using ethnographic and experimental data. In R. Iovita & K. Sano (Eds.), *Multidisciplinary approaches to the study of stone age weaponry* (pp. 189–201). Dordrecht: Springer.

Cotterell, B., Hayden, B., Kamminga, J., Kleindienst, M. R., Knudson, R., & Lawrence, R. A. (1979). The Ho Ho Classification and Nomenclature Committee report. In B. Hayden (Ed.), *Lithic use-wear analysis* (pp. 133–135). New York: Academic Press.

d'Errico, F., Backwell, L., Villa, P., Degano, I., Lucejko, J. J., Bamford, M. K., et al. (2012). Early evidence of San material culture represented by organic artifacts from Border Cave, South Africa. *Proceedings of the National Academy of Sciences of the United States of America, 109*, 13214–13219.

Dockall, J. E. (1997). Wear traces and projectile impact: A review of the experimental and archaeological evidence. *Journal of Field Archaeology, 24*, 321–331.

Farizy, C., David, F., & Jaubert, J. (1994). *Hommes et bisons du Paléolithique moyen à Mauran (Haute-Garonne)*. Paris: CNRS Editions.

Fischer, A., Hansen, P. V., & Rasmussen, P. (1984). Macro and micro wear traces on lithic projectile points: Experimental results and prehistoric examples. *Journal of Danish Archaeology, 3*, 19–46.

Fischer, A., & Tauber, H. (1986). New C-14 datings of late palaeolithic cultures from Northwestern Europe. *Journal of Danish Archaeology, 5*, 7–13.

Frison, G. C., Wilson, M., & Wilson, D. J. (1976). Fossil bison and artifacts from an early altithermal period Arroyo trap in Wyoming. *American Antiquity, 41*, 28–57.

Gaudzinski, S. (1995). Wallertheim revisited: A re-analysis of the fauna from the middle Paleolithic site of Wallertheim (Rheinhessen/Germany). *Journal of Archaeological Science, 22*, 51–66.

Gaudzinski, S. (2005). Monospecific or species-dominated faunal assemblages during the middle Paleolithic in Europe. In E. Hovers & S. L. Kuhn (Eds.), *Transitions before the Transition: Evolution and stability in the middle Paleolithic and middle stone age* (pp. 137–147). New York: Springer.

Gaudzinski, S., & Roebroeks, W. (2000). Adults only. Reindeer hunting at the middle Paleolithic site Salzgitter Lebenstedt, Northern Germany. *Journal of Human Evolution, 38*, 497–521.

Gaudzinski-Windheuser, S. (2005). Subsistenzstrategien frühpleistozäner Hominiden in Eurasien. Taphonomische Faunenbetrachtungen der Fundstellen der 'Ubeidiya Formation (Israel) (Vol. 61). Mainz and Bonn: Habelt.

Gaudzinski-Windheuser, S. (2016). Hunting lesions in pleistocene and early holocene European bone assemblages and their implications for our knowledge on the use and timing of lithic projectile technology. In R. Iovita & K. Sano (Eds.), *Multidisciplinary approaches to the study of stone age weaponry* (pp. 77–100). Dordrecht: Springer.

Geneste, J. M., & Plisson, H. (1989). Analyse technologique des pointes à cran solutréennes du Placard (Charente), du Fourneau du Diable, du Pech de la Boissière et de Combe Saunière (Dordogne). *Paléo, 1*, 65–106.

Griffith, A. A. (1921). The phenomena of rupture and flow in solids. *Philosophical Transactions of the Royal Society of London Series A, Containing Papers of a Mathematical or Physical Character, 221*, 163–198.

Horsfall, I., Prosser, P., Watson, C., & Champion, S. (1999). An assessment of human performance in stabbing. *Forensic Science International, 102*, 79–89.

Huckell, B. B. (1982). The Denver Elephant Project: A report on experimentation with thrusting spears. *Plains Anthropologist, 27*, 217–224.

Hughes, S. S. (1998). Getting to the point: Evolutionary change in prehistoric weaponry. *Journal of Archaeological Method and Theory, 5*, 345–408.

Hutchings, W. K. (1999). Quantification of fracture propagation velocity employing a sample of Clovis channel flakes. *Journal of Archaeological Science, 26*, 1437–1447.

Hutchings, W. K. (2011). Measuring use-related fracture velocity in lithic armatures to identify spears, javelins, darts, and arrows. *Journal of Archaeological Science, 38*, 1737–1746.

Hutchings, K. (2016). When is a point a projectile? Morphology, impact fractures, scientific rigor, and the limits of inference. In R. Iovita & K. Sano (Eds.), *Multidisciplinary approaches to the study of Stone Age weaponry* (pp. 3–12). Dordrecht: Springer.

Hutchings, W. K., & Brüchert, L. W. (1997). Spearthrower performance: Ethnographic and experimental research. *Antiquity, 71*, 890–897.

Iovita, R., Schönekeß, H., Gaudzinski-Windheuser, S., & Jäger, F. (2014). Projectile impact fractures and launching mechanisms: Results of a controlled ballistic experiment using replica Levallois points. *Journal of Archaeological Science, 48*, 73–83.

Kerkhof, F. (1975). Bruchmechanische Analyse von Schadensfällen an Gläsern. *Glastechnische Berichte, 48*, 112–124.

Kerkhof, F., & Müller-Beck, H. (1969). Zur bruchmechanischen Deutung der Schlagmarken an Steingeräten. *Glastechnische Berichte, 42*, 439–448.

Kindler, L. (2012). *Die Rolle von Raubtieren in der Einnischung und Subsistenz jungpleistozäner Neandertaler*. Mainz: Monographien des Römisch-Germanischen Zentralmuseums.

Lazuén, T. (2012). European Neanderthal stone hunting weapons reveal complex behaviour long before the appearance of modern humans. *Journal of Archaeological Science, 39*, 2304–2311.

Lee, R. B., & DeVore, I. (Eds.). (1968). *Man the hunter*. New York: Aldine de Gruyter.

Lombard, M. (2005). Evidence of hunting and hafting during the Middle Stone Age at Sibudu Cave, KwaZulu-Natal, South Africa: A multianalytical approach. *Journal of Human Evolution, 48*, 279–300.

Lombard, M. (2011). Quartz-tipped arrows older than 60 ka: Further use-trace evidence from Sibudu, KwaZulu-Natal, South Africa. *Journal of Archaeological Science, 38*, 1918–1930.

Lombard, M., & Haidle, M. N. (2012). Thinking a bow-and-arrow set: Cognitive implications of middle stone age bow and stone-tipped arrow technology. *Cambridge Archaeological Journal, 22*, 237–264.

Lombard, M., & Pargeter, J. (2008). Hunting with howiesons poort segments: Pilot experimental study and the functional interpretation of archaeological tools. *Journal of Archaeological Science, 35*, 2523–2531.

Lombard, M., Parsons, I., & Van der Ryst, M. (2004). Middle Stone Age lithic point experimentation for macro-fracture and residue analyses: The process and preliminary results with reference to Sibudu Cave points. *South African Journal of Science, 100*, 159–166.

Lombard, M., & Phillipson, L. (2010). Indications of bow and stone-tipped arrow use 64,000 years ago in KwaZulu-Natal, South Africa. *Antiquity, 84*, 635–648.

Lombard, M., & Wadley, L. (2016). Hunting technologies during the holwiesons poort at Sibudu Cave: What they Reveal about human cognition in KwaZulu-Natal, South Africa, between ~65 and 62 ka. In R. Iovita & K. Sano (Eds.), *Multidisciplinary approaches to the study of Stone Age weaponry* (pp. 273–286). Dordrecht: Springer.

Marean, C. W., & Kim, S. Y. (1998). Mousterian large-mammal remains from Kobeh Cave: Behavioral implications for Neanderthals and early modern humans. *Current Anthropology, 39* (supplement), 79–113.

McBrearty, S., & Brooks, A. S. (2000). The revolution that wasn't: A new interpretation of the origin of modern human behavior. *Journal of Human Evolution, 39*, 453–563.

Newman, K., & Moore, M. W. (2013). Ballistically anomalous stone projectile points in Australia. *Journal of Archaeological Science, 40*, 2614–2620.

Odell, G. H., & Cowan, F. (1986). Experiments with spears and arrows on animal targets. *Journal of Field Archaeology, 13*, 195–212.

Pargeter, J. (2007). Howiesons poort segments as hunting weapons: Experiments with replicated projectiles. *South African Archaeological Bulletin, 62*, 147–153.

Pargeter, J. (2011). Assessing the macrofracture method for identifying stone age hunting weaponry. *Journal of Archaeological Science, 38*, 2882–2888.

Plisson, H., & Beyries, S. (1998). Pointes ou outils triangulaires? Données fonctionnelles dans le Moustérien levantin. *Paléorient, 24*, 5–24.

Raymond, A. (1986). Experiments in the function and performance of the weighted atlatl. *World Archaeology, 18*, 153–177.

Rhodes, J., & Churchill, S. E. (2009). Throwing in the middle and upper Paleolithic: Inferences from an analysis of humeral retroversion. *Journal of Human Evolution, 56*, 1–10.

Richter, H. (1974). *Experimentelle Untersuchungen zur Rißausbreitung in Spiegelglas im Geschwindigkeitsbereich 10^{-3} bis $5 \cdot 10^{3}$ mm/s*. Karlsruhe: TH Karlsruhe.

Rieder, H. (2003). Der Große Wurf der frühen Jäger: Nachbau altsteinzeitlicher Speere. *Biologie in unserer Zeit, 33*, 156–160.

Roque, C., Guibert, P., Vartanian, E., Bechtel, F., & Schvoerer, M. (2001). Thermoluminescence-dating of calcite: Study of heated limestone fragments from upper Paleolithic layers at Combe Sauniere, Dordogne, France. *Quaternary Science Reviews, 20*, 935–938.

Rots, V., Van Peer, P., & Vermeersch, P. M. (2011). Aspects of tool production, use, and hafting in palaeolithic assemblages from Northeast Africa. *Journal of Human Evolution, 60*, 637–664.

Rust, A. (1943). *Die alt- und mittelsteinzeitlichen Funde von Stellmoor*. Neumünster: Karl Wachholtz Verlag.

Sahle, Y., Morgan, L. E., Braun, D. R., Atnafu, B., & Hutchings, W. K. (2013). Chronological and behavioral contexts of the earliest middle stone age in the Gademotta formation, Main Ethiopian rift. *Quaternary International, 331*, 6–19.

Sano, K. (2009). Hunting evidence from stone artefacts from the Magdalenian cave site Bois Laiterie, Belgium: A fracture analysis. *Quartär, 56*, 67–86.

Schmitt, D., Churchill, S. E., & Hylander, W. L. (2003). Experimental evidence concerning spear use in neandertals and early modern humans. *Journal of Archaeological Science, 30*, 103–114.

Schoville, B. J. (2010). Frequency and distribution of edge damage on middle stone age lithic points, Pinnacle point 13B, South Africa. *Journal of Human Evolution, 59*, 378–391.

Shea, J. J. (1988). Spear points from the middle Paleolithic of the Levant. *Journal of Field Archaeology, 15*, 441–450.

Shea, J. J. (2006). The origins of lithic projectile point technology: Evidence from Africa, the Levant, and Europe. *Journal of Archaeological Science, 33*, 823–846.

Shea, J., Davis, Z. J., & Brown, K. S. (2001). Experimental tests of middle palaeolithic spear points using a calibrated crossbow. *Journal of Archaeological Science, 28*, 807–816.

Shea, J. J., & Sisk, M. L. (2010). Complex projectile technology and *Homo sapiens* dispersal into western Eurasia. PaleoAnthropology, 100–122.

Shott, M. J. (1997). Stones and shafts redux: The metric discrimination of chipped-stone dart and arrow points. *American Antiquity, 62*, 86–101.

Sisk, M. L., & Shea, J. J. (2009). Experimental use and quantitative performance analysis of triangular flakes (Levallois points) used as arrowheads. *Journal of Archaeological Science, 36*, 2039–2047.

Sisk, M. L., & Shea, J. J. (2011). The African origin of complex projectile technology: An analysis using tip cross-sectional area and perimeter. *International Journal of Evolutionary Biology, 2011*, 1–8. doi:10.4061/2011/968012.

Smith, G. M. (2003). Damage inflicted on animal bone by wooden projectiles: Experimental results and archaeological implications. *Journal of Taphonomy, 1*, 105–113.

Stodiek, U. (1993). *Zur Technologie der jungpaläolithischen Speerschleudern* (Vol. 9). Tübingen: Institut für Ur- und Frühgeschichte der Universität Tübingen.

Thieme, H. (1997). Lower palaeolithic hunting spears from Germany. *Nature, 385*(6619), 807–810. doi:10.1038/385807a0.

Tomenchuk, J. (1985). The development of a wholly parametric use-wear methodology and its application to two selected samples of Epipaleolithic chipped stone tools from Hayonim Cave, Israel. Ph. D. Dissertation, University of Toronto.

Trinkaus, E. (2012). Neandertals, early modern humans, and rodeo riders. *Journal of Archaeological Science, 39*, 3691–3693.

Villa, P., Boscato, P., Ranaldo, F., & Ronchitelli, A. (2009a). Stone tools for the hunt: Points with impact scars from a middle Paleolithic site in southern Italy. Journal of Archaeological Science, 36, 850–859.

Villa, P., Soriano, S., Teyssandier, N., & Wurz, S. (2009b). The howiesons poort and MSA III at Klasies river main site, Cave 1A. Journal of Archaeological Science, 37, 630–655.

Wallner, H. (1939). Linienstrukturen an Bruchflächen. *Zeitschrift für Physik, 114*, 368–378.

Weber, M.-J., Grimm, S. B., & Baales, M. (2011). Between warm and cold: Impact of the younger dryas on human behavior in Central Europe. *Quaternary International, 242*, 277–301.

Wilkins, J., Schoville, B. J., Brown, K. S., & Chazan, M. (2012). Evidence for early hafted hunting technology. *Science, 338*, 942–946.

Witthoft, J. (1968). Flint arrowheads from the eskimo of Northwestern Alaska. *Expedition, 10*, 10–37.

Chapter 3
Experiments in Fracture Patterns and Impact Velocity with Replica Hunting Weapons from Japan

Katsuhiro Sano, Yoshitaka Denda, and Masayoshi Oba

Abstract Recent anthropological and archaeological studies in western Eurasia indicate that long-range projectile hunting was innovated by modern humans, and that complex projectile technology, such as using spearthrowers or bows (Shea and Sisk 2010), was an important component of behavioral modernity. The morphometric analysis of stone tips, including tip cross-sectional area (TCSA) and tip cross-sectional perimeter (TCSP), may facilitate suggestions for an optimum delivery method of stone tips as hunting weaponry. However, the suggested method does not always coincide with the true functions of the stone tips. Thus, this study developed a projectile experiment project to confirm additional indicators for identifying the delivery methods of prehistoric hunting armatures and to detect the emergence of spearthrower darts and bows and arrows in East Asia. Furthermore, macroscopic and microscopic analyses of the experimental specimens reveal a correlation between both the formation patterns of impact fractures as well as microscopic linear impact traces (MLIT) and impact velocities. This paper presents results of the projectile experiments, which provide indices to examine spearthrower darts and arrowheads in archaeological assemblages.

Keywords Delivery modes • Long-range projectiles • Projectile experiments • Impact fractures • MLIT • Trapezoids • Japanese Paleolithic

K. Sano (✉)
The University Museum, The University of Tokyo, Hongo 7-3-1, Bunkyo-Ku, Tokyo 113-0033, Japan
e-mail: sano@um.u-tokyo.ac.jp

Y. Denda
Miyagi Prefectural Board of Education, Honcho 3-8-1, Aoba-Ku, Sendai 980-8423, Japan
e-mail: denda-yo922@pref.miyagi.jp

M. Oba
Yamagata Prefectural Center for Archaeological Research, Nakayama-Kabeyashiki 5608, Kaminoyama, Yamagata 999-3246, Japan
e-mail: embrace@zb4.so-net.ne.jp

Introduction

The earliest clear evidence of hunting weaponry is wooden spears discovered at a Lower Paleolithic site in Schöningen, Germany. The objects are dated at c. 400 ka (Thieme 1996, 1997) or c. 310 ka (Jöris and Baales 2003), and the new U/Th data ranging from 348 to 280 ka (Urban et al. 2011) supports the latter. Although O'Brien (1981) concluded an experimental study by claiming that an Acheulian handaxe was used as a projectile weapon, the hypothesis was challenged because of the lack of impact damage on handaxes (Whittaker and McCall 2001). As the weight and position of maximum thickness of the Schöningen spears are similar to those of modern athletic javelins, Thieme (2005) suggested that the spears were utilized as hand-casting spears; however, this remains debatable. The Middle Paleolithic humans probably began using stone-tipped weapons, such as Levallois points (Boëda et al. 1999, 2008), which increased impact energy. However, their hunting included frequent close encounters with prey, based on the observation of scars from hunting wounds on several Neanderthal fossils (Berger and Trinkaus 1995). In addition, while marked asymmetry humeral retroversion of anatomically modern humans suggests habitual throwing, investigations of Neanderthal skeletons demonstrate a lack of regular throwing (Rhodes and Churchill 2009). This anthropological evidence suggests that modern humans would have been the first to innovate long-range projectile hunting.

On the other hand, the direct archaeological evidence for true long-range projectile hunting using spearthrowers or bows (Churchill 1993) emerged not from the initial Upper Paleolithic, but from the middle Upper Paleolithic period in Europe, as evidenced by the spearthrower hook discovered at the Solutrean layer in Combe Saunière, France, which was dated at between 19 and 17 ^{14}C kBP (Geneste and Plisson 1986; Cattelain 1989). However, studies on the tip cross-sectional area (TCSA) of hunting armatures indicated that stone tips, including darts propelled by spearthrowers,

Radu Iovita and Katsuhiro Sano (eds.), *Multidisciplinary Approaches to the Study of Stone Age Weaponry*, Vertebrate Paleobiology and Paleoanthropology, DOI 10.1007/978-94-017-7602-8_3

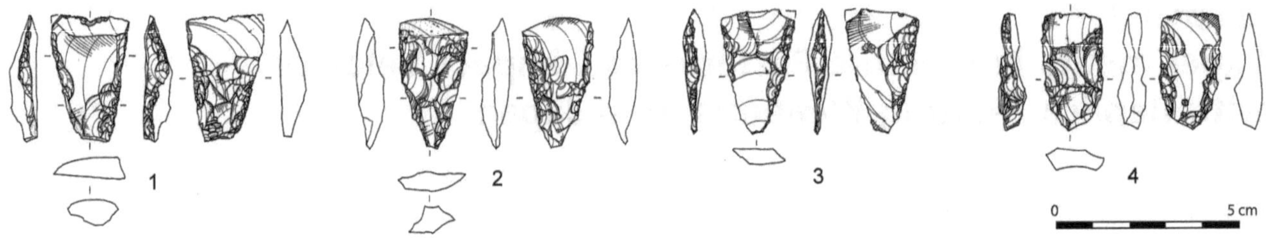

Fig. 3.1 Trapezoids from Layer Vb at the Hinatabayashi B site in Japan dated at between 31.4 and 28.2 ^{14}C kBP (after Tani 2000)

may have appeared after 50 ka in western Eurasia, which coincides with when modern humans expanded out of Africa and to the Old World (Shea 2006; Shea and Sisk 2010). Moreover, a tip cross-sectional perimeter (TCSP) analysis of several samples of African Middle Stone Age points, such as bifacial points from Porc Epic and Aterian tanged points from Aoulef, also suggested that they could have been used as spearthrower darts (Sisk and Shea 2011).

The TCSA and TCSP are practical indicators for suggesting the capability of hunting weaponry and for assuming the potential projectile systems. Nevertheless, the TCSA and TCSP values are not absolute proxies for identifying projectile delivery methods and for reconstructing actual functions (see Newman and Moore 2013; Clarkson 2016), and other indicators are therefore required to accurately detect the types of projectile systems for which the stone tips were actually employed. Thus, this study developed a projectile experiment project to establish criteria for identifying the employed hunting methods through formation patterns of impact fractures and MLIT related to delivery modes such as spear-thrusting, javelin-throwing, as well as the use of spearthrower darts and bows and arrows.

Projectile experiments regarding impact fracture formation patterns have been conducted to identify hunting weapons (e.g., Barton and Bergman 1982; Moss and Newcomer 1982; Bergman and Newcomer 1983; Fischer et al. 1984; Shea 1988; Midoshima 1991, 1996; Geneste and Plisson 1993; Caspar and De Bie 1996), and in recent decades, a variety of projectile experiments have been performed to understand prehistoric hunting technologies (Shea et al. 2001; Lombard et al. 2004; Lombard and Pargeter 2008; Sisk and Shea 2009; Yaroshevich et al. 2010; Pétillon et al. 2011; Sano and Oba 2015). One author (K. Sano) compared the formation patterns of the "diagnostic impact fractures" (DIF) with accidental fractures, which can occur during lithic production and syn-/post-depositional processes, and presented more reliable DIF exclusive to the hunting context (Sano 2009).

MLIT are another distinctive impact scar; they are microscopically observable at magnifications from 50× to 500× (Moss and Newcomer 1982; Fischer et al. 1984; Geneste and Plisson 1993; Caspar and De Bie 1996; Crombé et al. 2001; Yaroshevich et al. 2010; Sano and Oba 2015). MLIT are comprised of clusters of linear polishes running parallel to one another, which give them their striped appearance. MLIT are most likely formed due to contact with bone or fragments of stone tips (Moss and Newcomer 1982; Fischer et al. 1984). Since little is known regarding the formation patterns of MLIT, further experiments are required to better understand the formation mechanics.

If the hypothesis that complex projectile technology appeared after 50 ka, when *Homo sapiens* expanded to the Old World (Shea and Sisk 2010), is true, there must also be evidence of the use of spearthrowers or bows at early Upper Paleolithic sites in East Asia. This project performs projectile experiments with representative hunting armatures from the Japanese islands, including trapezoids, backed points, leaf-shaped points, and antler points in which microblades have been inserted, and reveals when the use of spearthrower darts and bows and arrows began in East Asia. This paper presents the results of the experiments centered on trapezoids (Fig. 3.1) that emerged between c. 38 and c. 30 cal kBP in early Upper Paleolithic Japan (Kudo and Kumon 2012), and some of which were probably hunting armatures (Yamaoka 2012). Furthermore, we discuss the possibility of reconstructing hunting delivery modes on the basis of the formation patterns of impact fractures and MLIT.

Methods

A calibrated crossbow was employed to accurately control loading conditions according to the estimated impact velocities of throwing, spearthrowers, and bows and arrows (Fig. 3.2). For thrusting, a realistic experiment was conducted because the kinematic mechanics of thrusting is difficult to reconstruct using the crossbow; one male student (1.81 m tall and weighing 76 kg) performed the required actions.

Ethnographic data indicate that spearthrowers enabled hunting at a distance of over 30 m (Churchill 1993; Stodiek 1993; Cattelain 1997). However, the effective hunting range of spearthrowers was 15–30 m (Stodiek 1993). Furthermore,

Fig. 3.2 Crossbow used for the projectile experiments

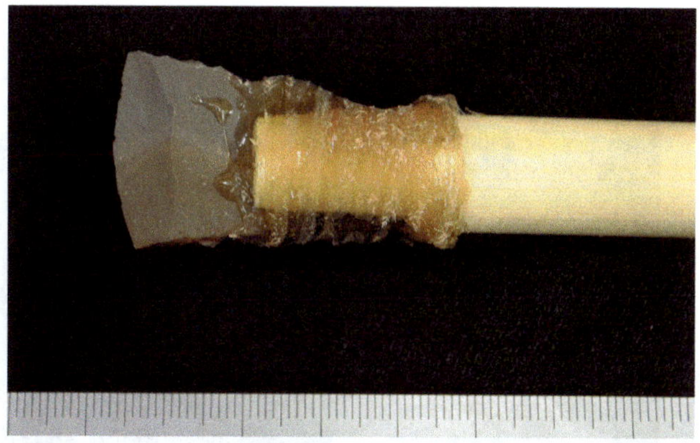

Fig. 3.3 Example of hafting a stone tip to a foreshaft

ethnographic and experimental studies indicated that its accuracy decreased between 20 and 30 m (Stodiek 1993; Cattelain 1997). The bow and arrow was effective between 20 and 30 m in the majority of cases and the "successful shots with high-performance equipment are taken at distances from 10 to 20 m on average" (Cattelain 1997). Based on these findings, we can estimate that the most effective and average range was approximately 20 m for both spearthrowers and bows. Stodiek (1993) recorded the velocities of spearthrower darts and bows and arrows using high-speed film and reported that the average velocity of spearthrowers from 20 m was 21.7 m/s while that of bows was 31.4 m/s. Regarding the throwing hunting, as there was no available data on the decline rate of the velocity according to distance, we employed the average velocity of 17.8 m/s presented by Hughes (1998). Thus, we calibrated the crossbow to shoot spears at impact velocities of 31.4 m/s for bows, 21.7 m/s for spearthrowers, and 17.8 m/s for throwing, with ±1.0 m/s deviation (Table 3.1).

Table 3.1 Velocities at a range of 20 m by bows and spearthrowers and the average throwing velocity

Delivery modes	Velocity (m/s)	Range (m)	References
Bow and arrow	31.4	20	Stodiek (1993)
Spearthrower dart	21.7	20	Stodiek (1993)
Throwing spear	17.8	–	Hughes (1998)

The lithic tips were first hafted to wooden foreshafts using glue (Fig. 3.3) before being fastened to the main shafts. A skillful knapper (M. Oba) produced lithic replicas of trapezoids made on siliceous shale from the Yamagata Prefecture in Japan. This shale was a high-quality raw material most frequently recovered at Paleolithic sites in the Tohoku region, which we are currently investigating. Forty trapezoid specimens were prepared for the experiments (Fig. 3.4), 10 of which were used for the experiments of thrusting, throwing, spearthrowers, and bows. A joint made from stainless steel, used to connect the foreshafts with the main shaft, weighed 16.8 g, and the wooden main shaft weighed 120.0 g. Each specimen was shot only once at an undamaged target assembled from deer hide, pig meat, and cattle scapulae. The target was set at a distance of 1.5 m from the crossbow to ensure that the impact and initial velocities were almost identical. The specimens were then macroscopically and microscopically observed. For the microscopic analysis, we utilized a digital microscope (KEYENCE VHX-1000) at magnifications from 100× to 500×.

Before the experiments, we examined whether the morphological variability of the trapezoid replicas can influence impact fracture formation patterns. The measured attributes of the trapezoid specimens, including length/width, length/thickness, TCSA, TCSP, weights, average angle at three parts of edges, a/b, c^l/b^l, and c^r/d^r (Fig. 3.5), are displayed in Table 3.2. The statistical significance of the difference among the attributes for thrusting, throwing, spearthrowers, and bows was assessed by using the

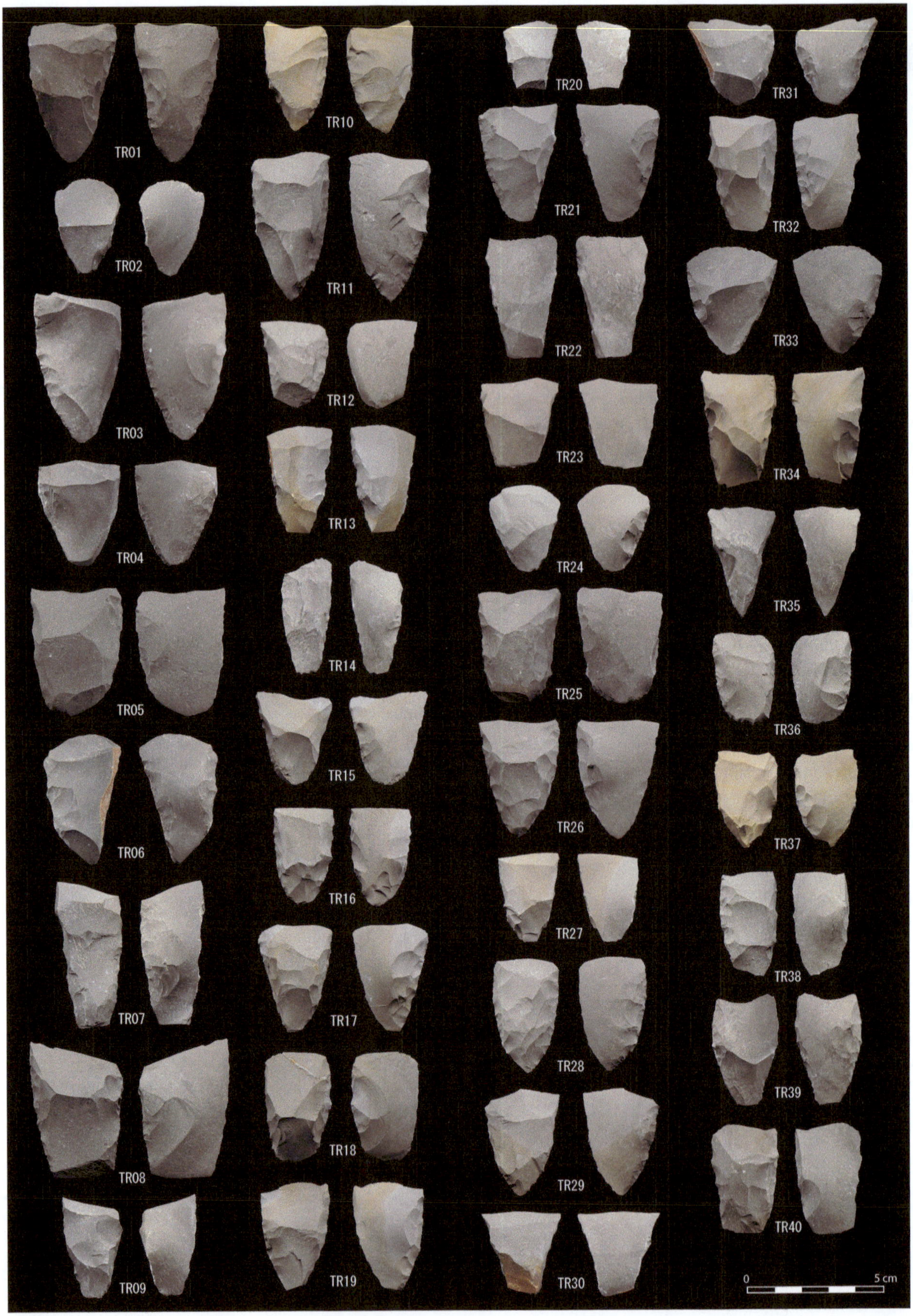

Fig. 3.4 Lithic replicas of trapezoids used for the experiments

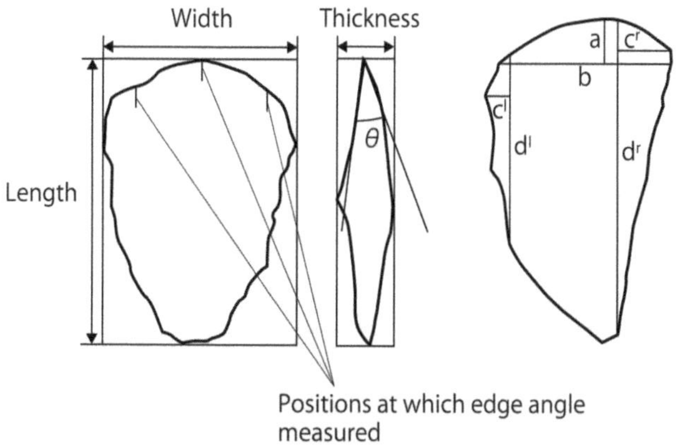

Fig. 3.5 Measured attributes for examining morphological variability among the experimental specimens. b is the maximum width of the distal portion. a is the distance from the distal end to the line b. d^l is the distance between the left end of the base and the point of contact between the line from the left base end parallel to the long axe and the outline. c^l is the distance from the point of the left maximum curvature to d^l. d^r is a distance between the right end of the base and the point of contact between the line from the right base end parallel to the long axe and the outline. c^r is the distance from the point of the right maximum curvature to d^r

Steel-Dwass multiple comparison test at a significance level of 0.05. As all the values were less than the critical value of 2.569 (Table 3.3), the null hypothesis that there were no significant differences between the specimens was not rejected. Hence, we cannot conclude that the morphological variance was sufficiently significant to influence the fracture formation patterns.

Furthermore, the TCSA and TCSP values of the trapezoid replicas were compared with those of North American ethnographic dart tips and arrowheads presented by Thomas (1978) and Shott (1997) (Fig. 3.6). The size and morphology of the replicas were based on the trapezoids unearthed at the Hinatabayashi B site in Japan. Both the TCSA and TCSP values of the replicas were larger than those of the ethnographic dart tips and arrowheads. However, because of the trapezoidal morphology, most trapezoids have their maximum width at the tip. Therefore, the TCSA and TCSP values of the trapezoids should not be directly compared to those of the ethnographic dart tips and arrowheads.

Table 3.2 Attributes of the trapezoid replicas

		L/W	L/Th	Weight (g)	Angle	a/b	c^l/d^l	c^r/d^r	TCSA	TCSP
Thrusting	Mean	1.59	4.97	11.0	43.9	0.15	0.20	0.17	132.1	59.3
	Std dev.	0.26	1.42	5.11	8.78	0.16	0.06	0.08	47.5	8.98
	Min.	1.25	3.64	4.56	28.7	−0.12	0.08	0.05	87.3	87.3
	Max	2.20	8.51	18.1	55.3	0.39	0.30	0.29	210.6	210.6
Throwing	Mean	1.62	3.89	8.15	42.8	0.13	0.19	0.19	116.8	51.2
	Std dev.	0.38	0.56	3.48	7.36	0.11	0.06	0.06	39.4	11.36
	Min.	1.15	3.10	2.52	35.0	−0.11	0.12	0.10	28.7	23.2
	Max	2.39	4.92	16.1	54.7	0.35	0.27	0.28	167.3	62.6
Spearthrower	Mean	1.34	4.50	8.35	41.7	0.11	0.21	0.19	118.4	58.0
	Std dev.	0.21	0.98	3.22	5.89	0.08	0.09	0.08	36.8	5.76
	Min.	1.01	3.19	4.37	33.3	0.04	0.12	0.06	78.3	47.5
	Max	1.66	6.63	12.96	50.3	0.26	0.36	0.29	197.1	65.7
Bow	Mean	1.45	4.21	7.47	37.7	0.05	0.19	0.15	115.7	55.1
	Std dev.	0.24	0.63	2.15	7.11	0.12	0.08	0.14	22.2	7.15
	Min.	0.85	3.55	4.86	30.0	−0.14	0.09	−0.06	91.4	45.5
	Max	1.75	5.24	10.8	50.7	0.18	0.35	0.40	151.4	67.3

Table 3.3 Multiple comparisons of attributes of the trapezoid replicas using Steel-Dwass test. Critical value = 2.569

		L/W	L/Th	Weight	Angle	a/b	c^l/d^l	c^r/d^r	TCSA	TCSP
Thrusting	Throwing	0.076	2.343	1.172	0.378	0.680	0.454	0.680	0.151	1.209
	Spearthrower	1.512	0.454	0.983	0.718	0.832	0.227	0.529	0.529	0.378
	Bow	0.756	1.436	1.512	1.512	1.663	0.529	0.605	0.454	0.983
Throwing	Spearthrower	1.739	1.436	0.151	0.189	1.058	0.151	0.076	0.302	1.663
	Bow	0.756	1.134	0.529	1.776	1.285	0.227	0.983	0.378	0.227
Spearthrower	Bow	1.285	0.529	0.340	1.512	0.529	0.454	0.983	0.076	1.134

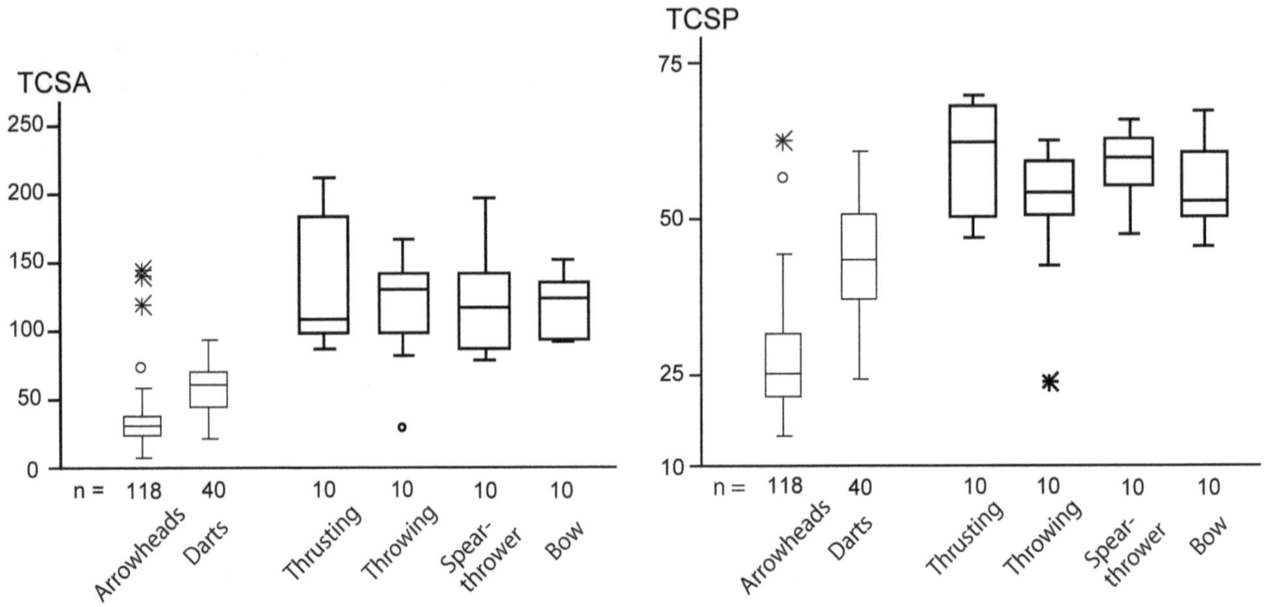

Fig. 3.6 Boxplot of TCSA and TCSP values for the experimental replicas compared with those of the ethnographic arrowheads and dart tips

Table 3.4 Frequency of impact fractures and MLIT. Impact fractures[1] = number of specimens with impact fractures, Impact fractures[2] = total number of the impact fractures, MLIT[1] = number of specimens with MLIT, MLIT[2] = total number of MLIT

	Impact fractures[1]	Impact fractures[2]	MLIT[1]	MLIT[2]
Thrusting	2	2	0	0
Throwing	6	23	4	8
Spearthrower	10	39	7	22
Bow	10	63	9	45

Results

Thrusting

The thrusting experiments produced just two impact fractures and no MLIT (Table 3.4; Fig. 3.7). The impact fractures were too small, making it difficult to distinguish them from micro-flaking formed by trampling or other accidental agencies (Fig. 3.8). Little or no morphological reduction of the specimens occurred due to impact damage. If the same traces are observed on archaeological stone tips, we are unable to determine whether the trapezoids were used as thrusting spear points.

Throwing

Regarding throwing velocity, several distinctive impact fractures were formed (Fig. 3.9). Six out of the 10 specimens included impact fractures and a total of 23 impact fractures were observed (Table 3.4). In addition to the typical DIF, such as flute-like fractures (Fig. 3.10b) and burin-like fractures, evidence of crushing (Odell and Cowan 1986) was frequently found (Fig. 3.10c). The dimension of the impact fractures was larger than that of the thrusting specimens, although half of them were extremely small.

Along with impact fractures, the throwing experiment induced MLIT on four trapezoids (Fig. 3.10a). Although the MLIT on the throwing specimens were generally faint and difficult to recognize, there are specimens bearing MLIT on several parts. Eight MLIT were observed on the throwing spear replicas.

Spearthrowers

The frequency of impact fractures in the spearthrower experiment was dramatically higher than that in the previous two experiments. All the trapezoids shot at the velocity of a

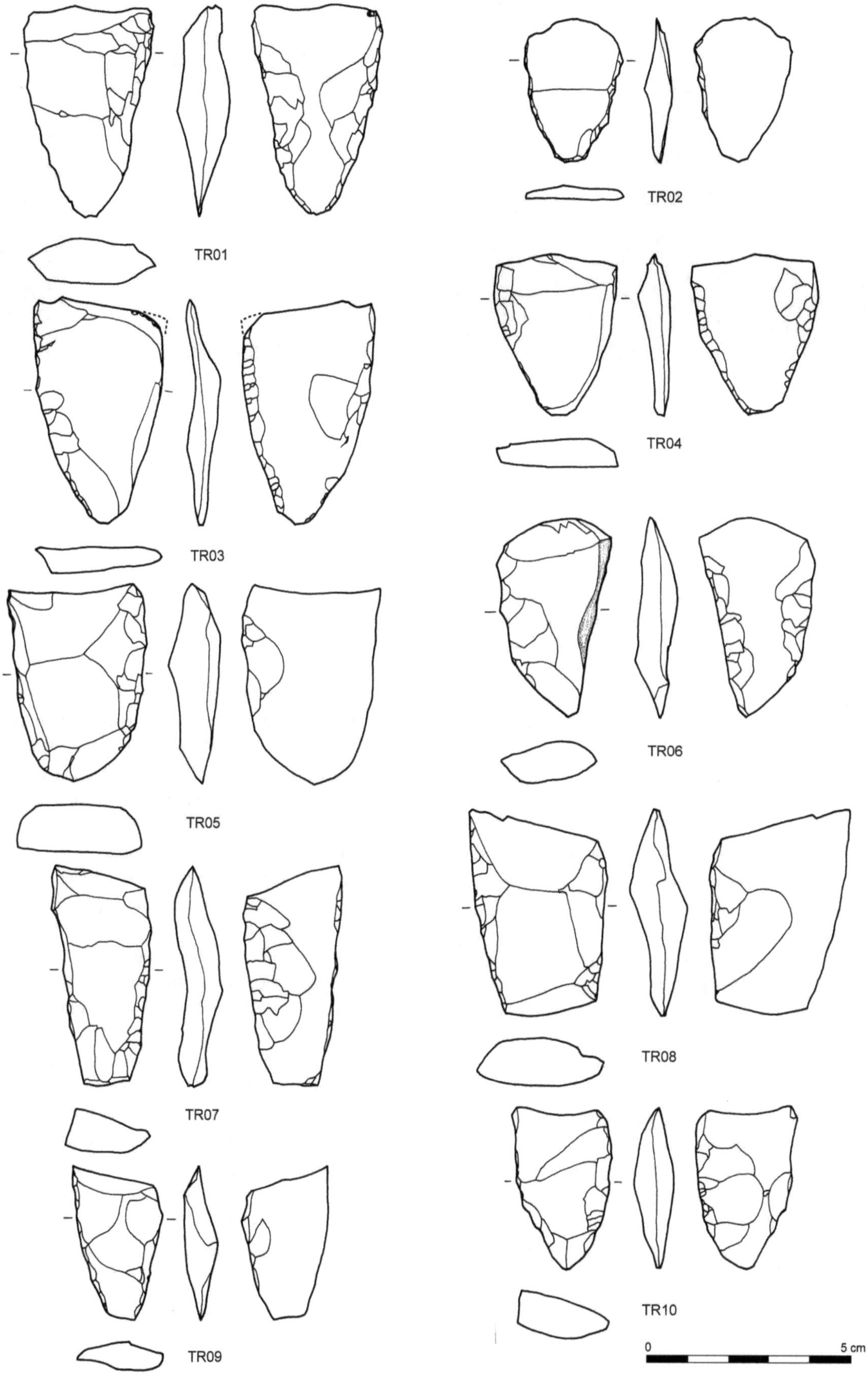

Fig. 3.7 Trapezoids after the thrusting experiment

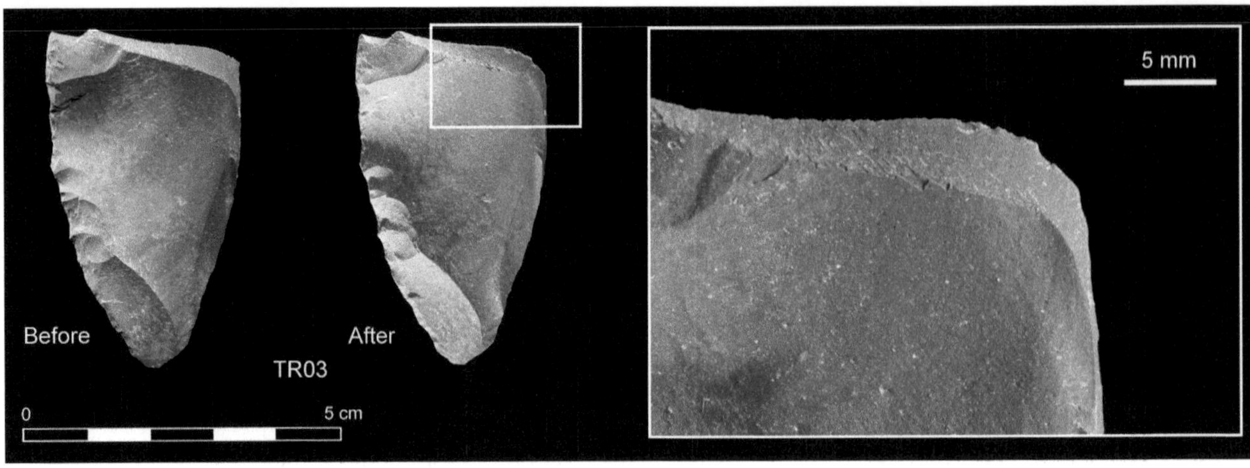

Fig. 3.8 Specimen with small impact fractures after the thrusting experiment

spearthrower exhibited impact fractures and a total of 39 fractures were observed (Table 3.4; Fig. 3.11). The dimensions of the flute- and burin-like fractures were larger than those of the throwing specimens, and most of them included step or hinge terminations (Fig. 3.12a, b).

The MLIT were formed on seven trapezoids (Fig. 3.12c) and a total of 22 MLIT were observed, more than twice the amount in the throwing experiment. One specimen exhibited a removal on the middle part of the ventral surface, which probably occurred due to hafting (Fig. 3.11: TR26).

Bows

The shooting velocity of bows also generated impact fractures on all the specimens (Table 3.4; Fig. 3.13). There was almost twice the number of impact fractures than that for the spearthrowers. Transverse fractures, which break specimens into two or more pieces, occurred due to the bow's high impact energy (Fig. 3.13: TR35, TR39). Several trapezoids exhibited complex fractures, including transverse, flute-like, burin-like, and spin-off fractures, as well as crushing. Furthermore, most specimens did not maintain their original morphology and broke into several pieces (Fig. 3.14: TR39) with fragments that were too small to be recovered.

MLIT were formed on nine specimens and a total of 45 MLIT were observed (Fig. 3.14a). The numbers of MLIT were larger than that for the spearthrowers. Hafting removals on the ventral surfaces were found on three specimens (Fig. 3.14c), and such removals on the middle surfaces were dissimilar to the hafting traces presented by Rots (2010). This may be a unique hafting scar exclusively formed by a projectile impact.

Discussion

The experiments of thrusting, throwing, spearthrowers, and bows exhibited distinctive results in formation patterns of impact fractures and MLIT. Currently, we discuss the frequency, MLIT, types, and dimension of impact fractures, as well as the volume reduction rate of the specimens to examine whether they provide new indicators for identifying the delivery modes of hunting weaponry.

The ratio of the specimens with impact fractures rose according to the delivery modes, and more impact fractures occurred when the specimens were shot at a higher velocity (Fig. 3.15). In addition, there were positive correlations between impact velocity and the frequencies of the MLIT.

Flute-like fractures and crushing occurred with high frequency in the experiments (Fig. 3.16). The high ratios resulted from the morphological features of trapezoids with vertical edges to the direction of the projectile movement. It is noteworthy that the transverse fractures were formed exclusively when the tips were shot at the velocity of a bow. Trapezoids are generally shorter and thicker than backed points, leaf-shaped points, and microblades, and are thus rarely broken transversely. Consequently, the presence

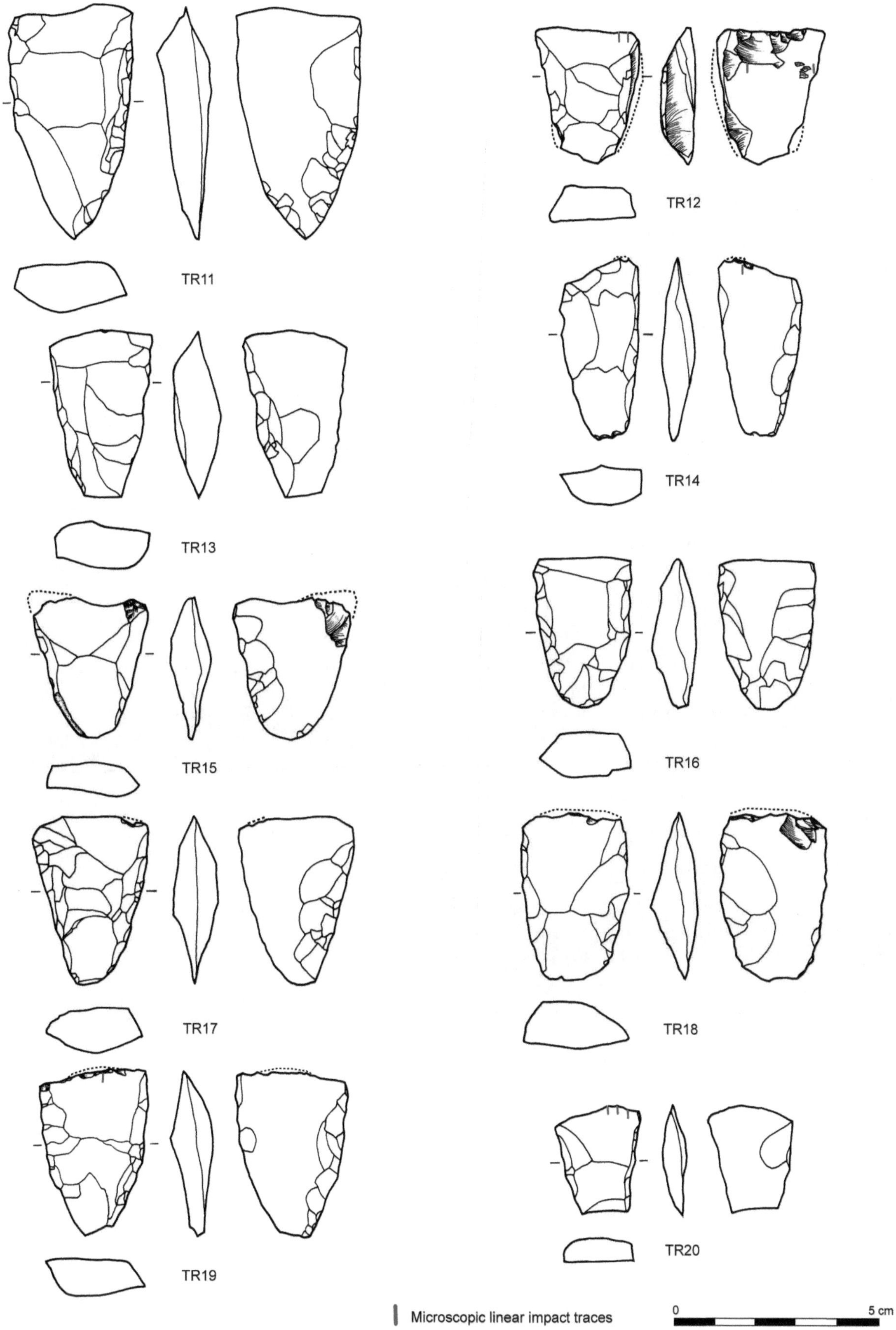

Fig. 3.9 Trapezoids after the projectile throwing velocity experiment

Fig. 3.10 Specimens with impact fractures and MLIT after the throwing velocity projectile experiment: **a** MLIT; **b** flute-like fracture; and **c** crushing

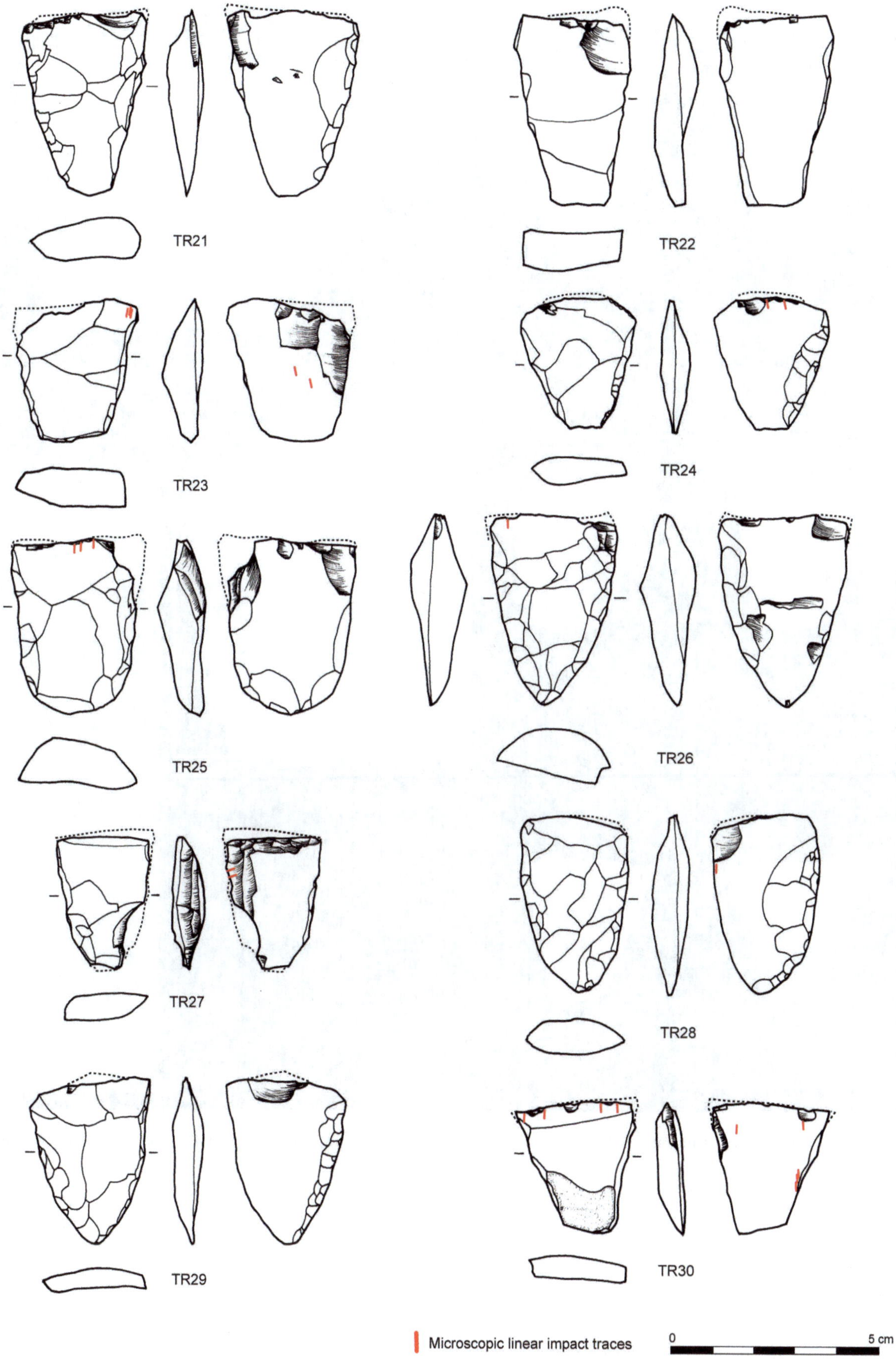

Fig. 3.11 Trapezoids after the spearthrower velocity experiment

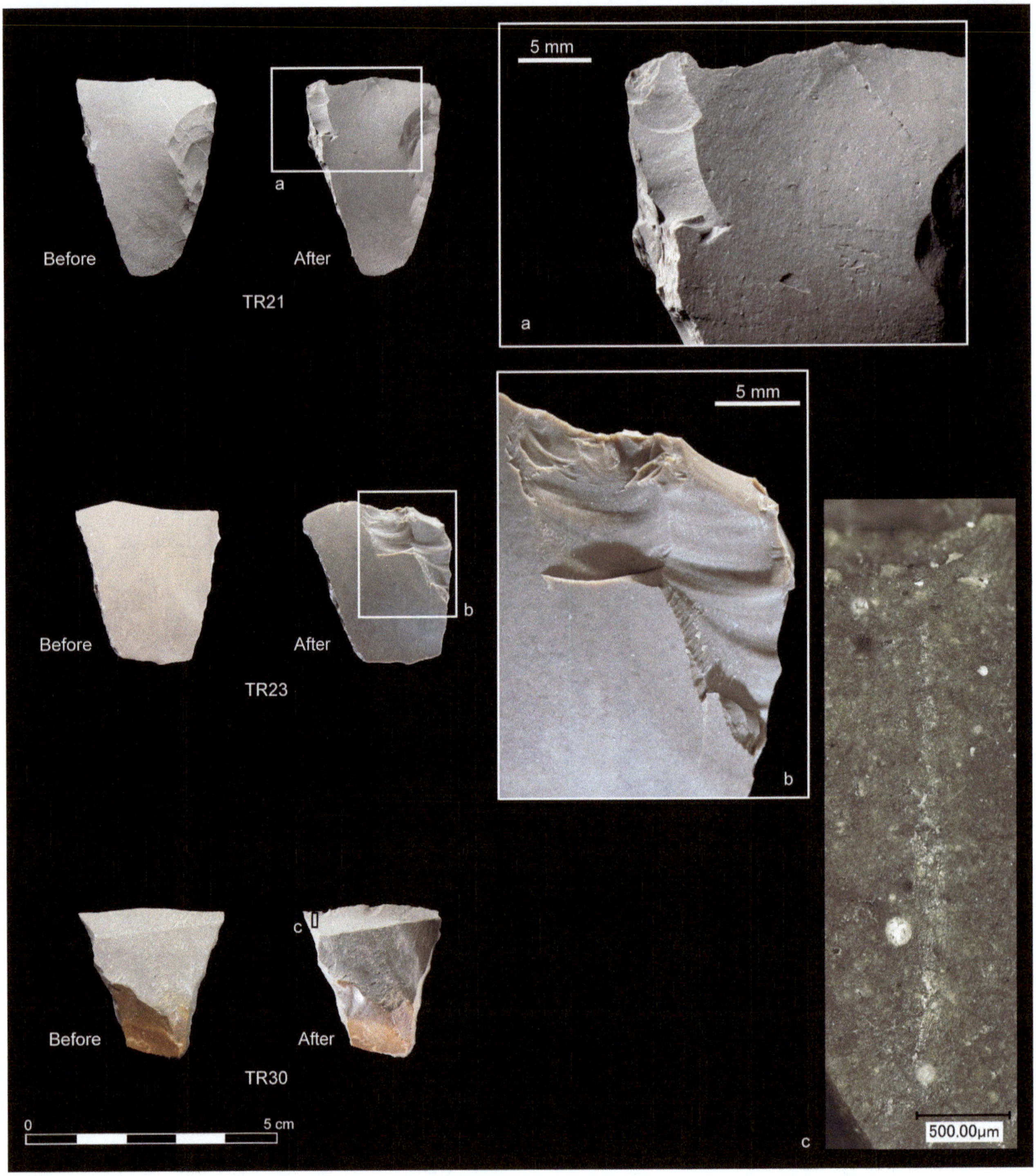

Fig. 3.12 Specimens with impact fractures and MLIT after the spearthrower velocity experiment: **a** burin-like fracture; **b** flute- and burin-like fractures; and **c** MLIT

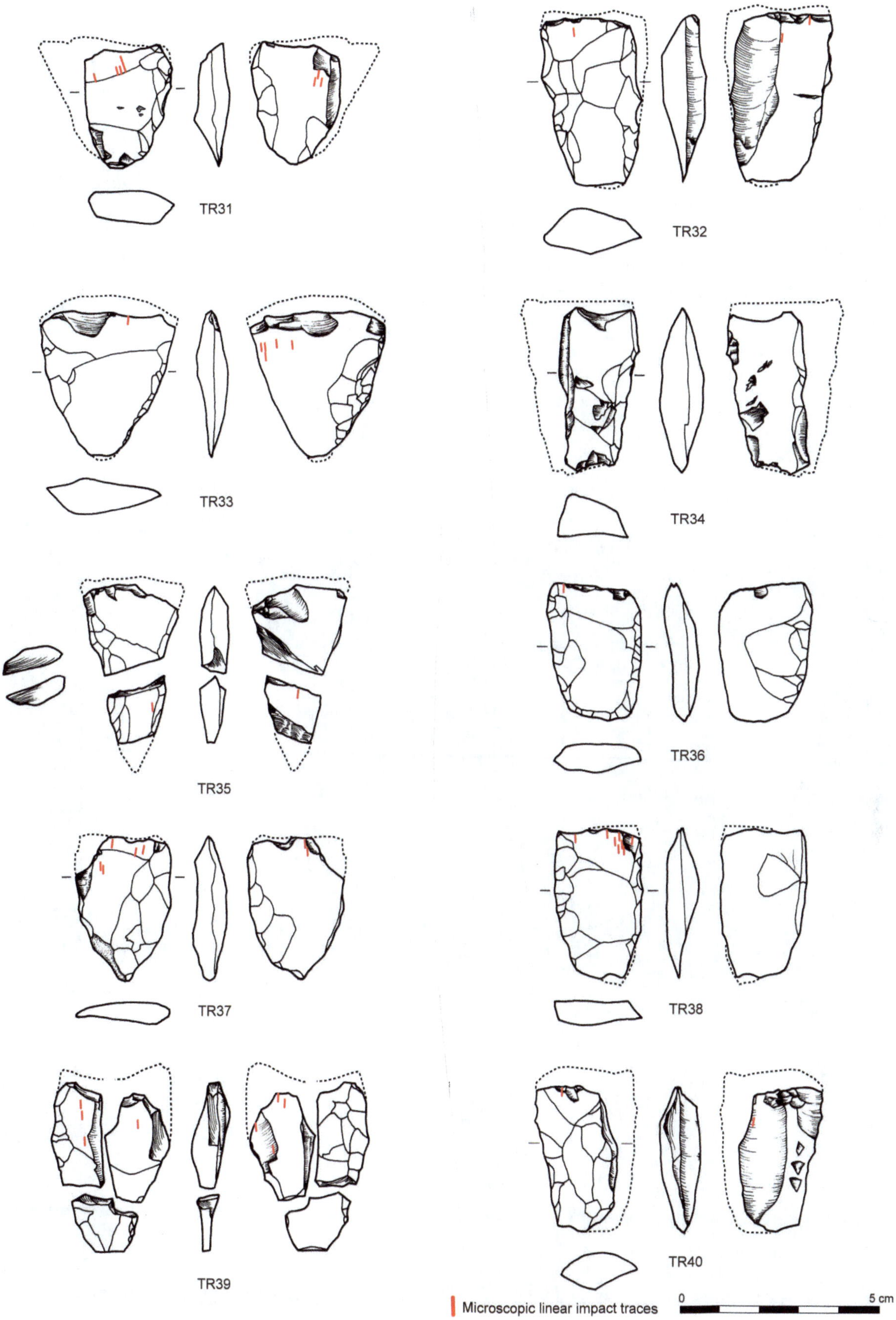

Fig. 3.13 Trapezoids after the bow velocity experiment

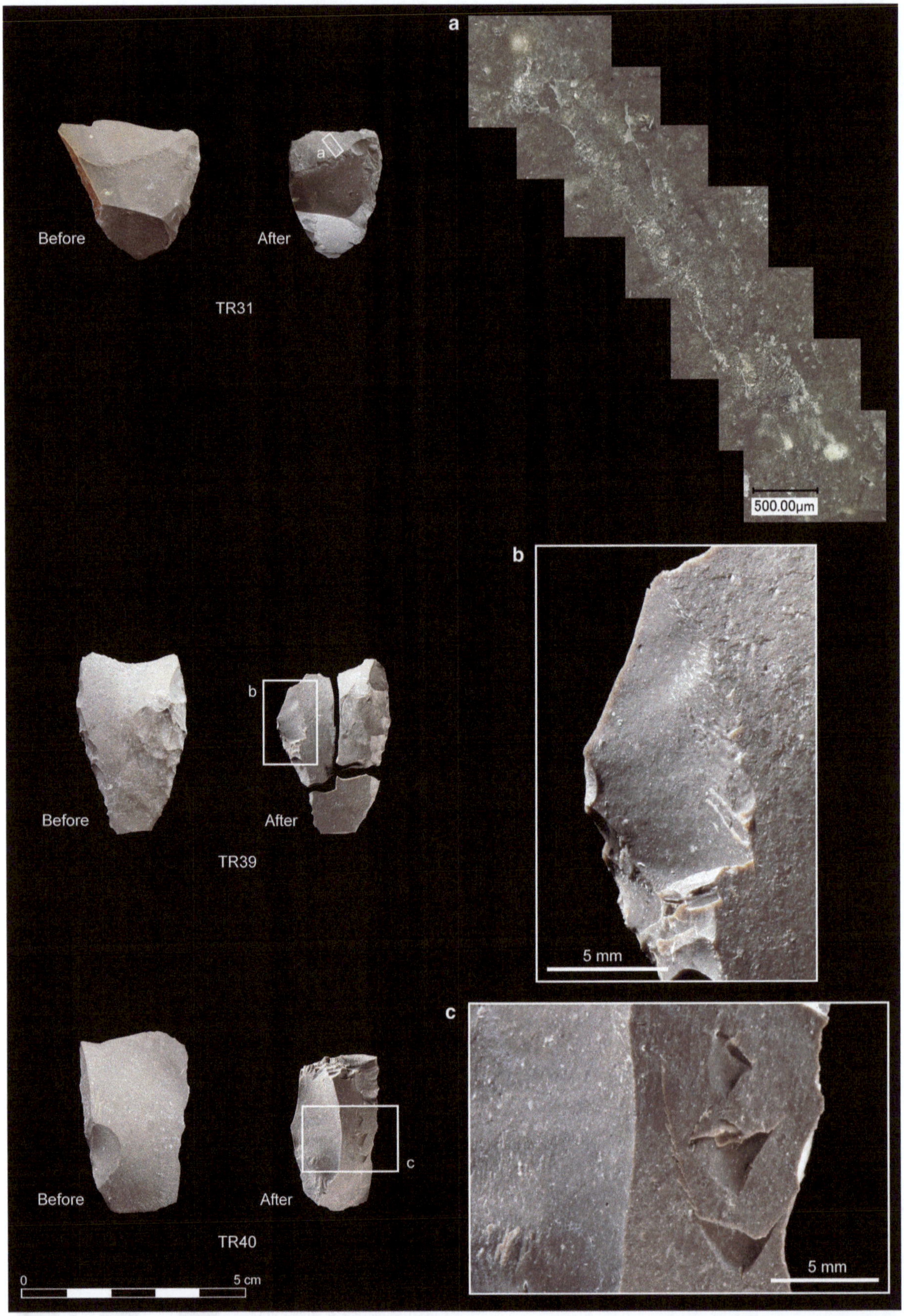

Fig. 3.14 Specimens with impact fractures and MLIT after the bow velocity experiment: **a** MLIT; **b** burin-like fracture; and **c** removals due to hafting

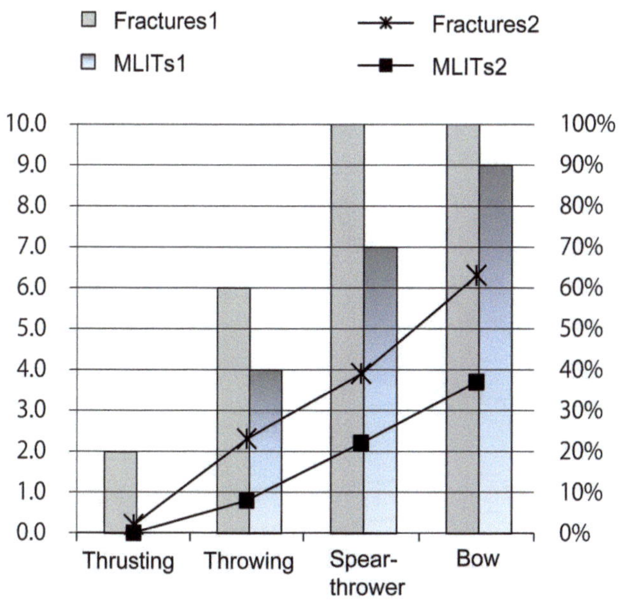

Fig. 3.15 Correlations between the delivery modes and the impact fractures as well as MLIT. Fractures^1 = ratio of the specimens with impact fractures; Fractures^2 = number of impact fractures per specimen; MLIT^1 = ratio of the specimens with MLIT; and MLIT^2 = number of MLIT per specimen

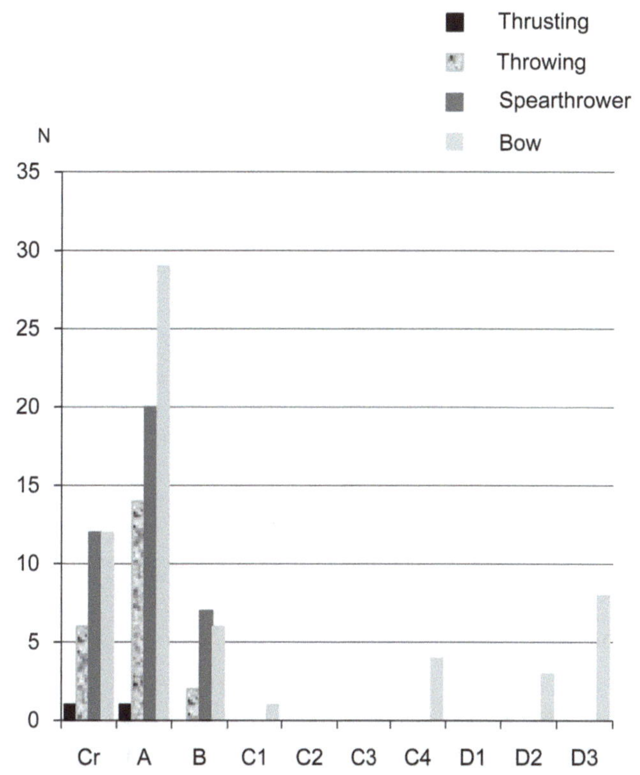

Fig. 3.16 Frequency of the impact fracture types for different delivery modes. Cr: crushing; A: flute-like fracture; B: burin-like fracture; C: transverse fracture; C1: feather termination; C2: hinge termination; C3: step termination; C4: snap termination; D1: bifacial spin-off fractures; D2: spin-off fracture > 6 mm; and D3: spin-off fracture < 6 mm. The fracture types are according to Sano (2009)

of transverse fractures on trapezoids may indicate that the stone tips were fired with high energy by a projectile system.

In addition, there is a correlation between the dimension of impact fractures and the delivery modes (Fig. 3.17; Table 3.5). Thrusting produced impact fractures shorter than 5 mm, while those produced by throwing were no larger than 10 mm, except for one outlier. Conversely, spearthrowers created impact fractures larger than 10 mm, while those created by bows were more than 30 mm, almost as large as the specimens themselves. Therefore, if impact fractures larger than 10 mm were observed, it could be concluded that the stone tips were delivered by either spearthrowers or bows.

Since we confirmed that certain specimens were substantially reduced due to impact damage, the specimen weights were compared before and after the experiments to evaluate the reduction ratio of the pieces (Fig. 3.18). Thrusting created minimal reductions on the specimens. All 10 specimens maintained 100–95% of their original volume. Regarding throwing and spearthrowers, while several trapezoids reduced in volume by over 25%, the majority maintained more than 95% of their original volume. The morphology of tips shot at the velocity of a bow was considerably altered, and two specimens lost over half of their volume, four lost 50–25%, two lost 25–5%, and two lost 5–0%. Accordingly, the high ratio of reduction due to impact damage enabled us to distinguish arrowheads from other hunting weapon tips.

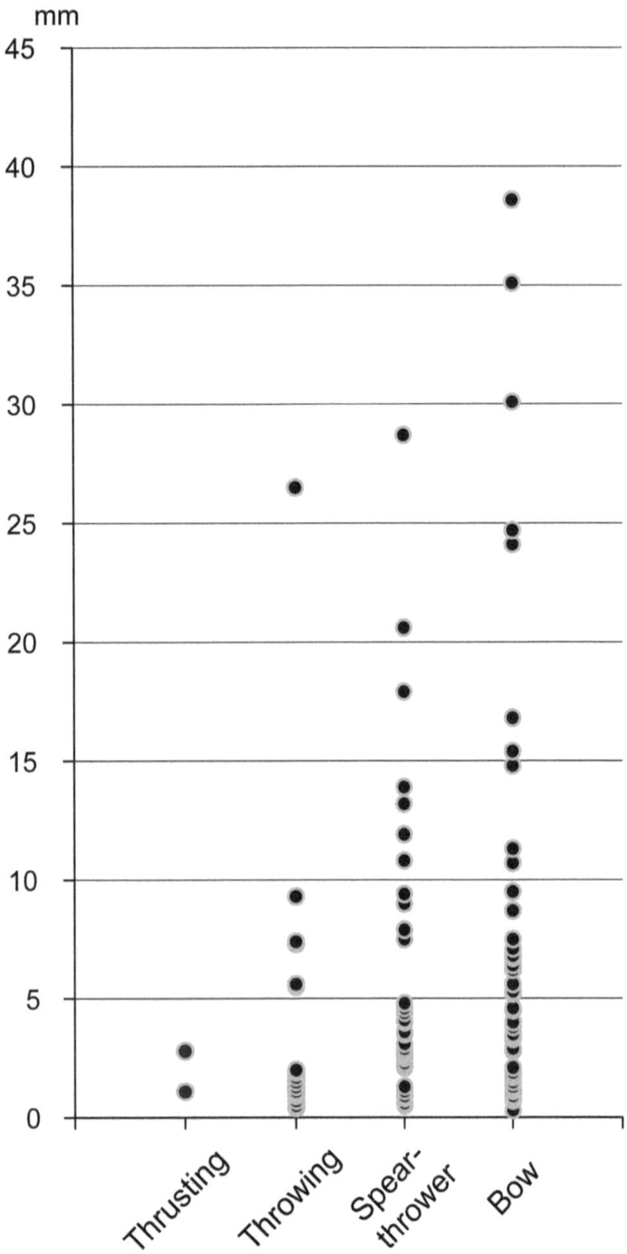

Fig. 3.17 Lengths of the impact fractures for different delivery modes

Table 3.5 Summary of the length (mm) of the impact fractures

	Thrusting	Throwing	Spearthrower	Bow
Number	2	22	39	47
Mean	1.95	3.67	5.74	7.61
Median	1.95	1.5	3.1	4.0
Stdev.	1.20	5.73	6.14	9.08
Min.	1.1	0.4	0.5	0.3
Max.	2.8	26.5	28.7	38.6

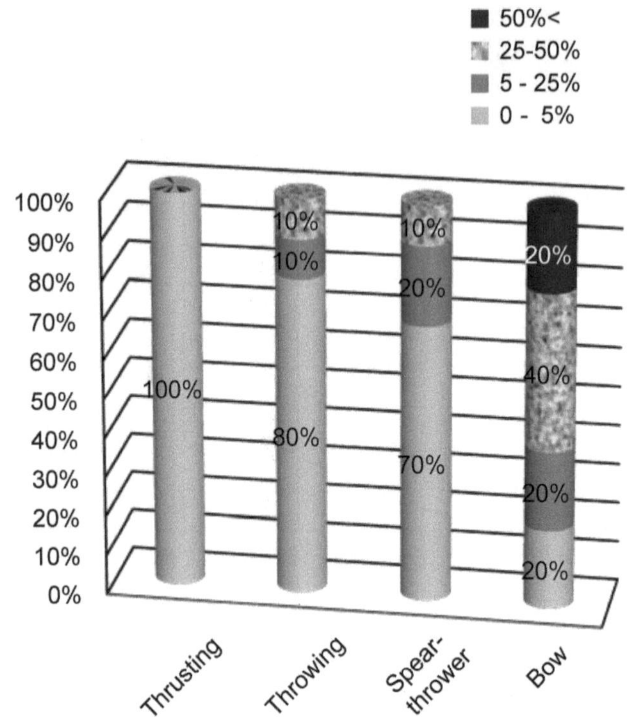

Fig. 3.18 Volume reduction rate of the specimens after the thrusting, throwing, spearthrower, and bow experiments

Conclusions

The results of the aforementioned experiments offered the following conclusions. First, trapezoids rarely experienced transverse fractures owing to their morphological feature, and such fractures occurred only on the bow specimens. Hence, if transverse fractures were observed on trapezoids in archaeological assemblages, we should consider that they may have been shot with bows. In addition, the high reduction ratio of the specimens was an important indicator for the use of bows. It is difficult to accurately estimate the reduction ratio of archaeological tips owing to impact damage. However, if several tips were transversely broken and fragmented, no longer retaining their original morphology, we can assume that these stone tips were used as arrowheads.

Yet, it is worth noting that if the transverse fractures terminated in a snap, they could frequently occur through other agencies such as retouching or trampling (Sano 2009). Therefore, without association with the DIF, including flute-like fractures, burin-like fractures, bifacial spin-offs, and unifacial spin-offs larger than 6 mm (Sano 2009), we cannot conclude that the transverse fractures with snap termination occurred due to hunting. The transverse fractures terminating in a feather, hinge, or step have been recognized as DIF (Fischer et al. 1984; Caspar and De Bie 1996), and they are rarely caused by retouching and trampling (Sano 2009). However, they could accidentally occur from knapping blades (Crabtree 1968; Roche and Tixier 1982; Sano 2009). Therefore, it is necessary to confirm whether the transverse fractures with feather, hinge or step terminations occurred before or after retouching on the lateral sides.

There were significant dimensional differences in impact fractures between delivery modes (see also Clarkson 2016). While thrusting and throwing produced small impact fractures, shooting by spearthrowers and bows frequently yielded impact fractures larger than 10 mm. Thus, impact fractures larger than 10 cm signify that the tips were delivered by spearthrowers or bows.

The frequencies of impact fractures and MLIT were positively correlated with impact velocities. Nevertheless, this cannot be directly used as criteria to evaluate the delivery modes for specific archaeological tips, as we are unaware of the ratio of analyzed archaeological specimens that included stone tips, which were already being utilized as hunting armatures. If the frequencies of impact fractures and MLIT are as low as in the thrusting or throwing experiments, it is difficult to conclude whether this was due to the delivery modes or because most of the analyzed specimens have yet to be used.

If the ratios of the archaeological tips with impact fractures and MLIT without them are similar to those in the throwing experiments, it implies that the use of stone tips comprised primarily of thrusting spears may have been low and that other projectile systems may have existed. The similar ratios of impact fracture occurrences and MLIT to those of the spearthrower experiment suggest that these pieces were shot with either spearthrowers or bows.

The projectile experiments indicated that the formation patterns of the impact fractures and MLIT provide an opportunity to estimate the employed delivery modes. Especially, the presence of the transverse fractures, the dimension of the impact fractures, and the volume reduction ratio are good indicators for distinguishing spearthrower darts and bows and arrows from javelins and thrusting-spears. As this is not the only index for identifying delivery modes, it is important to investigate archaeological specimens by analyzing the frequency, dimension, and types of impact fractures, as well as the volume reduction ratio of the specimens.

However, the results presented in this paper may only be valid for trapezoids (see Iovita et al. 2016 and Clarkson 2016 for comparison), as experiments with backed points showed results different from those for trapezoids (Sano and Oba 2014, 2015). Moreover, the hardness and fragility of raw materials influence fracture formation patterns. In addition, the siliceous shale used in this project is similar to flint, but much harder and less fragile than obsidian. In the future, we will investigate other types of stone tips (such as leaf-shaped points and microblades) to confirm the criteria for identifying the delivery modes within and beyond the variety of tip types. Furthermore, we will examine the influence of different raw materials.

Acknowledgements We would like to thank Radu Iovita for discussion on this study, Karl Hutchings and Chris Clarkson for their helpful comments at the Workshop of Stone Age Weaponry, and Walter Mehlem for creating the calibrated crossbow. We would also like to thank Kaoru Akoshima and Yoshitaka Kanomata for their constant support and helpful advice as well as the students of Tohoku University for their help with the experiments. In addition, this paper profited from the valuable comments of three anonymous reviewers, to whom we are grateful. Finally, this project was supported by grants from the Japan Society for the Promotion of Science (Kakenhi grant numbers 22820005 and 24720350).

References

Barton, R. N. E., & Bergman, C. A. (1982). Hunters at Hengistbury: Some evidence from experimental archaeology. *World Archaeology, 14*, 237–248.

Berger, T. D., & Trinkaus, E. (1995). Patterns of trauma among the Neandertals. *Journal of Archaeological Science, 22*, 841–852.

Bergman, C. A., & Newcomer, M. H. (1983). Flint arrowhead breakage: Examples from Ksar Akil, Lebanon. *Journal of Field Archaeology, 10*, 238–243.

Boëda, E., Bonilauri, S., Connan, J., Jarvie, D., Mercier, N., Tobey, M., et al. (2008). Middle Palaeolithic bitumen use at Umm el Tlel around 70,000 BP. *Antiquity, 82*, 853–861.

Boëda, E., Geneste, J. M., Griggo, C., Mercier, N., Muhesen, S., Reyss, J., et al. (1999). A Levallois point embedded in the vertebra of a wild ass Equus africanus: Hafting, projectiles and Mousterian hunting weapons. *Antiquity, 73*, 394–402.

Caspar, J.-P., & De Bie, M. (1996). Preparing for the hunt in the Late Paleolithic Camp at Rekem, Belgium. *Journal of Field Archaeology, 23*, 437–460.

Cattelain, P. (1989). Un crochet de propulseur solutréen de la grotte de Combe-Saunière 1 (Dordogne). *Bulletin de la Société préhistorique française, 86*, 213–216.

Cattelain, P. (1997). Hunting during the Upper Paleolithic: Bow, spearthrower, or both. In H. Knecht (Ed.), *Projectile technology* (pp. 213–240). New York and London: Plenum Press.

Churchill, S. E. (1993). Weapon technology, prey size selection, and hunting methods in modern hunter-gatherers: Implications for hunting in the Palaeolithic and Mesolithic. *Archeological Papers of the American Anthropological Association, 4*, 11–24.

Clarkson, C. (2016). Testing archaeological approaches to determining past projectile delivery systems using ethnographic and experimental data. In R. Iovita & K. Sano (Eds.), *Multidisciplinary approaches to the study of Stone Age weaponry* (pp. 189–201). Dordrecht: Springer.

Crabtree, D. E. (1968). Mesoamerican polyhedral cores and prismatic blades. *American Antiquity, 33*, 446–478.

Crombé, P., Perdaen, Y., Sergant, J., & Caspar, J.-P. (2001). Wear analysis on Early Mesolithic Microliths from the Verrebroek Site, East Flanders, Belgium. *Journal of Field Archaeology, 28*, 253–269.

Fischer, A., Hansen, P. V., & Rasmussen, P. (1984). Macro and micro wear traces on lithic projectile points. Experimental results and prehistoric examples. *Journal of Danish Archaeology, 3*, 19–46.

Geneste, J.-M., & Plisson, H. (1986). Le Solutréen de la grotte de Combe Saunière 1 (Dordogne). *Première approche paleth-nologique. Gallia préhistoire, 29*, 9–27.

Geneste, J. M., & Plisson, H. (1993). Hunting technologies and human behavior: Lithic analysis of Solutrean shouldered points. In H. Knecht, A. Pike-Tay & R. White (Eds.), *Before Lascaux: The complex record of the Early Upper Paleolithic* (pp. 117–135). Boca Raton, Ann Arbor, London, and Tokyo: CRC Press.

Hughes, S. S. (1998). Getting to the point: Evolutionary change in prehistoric weaponry. *Journal of Archaeological Method and Theory, 5*, 345–408.

Iovita, R., Schönekeß, H., Gaudzinski-Windheuser, S., & Jäger, F. (2016). Identifying weapon delivery systems using macrofracture analysis and fracture propagation velocity: A controlled experiment. In R. Iovita & K. Sano (Eds.), *Multidisciplinary approaches to the study of Stone Age weaponry* (pp. 13–27). Dordrecht: Springer.

Jöris, O., & Baales, M. (2003). Zur Altersstellung der Schöninger Speere. In L. Fiedler, W.-D. Heinrich, A. Justus & E. Brühl (Eds.), *Erkenntnisjäger. Kultur und Umwelt des frühen Menschen. Festschrift für Dietrich Mania. Veröffentlichungen des Landesamtes für Archäologie Sachsen-Anhalt-Landesmuseum für Vorgeschichte 57* (pp. 281–288). Halle: Landesamt für Denkmalpflege und Archäologie Sachsen-Anhalt.

Kudo, Y., & Kumon, F. (2012). Paleolithic cultures of MIS 3 to MIS 1 in relation to climate changes in the central Japanese islands. *Quaternary International, 248*, 22–31.

Lombard, M., & Pargeter, J. (2008). Hunting with Howiesons Poort segments: Pilot experimental study and the functional interpretation of archaeological tools. *Journal of Archaeological Science, 35*, 2523–2531.

Lombard, M., Parsons, I., & Van der Ryst, M. M. (2004). Middle Stone Age lithic point experimentation for macro-fracture and residue analyses: The process and preliminary results with reference to Sibudu Cave points. *South African Journal of Science, 100*, 159–166.

Midoshima, T. (1991). Collisional flaking of the stone arrowhead and tanged point. *Kodai, 92*, 79–97. (In Japanese).

Midoshima, T. (1996). Projectile experiments with backed points. *Kanagawa Koko, 32*, 79–97. (In Japanese).

Moss, E. H., & Newcomer, M. H. (1982). Reconstruction of tool use at Pincevent: Microwear and experiments. *Studia Praehistorica Belgica, 2*, 289–312.

Newman, K., & Moore, M. W. (2013). Ballistically anomalous stone projectile points in Australia. *Journal of Archaeological Science, 40*, 2614–2620.

O'Brien, E. M. (1981). The projectile capabilities of an Acheulian handaxe from Olorgesailie. *Current Anthropology, 22*, 76–79.

Odell, G. H., & Cowan, F. (1986). Experiments with spears and arrows on animal targets. *Journal of Field Archaeology, 13*, 195–212.

Pétillon, J.-M., Bignon, O., Bodu, P., Cattelain, P., Debout, G., Langlais, M., et al. (2011). Hard core and cutting edge: Experimental manufacture and use of Magdalenian composite projectile tips. *Journal of Archaeological Science, 38*, 1266–1283.

Rhodes, J. A., & Churchill, S. E. (2009). Throwing in the Middle and Upper Paleolithic: Inferences from an analysis of humeral retroversion. *Journal of Human Evolution, 56*, 1–10.

Roche, H., & Tixier, J. (1982). Les accidents de taille. *Studia Praehistorica Belgica, 2*, 65–76.

Rots, V. (2010). *Prehension and hafting traces on flint tools: A methodology*. Leuven: Leuven University Press.

Sano, K. (2009). Hunting evidence from stone artefacts from the Magdalenian cave site Bois Laiterie, Belgium: A fracture analysis. *Quartär, 56*, 67–86.

Sano, K., & Oba, M. (2014). Projectile experimentation for identifying hunting methods with replicas of Upper Palaeolithic weaponry from Japan. In J. Marreiros, N. Bicho & J. Gibaja Bao (Eds.), *International conference on use-wear analysis, Faro, Portugal. Use-wear 2012* (pp. 474–486). Newcastle upon Tyne: Cambridge Scholars Publishing.

Sano, K., & Oba, M. (2015). Backed point experiments for identifying mechanically-delivered armatures. *Journal of Archaeological Science, 63*, 13–23.

Shea, J. J. (1988). Spear points from the Middle Paleolithic of the Levant. *Journal of Field Archaeology, 15*, 441–450.

Shea, J. J. (2006). The origins of lithic projectile point technology: Evidence from Africa, the Levant, and Europe. *Journal of Archaeological Science, 33*, 823–846.

Shea, J., Davis, Z., & Brown, K. (2001). Experimental tests of Middle Palaeolithic spear points using a calibrated crossbow. *Journal of Archaeological Science, 28*, 807–816.

Shea, J. J., & Sisk, M. L. (2010). Complex projectile technology and *Homo sapiens* dispersal into Western Eurasia. *PaleoAnthropology, 2010*, 100–122.

Shott, M. J. (1997). Stones and shafts redux: The metric discrimination of chipped-stone dart and arrow points. *American Antiquity, 62*, 86–101.

Sisk, M. L., & Shea, J. J. (2009). Experimental use and quantitative performance analysis of triangular flakes (Levallois points) used as arrowheads. *Journal of Archaeological Science, 36*, 2039–2047.

Sisk, M. L., & Shea, J. J. (2011). The African origin of complex projectile technology: An analysis using tip cross-sectional area and perimeter. *International Journal of Evolutionary Biology, 2011*, Article ID 968012, doi: 10.4061/2011/968012.

Stodiek, U. (1993). *Zur Technologie der jungpaläolithischen Speerschleuder: Eine Studie auf der Basis archäologischer, ethnologischer und experimenteller Erkenntnis*. Tübingen: Verlag Archaeologica Venatoria, Institut fur Ur- und Fruhgeschichte der Universitat Tübingen.

Tani, K. (Ed.). (2000). *Excavation report of Nagano Prefectural Center for Archaeological Research 48: Hinatabayashi B Site, Hinatabayashi A Site, Nanatsuguri Site, Ohira B Site*. Nagano: Nagano Prefectural Center for Archaeological Research. (In Japanese).

Thieme, H. (1996). Altpaläolithische Wurfspeere aus Schöningen, Niedersachsen: Ein Vorbericht. *Archäologisches Korrespondenzblatt, 26*, 377–393.

Thieme, H. (1997). Lower Palaeolithic hunting spears from Germany. *Nature, 385*, 807–810.

Thieme, H. (2005). The Lower Palaeolithic art of hunting. In C. Gamble & M. Porr (Eds.), *The hominid individual in context: Archaeological investigations of Lower and Middle Palaeolithic landscapes, locales and artefacts* (pp. 115–132). London and New York: Routledge.

Thomas, D. H. (1978). Arrowheads and atlatl darts: How the stones got the shaft. *American Antiquity, 43*, 461–472.

Urban, B., Sierralta, M., & Frechen, M. (2011). New evidence for vegetation development and timing of Upper Middle Pleistocene interglacials in Northern Germany and tentative correlations. *Quaternary International, 241*, 125–142.

Whittaker, J. C., & McCall, G. (2001). Handaxe-hurling hominids: An unlikely story. *Current Anthropology, 42*, 566–572.

Yamaoka, T. (2012). Use and maintenance of trapezoids in the initial Early Upper Paleolithic of the Japanese Islands. *Quaternary International, 248*, 32–42.

Yaroshevich, A., Kaufman, D., Nuzhnyy, D., Bar-Yosef, O., & Weinstein-Evron, M. (2010). Design and performance of microlith implemented projectiles during the Middle and the Late Epipaleolithic of the Levant: Experimental and archaeological evidence. *Journal of Archaeological Science, 37*, 368–388.

Chapter 4
Thirty Years of Experimental Research on the Breakage Patterns of Stone Age Osseous Points. Overview, Methodological Problems and Current Perspectives

Jean-Marc Pétillon, Hugues Plisson, and Pierre Cattelain

Abstract Numerous projectile experiments focusing on the replication and use of Stone Age spearheads and arrowheads made of bone or antler have been undertaken since the early 1980s. A survey of this literature is presented here, focusing on aspects of point resistance and breakage patterns, in order to provide a synthetic view of the experimentally-attested macroscopic use-wear traces on this type of implements. Emerging from this general overview, a consistent discrepancy in the extent of fracture damage between the experimental results and the archeological record is pointed out. A first explanation for this situation is suggested, based on recent experimental results. Finally, several directions for further research on this topic are proposed.

Keywords Fracture • Osseous industry • Point • Projectile experiment • Upper Paleolithic

Introduction

Beginning in the early 1980s, a steady flow of experimental projectile studies have addressed the question of the design, performance and use-wear of Stone Age weapon tips (Iovita and Sano 2016; for a review in the 1990s see Dockall 1997; Knecht 1997). In this domain, over 30 years, lithic projectile elements have attracted much more attention than their osseous counterparts – a situation that is also exemplified in the contents of this volume (but see Langley 2016). Nevertheless, today, more than 20 sessions of projectile experiments focusing on the replication and use of prehistoric spearheads and arrowheads made of bone, antler or ivory have been published. Together they form a quite extensive but somewhat scattered body of data. In this article we present a survey of this literature, focusing on aspects of point resistance and breakage patterns, in order to provide a synthetic view of the experimentally-attested macroscopic use-wear traces on this type of implements. Emerging from this general overview, a consistent discrepancy between the experimental results and the archeological record is then pointed out, and a first explanation for this situation is suggested, based on recent experimental results.

Overview of Projectile Experiments with Osseous Weapon Tips

Twenty-five references describing altogether 29 distinct sessions of projectile experiments have been recorded (Table 4.1). Of course, this survey ignores experiments that are still pending publication or remained unpublished. Furthermore, experiments by Bradfield and Lombard (2011) and Foletti (2012) are unfortunately not included here because the results were made available after this article was written.

Most of the 29 studies are centered on material from the Upper Paleolithic of western Europe (from the Aurignacian to the Magdalenian); almost all types of points known from this context have been the subject of at least one experimental test. Prehistoric material from northern Europe and Russia (Ikäheimo et al. 2004; Nuzhnyi 2007), from the Near East (Bergman 1987) and from the Americas (Tyzzer 1936; Frison and Zeimens 1980; Buc 2011) has also been replicated and used. For the manufacture of the experimental

J.-M. Pétillon (✉)
TRACES, CNRS, Université Toulouse 2, Maison de la recherche, 5 allées A.-Machado, Toulouse 31058, France
e-mail: petillon@univ-tlse2.fr

H. Plisson
University of Bordeaux, CNRS, MCC, INRAP, PACEA, UMR 5199, Bâtiment B8, Allée Geoffroy Saint-Hilaire, CS 50023, Pessac 33615, France
e-mail: hugues.plisson@u-bordeaux1.fr

P. Cattelain
Cedarc/Musée du Malgré-Tout, GDRE Préhistos, CReA-Patrimoine, Université Libre de Bruxelles, ULB – Site de Treignes, 28, rue de la Gare, 5670 Treignes, Belgium
e-mail: pcattela@ulb.ac.be

Radu Iovita and Katsuhiro Sano (eds.), *Multidisciplinary Approaches to the Study of Stone Age Weaponry*, Vertebrate Paleobiology and Paleoanthropology, DOI 10.1007/978-94-017-7602-8_4

Table 4.1 Published experiments with osseous weapon tips (distances in meters). See also Bradfield and Lombard (2011) and Foletti (2012)

References	n of points	Point type	Point raw material	Projectile delivery	Target	Shooting distance
Tyzzer (1936)	9	Algonkian simple p.[a]	Ox bone	Ironwood bow	Rocky loam, gravel	37 (loam), ≈8 (gravel)
Tyzzer (1936)	1	Algonkian simple p. (multipronged)	Ox bone	Hand-thrust, bow	Rocks, wooden board	6 (board)
Frison and Zeimens (1980)	1	Folsom p.	Bison femur	Hand-thrust?	Freshly killed elk	0?
Guthrie (1983)	?	Single-beveled p.	Bone, antler (various species)	Compound bow	Freshly killed moose	5
Arndt and Newcomer (1986)	20	Double-beveled p., bipoints	Bone, antler, ivory (various species)	Recurved composite bow	Freshly killed ewe, artificial carcass	5–7
Bergman (1987)	22?	Bipoints	Bone, antler (various species)	Copy of Mesolithic bow	Artificial carcass	5–15
Carrère and Lepetz (1988)	8?	Split-based p.	Reindeer antler	Calibrated crossbow	Gelatin, polystyrene	2
Knecht (1991, 1993)	23	Split-based p., bipoints, single- and double-beveled p.	Bone, antler (various species)	Calibrated crossbow	Freshly killed goats	?
Rozoy (1992)	9	Single-beveled p.	Ox femur	Spearthrower	Competition target, ground	10–30
Stodiek (1991, 1993, 2000), Stodiek and Paulsen (1996)	28	Single- and double-beveled p.	Reindeer antler	Calibrated crossbow	Freshly killed fallow deer	10–15
Stodiek (1993), Stodiek and Paulsen (1996)	5	Late Upper Paleolithic p.	Antler	Spearthrower	Freshly killed bison	12
Knecht (1997)	90	Split-based p.?	Antler	Calibrated crossbow	Freshly killed cow	?
Pokines and Krupa (1997)	6	Self-barbed p., harpoon	Moose antler	Hand-thrown	Buffalofish, half goat carcass	0 (fish), 5 (goat)
Nuzhnyi (1998)	3	Split-based p.	Reindeer antler	Bow	Freshly killed wild boar	?
Pokines (1998)	20	Single-beveled p.	Moose antler	Hand-thrown	Goat	3–5
Bertrand (1995, 1999)	10?	Fork-based p., bipoints, single- and double-beveled p.	Bone, antler	Spearthrower	Ground?	?
Ikäheimo et al. (2004)	4	Bronze age Fenni p.	Elk bone, reindeer antler	Copy of Bronze Age bow	Freshly killed reindeer	8
Pétillon (2004a, 2005, 2006)	42	Fork-based p.	Reindeer antler	Copy of Neolithic bow, spearthrower	Freshly killed calves	10–13
Pétillon (2004a, 2006)	54	Fork-based p., double-beveled p.	Reindeer antler	Copy of Neolithic bow, spearthrower	Freshly killed fallow deer	10–13
Nuzhnyi (2007)	?	Mesolithic p.	Bone	Bow	Freshly killed goat	?
Buc (2011)	2	Harpoon	Red deer antler	Hand-thrust	Fish	0
Buc (2011)	2	Drilled p.	Red deer antler, sheep bone	Hand-thrust	Sheep carcass	0
Buc (2011)	4	Bipoints	Red deer antler, sheep bone	Bow	Sheep carcass	8
Pétillon et al. (2011)	34	p. with massive base, single- and double-beveled p.	Reindeer antler	Spearthrower	Freshly killed deer	12

[a]p. = Points

Table 4.2 Experimental settings described in the references listed in Table 4.1

Target	Projectile delivery				
	Bow	Spearthrower	Crossbow	Hand	Total
Caprine	3	–	1	3	7
Cervid	3	2	1	1	7
Bovine	1	2	1	–	4
Suid	1	–	–	–	1
Fish	–	–	–	2	2
Artificial carcass	2	–	–	–	2
Vegetal or non-organic	2	2	1	1	6
Total	12	6	4	7	29

replicas, antler is the raw material most commonly used, followed by bone, little work having been devoted to ivory points so far.

All these experiments are "replicative experiments" as understood by Iovita et al. (2016). The projectiles are either thrown by hand, thrown with a spearthrower, shot with a bow, or shot with a calibrated crossbow designed to replicate the average speed of a spearthrower throw (Carrère and Lepetz 1988; Knecht 1991; Stodiek 1993). Shooting distances are usually short, between 5 and 15 m, assuming that prehistoric hunters in "normal" hunting conditions would attempt to approach game as closely as possible before shooting (Cattelain 1994, 1997). Most of the targets are bodies of freshly killed ungulates of medium to large size – from fallow deer, sheep and goat to adult cow and moose. Artificial carcasses (i.e., arrangements of bones and meat), vegetal and non-organic targets were also used (Table 4.2). Some of these experiments mostly aimed at assessing the overall performance of the points (efficiency, penetration and wounding power); others were concerned with the study of impact resistance, fractures and the identification of breakage patterns. This is the data that we are going to focus on.

High Impact Resistance

Whatever the means of projectile delivery, almost all experimenters stress the high impact resistance of the bone and antler points. Impacts into soft soil – loam, grassy field, etc. – usually cause no damage to the points (Tyzzer 1936; Rozoy 1992; Bertrand 1995, 1999; Pétillon 2006: 139–140). Shots directed at small animals, such as fish, or shots hitting the soft tissues and the small bones of larger animals – muscles, viscera, ribs and sternum, but also the thoracic vertebrae in medium-sized ungulate species – also leave the point intact in the large majority of cases (Knecht 1991, 1993; Pokines and Krupa 1997; Pokines 1998; Ikäheimo et al. 2004; Pétillon 2006; Buc 2011). The ability of osseous points to break through thin bones such as ribs without taking macroscopic damage was repeatedly noted (Nuzhnyi 1998; Pokines 1998; Pétillon 2006; Pétillon et al. 2011). Higher frequencies of macroscopic damage were obtained only with impacts on hard and thick obstacles, such as stones, frozen soil, and large bones – notably the scapula, innominate and long bones – of medium and large ungulates (Tyzzer 1936; Arndt and Newcomer 1986; Bergman 1987; Rozoy 1992; Stodiek 1993, 2000; Pokines 1998; Pétillon 2006; Pétillon et al. 2011). This good impact resistance clearly contrasts with what was observed in projectile experiments involving lithic points; this difference is of course linked to the much higher elasticity of osseous tissues as compared to cryptocrystalline stones (Ellis 1997; Knecht 1997). As noted by Arndt and Newcomer (1986), Knecht (1997) and Pokines (1998), the low durability of antler points in the experiment by Guthrie (1983) is at odds with all subsequent results and is probably due to his experimental protocol that applies a much higher kinetic energy to the projectiles: heavy, 500 g fiberglass shafts fired from a compound bow at a range of only 5 m.

Tip Damage: Crushing and Mushrooming

When impact damage occurs on osseous points, it is often in the form of a simple crushing of the tip, destroying only the last few millimeters of the point (Fig. 4.1). This type of damage was mentioned, described and/or illustrated by Tyzzer (1936) – who terms it "chipping" –, by Arndt and Newcomer (1986), Bergman (1987), Stodiek (1993), Nuzhnyi (1998), Pétillon (2006) and Pétillon et al. (2011). In the experiments made in 2004 at the Cedarc/Musée du Malgré-Tout by Pétillon (2006), tip crushing is the most common type of damage recorded (Table 4.3). In an unpublished projectile experiment organized by V. Guillomet-Malmassari within the TFPS project (Chadelle et al. 1995, 1996, 1997; Geneste and Magontier 1998) with replicas of Solutrean antler points, crushing is the only type of damage that could be obtained on the points despite long shooting sessions at bodies of large ungulates (H. Plisson, personal observation).

Fig. 4.1 Examples of tip damage on replicas of Solutrean antler points from the experiments organized by V. Guillomet-Malmassari within the TFPS project (April 2000). 1–4: tip crushing

The flattening of the tip of the point ("mushrooming") was observed by Stodiek (1993) on some antler spear points shot at an unprotected ox scapula at a range of 10 m. Mushrooming is not illustrated in this article but examples are given in Stodiek (1993), Plate 113, 4; Arndt and Newcomer (1986), Plate 24. Mushrooming is also described by Arndt and Newcomer (1986) and Bergman (1987). Bergman stresses that this type of damage occurs only on antler points and "never happens with bone which splinters under stress" (see below).

Beveled Breaks

Beveled breaks, or bending fractures, are oblique fractures that start on the side of the point, break off the distal part of it, and show a step-, hinge- or bevel-shaped proximal termination (Fig. 4.2). Beveled breaks are described and illustrated by Arndt and Newcomer (1986), Stodiek (1993), Pétillon (2005, 2006) and Pétillon et al. (2011). Tyzzer (1936), Bergman (1987) and Pokines (1998) do not describe them explicitly but show several experimental points that obviously suffered this kind of breakage. These fractures represent an important proportion of the macroscopic breaks observed experimentally on osseous points: in the experiments by Pétillon et al. at the Cedarc/Musée du Malgré-Tout in 2003 and 2008, they are the most common type of distal damage (Table 4.3; Pétillon 2006; Pétillon et al. 2011). These fractures are almost always the result of an impact on bone, or the result of a missed shot hitting a hard obstacle.

Arndt and Newcomer note that "rounding (...) may be visible on (...) the uppermost edge of a beveled break, extending in some cases down part of the fracture surface. This rounding probably occurs in the instant after the tip breaks away, as the [projectile's] momentum pushes the broken tip against the target" (Arndt and Newcomer 1986). Stodiek (1993) describes a similar phenomenon and suggests the same explanation. Beveled breaks with spin-off fractures are also documented (e.g., Stodiek 1993, Plate 114, 2) – spin-off fractures being defined here as shorter fracture

Table 4.3 Types of distal damage on the antler points in the experiments by Pétillon et al. at the Cedarc/Musée du Malgré-Tout (Pétillon 2006; Pétillon et al. 2011). 2003: Projectiles delivered with spearthrower and bow at bodies of ox calves; 2004: projectiles delivered with spearthrower and bow at bodies of fallow deer; 2008: projectiles delivered with spearthrower at bodies of young deer

Area impacted	Experimental session							
	2003		2004		2008		Total	
	c.[a]	b.b.[b]	c.	b.b.	c.	b.b.	c.	b.b.
Skull	–	–	1	–	–	–	1	–
Vertebrae	1	2	1	1	1	–	3	3
Thorax	1	–	7	3	1	–		3
Scapula and innominate	1	–	3	4	–	1	9	5
Long bones	–	–	5	2	–	1	4	3
Soft tissue	–	–	5	1	–	–	5	1
Off target	–	4	4	–	5	15	9	19
Total	3	6	26	11	7	17	36	34

[a]c. = Crushing
[b]b.b. = Beveled breaks

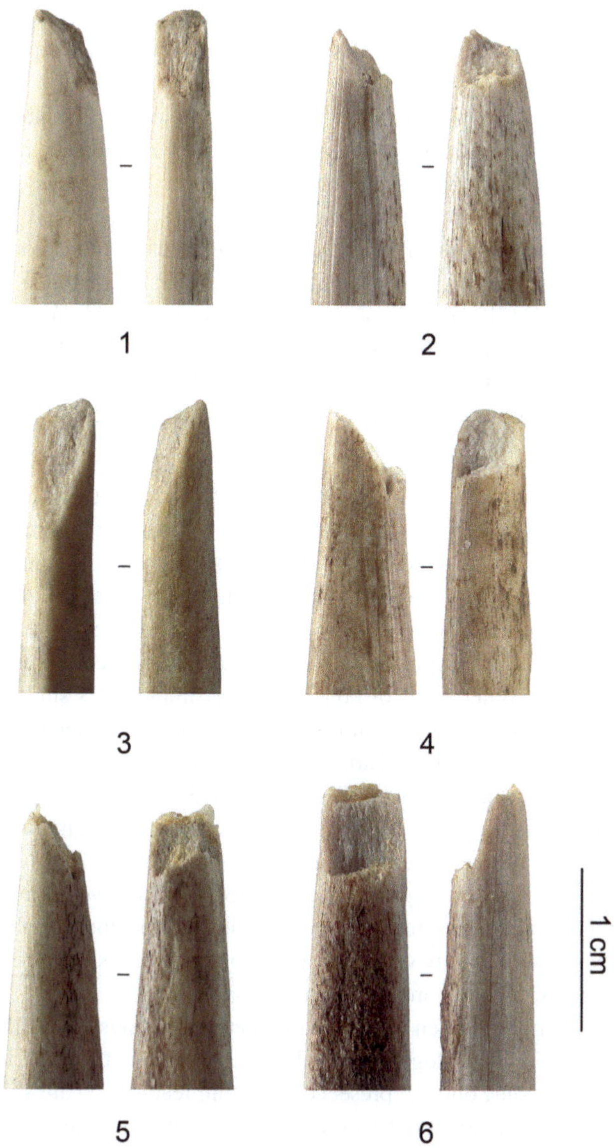

Fig. 4.2 Experimental beveled breaks on antler projectile points, from the experiments at the Cedarc/Musée du Malgré-Tout in 2003 and 2004. 1–2: Step-terminating beveled breaks. 3: Bevel-terminating beveled break. 4–6: Hinge-terminating beveled breaks. All are from spearthrower shots except 2 and 6 (bow shots)

Fig. 4.3 Archeological examples of beveled breaks with spin-off fractures. Antler points from the Upper Magdalenian layer in Isturitz

surfaces, located on the side opposite the main beveled break, and likely resulting from the breakage of the distal part of the point in several fragments upon impact (Fig. 4.3).

"Catastrophic" Damage: Splitting and Shattering

Longitudinal breaks, originating from the tip and dividing the point in two halves, are shown by Bergman (1987) and mentioned by Tyzzer (1936) as "splitting from the tip" (Fig. 1, 4 in Bergman 1987). In both cases, the points affected are made of bone, and this type of fracture does not seem to occur with antler projectile heads.

The shattering of the point in multiple fragments – either through multiple beveled breaks, or through transverse jagged breaks – has been recorded by Rozoy (1992) on one bone point and by Stodiek (1993) on two antler points (Stodiek 1993, Plate 114). Another case is shown by Arndt and Newcomer (1986, Plate 19, left).

Both splitting and shattering cause irreparable damage to the points. These fractures, however, seem infrequent. They are rarely reported and, when they are, they always seem to be the result of a particularly misplaced shot: point-blank bow shots into a bank of coarse and sharp gravel (Tyzzer 1936); an accidental hit on a stone jar (Rozoy 1992); a direct impact on an unprotected ox shoulder blade; and an impact on a fallow deer shoulder blade at acute angle (both in Stodiek 1993; on the influence of acute impact angles on the formation of fractures see Iovita et al. 2016).

Breaks at the Base

According to the experimental data, projectile impacts seldom produce breaks on the proximal, hafted part of the points. Bergman (1987) observes that "damage [to the points] is mostly confined to the tip and in only one case did a break occur at the base". Similarly, Arndt and Newcomer (1986) record a single break at the base, likely due to a manufacturing defect of the point ("this break, the only one in this position, was almost certainly caused by the slightly asymmetrical nature of the bevel, which encouraged this end to bend and finally snap"). Nuzhnyi (1998) also reports one case of (double) base break during an experiment with split-based points. Pokines (1998) writes that, after 249 throws launched at a goat target, only two of the 20 antler points in his experimental sample snapped "where the haft binding ends at the top of the foreshaft". After 74 spear-thrower throws at deer targets, Pétillon et al. (2011) report that only one of their 34 experimental antler points broke in a similar manner at the joint between the point and the shaft. The other experimenters report no breaks in this area; thus the scarcity of damage to the proximal part seems to be the rule.

One of the rare exceptions to this trend is the experiment by Guthrie (1983) in which a significant proportion of breaks occurred at the level of the hafted part. As suggested by Knecht (1997), this discrepancy between Guthrie's results and observations made by other authors might be due to the hafting technique he used.

The only other case of recurrent breaks at the base is the series of experiments by Pétillon (2005, 2006): in several cases, antler fork-based points mounted on spears and launched with a spearthrower suffered proximal fractures when hitting the bones of the target or the frozen ground behind it. These fractures never occurred with similar points used as arrowheads and shot with a bow in identical conditions. Thus, here again, the unusually high frequency of base breaks in this experiment is obviously due to the combination of an especially fragile hafting shape (a two-pronged fork) and a system of projectile delivery (the spearthrower) that places the point under greater flexing forces upon impact [Whittaker 2016; and the large number of similar proximal fractures on fork-based points from Magdalenian assemblages indicates that the spearthrower was the most likely mode of projectile delivery used with these points in the Magdalenian: Pétillon (2005, 2006)].

Summary

The diversity of the protocols used (types of point, raw material, mode of projectile delivery, targets, etc.) makes it difficult to directly compare the results of these different studies. However, several consistent trends emerge from this survey.

- Bone and antler points are very tough projectile tips, especially compared to their lithic counterparts; they often sustain no damage even when hitting hard materials.
- When damage occurs on an osseous point, it is generally located on the distal part of the point and usually takes the form of tip crushing or bending breaks. Experimenters consistently describe these two types of damage as "easily repaired".
- More extensive damage does happen (splitting, shattering, base break) and renders the point irreparable, but these accidents seem rather infrequent and linked to especially violent impacts.
- Several experimenters report that bone points are more prone to breakage than antler points (Guthrie 1983; Bergman 1987; Knecht 1997). Although this parameter deserves further systematic testing and quantification, this observation is consistent with the difference in toughness and elasticity measured between bone and antler tissues (Albrecht 1977; Currey 1979; MacGregor and Currey 1983).

It must be kept in mind that the fracture patterns presented here represent only the range of macroscopic damage that one might expect to find on osseous projectile points identified as such. These fracture patterns must not be over interpreted as diagnostic traces of projectile use. Indeed, several of them, notably beveled breaks, are commonly found on other types of bone and antler implements subject to percussion and flexion constraints: bone and antler wedges (Legrand 2000; Tartar 2009), piercing tools such as needles (Stordeur 1979; Chauvière 2003), etc. Such as on flint artifacts, beveled breaks are the result of a longitudinal stress, whatever its circumstances, and cannot in themselves be linked to a unique cause. Detailed comparative experiments on the fracture patterns of different osseous implements showing beveled breaks (projectile heads, wedges, needles, awls…) are still largely missing.

Therefore, when attempting to identify osseous projectile points in archeological assemblages, morphometric criteria (dimensions, outline, presence of hafting features…) are still the primary arguments; in most cases, impact fractures can only come as supporting arguments ensuring that the traces on the artifacts are compatible with a projectile function. Thus, in the absence of diagnostic morphometric criteria, no identification as projectile point can be done securely (see Hutchings 2016 for a comparable statement regarding lithic points).

Experimental and Archeological Beveled Breaks: An Attempt at Quantitative Traceology

The large majority of experimental breaks described here are in the range of damage that could be easily repaired by resharpening the distal part of the point. However, in archeological assemblages of osseous points, a significant proportion of the artifacts usually show more extensive damage. This damage notably takes the form of fractures occurring "low" on the mesial part, close to the base, thus breaking off a large part of the point and preventing its reworking into a new point (Fig. 4.4). In other words, there seems to be a discrepancy between the experimental and the archeological records as to the extent of damage encountered on osseous projectile points. In order to move beyond this first intuitive statement, we assessed and quantified this discrepancy by using data from an extensively studied assemblage: the antler projectile points from the Upper Magdalenian layer in the Isturitz cave (Pyrénées-Atlantiques, France).

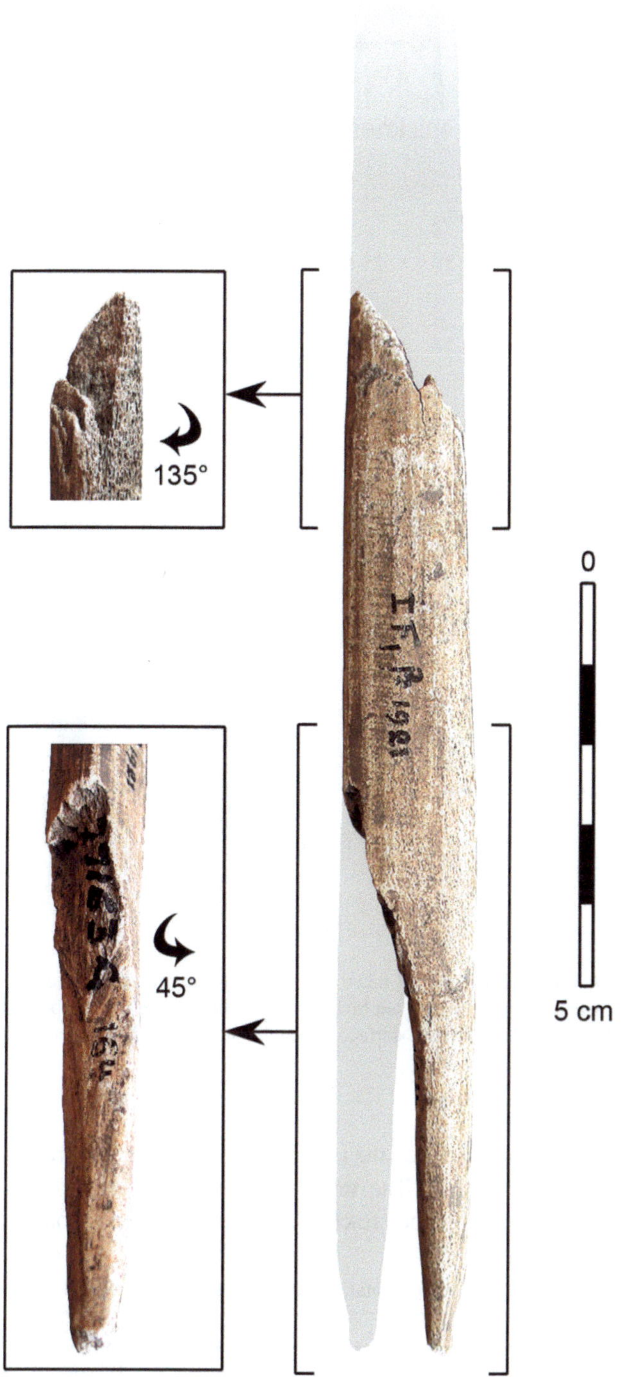

Fig. 4.4 Large fork-based point from Isturitz with extensive, irreparable proximal damage (fracture of the left tine of the fork) and distal damage (bending fracture occurring low on the mesial part). Detail of the proximal fracture as seen from the three-quarters of the upper side; detail of the distal fracture as seen from the three-quarters of the lower side

Materials and Methods

The Isturitz cave was intensively occupied by human groups during most of the Upper Paleolithic. The Upper Magdalenian layer, labeled "I/F1" and dated ca. 16,000–15,000 cal BP (Pétillon 2004b; Szmidt et al. 2009), was completely excavated between 1912 and 1937 (Passemard 1924, 1944; Saint-Périer 1936, 1947). Among the very rich osseous industries, points made of reindeer antler are the most common artifact type. The 662 pieces include 419 fork-based

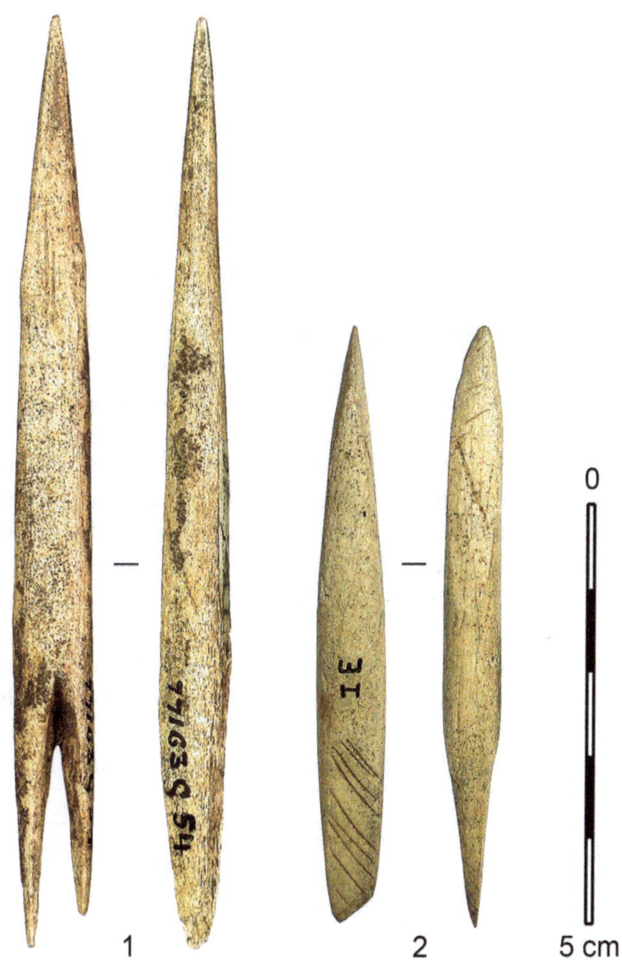

Fig. 4.5 Types of antler points from the Upper Magdalenian in Isturitz. 1: Fork-based point. 2: Double-beveled point

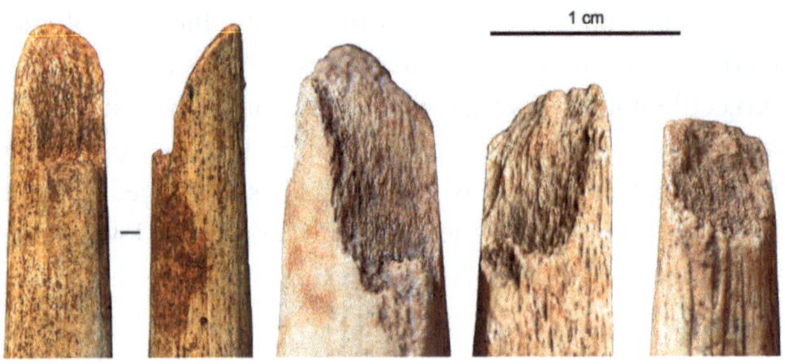

Fig. 4.6 Examples of distal beveled breaks on antler points from the Upper Magdalenian in Isturitz

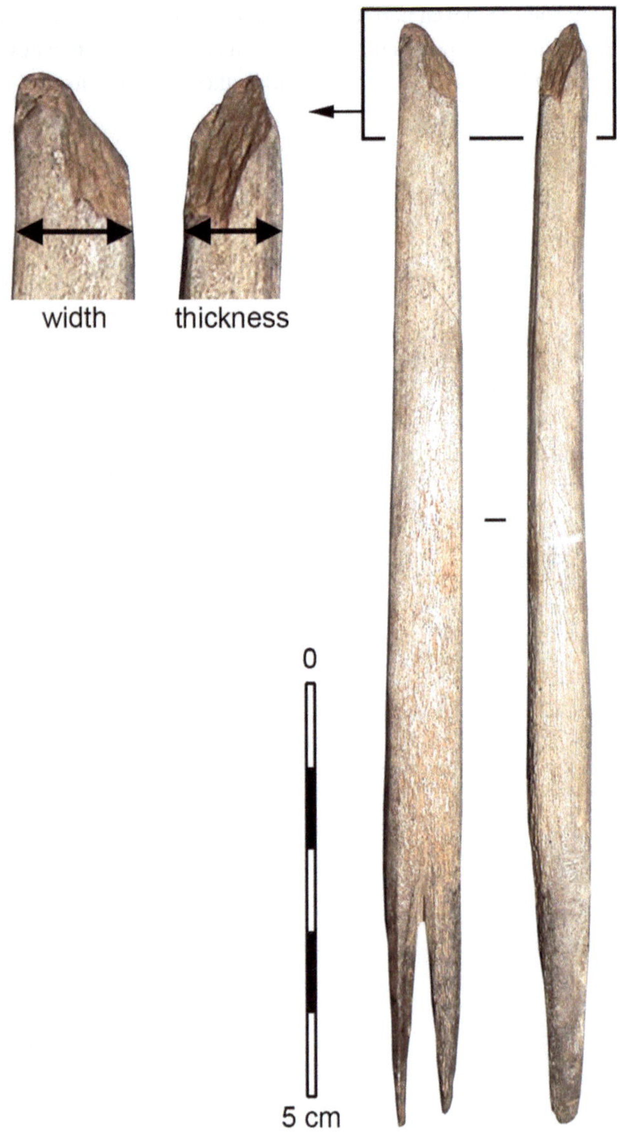

Fig. 4.7 Measurements taken on the archaeological and the experimental points with beveled breaks: width and thickness of the point at the base of the fracture surface (the point shown is a fork-based point from Isturitz)

points, 122 double-beveled points and 121 typologically unidentified fragments (Fig. 4.5). In this assemblage, 155 artifacts show distal beveled breaks (Fig. 4.6; Table 4.4).

In 2003–2004 two projectile experiments were organized at the Cedarc/Musée du Malgré-Tout (Treignes, Belgium). Replicas of the Isturitz points were manufactured and used in order to generate a sample of experimental impact traces and to explore the characteristics and probable operating conditions of these points during the Upper Magdalenian. The experimental protocol has been published in detail elsewhere (Pétillon 2006; Letourneux and Pétillon 2008) and will only be outlined here. A total of 78 fork-based points and 18 double-beveled points, with morphology and dimensions similar to those of the archeological ones, were manufactured from reindeer antler. Half of them were then hafted to arrows shot with a bow and the other half to spears thrown with a spearthrower (the nature of the weapon used in the Upper Magdalenian being first undetermined). The projectiles were launched at bodies of two ox calves and two adult female fallow deer; the shooting distance was 10–13 m. Each arrow or spear was shot repeatedly until the point, shaft or hafting was damaged. A total of 618 shots were performed, 455 of which hit the target (the rather high proportion of missed shots – 26% – is due to the fact that certain shots were specifically aimed at precise body parts such as the head, limbs and vertebrae). At the end of the experiments, 17 points showed distal beveled breaks (Table 4.4).

We quantified the extent of damage inflicted by the beveled breaks by measuring the width and thickness of each point at the base (i.e., the proximal extremity) of the fracture surface (Fig. 4.7). The rationale was that, contrary to the morphology of the fracture – which depends on the

Table 4.4 Antler points from the Upper Magdalenian layer in Isturitz: archeological and experimental samples and number of beveled breaks

		Number of p.[a]	Number of p. with b.b.[b]	Percentage of p. with b.b. (%)
Archeological	Fork-based p.	419	86	21
	Double-beveled p.	122	31	25
	Indeterminate	121	38	31
Experimental	Fork-based p.	78	14	18
	Double-beveled p.	18	3	17

[a]p. = Points
[b]b.b. = Beveled breaks

orientation of the mechanical stress –, the location of the fracture is significant of the level of energy involved. The points having a roughly conical shape, resistance to breakage increases along the shaft; hence for a given energy there is a maximal thickness of possible break.

Results

One of the 17 experimental beveled breaks was excluded from the sample because of its atypical morphology; all others were measured. Of the 155 archeological points with beveled breaks, 16 were excluded from the sample because the relevant measures could not be taken (because of bad preservation, fragmentation, etc.). The graph (Fig. 4.8) shows that the location of the experimental beveled breaks along the shaft of the point is compatible with that of the archeological ones, but clearly restricted to the low width and thickness values. The distribution of the experimental breaks seems bound by a 6 mm threshold: the experimental points never broke at a level where the shaft is more than 6 mm wide and/or 6 mm thick. In the archeological sample, this 6 mm threshold does not exist: 39.6% of the fractures (n = 55) occur on the shaft at a level where the point is wider and thicker than 6 mm, up to 12 and 10 mm respectively. The mean location of the experimental beveled breaks is at 4.4 mm width by 3.8 mm thickness, while these values are 50 and 42% higher for the archeological sample (6.6 × 5.4 mm). This contrast cannot be the result of a morphometric difference between the archeological and experimental samples, since the Isturitz points were precisely used as the model for the manufacture of the experimental ones (Pétillon 2006).

Comparison Data

Before discussing the possible reasons for this contrast, it was necessary to determine if it was specific to our experimental and archeological samples.

Experimental comparison data was sought in the references listed in Table 4.1. But since the necessary measures are not given by the authors, the only way to collect quantified data was to take measurements as precise as possible on the illustrations (pictures and drawings) of the original publications, concentrating on antler points only to ensure

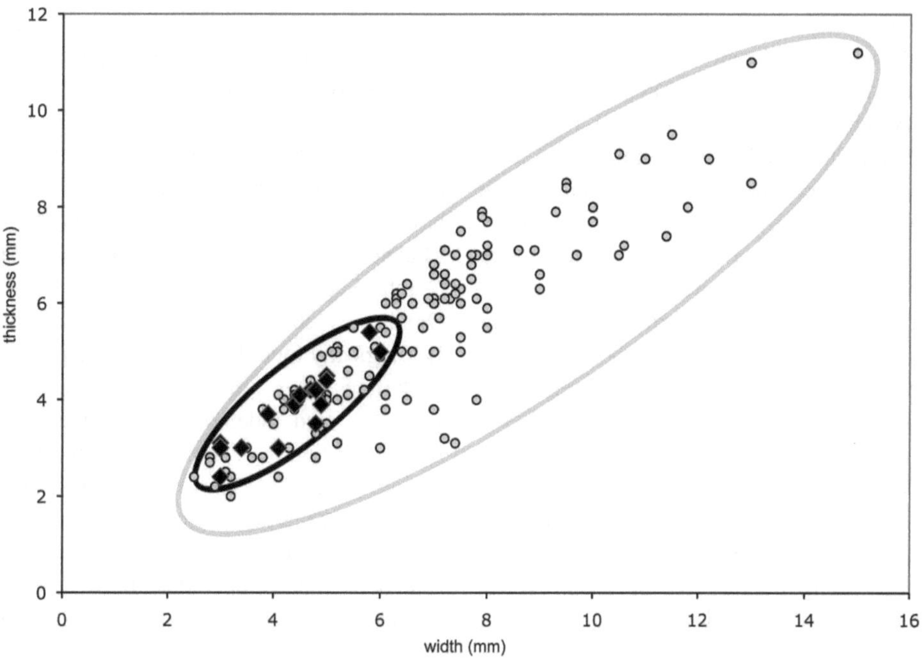

Fig. 4.8 Width and thickness of the point at the base of the fracture surface on points showing beveled breaks. Grey circles: archeological points from the Upper Magdalenian in Isturitz (n = 139). Black diamonds: experimental points from the experiments at the Cedarc/Musée du Malgré-Tout in 2003 and 2004 (n = 16)

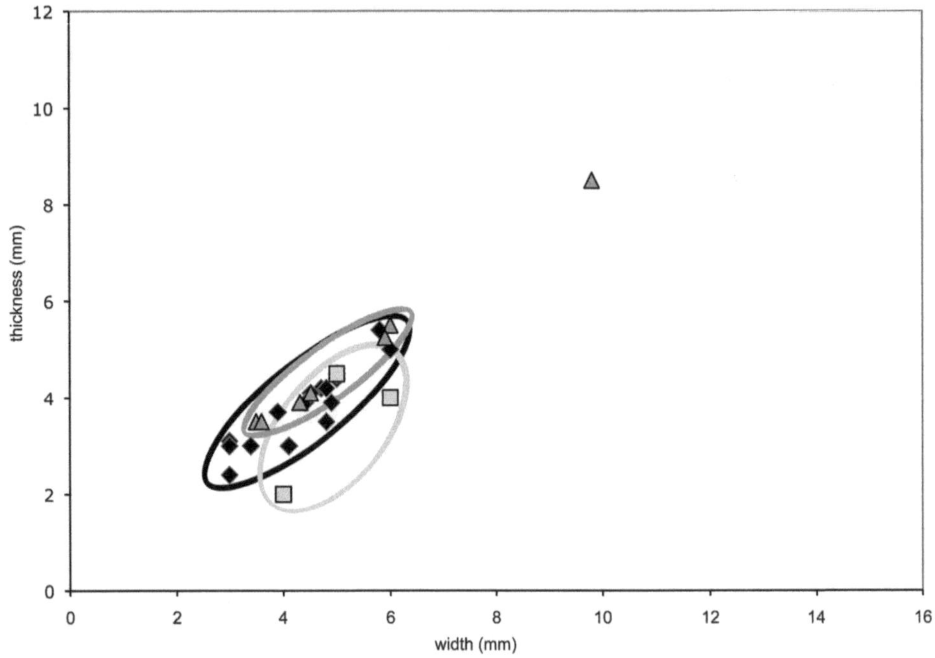

Fig. 4.9 Width and thickness of the point at the base of the fracture surface on points showing beveled breaks. Black diamonds: experimental points from the experiments at the Cedarc/Musée du Malgré-Tout in 2003 and 2004 (n = 16). Dark grey triangles: experimental points measured on the publication by Stodiek (1993; n = 7). Light grey boxes: experimental points measured on the publication by Bergman (1987; n = 3)

comparability with the Isturitz samples. These measurements could be made only on the publications by Bergman (1987: three antler points with beveled breaks on Fig. 1) and by Stodiek (1993: six antler points with beveled breaks on Plate 112, and one on Plate 114, 2). Most of the measures are fully compatible with the Isturitz experimental sample (Fig. 4.9). The bending breaks shown by Pokines (1998, Fig. 2, number 15 to 17) also seem to fall in the same range, although only the width of the fracture base (ca. 4–5 mm) can be measured on the illustration. The only exception is an antler point from Stodiek's sample, with a bending break located where the shaft is 9.8 mm wide and 8.5 mm thick. This break corresponds to the single case reported by Stodiek of point shattering through multiple beveled breaks, after a direct impact on an unprotected ox shoulder blade. A similar case of a point shattering is shown by Arndt and Newcomer (1986, Plate 19, left), with a width of ca. 11 mm at the base of the bending fracture. To sum up, even if certain very violent impacts produced more extensive breaks, the general trend in at least some of the other published projectile experiments appears similar to what was observed on the Isturitz experimental sample.

A small set of archeological comparison data was obtained through the study of collections from four sites in Dordogne. Measurements were taken on antler points with beveled breaks from the Middle Magdalenian levels of La Madeleine (n = 20) and from the Upper Solutrean levels of Combe Saunière, Pech de la Boissière and Le Fourneau du Diable (n = 26 for the total of the three sites). Only artifacts typologically identified as projectile points were considered, excluding other types of implements – especially antler wedges – and indeterminate fragments. The two samples show distributions comparable to the Isturitz archeological sample (Fig. 4.10). Thus the important extent of the bending fractures on the points from the Isturitz Upper Magdalenian does not seem to be a local specificity: this trait is apparently shared by other Upper Paleolithic assemblages from older periods in another region of southwest France.

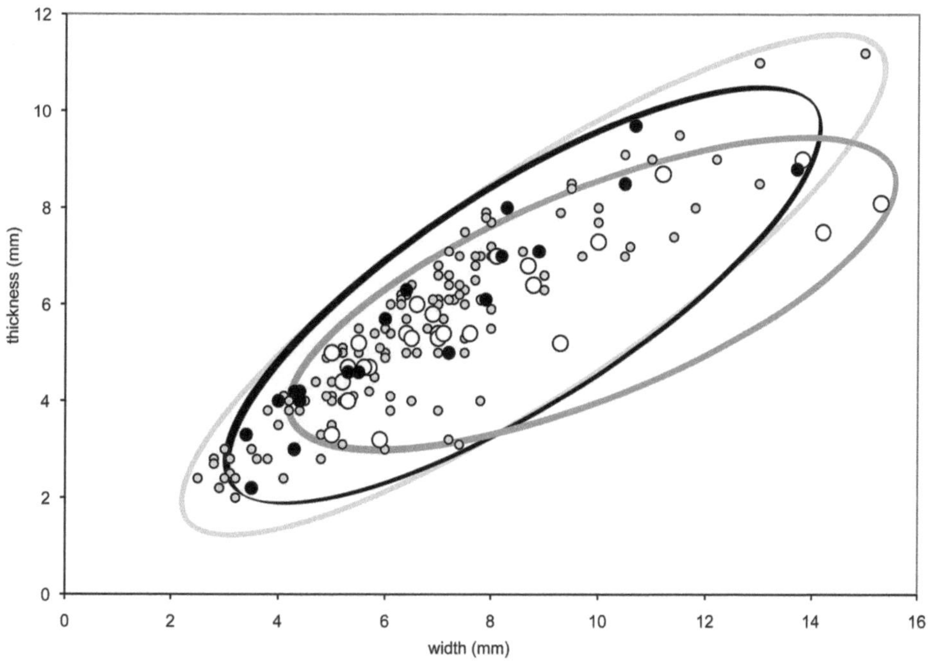

Fig. 4.10 Width and thickness of the point at the base of the fracture surface on points showing beveled breaks. Grey circles: archeological points from the Upper Magdalenian in Isturitz (n = 139). Black circles: archeological points from the Middle Magdalenian in La Madeleine (n = 20). White circles: archeological points from the Upper Solutrean in Combe-Saunière, Pech de la Boissière and Le Fourneau du Diable (n = 26)

Discussion

In the present state of research, and in the context of this article, we can only briefly hint at several possible causes for this phenomenon.

The first line of investigation is that the Paleolithic points could have been used in specific conditions that made their material less resistant to fracture:

- One possible reason for this greater vulnerability is the fatigue of the material. The length of the use life of an osseous projectile point is an issue that has rarely been addressed (see Langley 2016 and references therein); and the consequences of the "aging" of the material on the performance of the implements are poorly known. A progressive loss of elasticity and the formation of microscopic cracks after impacts are two factors that could result in the fact that "old" points eventually break in ways that are not seen with "new" points such as the ones used in the experiments (Shott 2016).
- A second possible reason for a lower impact resistance of the Paleolithic points is the greater brittleness of materials under low temperatures. The environmental record of the Upper Pleistocene in Western Europe indicates very rigorous winters, and thus, during a part of the year at least, a context of operation much more severe for the projectiles than the conditions of modern experiments. Ethnographic evidence shows that lithic points break more easily under low temperatures (Ellis 1997). The existence of a similar liability for bone and antler points remains to be explored, but Khlopachev and Girya (2010) already demonstrated that below −25 °C the mechanical properties of mammoth ivory change considerably, with an increase in brittleness.

The second line of investigation is that the Paleolithic points could have been subject to a greater mechanical stress than the experimental replicas:

- This could first be due to the fact that the Paleolithic projectiles had a higher kinetic energy, that is, a bigger mass and/or a higher speed. This question particularly arises for projectiles thrown with a spearthrower: it is possible that most of the experimental reconstructions of spearthrower hunting are in fact "below" the plausible conditions of use of this weapon in the Upper Paleolithic. Indeed, the spears used in these experiments usually weigh between 80 and 240 g (Carrère and Lepetz 1988; Rozoy 1992; Stodiek 1993; Pétillon 2006; Pétillon et al. 2011), while spearthrower projectiles known from ethnographic contexts can exceed 300 or even 500 g (e.g., Cattelain 1994, 1997; similarly Clarkson (2016) gives a weight of 274 ± 45 g for a sample of 30 Australian spears). Similarly, the calibrated crossbows used by several experimenters to simulate spearthrower use were set to propel the spears at an initial speed of 30 m/s (Stodiek 1993) or an impact speed of 21 m/s (Carrère and Lepetz 1988; settings also used in the experiments by H. Knecht in the TFPS project); but speed measurements by Hutchings and Brüchert (1997) yielded higher values for spearthrower throws (average initial speeds between 31.8 and 47.5 m/s; and see discussion of crossbow calibration in Sano et al. 2016). These experimental biases – too light spears, too slow projectiles – probably had an influence on the point's behavior upon impact. Actually, we believe it to be the most likely hypothesis to explain the very small extent of damage on the antler points in the experiments organized by Guillomet-Malmassari (see above) and by Knecht (1991, 1993, 1997) in the context of the TFPS project. Regarding the experiments by Knecht specifically, the wide outline and large dimensions of many archeological Aurignacian split-based points (Knecht 1993) suggest that they were designed to resist high-energy impacts, probably more violent than those to which they were exposed in the experiments, and that they were attached to much heavier shafts, maybe thrown by hand, since the earliest archeological evidence of the spearthrower is younger than the Aurignacian.

- The necessary use of dead animals as targets, instead of live game, can also be considered as an experimental bias, and Knecht (1997) adequately noted that "stress in bending and torsion will be greater with live animals". The lack of muscular tonus, the higher inertia of a suspended carcass as compared to a standing animal, the absence of body movements (escape attempts, struggle, fall, etc.) are all factors that probably tend to limit the opportunities of damage being inflicted to the projectile points.
- Finally, another cause of higher mechanical stress on the Paleolithic points could be a much more resistant target. In "real life" situations of hunting with bow, spearthrower or hand-thrown spears, missed shots are likely to have been quite frequent, and might have hit hard natural obstacles – rocks, pebbles, etc. The few experimental data available on these impacts suggest that they are particularly damaging to bone and antler points (Tyzzer 1936; Rozoy 1992; Pétillon 2005). However, in most projectile experiments, either missed shots were completely avoided thanks to the use of a precise delivery machine such as the crossbow, or the targets were set against a "soft" background (hay balls, grassy field…) to minimize damage to the projectiles (Fig. 4.11). With this bias in mind, the experiment presented below was organized.

Testing the "Hard Obstacle" Hypothesis

In early 2008 a projectile experiment was held at the Cedarc/Musée du Malgré-Tout. The experimental protocol

Fig. 4.11 General view of three experimental settings. 1: TFPS 1995. 2: Cedarc/Musée du Malgré-Tout 2008. 3: TFPS 2000. The spears were shot with a crossbow in the TFPS experiments and with a spearthrower in the Cedarc/Musée du Malgré-Tout experiments. In the Cedarc/Musée du Malgré-Tout experiments the target is set before a rocky slope covered with vegetation. Pictures by E. Demoulin (Cedarc) and H. Plisson (TFPS)

has been described by Pétillon et al. (2011) and is only summarized here. Thirty-four points were manufactured from reindeer antler, based on various Lower Magdalenian and Upper Magdalenian designs; 24 of them were equipped with lithic inserts (side-hafted flint bladelets and microbladelets). All points were hafted to spears launched with a spearthrower at bodies of two young deer. The shooting distance was 12 m and the targets were deliberately set before a rocky slope covered with vegetation (Fig. 4.11). Each spear was shot repeatedly until the point, shaft or hafting was damaged. Of the 74 shots, 44 hit the target and the others struck the surrounding landscape.

At the end of the experiment, 25 points showed macroscopic distal damage (Table 4.5). Damage to the points occurred, on average, after 2.2 shots. As in previous experiments, impacts in the soft parts of the body, and even against small bones (ribs, sternum, thoracic vertebra), were usually harmless for the points; only the impacts on the humerus and the scapula caused beveled breaks. However, the 30 shots that missed the target and impacted the ground, rocks and vegetation behind the

Table 4.5 Types of distal damage on the antler points in the experiments by Pétillon et al. (2011) at the Cedarc/Musée du Malgré-Tout

Area impacted	Number of hits	Number of hits causing damage	Percentage of hits causing damage (%)	Type of damage		
				c.[a]	b.b.[b]	j.b.[c]
Vertebrae	8	1	12.5	1	–	–
Thorax	29	1	3.4	1	–	–
Scapula and innominate	2	1	50	–	1	–
Long bones (humerus)	1	1	100	–	1	–
Soft tissue	4	0	0	–	–	–
Off target	30	21	70	5	14	2
Total	74	25	33.8	7	16	2

[a]c. = Crushing
[b]b.b. = Beveled breaks
[c]j.b. = Jagged breaks

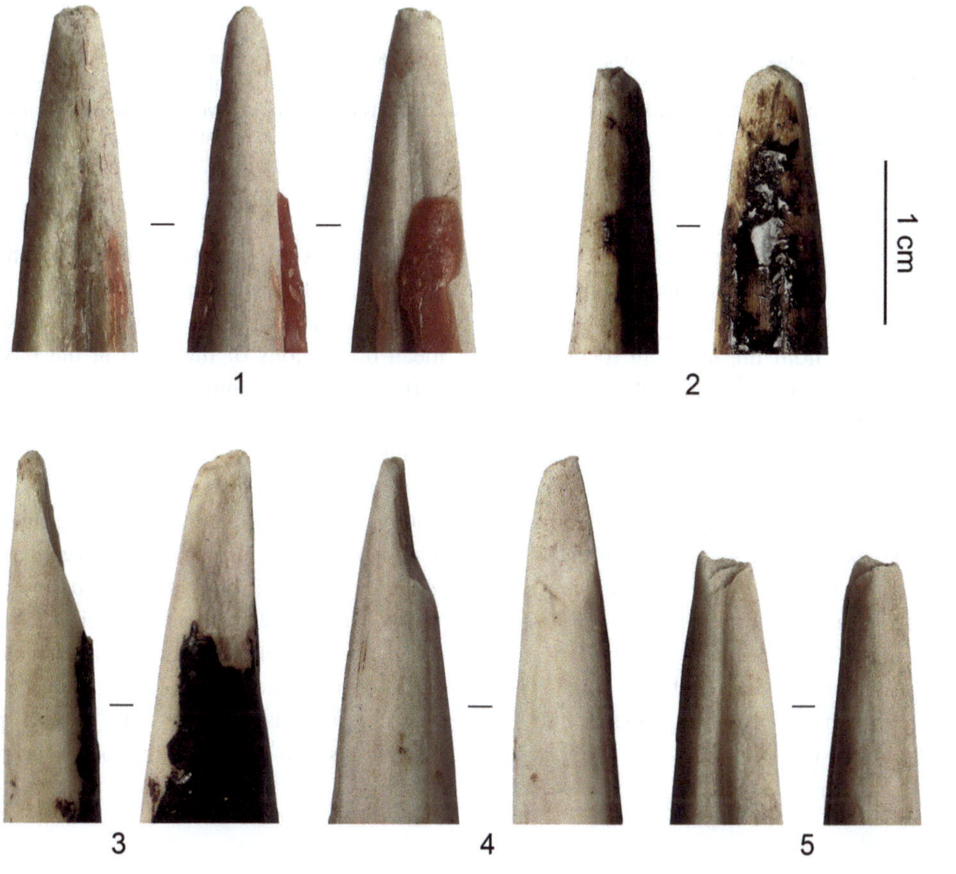

Fig. 4.12 Experimental antler points showing impact damage of limited extent (projectile experiment, Cedarc/Musée du Malgré-Tout, 2008). 1, 2: Tip crushing. 3, 4, 5: Bending fractures. All crushings and fractures correspond to off-target impacts. The red and black stains on points 1, 2 and 3 are remains of adhesives used to attach flint backed bladelets

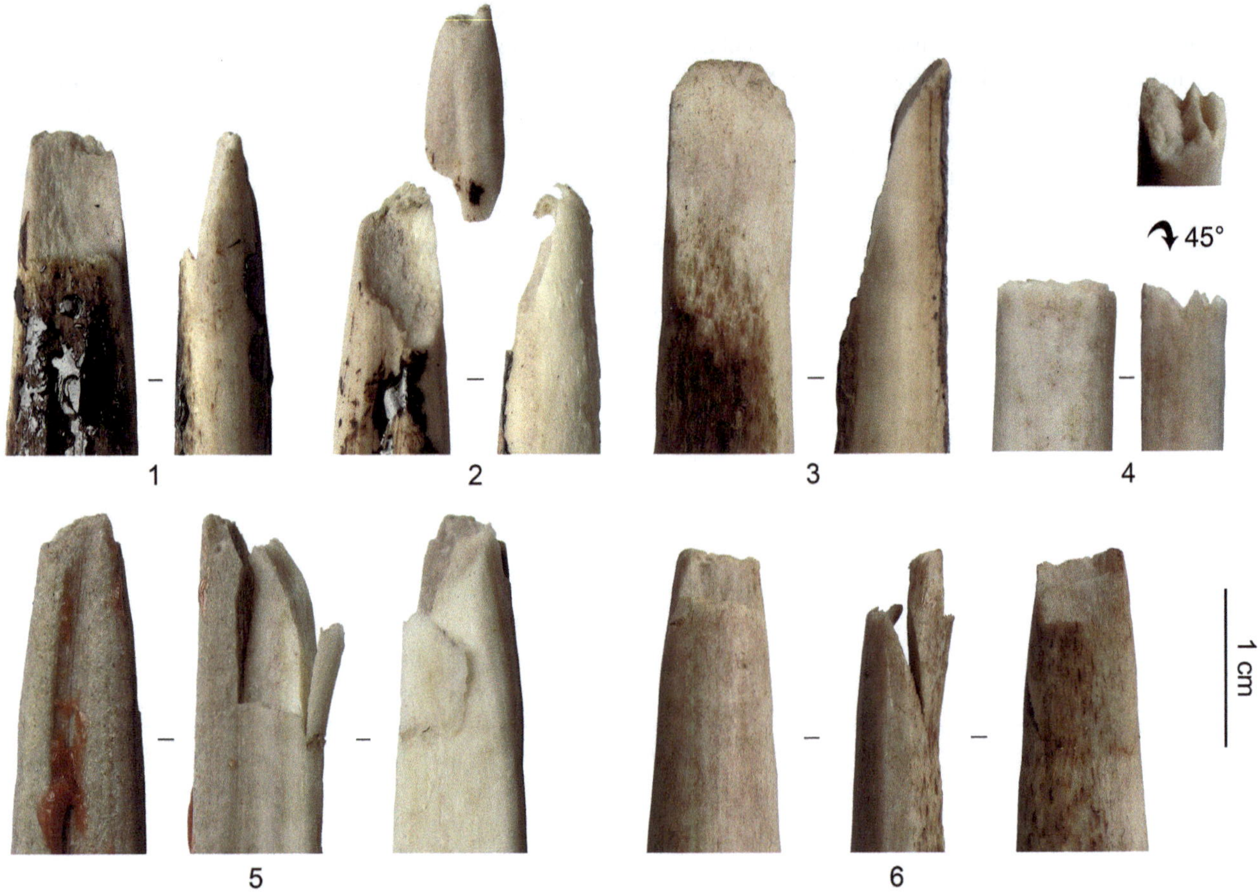

Fig. 4.13 Experimental antler points showing extensive impact damage (projectile experiment, Cedarc/Musée du Malgré-Tout, 2008). 1: Beveled break; 2: Beveled break with rounding (and broken distal fragment); 3: beveled break; 4: transversal jagged break at the limit of the hafted part (detail of the fracture surface as seen from the three-quarters of the distal extremity); 5: multiple beveled break with rounding; 6: beveled break with spin-off. All fractures correspond to off-target impacts. The red and black stains on points 1, 2, 3 and 5 are remains of adhesives used to attach flint backed bladelets

animal caused distal damage in 21 instances (70%). Among these 21 damaged points, beside tip crushing and beveled breaks of limited extent (Fig. 4.12), more important damage occurred, likely resulting from impacts with rocks: "large" beveled breaks, beveled breaks with spin-offs, and, as mentioned above, one case of transversal jagged break at the limit of the hafted part (Fig. 4.13). At least three of these broken points were deemed irreparable.

The same measures were taken on the 16 beveled breaks of the 2008 experiment as on the previous ones (Fig. 4.14). The result is that, in this sample, the 6 mm threshold observed in the other experimental samples does not exist: 6 fractures (37.5%) occur on the shaft at a level where the point is wider and thicker than 6 mm. Five of these 6 breaks correspond to off target impacts (the sixth one being a shot against the humerus). The proportion of 37.5% is close to that of the Isturitz archeological sample (39.6%), and the general distribution of the measures is quite similar given the difference in sample size. The mean dimensions of the shaft at the base of the fracture surface are also comparable: 6.1 × 4.9 mm for the 2008 experiment and 6.6 × 5.4 mm for the Isturitz archeological sample.

Although other parameters certainly deserve systematic testing, we believe that this pilot experiment has pointed out a bias of previous experimental projectile studies. Given the trend evidenced in this small experimental sample, it is highly probable that, in archeological assemblages, at least one part of the antler points showing beveled breaks of large extent correspond to off target impacts against a hard obstacle such as a rock – the broken point being then brought back to the site at the top of the projectile.

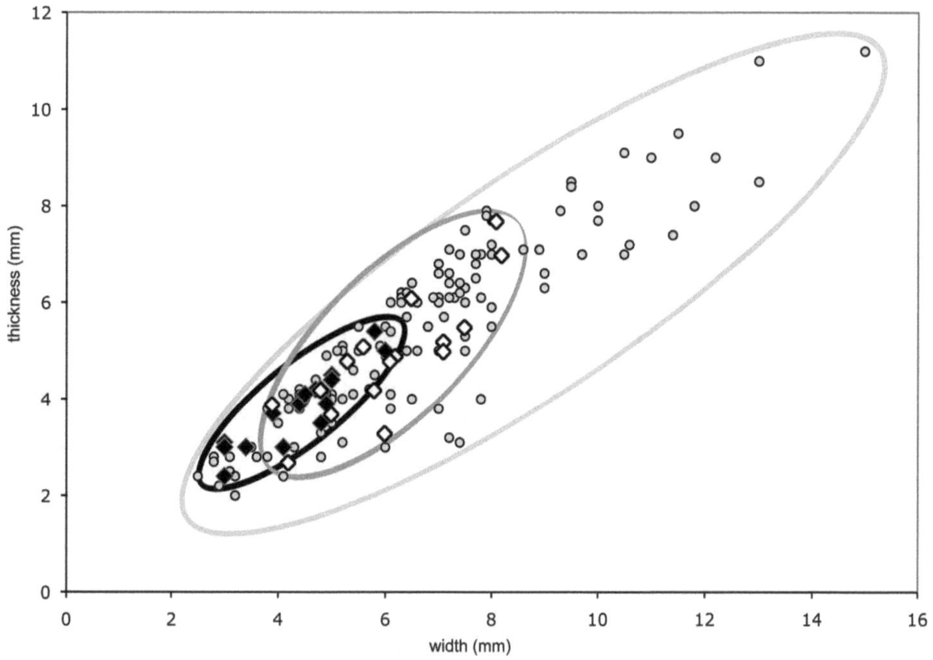

Fig. 4.14 Width and thickness of the point at the base of the fracture surface on points showing beveled breaks. Grey circles: archeological points from the Upper Magdalenian in Isturitz (n = 139). Black diamonds: experimental points from the experiments at the Cedarc/Musée du Malgré-Tout in 2003 and 2004 (n = 16). White diamonds: experimental points from the experiments at the Cedarc/Musée du Malgré-Tout in 2008 (n = 16)

Conclusion

The aim of this article was to summarize the existing knowledge on the experimental fracture patterns of osseous points, and to identify possible perspectives for future research. These perspectives can be laid out in three points:

- Even if recurrent patterns of breakage have been noted and described in several experiments, detailed comparative studies with other types of osseous implements are still needed to identify fractures that are truly projectile diagnostic. This condition must be fulfilled if traceology is to become more than a useful help to morphometric analyses when attempting to identify potential projectile heads among bone and antler implements.
- The most informative experiments are type specific. The proximal fractures of the fork-based points used as spearheads and the specific breakage patterns of bone points as opposed to antler points are two examples of issues that require ad hoc experiments and cannot be addressed through a generic catalogue of fracture patterns on osseous points.
- The experimental conditions do not always adequately reflect the original context of use of the implements, and this can produce experimental results that are not fully compatible with the archeological record. One of these biases – the importance of damage caused by missed shots – has been identified by a recent experiment. Other parameters – including, but not limited to, the fatigue of the osseous material, the influence of extreme temperatures and the speed and mass of the projectiles (at least for the spearthrower) – remain to be tested.

Acknowledgements The authors most strongly thank R.P. Iovita and K. Sano for their invitation to the Stone Age Weaponry workshop and their perfect organization and editing work. Sincere thanks are extended to all participants in the workshop for their very useful comments. Previous versions of this research were presented at the second International Northern Archaeological Congress (Khanty-Mansiisk, Russia, 2006), the Seminario de Tecnología Prehistórica of the Laboratorio de Tecnología Prehistórica, IMF-CSIC in Barcelona, Spain (2009) and the Séminaire de Technologie Osseuse of Université Paris X Nanterre, France (2009); we are grateful to the organizers and attendants of these events. We also want to express our deepest gratitude to all the participants in the Cedarc and TFPS experiments presented in this article, especially to V. Guillomet-Malmassari for kindly allowing us to use her unpublished data. Funding and logistical support to these experiments were provided by associations Cedarc/Musée du Malgré-Tout and L'Homme Retrouvé, by the collective research projects "Technologie fonctionnelle des pointes solutréennes" and "Habitats et peuplements tardiglaciaires du Bassin parisien", by laboratories ArScAn (UMR 7041) and TRACES (UMR 5608), and by universities Paris I and ULB.

References

Albrecht, G. (1977). Testing of materials as used for bone points in the Upper Palaeolithic. In H. Camps-Fabrer (Ed.), *Méthodologie appliquée à l'industrie de l'os préhistorique* (pp. 119–124). Paris: CNRS Éditions.

Arndt, S., & Newcomer, M. (1986). Breakage patterns on prehistoric bone points. In D. A. Roe (Ed.), *Studies in the Upper Palaeolithic of Britain and Northwest Europe* (pp. 165–173). Oxford: Archaeopress (BAR International Series 296).

Bergman, C. A. (1987). Hafting and use of bone and antler points from Ksar Akil, Lebanon. In D. Stordeur (Ed.), *la main et l'outil. Manches et emmanchements préhistoriques, actes du Colloque de Lyon, 1984* (pp. 117–126). Paris: CNRS Éditions.

Bertrand, A. (1995). *Les armatures de sagaies magdaléniennes en matière dure animale dans les Pyrénées*. PhD dissertation, Université Paris I.

Bertrand, A. (1999). *Les armatures de sagaies magdaléniennes en matière dure animale dans les Pyrénées*. Oxford: Archaeopress (BAR International Series 773).

Bradfield, J., & Lombard, M. (2011). A macrofracture study of bone points used in experimental hunting with reference to the South African Middle Stone Age. *South African Archaeological Bulletin, 66*, 67–76.

Buc, N. (2011). Experimental series and use-wear in bone tools. *Journal of Archaeological Science, 38*, 546–557.

Carrère, P., & Lepetz, S. (1988). *Étude de la dynamique des pointes de projectile: élaboration d'une méthode*. M.A. thesis, Université Paris I.

Cattelain, P. (1994). La chasse au Paléolithique supérieur: Arc ou propulseur, ou les deux? *Archéo-Situla, 21–24*, 5–26.

Cattelain, P. (1997). Hunting during the Upper Paleolithic: Bow, spearthrower, or both? In H. Knecht (Ed.), *Projectile technology* (pp. 213–240). New York: Plenum press.

Chadelle, J.-P., Geneste, J.-M., & Plisson, H. (1995). Technologie fonctionnelle des pointes de projectiles solutréennes. In Service Régional de l'Archéologie (Ed.), *Bilan scientifique Aquitaine 1994* (p. 130). Bordeaux: ministère de la Culture.

Chadelle, J.-P., Geneste, J.-M., & Plisson, H. (1996). Technologie fonctionnelle des pointes de projectiles solutréennes. In Service Régional de l'Archéologie (Ed.), *Bilan scientifique Aquitaine 1995* (pp. 131–132). Bordeaux: ministère de la Culture.

Chadelle, J.-P., Geneste, J.-M., & Plisson, H. (1997). Technologie fonctionnelle des pointes de projectiles solutréennes. In Service Régional de l'Archéologie (Ed.), *Bilan scientifique Aquitaine 1997* (pp. 142–143). Bordeaux: ministère de la Culture.

Chauvière, F.-X. (2003). Les aiguilles à chas. In J. Clottes & H. Delporte (Eds.), *La Grotte de la Vache (Ariège), I: Les occupations du Magdalénien* (pp. 325–330). Paris: CTHS—RMN.

Clarkson, C. (2016). Testing archaeological approaches to determining past projectile delivery systems using ethnographic and Experimental data. In R. Iovita & K. Sano (Eds.), *Multidisciplinary approaches to the study of Stone Age weaponry* (pp. 189–201). Dordrecht: Springer.

Currey, J. D. (1979). Mechanical properties of bone tissues with greatly differing functions. *Journal of Biomechanics, 12*, 313–319.

Dockall, J. E. (1997). Wear traces and projectile impact: A review of the experimental and archaeological evidence. *Journal of Field Archaeology, 24*, 321–331.

Ellis, C. J. (1997). Factors influencing the use of stone projectile tips: An ethnographic perspective. In H. Knecht (Ed.), *Projectile technology* (pp. 37–74). New York: Plenum press.

Foletti, G. (2012). Doubles pointes en matières dures animales et armatures de projectiles à la fin du Néolithique moyen: étude fonctionnelle et tentative d'interprétation à partir du site de Marin-Les Piécettes (NE, Suisse). M.A. thesis, Université de Neuchâtel.

Frison, G. C., & Zeimens, G. M. (1980). Bone projectile points: An addition to the Folsom cultural complex. *American Antiquity, 45*, 231–237.

Geneste, J.-M., & Magontier, P. (1998). *Le Temps des chasseurs solutréens*, Scientific documentary, ADCC Archéolud, France.

Guthrie, R. D. (1983). Osseous projectile points: Biological considerations affecting raw material selection and design among Paleolithic and Paleoindian people. In J. Clutton-Brock & C. Gribson (Eds.), *Animals and archaeology, 2 Hunters and their prey* (pp. 273–294). Oxford: Archaeopress (BAR International Series 165).

Hutchings, K. (2016). When is a point a projectile? Morphology, impact fractures, scientific rigor, and the limits of inference. In R. Iovita & K. Sano (Eds.), *Multidisciplinary approaches to the study of Stone Age weaponry* (pp. 3–12). Dordrecht: Springer.

Hutchings, W. K., & Brüchert, L. W. (1997). Spearthrower performance: ethnographic and experimental research. *Antiquity, 71*, 890–897.

Ikäheimo, J. P., Joona, J. P., & Hietala, M. (2004). Wretchedly poor, but amazingly practical: Archaeological and experimental evidence on the bone arrowheads of the Fenni. *Acta Borealia, 2004*, 3–20.

Iovita, R., & Sano, K. (Eds.). (2016). *Multidisciplinary approaches to the study of Stone Age weaponry*. Dordrecht: Springer.

Iovita, R., Schönekeß, H., Gaudzinski-Windheuser, S., & Jäger, F. (2016). Identifying weapon delivery systems using macrofracture analysis and fracture propagation velocity: A controlled experiment. In R. Iovita & K. Sano (Eds.), *Multidisciplinary approaches to the study of Stone Age weaponry* (pp. 13–27). Dordrecht: Springer.

Khlopachev, G. A., & Girya, E. Y. (2010). Секреты древних косторезов Восточной Европы и Сибири. Приемы обработки бивня мамонта и рога северного оленя в каменном веке по археологическами и экспериментальным данным. [*Secrets of ancient carvers of Eastern Europe and Siberia: Treatment techniques of ivory and reindeer antler in the Stone Age according to archaeological and axperimental data*]. Saint Petersburg: Nauka.

Knecht, H. (1991). *Technological innovation and design during the Early Upper Paleolithic: A study of organic projectile technologies*. PhD dissertation, New York University.

Knecht, H. (1993). Early Upper Palaeolithic approaches to bone and antler projectile technology. In G. L. Peterkin, H. M. Bricker & P. Mellars (Eds.), *Hunting and animal exploitation in the Later Palaeolithic and Mesolithic of Eurasia* (pp. 33–47). Washington, DC: APAAA.

Knecht, H. (1997). Projectile points of bone, antler and stone: Experimental explorations of manufacture and use. In H. Knecht (Ed.), *Projectile technology* (pp. 191–212). New York: Plenum press.

Langley, M. C. (2016). More to the point: Developing an Multi-faceted approach to investigating the curation of Magdalenian osseous projectile points. In R. Iovita & K. Sano (Eds.), *Multidisciplinary approaches to the study of Stone Age weaponry* (pp. 229–244). Dordrecht: Springer.

Legrand, A. (2000). *Vers une identification technologique et fonctionnelle des outils biseautés en matière osseuse: le site magdalénien de la Garenne Saint-Marcel (Indre)*. Master's thesis, Université Paris I.

Letourneux, C., & Pétillon, J.-M. (2008). Hunting lesions caused by osseous projectile points: Experimental results and archaeological implications. *Journal of Archaeological Science, 35*, 2849–2862.

MacGregor, A. G., & Currey, J. D. (1983). Mechanical properties as conditioning factors in the bone and antler industry of the 3rd to the 13th century AD. *Journal of Archaeological Science, 10*, 71–77.

Nuzhnyi, D. (1998). The preliminary results of experiments with Aurignacian split based points production, hafting and usage. *Préhistoire européenne, 13*, 117–132.

Nuzhnyi, D. (2007). Розвиток мікролітичної техніки в кам'яному віці: удосконалення зброї первісних мисливців [*Development of the microlithic technique in the Stone Age: Improvement of the weapons of the primitive hunters*]. Kiev: KNT Press.

Passemard, E. (1924). *Les stations paléolithiques du Pays Basque et leurs relations avec les terrasses d'alluvions*. Bayonne: Bodiou.

Passemard, E. (1944). La caverne d'Isturitz en Pays Basque. *Préhistoire, 9*, 7–95.

Pétillon, J.-M. (2004a). *Des Magdaléniens en armes. Technologie des armatures de projectiles en bois de Cervidé du Magdalénien supérieur de la grotte d'Isturitz (Pyrénées-Atlantiques)*. PhD dissertation, Université de Paris I.

Pétillon, J.-M. (2004b). Lecture critique de la stratigraphie magdalénienne de la Grande Salle d'Isturitz (Pyrénées-Atlantiques). *Antiquités Nationales, 36*, 105–131.

Pétillon, J.-M. (2005). Tir expérimental de pointes à base fourchue en bois de renne. In V. Dujardin (Ed.), *Industrie osseuse et parures du Solutréen au Magdalénien en Europe* (pp. 243–256). Paris: Société préhistorique française.

Pétillon, J.-M. (2006). *Des Magdaléniens en armes: Technologie des armatures de projectile en bois de Cervidé du Magdalénien supérieur de la Grotte d'Isturitz (Pyrénées-Atlantiques)*. Treignes: CEDARC.

Pétillon, J.-M., Bignon, O., Bodu, P., Cattelain, P., Debout, G., Langlais, M., et al. (2011). Hard core and cutting edge: Experimental manufacture and use of Magdalenian composite projectile tips. *Journal of Archaeological Science, 38*, 1266–1283.

Pokines, J. T. (1998). Experimental replication and use of Cantabrian lower Magdalenian antler projectile points. *Journal of Archaeological Science, 25*, 875–886.

Pokines, J. T., & Krupa, M. (1997). Self-barbed antler spearpoints and evidence of fishing in the late Upper Paleolithic of Cantabrian Spain. In H. Knecht (Ed.), *Projectile technology* (pp. 241–262). New York: Plenum Press.

Rozoy, J.-G. (1992). Expérimentation de lancer de sagaies avec le propulseur. *Bulletin de la société royale belge d'études géologiques et archéologiques, les chercheurs de la Wallonie, 32*, 169–184.

de Saint-Périer, R. (1936). *La Grotte d'Isturitz, II: Le Magdalénien de la Grande Salle*. Paris: Masson.

de Saint-Périer, R. (1947). Les derniers objets magdaléniens d'Isturitz. *L'Anthropologie, 51*, 393–415.

Sano, K., Denda, Y., & Oba, M. (2016). Experiments in fracture patterns and impact velocity with replica hunting weapons from Japan. In R. Iovita & K. Sano (Eds.), *Multidisciplinary approaches to the study of Stone Age weaponry* (pp. 29–46). Dordrecht: Springer.

Shott, M. J. (2016). Survivorship distribution in experimental spear points: Implications for tool design and assemblage formation. In R. Iovita & K. Sano (Eds.), *Multidisciplinary approaches to the study of Stone Age weaponry* (pp. 245–258). Dordrecht: Springer.

Stodiek, U. (1991). Erste Ergebnisse experimenteller Untersuchungen von Geweihgeschoßspitzen des Magdalénien. In M. Fansa (Ed.), *Experimentelle Archäologie, Bilanz 1991* (pp. 245–256). Isensee: Oldenburg.

Stodiek, U. (1993). *Zur Technologie der Jungpaläolithischen Speerschleuder: eine Studie auf der Basis Archäologischer, Ethnologischer und Experimenteller Erkenntnisse*. Tübingen: Archaeologia Venatoria.

Stodiek, U. (2000). Preliminary results of an experimental investigation of Magdalenian antler points. In C. Bellier, P. Cattelain & M. Otte (Eds.), *La chasse dans la préhistoire/Hunting in prehistory* (pp. 70–78). Bruxelles: SRBAP – Université de Liège – CEDARC.

Stodiek, U., & Paulsen, H. (1996). *Mit dem Pfeil, dem Bogen... Technik der Steinzeitlichen Jagd*. Isensee: Oldenburg.

Stordeur, D. (1979). *Les Aiguilles à Chas au Paléolithique*. Paris: CNRS Éditions.

Szmidt, C., Pétillon, J.-M., Cattelain, P., Normand, C., & Schwab, C. (2009). Premières dates radiocarbone pour le Magdalénien d'Isturitz (Pyrénées-Atlantiques). *Bulletin de la Société Préhistorique Française, 106*, 588–592.

Tartar, É. (2009). *De l'os à l'outil: caractérisation technique, économique et sociale de l'utilisation de l'os à l'Aurignacien ancien*. PhD dissertation, Université Paris I.

Tyzzer, E. E. (1936). The "simple bone point" of the shell-heaps of the northeastern Algonkian area and its probable significance. *American Antiquity, 1*, 261–279.

Whittaker, J. C. (2016). Levers, not springs: how a spearthrower works and why it matters. In R. Iovita & K. Sano (Eds.), *Multidisciplinary approaches to the study of Stone Age weaponry* (pp. 65–74). Dordrecht: Springer.

Chapter 5
Levers, Not Springs: How a Spearthrower Works and Why It Matters

John C. Whittaker

Abstract A spearthrower, or atlatl, works as a lever to propel a light spear or dart, but there are still alternative theories about the mechanical principles. Howard proposed that atlatls work by extending the time force can be applied to a spear. Others suggest that the flex of the atlatl or the dart, or both, stores energy to propel the dart as from a spring. Both of these theories can be demonstrated to be wrong by a variety of evidence, including slow motion images. Those who believe that spearthrowers work by spring power often see them as ancestral to bows. Because they work by different principles, this is highly unlikely. Understanding how a spearthrower works is important in examining its capabilities and place in the evolution of technology, and both practical experimentation and theoretical understanding are necessary.

Keywords Atlatl • Spearthrower • Bow • Lever • Mechanical principles • Experiment • Evolution of technology

Introduction

The simple message of this paper is that it takes both practical experience and theoretical understanding of a tool to evaluate its evolutionary implications. The atlatl, or spearthrower, is familiar in a variety of forms from many cultures, prehistoric and ethnographic (Fig. 5.1). It was one of humankind's earliest mechanical inventions, but the timing of this invention remains in dispute. One reason for this uncertainty is the difficulty of recognizing the weapon that propelled a projectile, when that weapon has not survived in the archaeological record, or is only present as fragments. Most often, all we have are fragments of the associated projectile, usually in the form of stone points. The spearthrower is seen as a considerable advance over a spear thrust or thrown by hand. Heavy spears can do a lot of damage, but require getting close to the target. A spearthrower provides good killing power at a safer distance: When pursuing pachyderms, prudent primitives prefer projectiles. However, to assess the efficiency of a weapon and understand its place in the history of human and technological evolution, we need know how it works.

A spearthrower is a device with a handle at one end and a hook or a socket at the other to engage the butt end of a light spear. Using an atlatl (the word preferred by North American archaeologists), one can throw a light spear (commonly called a dart) much farther and faster than by hand alone. However, despite more than 10,000 years of successful use and the current abilities and experiments of dozens of academic and sporting users (Whittaker and Kamp 2006; Whittaker 2010b), there are still competing theories about how atlatls work.

A bit of history helps to illustrate these theories and to demonstrate the ill effects of misunderstanding how a weapon works. Fortunately, the ill effects are mostly theoretical rather than fatal – with an atlatl, no matter how badly you misunderstand it, you cannot forget it is loaded and shoot yourself. But it is always best to know your tools.

The atlatl is a lever, or more correctly, it is used with the human body as one of a series of levers operating together and in sequence in the course of throwing something (Cundy 1989; Cattelain 1997). This is the dominant understanding today, one that is supported by both theoretical consideration and years of practical experiment by many atlatlists, myself included.

The basic throwing motion, visible on films and ethnographic photos, and described by many others, is the same with an atlatl or a ball (Cundy 1989; Stodiek 1993; Vanderhoek 1998; Whittaker 2010b) (Fig. 5.2). With the dart raised and aimed at the target, the atlatlist begins by shifting

J.C. Whittaker (✉)
Department of Anthropology, Grinnell College, Grinnell, IA 50112, USA
e-mail: whittake@grinnell.edu

Radu Iovita and Katsuhiro Sano (eds.), *Multidisciplinary Approaches to the Study of Stone Age Weaponry*, Vertebrate Paleobiology and Paleoanthropology, DOI 10.1007/978-94-017-7602-8_5

Fig. 5.1 A selection of atlatl forms, Bottom left to right Replica of ancient Peruvian atlatl (JW); reconstruction of Indian Knoll, Kentucky, Archaic Period atlatl type (JW); reconstruction of Upper Paleolithic spearthrower, France (Pascal Chavaux); reconstruction of prehistoric Key Marco, Florida form (JW); replica of Basketmaker atlatl, Arizona (JW); replicas of spearthrowers from ethnographic Lake Patzcuaro (Mexico), Australia, and Inuit (all JW); replica and modern PVC interpretation of ethnographic Australian varieties (Ray Madden, Chris Oberg). Top left to right two modern atlatls (Richard Lyons, JW); and a plastic "Chuck-it" ® dog ball thrower

weight from the rear foot to the front foot, or stepping forward, which brings body, arm, and dart forward. As the step is completed, the torso rotates and the throwing arm flexes at the shoulder, bringing the hand and atlatl forward. The atlatl stays level and the dart on target throughout this motion. To complete the throw, the wrist flexes violently, swinging the atlatl up to vertical and flicking the dart away. Finally the arm and body follow through as the dart flies toward the target. The motion is essentially the same as in throwing a rock or a ball, and it is the "snap" of the wrist at the end that imparts much of the velocity to the throw. The atlatl simply makes a longer lever arm at the wrist – and by flexing the wrist rapidly a small distance, the distal end of the atlatl moves a much greater distance, acting as a lever to impart energy to the dart (Cotterell and Kaminga 1990: 163–170; Baugh 2003).

A simple test of the atlatl's lever action is to compare atlatls that differ only in length (Whittaker 2011; Whittaker and Kamp 2011; Taylor 2012). Using two replicas of an atlatl found by Cushing in his Key Marcos excavations, one of my students and I threw for distance, and found, not surprisingly, that the longer atlatl consistently achieved longer throws for both subjects and different sets of darts. Taylor (2012) produced even clearer results using dog ball throwers of different lengths. To understand why, a quick explanation of lever principles as applied to the spearthrower is in order.

The spearthrower is best thought of as a class 1 lever, meaning that force is applied to one end to move an object at the other end, with the fulcrum in between. For an atlatl, force is applied by the hand to the short arm of the lever, with the wrist as the fulcrum, to move the dart at the long end of the lever. Levers can either operate as force multipliers or velocity multipliers. In other words, the distal end of the atlatl with the hook and dart travels a much greater distance than the grip at the other end, but in exactly the same amount of time, and therefore it must be traveling at higher velocity. But there is a compromise: a single lever system cannot be optimized to enhance both force output and speed output. If you think of the more common use of a class 1 lever, where you can apply a small amount of force to a long lever arm to move a heavy object (a greater force) on the shorter arm, the reverse must also be true: The long arm provides the mechanical advantage, and greater force must be used on the short arm to move a lesser weight on the long arm. You can increase the speed with which the dart is

Fig. 5.2 Six frames from a video of the author throwing with an atlatl. Note the flipping motion (lever action) of the atlatl, and the flex of the dart. Even at 240 frames per second there is some blurring of the fast tip of the atlatl. Photos by Mike Conner, Grinnell College Instructional Technology

thrown by increasing the force applied by your hand, or the length of the atlatl, or you can increase the weight of the projectile to do more damage to the target. In either case, you rapidly reach the limits of the force your hand can apply to swing the atlatl, so atlatl darts tend to be lighter than the kind of spear you can throw by hand, and optimal atlatl length is also limited.

In parts of North America, "bannerstones" or other stone weights were attached to the shaft of an atlatl. Their function remains controversial, but the mechanical principles above tell us that adding weight to the long shaft of a spearthrower must decrease its efficiency (Cundy 1989; Cotterell and Kaminga 1990; Baugh 2003), adding to the force necessary to swing it and thus usually slowing it down. Furthermore, the farther from the fulcrum at the wrist the weight is attached, the more it affects the lever. Added weight cannot possibly increase the velocity of the dart without greater force input, contrary to some theories (Webb 1957). Webb and others (e.g., Perkins 1993) argue that adding a weight increases the flex of the atlatl and thus its spring force. As we will see, the spring theory of atlatl operation is incorrect, so we must dismiss this interpretation of weights. Various other reasons for atlatl weights (which are relatively scarce) have been suggested. Two more plausible ideas are that they act as a flywheel to help stabilize the motion of the atlatl during the throw, and that they balance a spear when the atlatl is held at rest in a "cocked" position waiting to throw (Peets 1960; Cundy 1989; Kinsella 2013).

It is clear that spearthrowers should be understood as lever systems. However there are two competing ideas about how the atlatl works. (1) It extends the application of force during a throw. (2) Either the atlatl or the dart, or both, flex and store spring energy like a bow to propel the dart forward.

Prolonged Thrust Theories

In 1976, as a young archaeology student, I made my first atlatl. It was an abject failure, and I gave up on atlatls for twenty years. Some of the problems are visible in Fig. 5.3. Not knowing much about spearthrowers, I made a clumsy, oversized, heavy atlatl, and a short, rigid spear. Neither is conducive to good throwing. However, I will lay much of the blame on Calvin Howard. I do so with respect and entirely without animosity; his 1974 paper was one of the few coherent descriptions of how an atlatl worked, and inspired me at the time. Unfortunately, it was largely incorrect.

Howard (1974, 1976) believed that the spearthrower is not a catapult, flipping device, or lever arm. As he described atlatl action, spur and handle remain level throughout a throw, the atlatl providing greater thrust because the spur remains in contact with the spear longer than the hand would (Fig. 5.4). Others have suggested similar theories (Mason 1885, 1895; Raymond 1986). As Howard (1974: 102) described his prolonged thrust model, "…during a proper throw, the spur…reaches no greater elevation than that reached by the handle… The spur does not swing upward in an arc, but merely 'follows through' in the original portion of the spear's flight path…[You] throw the spear with the atlatl exactly the way it is thrown without it…The atlatl provides greater thrust than the unaided hand simply because it remains in contact with the spear during a greater proportion of the total thrust than does the hand." At the point where the hand releases the spear and starts to swing down, the atlatl handle goes down a bit, but the "spur" (called "hook" by most American atlatlists) continues to propel the spear. "Hooking results when the thrower fails to keep the atlatl level during the thrust. Any attempt to use the atlatl in a catapult or whipping fashion will hook the end of the spear, forcing it down, and resulting in a completely uncontrolled flight" (Howard 1974: 103).

Fig. 5.3 The author's first experiments with spearthrowers, 1976. The atlatl was too long and heavy, the dart was too short and rigid, and my throwing motion was incorrect

The hooking Howard hated happened to me. It was exactly how my first spears behaved, and today I can demonstrate it at will by throwing a rigid spear with any atlatl. Even adding fletching to a rigid dart will only stabilize it partially after a wild flight with the tail down, or even end-over-end tumbling. A successful spearthrower dart must flex some, a topic to which we will return. According to Howard's notions, the flexibility of the dart is not an issue. However, no matter how hard I tried, I could not make my atlatl work well the way Howard said it should be used, keeping it level throughout the throw.

In fact, I don't think Howard could either. This is not an accusation of dishonesty. We will see more than once in this paper that models of how atlatls work affect our observations, and I believe that Howard was just not seeing what he actually did during a successful throw. Even at the time of his publication, there were others who refuted his model (Butler 1975; Patterson 1975), and there were available ethnographic photos clearly showing that a spearthrower does not stay level during a throw (e.g., Hermann 1967: 111 in Cotterell and Kaminga 1990: 169 and Stodiek 1993, Plate 31).

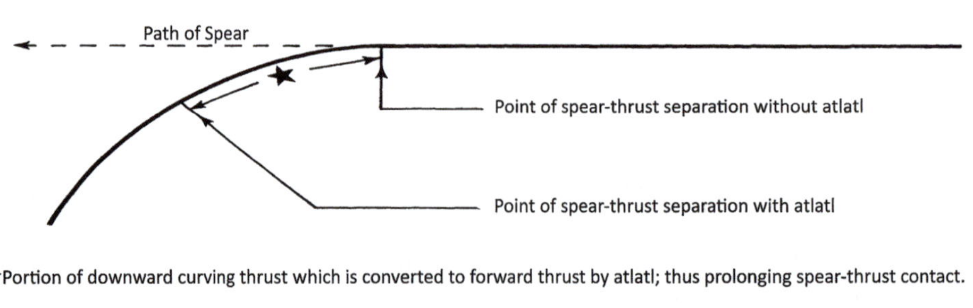

Fig. 5.4 Howard's figure illustrating his theory of atlatl function (Howard 1974, Fig. 5.1). It is not clear to me where the hand and spear actually are, but it is how he expressed his ideas

But here is the power of a mistaken idea: Louis Brennan, writing in 1975, explained atlatls in Howard's terms, saying "Although it appears that the dart is about to be catapulted, the proper throwing motion is to keep the dart and atlatl in contact on a straight horizontal line throughout the entire casting action; the atlatl adds to the length of time of this contact, in effect lengthening the arm" (Brennan 1975: 31). At the same time, he illustrated this discussion with a nice series of photographs showing archaeologist Richard Regensburg using an atlatl. The fourth photo clearly shows Regensburg throwing correctly: the atlatl is vertical as the dart departs, completely contradicting what Brennan has just said. Brennan, like Howard, was evidently blinded by his theory.

Spring Power Theories

Frank Hamilton Cushing provides a second example of mistaken atlatl theory. Cushing is one of my heroes, one of the most colorful figures in all of anthropology. At the age of 17 he spent a semester at Cornell University (my alma mater) where he so impressed his mentors that he was sent to Washington and shortly appointed a curator at the Smithsonian. He went with an expedition to the pueblo of Zuni, where he more or less forced himself on the tribe until they had to accept him, and he stayed for several years as one of the first real participant observer ethnographers. He wrote both popular and scholarly pieces from this experience, became an advocate for the tribe in Washington, liked to see himself as a sort of romantic modern savage, and was famously portrayed as such by Thomas Eakins, a well-known artist of the time. Cushing was fond of replicating prehistoric artifacts, and was among the first archaeologists to figure out some flintknapping. He went on to do archaeological work in Arizona and Florida before dying in 1900 at the unfortunately early age of 43. His ethnographic methods would not pass muster today, some of his colleagues thought he was mad, and some of his enemies accused him of fraud, but he undoubtedly had more experience with tribal peoples and pre-industrial technology than almost anyone else in the scholarly world at the time.

Cushing is relevant here because he was among the first archaeologists to recognize prehistoric spearthrowers in the late 1800s. Earlier accounts and specimens collected by explorers in the Arctic and Australia provided recognizable analogies and spearthrowers were beginning to be recognized among French Paleolithic specimens (Cattelain 1988, 2000; de Mortillet 1891; Lansac 2001), while Zelia Nuttall (1891) described Mesoamerican atlatls based on art and 3 surviving Mexican examples. Otis T. Mason, who also looked at Arctic spear throwers (1885), was the first to claim an archaeological specimen in North America (1893), recognizing that a southwestern "Cliff Dweller" (Basketmaker) artifact collected in the canyons of Colorado and displayed at the World Columbian Exhibition (Fig. 5.5) was equivalent to the atlatls described by Nuttall, and related to Mexican ethnographic specimens. Nuttal and Mason popularized the Nahuatl (Aztec) word *atlatl*, and are apparently responsible for the dominance of that term for spearthrowers in American archaeology.

Fig. 5.5 Cushing's illustration of the Basketmaker atlatl from the Columbian Exposition (Cushing 1895, Fig. 31), comparable to the replica in Fig. 5.1

Cushing and Mason were colleagues, and apparently Cushing prepared the labels for the Columbian Exposition atlatls when they were acquired by the University Museum of the University of Pennsylvania (Mason 1928: 305). Cushing was immediately fascinated by such an esoteric weapon, ancient and loaded with symbolism. Based on knowledge of southwestern ritual practices from his experiences at Zuni, he apparently guessed with some accuracy the objects obscured by bindings on the archaeological specimen (Culin 1898; Mason 1928). He claimed (Cushing 1895) that he made a Southwestern type atlatl, and it worked, so he may be the first archaeologist to actively experiment with spearthrowers, but we have no details. Atlatls remained a fascination for him, and he recognized some unique specimens a couple of years later when he excavated organic remains from the "Court of the Pile Dwellers" in the swamps of Florida (Cushing 1897; Whittaker 2011). In any case, when he learned about southwestern spearthrowers, he eagerly worked atlatls into a theory of technological evolution, illustrating the second faulty theory about atlatl operation: that they work by spring power.

Cushing's article *The Arrow* is a fascinating piece of 19th century scholarship. In it he explains his personal philosophy and the necessity of replicating prehistoric technologies to understand them and their makers, gives a reasonably detailed description of flintknapping based on his experiments, and provides a fanciful model of the evolution of weapon technology, leading to the essential Indian weapon, the bow and arrow.

According to Cushing (1895), you can throw a spear with your finger on the butt (Fig. 5.6a). (I have tried this, and it does work, but only for a weak throw with very light spears). An improvement is a strap on the spear shaft, as used by the ancient Greeks and others. Then it seems that the spearthrower was accidentally invented by the ancients (Fig. 5.6b, c, Cushing 1895: 337): "Let us suppose that a man holding an extra spear in the hand (point backward) with which he hurled another, happened now and then to catch the butt of the one thrown on the barb of the one held, he would not fail to find that this gave great additional force to his cast." The Southwestern atlatl (Fig. 5.6d) is an improvement on a simple stick atlatl. It is curved like a bow's limb in Cushing's figure, because he believed that the springing force of a flexed atlatl helped to propel the dart. In the next step in the evolutionary sequence, he combined atlatl and bow (Fig. 5.6e). This one is an adaptation of a Zuni prayer stick. Richard Lyons, one of my atlatlist friends, made a version of this. It does work, but not as Cushing imagined: you have to engage the dart on the top of the wooden curve, not on the string, and then it works like any other atlatl as a stick with a hook on the end. Continuing to play with the principle of the spring weapon, Cushing describes a flexible stick thrust into the ground and pulled back to propel a dart. He then presents us with a sort of slingshot for arrows (Fig. 5.6f), as the ancestor of reflexed bows. He gave this device the unfortunate name of "bow-crotch." I have not dared to try this one, so maybe I should not laugh at it, but I really do not think it could have been the effective weapon of war that Cushing depicts. Finally the ancients figured it out, connecting their various crotches and flexing limbs to produce a true bow. How one weapon led to the other in the evolutionary sequence that seems to have been on Cushing's mind is never quite clear, beyond the improbable "just so story" about the invention of the atlatl.

What's wrong with all this? Never mind the speculative reconstruction of dubious prehistoric weapons – Cushing started from a mistaken principle. The atlatl does not work by flexing like a bow.

Certainly, some atlatls flex as force is applied to them during a throw, and darts also flex visibly during a throw, sometimes quite dramatically (Fig. 5.7). So this is a good place to say that I am not making fun of what I consider mistaken theories – they mostly make intuitive sense to some users of atlatls, and indeed the question of whether the flex of atlatl or dart adds to the force of the throw continues to be debated. However, there are several pieces of evidence that quite clearly show that the stored energy of springy wood in a flexible atlatl and its dart do not add significantly to the velocity of the dart.

1. Many ethnographic atlatls are rigid, or essentially so. Everyone admits that atlatl flex is not necessary, but some say it is more efficient.

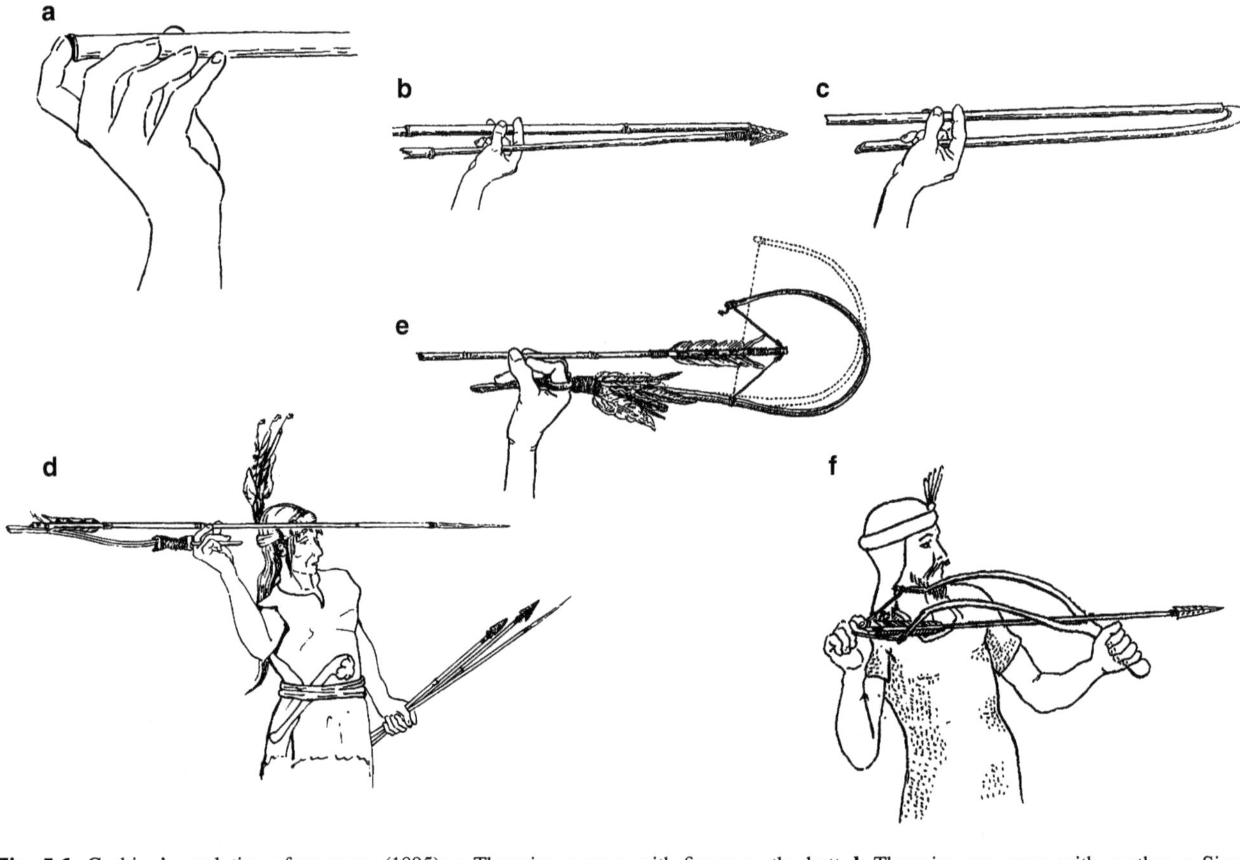

Fig. 5.6 Cushing's evolution of weapons (1895): **a** Throwing a spear with finger on the butt. **b** Throwing one spear with another. **c** Simple spearthrower derived from throwing one spear with another. **d** "Cliff-dweller atlatl or throwing stick in use." **e** "Restoration of ancient stringed spear-crook or throwing-bow (from Zuii prayer-stick of war)." **f** An unlikely weapon, the "bow-crotch."

Fig. 5.7 Dart flexing in flight shortly after launch. Photo by Mike Conner, Grinnell College Instructional Technology

2. Mathematical modeling: if you treat the spear thrower as a cantilevered spring, flexed during a throw by the resistance of the dart, in theory it could add 5–10% to a throw as that stored energy is released (Baugh 1998, 2003; Whittaker and Maginnis 2006). Baugh (1998) also got similar results by modeling the flexible atlatl as a spring at the hook. One can also test this simply and directly. Weathermon (2011) found that an atlatl fixed in place and flexed could propel a 50 g arrow 5.5 m. My own experiments have been less successful. The overall conclusion is usually that atlatl flex could add some force, but probably not a lot.
3. None of the above includes the possible force added by a flexing spear. Perkins (1992, 1993, 1995, 2000a), Perkins and Leininger (1989) has been the most vocal of modern atlatlists claiming that a flexed spear bounds off the spearthrower's hook. Because he is a dramatic publicist for atlatls, appearing in video documentaries and the like, his theories have a wide following among the public as well as atlatlists. Perkins feels that the springing action of a flexible dart both propels the projectile, and ensures a clean and timely separation from the hook of the atlatl. However, you can test at least the first idea very simply: flex a dart by pressing it against an immobile surface, the floor or wall, and release it. It may jump a few centimeters, but that is all. As various other observers have pointed out, most of the spring energy stored by a flexing dart is expended in side-to-side oscillation (Baugh 1998; Weathermon 2011).
4. But in fact, it turns out that the spring force in atlatl or in dart is irrelevant. One of my students and I tested three atlatls, identical except for differences in flexibility, filming them in action with a video camera and strobes (Whittaker and Maginnis 2006). First we noticed that velocity, as measured through our films, did not increase with increased atlatl flexibility. The strobe explained why. The intervals in Fig. 5.8 are 1/120th of a second. When the atlatl is flexed during a throw, by pressing against the resistance of the dart, it continues to move forward, still flexed, until the dart flies off the hook. Only after the dart has departed does the atlatl de-flex and rebound. Only if the atlatl slowed down or stopped would it be able to release the spring tension into propelling the dart. Figure 5.8 shows the flexed atlatl. Labels 1–3 add a straight line to show how much the atlatl has flexed. At 3, the dart has left, but the atlatl is still flexed and is only beginning to rebound at 4, well after the dart has gone down range. The dart is also flexing as it leaves the atlatl.

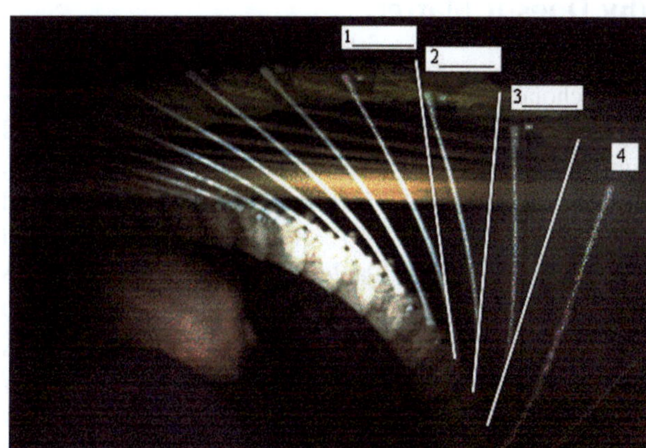

Fig. 5.8 Stroboscopic photograph of atlatl flex, intervals of 1/120th second. Straight lines in *1* to *3* show amount of atlatl flex, with the dart launched but atlatl still flexed at *3*. When this experiment was performed in 2005, our digital equipment was primitive, accounting for the mediocre photo quality. Photo by A. Maginniss

5. There are other demonstrations as well. One atlatlist (Spangler 1998) tested a spearthrower that was hinged in the middle, to provide yet one more added lever – if one wrist joint is good, two must be better. Of course the end of the atlatl did not snap forward until after the dart had left, anymore than it springs forward. And David Cain at Missouri State University has just built a mechanical dart thrower as a thesis project (Cain 2012; personal communication). It operates as a system of levers, and will let him test a number of ideas about the effects of flex, atlatl weights, and so on, that are difficult to examine because of the variability of human throws.

Atlatls and darts do flex (see Fig. 5.2), so it is still worth considering why. The spearthrower does not need to flex, but the dart usually does. The dart must flex in the same plane as the atlatl, to keep the dart tip pointed at the target as the atlatl rises. The flex is also necessary to compensate for the rotational motion of the end of the atlatl which would pull the butt of the spear down and make it tumble (Cotterell and Kaminga 1990; Baugh 1998). This is the "hooking" described by Howard.

But not all spears thrown with spear throwers are flexible. Arctic harpoons are essentially rigid. I have not experimented with them so my comments must be brief. First, photos of Inuit throwing harpoons show that the atlatl does rise and flip the spear away (e.g., Herbert 1981: 142, 148; Alexander 1988: 89 in Stodiek 1993: Plate 3). The inflexibility of the harpoon may be compensated by propelling it not from the end, but from pegs part way along the shaft. The accounts I have read suggest that the balance of the spear and the motion of the throw may be somewhat different from a conventional atlatl as well.

Why Does It Matter?

Many scholars have argued that projectile weaponry represented a significant advantage to those who had it, at several points in human evolutionary history (e.g., Bingham 2000; Crosby 2002; Thieme 2005; Brooks et al. 2006; McCall and Whittaker 2007; Perkins 2007; Shea and Sisk 2010; Sisk and Shea 2010; Whittaker 2010a; Iovita and Sano 2016). What is the technological path from shaking and tossing branches to throwing stones, to throwing spears, adding stone points, propelling them with a lever, and inventing the springy bow?

Cushing was not the only one to believe that the bow evolved from the atlatl. Practical atlatlists (Perkins 2000b; Lyons 2004), archaeologists (Hill 1948; Farmer 1994), and even novelists (Kjelgaard 1951) have said as much. But understanding how an atlatl works as a lever allows us to say that, contrary to popular opinion, it is not directly ancestral to the bow. They work on completely different principles. I suppose it is possible that prehistoric hunters, like some moderns, misunderstood the atlatl as a spring and were inspired to stick two of them together to make a bow. This seems unlikely to me, and is really unknowable. It is equally hard to know how the spearthrower was invented. Was it apparent to a Paleolithic inventor that the hand works as a lever in throwing, and that lever could be lengthened? More likely, someone noticed that you can flick a stone or blob of mud off a stick with some force. But applying that to a spear requires a lot of adjustment: not just any spear will work. Nor will just any stick with a hook – you really cannot throw one spear with another. For good reasons, it is not fashionable to write just-so stories about prehistoric inventions any more, but again, we can learn from practical experience that even the simple spearthrower is complex enough that prehistoric ingenuity, invention, and experiment were necessary to develop it. The variety of forms also shows this.

Since the path of development was probably complex, it is also likely that spearthrowers were invented more than once, and that they competed with and adopted features from other technologies. Certainly they survived in the Arctic, Mesoamerica, and elsewhere alongside the later bow and arrow, and with the earlier thrust and hurled spears and lances.

If we understand the spearthrower as a lever, we see that it is not directly ancestral to the bow, but there remains an interesting evolutionary trend, which continues today. Weapons increasingly distance the user from the target, increasing the hunter or warrior's safety and effect. In the case of thrust spear, javelin, atlatl, and bow there is a trade, a compromise. The two most relevant physical principles here are momentum and kinetic energy (Hrdlicka 2003 provides a detailed discussion). Momentum is the tendency of objects in motion to keep going, and is measured by mass times velocity. Heavier, or faster projectiles continue to penetrate the target longer (all else equal). An object in motion also has kinetic energy that is transferred to the target when it strikes, and can be thought of as force of impact. Kinetic energy is one half mass times velocity squared, thus increasing velocity increases kinetic energy much more than increasing mass. Thus while heavier projectiles have short range and low velocity (because it is difficult to accelerate them effectively), they can do a lot of damage. Lighter projectiles go farther and faster, to some extent increasing striking power through increase in velocity as mass is reduced. Most atlatl darts are too light to be very effective as slow hand-thrown spears, but the lever action of an atlatl allows the human arm to accelerate a dart enough to make it a deadly weapon.

Interpreting prehistoric weaponry is complex. Experimental experience is important. Theory is not enough. You do not have to be as expert as a Paleolithic hunter, but it is hard to understand a technology unless you use it at least a little.

Today we can also make observations that were impossible for earlier generations, using such things as slow motion cameras, microscopes, and measuring instruments of super-human precision (Hutchings 2016; Rots 2016; Yaroshevich et al. 2016). Artificial controlled experiments with standardized targets, weapon points, and projectile launching systems make it possible to sort out complex variables (Iovita et al. 2016; Sano et al. 2016). And we can understand what we are seeing with well-supported mechanical theories and mathematical models. With the conjunction of archaeological evidence, theoretical ideas, and practical experimentation, we can do better than most artists (e.g., Sattlern 1993), who do not know how the atlatl was really used. Like some archaeologists, their prehistoric hunters are in big trouble: they are often shown grasping the spear thrower with their hand wrapped tightly around both the atlatl and the dart, making it impossible to get off a shot.

References

Alexander, B., & Alexander, C. (1988). *Les Eskimos: Chasseurs du Grand Nord*. Courbevoi.

Baugh, R. A. (1998). Atlatl dynamics. *Lithic Technology, 23*, 31–41.

Baugh, R. A. (2003). Dynamics of spear throwing. *American Journal of Physics, 71*, 345–350.

Bingham, P. M. (2000). Human evolution and human history: A complete theory. *Evolutionary Anthropology, 9*, 248–257.

Brennan, L. A. (1975). *Artifacts of ancient America*. Harrisburg: Stackpole Books.

Brooks, A., Nevell, L., Yellen, J. E., & Hartman, G. (2006). Projectile technologies of the African MSA: Implications for modern human origins. In E. Hovers & S. L. Kuhn (Eds.), *Transitions before the transition: Evolution and stability in the Middle Paleolithic and Middle Stone Age* (pp. 233–255). New York: Springer.

Butler, W. B. (1975). The atlatl: Physics of function and performance. *Plains Anthropologist, 20*, 105–110.

Cain, D. (2012). *Sticks with stones: Controlled experimentation in the use of the weighted atlatl*. Unpublished M.A. Thesis, Missouri State University.

Cattelain, P. (1988). *Fiches typologiques de l'industrie osseuse préhistorique, cahier II: Propulseurs*. Aix en Provence: Publications de l'Université de Provence.

Cattelain, P. (1997). Hunting during the Upper Paleolithic: Bow, spearthrower, or both? In H. Knecht (Ed.), *Projectile technology* (pp. 213–240). New York: Plenum.

Cattelain, P. (2000). L'apport de la comparaison ethnographique á la connaisssance et aux tentatives de reconstitution des propulseurs paléolithiques. In C. Bellier, P. Cattelain & M. Otte (Eds.), *La chasse dans la préhistoire/ hunting in prehistory* (pp. 60–69). Bruxelles: Societé Royale Belge d'Anthropologie et Préhistoire.

Cotterell, B., & Kamminga, J. (1990). *The mechanics of preindustrial technology*. Cambridge: Cambridge University Press.

Crosby, A. W. (2002). *Throwing fire: Projectile technology through history*. Cambridge: Cambridge University Press.

Culin, S. (1898). An archaeological application of the Roentgen rays. *Free Museum of Science and Art (University Museum), Bulletin, 1*, 180–183. Philadelphia: University of Pennsylvania.

Cundy, B. J. (1989). *Formal variation in Australian spear and spearthrower technology*. Oxford: BAR International Series 546.

Cushing, F. H. (1895). The arrow. *American Anthropologist, 8*, 307–349.

Cushing, F. H. (1897). Explorations of ancient Key Dwellers' remains on the Gulf Coast of Florida. *Proceedings of the American Philosophical Society, 35*, 329–448.

de Mortillet, A. (1891). Les propulseurs à crochet: Modernes et préhistoriques. *Revue de l'École d'Anthropologie de Paris, 1*, 241–248.

Farmer, M. F. (1994). The origins of weapons systems. *Current Anthropology, 35*, 679–681.

Herbert, W. (1981). *Hunters of the Polar North: The eskimos*. Amsterdam: Time-Life Books.

Hermann, F. (1967). *Völkerkunde Australiens*. Mannheim: Bibliographisches Institut.

Hill, M. (1948). The atlatl, or throwing stick, a recent study of atlatls in use with darts of various sizes. *Tennessee Archaeologist, 4*, 37–44.

Howard, C. D. (1974). The atlatl: Function and performance. *American Antiquity, 39*, 102–104.

Howard, C. D. (1976). Atlatl function: A reply to Butler. *Plains Anthropologist, 21*, 313–314.

Hrdlicka, D. (2003). How hard does it hit? A revised study of atlatl and dart ballistics. *The Atlatl, 16*, 15–18.

Hutchings, K. (2016). When is a point a projectile? Morphology, impact fractures, scientific rigor, and the limits of inference. In R. Iovita & K. Sano (Eds.), *Multidisciplinary approaches to the study of Stone Age weaponry* (pp. 3–12). Dordrecht: Springer.

Iovita, R., & Sano, K. (Eds.). (2016). *Multidisciplinary approaches to the study of Stone Age weaponry*. Dordrecht: Springer.

Iovita, R., Schönekeß, H., Gaudzinski-Windheuser, S., & Jäger, F. (2016). Identifying weapon delivery systems using macrofracture analysis and fracture propagation velocity: A controlled experiment. In R. Iovita & K. Sano (Eds.), *Multidisciplinary approaches to the study of Stone Age weaponry* (pp. 13–27). Dordrecht: Springer.

Kinsella, L. (2013). The bannerstone: A prehistoric prey-specific artifact designed for use in the eastern woodlands of North America. *Ethnoarchaeology: Journal of Archaeological, Ethnographic, and Experimental Studies, 5*, 24–55.

Kjelgaard, J. (1951). *Fire hunter*. New York: Scholastic Book Services.

Lansac, J.-P. (2001). Discussion d'un cadre chronologique pour l'utilization du propulseur et de l'arc. Unpublished M.A. Thesis, University of Bordeaux. http://perso.wanadoo.fr/archeries/MEMOIRE.htm, Accessed August 20, 2001.

Lyons, R. B. (2004). Atlatl to bow. *The Atlatl, 17*, 12.

Mason, J. A. (1928). Some unusual spear throwers of ancient America. *The Museum Journal (University of Pennsylvania Museum), 19*, 290–324.

Mason, O. T. (1885). Throwing sticks in the National Museum. *Smithsonian Institution Annual Report for 1884, part 2*, 279–290, plates 1–16. Washington: Government Printing Office.

Mason, O. T. (1893). Throwing sticks [letter Sept 15]. *Science, 22*, 152–153.

Mason, O. T. (1895). *The origins of invention: A study of industry among primitive peoples*. London: Walter Scott Ltd. Reprinted (1966, M.I.T. Press, Cambridge).

McCall, G., & Whittaker, J. (2007). Handaxes still don't fly. *Lithic Technology, 32*, 195–202.

Nuttall, Z. (1891). The atlatl or spear-thrower of the ancient Mexicans. *Archaeological and Ethnographic Papers of the Peabody Museum, 1*, 171–198 (Cambridge).

Patterson, L. W. (1975). The atlatl function: Some comments. *The Record (Dallas Archaeological Society), 31*, 5–7.

Peets, O. (1960). Experiments in the use of atlatl weights. *American Antiquity, 26*, 108–110.

Perkins, W. R. (1992). The weighted atlatl and dart: A deceptively complicated mechanical system. *Archaeology in Montana, 31*, 65–77.

Perkins, W. R. (1993). Atlatl weights: Function and classification. *Bulletin of Primitive Technology, 1*, 58–61.

Perkins, W. R. (1995). Effects of stone projectile points as a mass within the atlatl and dart mechanical system. *Bulletin of Primitive Technology, 10*, 69–72.

Perkins, W. R. (2000a). Archeological, experimental, and mathematical evidence supporting the use of the atlatl as a primary big game procurement weapon of prehistoric Americans. *Bulletin of Primitive Technology, 20*, 69–72.

Perkins, W. R. (2000b). Effects of stone projectile points as a mass within the atlatl and dart mechanical system and its relationship to the bow and arrow. *Indian Artifact Magazine, 19*(8–9), 78–79.

Perkins, W. R. (2007). Throw this article. *Bulletin of Primitive Technology, 34*, 91–92.

Perkins, W. R., & Leininger, P. (1989). The weighted atlatl and dart: A deceptively complicated mechanical system. *The Atlatl, 2*,1–3; *2*,1–4; *3*,1–3.

Raymond, A. (1986). Experiments in the function and performance of the weighted atlatl. *World Archaeology, 18*, 153–177.

Rots, V. (2016). Projectiles and hafting technology. In R. Iovita & K. Sano (Eds.), *Multidisciplinary approaches to the study of Stone Age weaponry* (pp. 167–185). Dordrecht: Springer.

Sano, K., Denda, Y., & Oba, M. (2016). Experiments in fracture patterns and impact velocity with replica hunting weapons from Japan. In R. Iovita & K. Sano (Eds.), *Multidisciplinary approaches to the study of Stone Age weaponry* (pp. 29–46). Dordrecht: Springer.

Sattler, H. R. (1993). *The earliest Americans, illustrated by Jean Day Zallinger*. New York: Clarion Books.

Shea, J. J., & Sisk, M. L. (2010). Complex projectile technology and *Homo sapiens* dispersal into Western Eurasia. *PaleoAnthropology, 2010*, 100–122.

Sisk, M. L., & Shea, J. J. (2010). Defining complex projectile technology: A reply to Whittaker. *PaleoAnthropology, 2010*, L9.

Spangler, D. (1998). The broken atlatl. *The Atlatl, 11*, 8–9.

Stodiek, U. (1993). *Zur Technologie der jungpaläolithischen Speerschleuder: Eine Studie auf der Basis archäologischer, ethnologischer, und experimenteller Erkenntnisse*. Tübingen: Verlag Archaeologica Venatoria.

Taylor, H. (2012). The effectiveness of spear throwers. Unpublished paper, University of Manchester.

Thieme, H. (2005). The Lower Paleolithic art of hunting: The case of Schoningen 13 II-4, Lower Saxony, Germany. In C. Gamble & M. Porr (Eds.), *The hominid individual in context: Archaeological investigations of Lower and Middle Palaeolithic landscapes, locales, and artefacts* (pp. 115–132). London: Routledge.

VanderHoek, R. (1998). The atlatl and dart. Unpublished M.A. Thesis, Department of Anthropology, University of Illinois at Urbana-Champaign.

Weathermon, R. L. (2011). Late Archaic hunter-gatherer perishable technologies in the Black Hills: An investigation of the Crystal Cave location and its culturally modified wood. PhD dissertation, Department of Anthropology, University of Wyoming.

Webb, W. S. (1957). *The development of the spearthrower*. University of Kentucky Occasional Papers in Anthropology No. 2. Reprinted 1981, Program for Cultural Resource Assessment, Department of Anthropology, University of Kentucky.

Whittaker, J. C. (2010a). Comment on Shea and Sisk's "Complex projectile technology". *PaleoAnthropology, 2010*, L7–L8.

Whittaker, J. C. (2010b). Weapon trials: The atlatl and experiments in hunting technology. In J. Ferguson (Ed.), *Designing experimental research in archaeology: Examining technology through production and use* (pp. 195–224). Boulder: University of Colorado Press.

Whittaker, J. C. (2011). Cushing's Key Marco atlatls: Reconstructions and experiments. *Ethnoarchaeology: Journal of Archaeological, Ethnographic, and Experimental Studies, 3*, 139–161.

Whittaker, J., & Kamp, K. (2006). Primitive weapons and modern sport: Atlatl capabilities, learning, gender, and age. *Plains Anthropologist, 51*, 213–221.

Whittaker, J. C., & Kamp, K. A. (2011). Long and short: Reconstructing Key Marco atlatls. *The Atlatl, 24*, 21–22.

Whittaker, J., & Maginniss, A. (2006). Atlatl flex: Irrelevant. *The Atlatl, 19*, 1–3.

Yaroshevich, A., Zaidner, Y., & Weinstein-Evron, M. (2016). Projectile damage and point morphometry at the Early Middle Paleolithic Misliya Cave, Mount Carmel (Isral): Preliminary results and interpretations. In R. Iovita & K. Sano (Eds.), *Multidisciplinary approaches to the study of Stone Age weaponry* (pp. 119–134). Dordrecht: Springer.

Part II

Recognizing Weapons: Archaeological Applications

Chapter 6
Hunting Lesions in Pleistocene and Early Holocene European Bone Assemblages and Their Implications for Our Knowledge on the Use and Timing of Lithic Projectile Technology

Sabine Gaudzinski-Windheuser

Abstract This paper presents a review of our current state of knowledge about hunting lesions in faunal assemblages from Pleistocene and early Holocene contexts. Differences in the character of hunting lesions throughout the Pleistocene and early Holocene are described. This evidence is contextualized against the archaeological record and its potential for assertions on human hunting tactics is outlined. From the evidence considered here a relatively late onset of lithic projectile technology in human evolution can be implied, which was regularly in use no earlier than the Late Glacial period.

Keywords Hunting lesions • Faunal assemblages • Pleistocene • Early Holocene • Hunting tactics • Stone tipped projectile technology

Introduction

The hunting way of life is as old as mankind. Even though discussions on our ancestors' hunting abilities were challenged during the 1980s and 1990s, research using methods developed as a result of this debate has demonstrated that regular large mammal hunting was established already by at least 1.5 Ma (Dominguez-Rodrigo 2002; Gaudzinski-Windheuser 2005; Dominguez-Rodrigo et al. 2007; Rabinovich et al. 2011). Although we know that hunting formed an important aspect of hominins' behavioural repertoire, hunting techniques and strategies remain largely unknown for long phases of our past. From the zooarchaeological record we can infer that large cooperative hunts as well as ambush hunting of large herd animals (Gaudzinski 1995, 2000), territorial game (Valensi and Psathi 2004) and confrontational hunting (Gaudzinski-Windheuser and Roebroeks 2011) were already being employed by Neanderthals.

The zooarchaeological record, however, only provides a very coarse picture of the hunting tactics employed. Hunting methods and/or the organisation of hunting events remain invisible from this evidence. Even though we can demonstrate that hominins already hunted at a very early stage of human evolution the social organisation connected to these hunting events, remains enigmatic. It is these inferences which were once highly valued as beneficial for our understanding of hominin evolution and considered as the immediate benefits of having demonstrated hominin hunting in pre-Upper Palaeolithic times, an attitude that triggered 20 years of hunting/scavenging debate (e.g., Washburn and Lancaster 1968; Binford 1981, 1985). As long as zooarchaeological/taphonomical research does not envisage a more holistic perspective to develop the full potential of zooarchaeological studies (Gaudzinski-Windheuser and Kindler 2012), we have to focus on other aspects of the archaeological record such as lithics and/or organic implements in order to answer these questions. However, this also provides us with some difficulties.

In pre-Upper Palaeolithic contexts the identification of weaponry systems is particularly problematic. The most uncontested evidence is probably provided by wooden spears from German interglacial sites (Thieme and Veil 1985; Thieme 2007). At least 9 spears have been unearthed at the site of Schöningen, dated to MIS 9 or 7. Even though excavations in this particular exposure of the open cast lignite mine at Schöningen came to an end in layer 13II-4, the spear horizon, more than 10 years ago, it is to be expected that the number of spears will rise, as spear fragments have additionally been identified amidst the wooden debris associated with animal bones and lithics. For the spears the

S. Gaudzinski-Windheuser (✉)
MONREPOS Archaeological Research Centre and Museum for Human Behavioural Evolution, Schloss Monrepos, 56567 Neuwied, Germany
e-mail: gaudzinski@rgzm.de

and

Institut für Altertumswissenschaften, Arbeitsbereich Vor- und Frühgeschichtliche Archäologie, Johannes-Gutenberg-University of Mainz, Schönborner Hof, Schillerplatz 11, 55116 Mainz, Germany

Radu Iovita and Katsuhiro Sano (eds.), *Multidisciplinary Approaches to the Study of Stone Age Weaponry*,
Vertebrate Paleobiology and Paleoanthropology, DOI 10.1007/978-94-017-7602-8_6

majority of their weight ("taring") is forward towards the tip, as with modern javelins (Rieder 2007). Further important evidence comes from the MIS 5e site of Lehringen. Here a wooden weapon was found amidst the carcass of an elephant individual (Thieme and Veil 1985). In contrast to the javelins from Schöningen, the spear from Lehringen shows its taring in the base region and was thus interpreted as a lance, a thrusting weapon (Thieme and Veil 1985). A tipped wood fragment was additionally uncovered at Clacton (GB) (Oakley et al. 1977). Its provenience is unclear and as a tip fragment, it does not allow implications according to its handling. The fact that the Lehringen spear can clearly be interpreted as a thrusting weapon does not mean that the Schöningen spears are clearly projectile weapons.

Summarizing the above given evidence different wooden weapons can be outlined:

- Spears are pointed weapons delivered thrown by hand, their weight ("taring") being forward toward the tip.
- Lances are pointed weapons delivered thrusted by hand (bayonet style), their weight ("taring") being backwards toward the base.

According to studies undertaken by Churchill (1993), ethnographic sources show that throwing spears were primarily used to kill large and medium sized prey and are lethal at a range in the order of 8 m. The latter is in accordance with experimental studies undertaken with replicas of the Schöningen spears (Rieder 2007). Villa and Lenoir (2009) however argue based on historical and ethnographic analogues for larger target distances up to 20 m and beyond. Even though mutual agreement consists that the wooden implements from Germany must be interpreted as weapons, there seems to be no agreement as to their detailed use. What is clear however is that these spears were not equipped with lithic projectiles. Animals struck by projectiles without cutting edges, comparable to the Schöningen spears have a long time to die and could in certain circumstances require much tracking (Guthrie 2005). Taking into account that the Schöningen spears were found amidst the remains of more than 20 horses (Thieme 2007; Voormolen 2008), the detailed hunting tactics associated with these spears need to be addressed in more detail.

The lithic record does equally not provide us with information that brings us closer to answer questions about hunting tactics employed. Pointed lithic artefacts formally classified as Levallois and Mousterian points have been identified as projectiles due to the presence of flute scar impacts (Callow 1986; Villa and Lenoir 2009). In addition a mesial point fragment was found embedded in the cervical vertebra of a steppe ass (Boëda et al. 1999).

For Western Europe it was suggested that at least some of the Mousterian points were used to arm thrusting or throwing spears (Shea 2009; Yaroshevich et al. 2016). Arguments which give reason for the preference of either throwing- or thrusting spears cannot be provided (Villa and Lenoir 2009).

Shea (2009) argues that such thrusting and throwing sticks can only function in an effective way if they are launched in close proximity to their intended target, especially if large animals are preyed upon. As the risk of injury seems far too high to make this a viable strategy Shea suggests that much further functional analysis of these points is necessary.

He argues that projectile technology was invented no earlier than the "Transitional" industries at the very end of the Late Middle Palaeolithic. Given that the wooden spears from Germany were clearly weighted similarly to modern javelins this conclusion seems rather unfounded.

In sum, not only do we have problems to identify weaponry in the pre-Upper Palaeolithic record, the use of this weaponry is additionally highly disputed.

Recent discoveries could point to a different weaponry tradition in other parts of the world, i.e., the use of composite weaponry already at around 0.5 Ma. At Kathu Pan 1 (South Africa) tips of stone points are occasionally characterised by fractures which can also occur during use as weapon tips. Within this assemblage of stone points there are also some specimens showing basal thinning, interpreted to facilitate hafting. The interpretations provided for tip damages, basal thinning and studies of post-depositional modifications serve as major arguments to propose the use of spears with hafted stone points (Wilkins et al. 2012; Wilkins and Schoville 2016).

With the beginning of the Upper Palaeolithic in Western and Central Europe the body of source material changes. A huge variety of organic projectiles becomes a regular and reoccurring part of archaeological inventories (Knecht 1997). Their standardised variety could indicate that they were used both as tips for javelins and thrusting spears.

By the start of the Late Upper Palaeolithic at least the use of the spear thrower is evidenced by the preservation of crook ends. It has been proposed that during the Upper Palaeolithic early examples of spear throwers were entirely made of wood (Stodiek 1993). Thus this weapon might well have been invented long before the Upper Paleolithic but would not become visible until some of the diagnostic parts became made of less perishable organic materials such as bone or ivory.

The spear thrower functions as a lever arm that increases the initial range and velocity of a thrown spear by up to 5 times with a distant range of ca. 10–30 m (Stodiek 1993; Jungmanns 2001). The earliest known physical evidence for its use is represented by a reindeer antler hook from the Solutrean level of the Combe Saunière cave in France (Cattelain 1989; see also Geneste and Plisson 1993). This weapon technology prevails until the Upper Magdalenian in Western and Central Europe (Stodiek 1993).

One can infer the use of composite projectile technology from at least the earlier part of the Late Glacial Interstadial in Late Magdalenian contexts, based on the character of composite projectile points (Pétillon et al. 2011) and the presence of a spearthrower at Isturitz (F) (Szmidt et al. 2009). Moreover, the presence of shaft smoothers in Late Magdalenian assemblages (Ginter and Połtowicz 2007; Wojtal 2007) could indicate that bow-and-arrow technology was simultaneously known and/or used. It should be mentioned however, that based on the presence of small and light lithic points such as micro-gravette points or Solutrean shouldered points (Pericot Garcia 1942) interpreted as projectiles, the use of bow and arrow is already proposed for mid-Upper Palaeolithic contexts (Rozoy 1992).

The earliest direct evidence for the use of bow and arrow was discovered in the Stellmoor kettle hole in the Ahrensburg tunnel valley (Germany) (Rust 1943). The Ahrensburgian level at Stellmoor is assigned to the Younger Dryas. The site is famous for the preservation of wooden arrows and fragments thereof made from pine wood, for which a minimum number of 105 arrows were reconstructed. It has been suggested that the arrows were used in a composite form, with a long shaft at the base and a smaller foreshaft that was tipped with large Ahrensburgian tanged points. Some of these stone points were found still attached to fragments of arrow foreshafts (e.g., Rust 1943: Table 93, Fig. 2).

For the arrows Rust (1943) distinguishes several variants (Fig. 6.1) characterised by a flat rounded tip with a simple narrow (Form 1) or broader (Form 2) (up to 0.2 mm) incision parallel to the annual rings of the pine wood, serving the accommodation of the flint point. Form 3 shows an unmodified diameter of the arrow shaft in the area of the modification, which consisted of a flat indentation. This modification either ensured the utilization as part of a composite arrow or served the accommodation of the bowstring. Finally, the proximal end was sharpened to form a tip 1–3 cm in length (Form 4) (Fig. 6.1).

An arrow assigned to Form 4 was found pierced through the still articulated vertebra of a juvenile wolf (Rust 1943: Table 94, Fig. 9 and Table 106, Fig. 3). For the largest arrow a length of 85 cm was reconstructed with a diameter of ca. 1 cm, very rarely with a diameter of up to 1.7 cm.

In addition to the varying shape of the tip region it was reported that the arrows came in different sizes due to their utilization as composite tools. A small complete arrow possessed a length of only 15 cm (0.5 cm in diameter). The tip corresponded to Form 1. The taring of the arrows lies in their middle part (Fig. 6.1).

The Stellmoor assemblage is completed by two fragments of pine bows reconstructed to have had a length of 1.5 m. From the small size of the fragments it was not possible to reconstruct the shape of the bows diameter (Rust 1943) and thus this evidence remains doubtful, as is equally true for

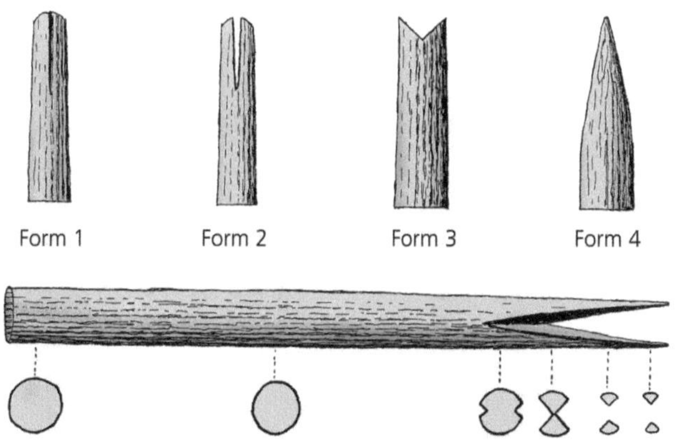

Fig. 6.1 Schematic depiction of different arrow tip modifications and cross section of the lower part of an arrow from the Ahrensburgian level at Stellmoor (D). For the detailed description of Forms 1–4 compare text (redrawn after Rust 1943: Fig. 20)

evidence from Mannheim (Germany), where a fragment of pine wood dated to the Magdalenian, was interpreted as a bow fragment (Rosendahl et al. 2006).

From the Younger Dryas onwards, the use of bow and arrow remains the most visible hunting equipment used. Straight bows of the "Holmegard Type" are known from Mesolithic contexts in Northern Europe. During the same time double-curved bows have been reported from Vis I (Russia) (Burow 1980). These well-engineered weapons projected a variety of arrows which varied according to the target species. Moreover, evidence uncovered from Final Palaeolithic (e.g., High Furlong, GB) (Hallam et al. 1973) and early Mesolithic sites (e.g., Tåderup, DK) (Ødum 1920) show that a variety of bone points were simultaneously in use during this period (see also Cziesla 2001).

Armaments were as varied as weaponry systems during the entire Upper Palaeolithic and early Mesolithic. Among the variety of organic points used during this period only some examples should be mentioned here. For the Early- and Mid-Upper Palaeolithic organic split based points (Fig. 6.2, 1–2), massive points (Fig. 6.2, 3), Lautscher points (Fig. 6.2, 4), and for the late Upper Palaeolithic biconical points (Fig. 6.2, 5), Baguettes demi-rondes (Fig. 6.2, 6), points with forked base (Fig. 6.2, 7) and barbed points (Fig. 6.2, 8, 9) occur.

During the Mid-Upper Palaeolithic thrusting spears made of ivory have also been in use (Nikolskij and Pitulko 2013).

Abundant evidence shows that the lateral edges of organic points were hafted with lithic elements, e.g., backed bladelets or chips that possessed sharp cutting edges to produce deep wounds with more rapid haemorrhaging compared to blunt organic tips. Impressive examples for wooden thrusting spears with inserted lithics were reported from the burials at Sungir (Russia) (Bader and Bader 2000), dated to between 33,300 and 36,300 cal BP (Marom et al. 2012). Their length was reconstructed to 1.7 m, their upper

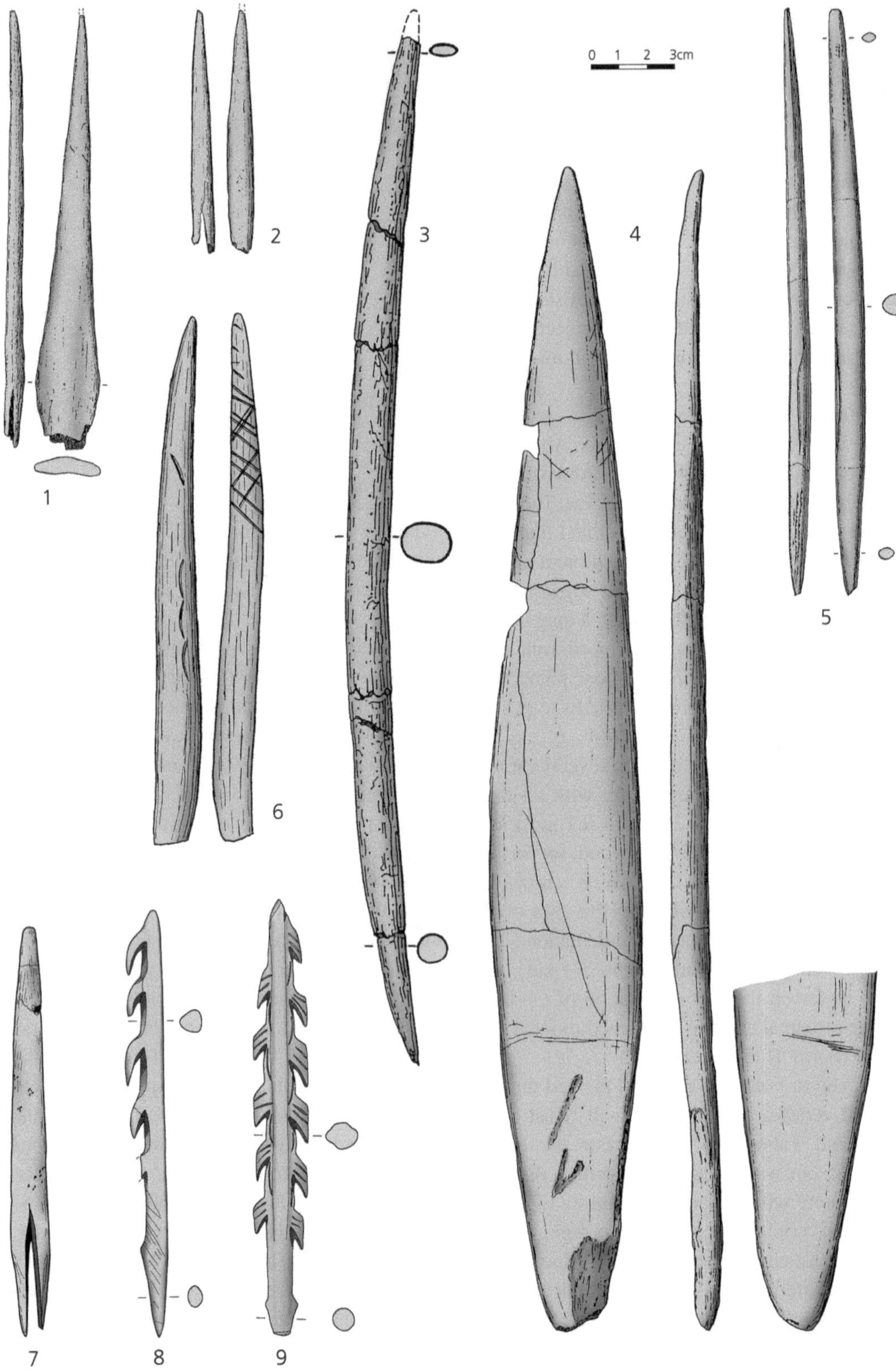

Fig. 6.2 Upper Palaeolithic organic weaponry. **1–2** Split based bone points from (1) Bocksteinhöhle (D), (2) Vogelherd V (D) (redrawn from Hahn 1977), **3** massive point from Amvrosievka (RUS) (redrawn after Bosinski 1987), **4** Lautscher point from Wildhaus Höhle (D) (redrawn from Hahn 1977), **5** biconical point from Lascaux (F) (redrawn after Allain 1979), **6** Baguette demi-ronde from Laugerie-Haute (F) (redrawn from Peyrony and Peyrony 1938), **7** point with forked base for composite use from Isturitz (F) (redrawn after Passemard 1944), **8** barbed point from Laugerie-Basse (F) (redrawn from Weniger 1995), **9** barbed point from La Madeleine (F) (redrawn from Weniger 1995)

part was laterally armed with a row of inserted/attached flint chips measuring more than 30 cm in length. Additional impressive examples for the sheathing of an organic base were unearthed at the Mid-Upper Palaeolithic Talicki-Station (Russia) (Fig. 6.3) and the Late Upper Palaeolithic site of Pincevent (France) where antler shafts were equipped with grooves inset with backed bladelets (Abramova 1982; Leroi-Gourhan 1983).

Among Upper Palaeolithic and Final Palaeolithic lithic composite elements backed bladelets dominate (Fig. 6.4, 7). Small symmetrical versions of Font-Robert points of the Mid-Upper Palaeolithic are generally considered to have functioned as lithic projectiles (Shea 2006) (Fig. 6.4, 8). In addition, microgravettian points (Fig. 6.4, 2), shouldered points (Fig. 6.4, 3), triangular microliths (Fig. 6.4, 4), curved backed points (Fig. 6.3, 5) and pen-knife points (Fig. 6.4, 6) are generally discussed as having been used as armament.

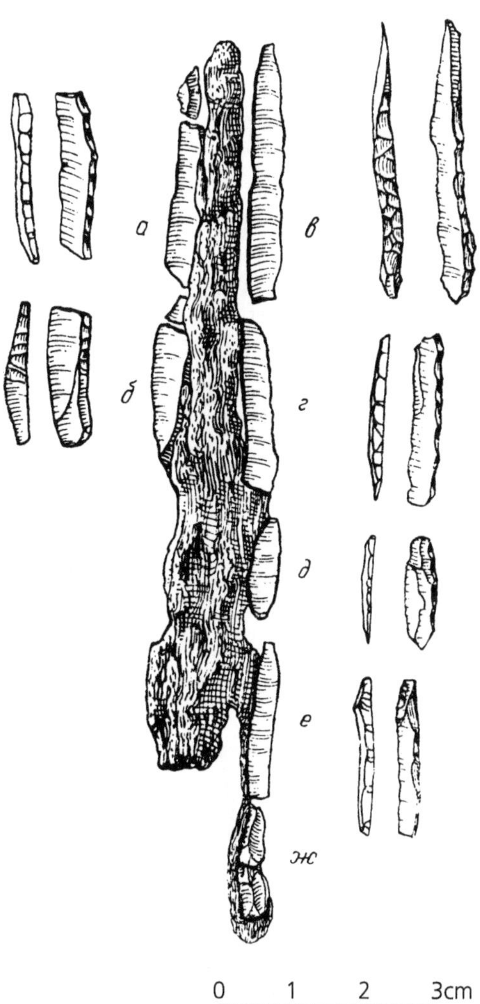

Fig. 6.3 Talicki-Station (RUS). Projectile point inserted with backed bladelets (after Abramova 1982)

For the Mesolithic, microliths and microblades prevail. A recent discovery of an Early Mesolithic arrow was reported from Rönneholms Moor in Sweden and provides us with just one example on how microliths have been hafted. The arrow shaft was made from a one-year old branch of a hazel which was modified by a v-shaped groove in which four triangular microliths were glued with resin to form barbs. It is considered possible that a further microlith functioned as an arrow tip (Larsson and Sjöström 2011).

It must be emphasised that we have no direct evidence that weapons were provided with a tip of flint for the entire Upper Palaeolithic: the direct evidence we do have is indicative only of lateral hafting of lithic elements. Laterally hafted weapons have an advantage over tip hafted weapons in that they can inflict wounds with larger diameter and penetration depth.

Thus, it could be assumed, that maybe with the exception of particular specimens of Font-Robert points, lateral hafting prevailed. It needs to be noted however that even though it is usually assumed that particular specimens of Font-Robert points were used as weapon tips, microwear studies of Belgian specimens indicate that some have been used as knives (Otte and Caspar 1987).

With the implementation of bow and arrow, we find numerous direct evidence for tip hafting e.g., from the Final Palaeolithic site of Stellmoor already mentioned, and for the early Mesolithic with finds from Lila Loshult (Sweden) (Malmer 1969), Vinkel (DK), Holmegaard IV (DK) or Rönneholm (S) (Larsson and Sjöström 2011).

Even though we have numerous indications for the equipment used for hunting, it is only from the later part of the Mid Upper Palaeolithic that we get an idea on how large mammal hunting was organised. Especially for the late-Upper Palaeolithic artistic depictions provide helpful indications here. Wounded herd animals with spears and arrows still protruding from their bodies are common within the artistic repertoire during this period. According to Leroi-Gourhan (1971) approximately 15% of large herd animals depicted in Palaeolithic art are depicted speared and/or bleeding from wounds, the animal's mouth or from nostrils. These images occur on portable art as well as in cave art. Concerning cave art, for the earlier part of the Late Upper Palaeolithic examples from Lascaux (F) (e.g., Leroi-Gourhan and Allain 1979: Fig. 232) and for the later part of the Upper Palaeolithic examples from Niaux (F) (e.g., Clottes 1995: Panneau 4, Fig. 129) illustrate this.

Moreover, a few engravings from Magdalenian contexts show large groups of people each person carrying at least one spear. The best example comes from Abri du Chateaux (Les Eyzies, Dordogne, F) (Fig. 6.5). This and comparable scenes could be interpreted in terms of hunting parties, indicating that large communal hunts were among the hunting tactics employed during the late Upper Palaeolithic.

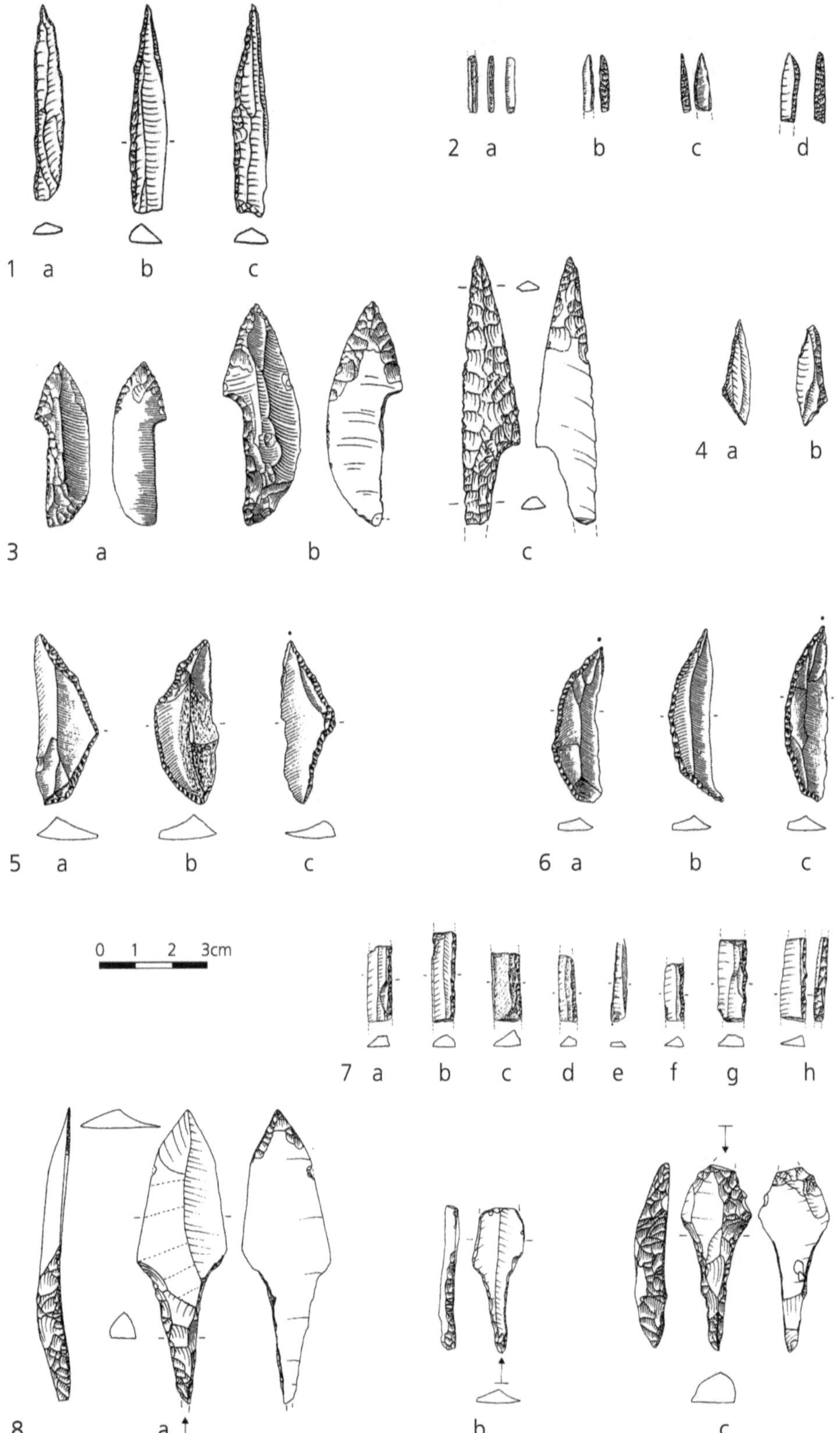

Fig. 6.4 Upper Palaeolithic and Final Palaeolithic lithic amendments. **1** a–c Gravettian points from Kostenki IV (RUS) (redrawn after Bosinski 1987), **2** a–d Microgravettian points from Mainz-Linsenberg (D) (redrawn after Hahn 1969), **3** a–c shouldered points, a–b from Kostenki I,1 (RUS), c from Placard (F) (redrawn after Bosinski 1987), **4** a–b triangular microliths from Kniegotte (D) (redrawn after Feustel 1974), **5** curved backed points from Petersfels (D) (from Bosinski 1987), **6** pen-knife points from Peterfels (D) (from Bosinski 1987), **7** backed bladelets from Gönnersdorf (D) (from Bosinski 1969), **8** Font-Robert points from Steinacker (D) (redrawn after Pasda 1998)

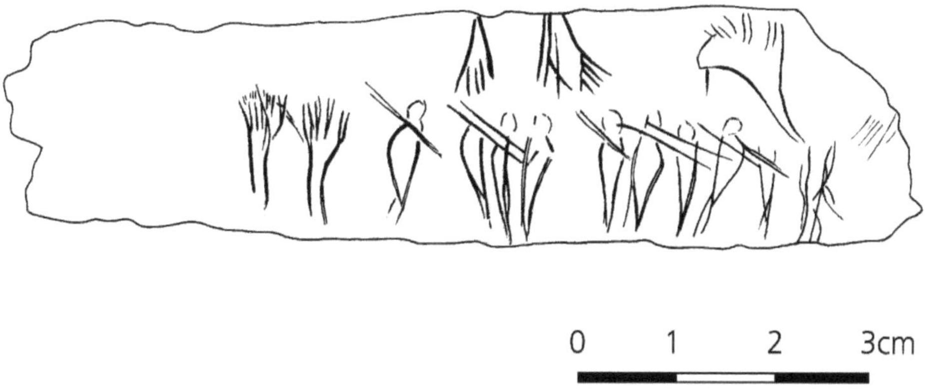

Fig. 6.5 Depiction of male hunters equipped with spears from Abri du Château (Les Eyzies, Dordogne, F) (Magdalenian). Redrawn from Cluzel and Cleyet-Merle (2011), Leroi-Gourhan (1971)

This is also underlined by the fact that large mammals are often depicted as hit by more than one spear. Engraved pebbles showing a speared horse from the late Magdalenian site of Oelknitz (Thuringia, Germany) serves as an example from portable art here (Fig. 6.6) (Gaudzinski-Windheuser 2013; Feustel 1970).

The short survey on weapon technology given above illustrates the huge mismatch between the evidence prior to and after the arrival of anatomically modern humans in Europe ca. 40,000 years ago even though during both epochs we are dealing with hunting communities. This mismatch might indicate that the concept of hunting and thus its social imbedding differed significantly. What also became clear is that the study of the weaponry itself only provides us with a limited perspective here. Therefore, the diachronous survey of hunting lesions might thus open up a complementary perspective to shed light on this highly complex topic. In the following sections evidence for hunting lesions from the Pleistocene and early Holocene are considered.

Hunting Lesions in the Pleistocene and Early Holocene

Experimental Set-Ups

In order to understand Pleistocene organic and lithic weaponry and hunting lesions on bones documented in the archaeological record, experimental studies have been undertaken. Maybe the most influential study that triggered and set the agenda for further research was a major study by Stodiek at the end of the 1980s (Stodiek 1993) on the technology of Upper Palaeolithic spear throwers. Since then experiments intended to help us identify hunting lesions in Pleistocene bone assemblages have been published. A short summary on our current stage of knowledge for studies relevant to the Upper Palaeolithic/Mesolithic can be found in Leduc (2012).

These experiments have been mostly undertaken using lithic shouldered points hafted as projectiles (e.g., Morel 2000; Smith et al. 2007; Castel 2008) and antler points (e.g., Letourneux and Péttilon 2008). According to what we know on the basis of experimental studies undertaken by Stodiek (1993), Morel (2000), Sudhues (2004), Letouneux and Pétillon (2008), Castel (2008) and Pétillon et al. (2011) lesions on bones caused by lithic and organic projectiles can generally be categorized in four damage types: notches and punctures, perforation and fragmentation. Definitions for the different damage types can be found in Letourneux and Pétillon (2008, see also Morel 1993):

- Notches are angular or v-shaped regular/or irregular indentations or slits on the edge of a bone resulting from grazing shots, where the weapon removed bone material.
- Punctures are depressions of various size, form and depths on a bone. Punctures result from inflictions by pointed weapons which did not penetrate the bone. Projectile inclusions or fragments thereof occur regularly.
- Perforations are holes of various size, form and depth in a bone. Perforations result from inflictions by pointed weapons which penetrated the bone. Projectile inclusions or fragments thereof occur regularly.

These damage types do not vary, independently of whether they have been inflicted by bow and arrow or spear thrower delivered javelin although spear impacts have approximately 2.5 times the kinetic energy of arrow impacts (Sudhues 2004).

The morphology of indirect hunting lesions (i.e., lesions without flint being embedded in the bone) however shows a high range of variation, depending on the targeted species, the age of the prey, bone morphology and density and of course on the shape of the projectile used. Spongy bones for

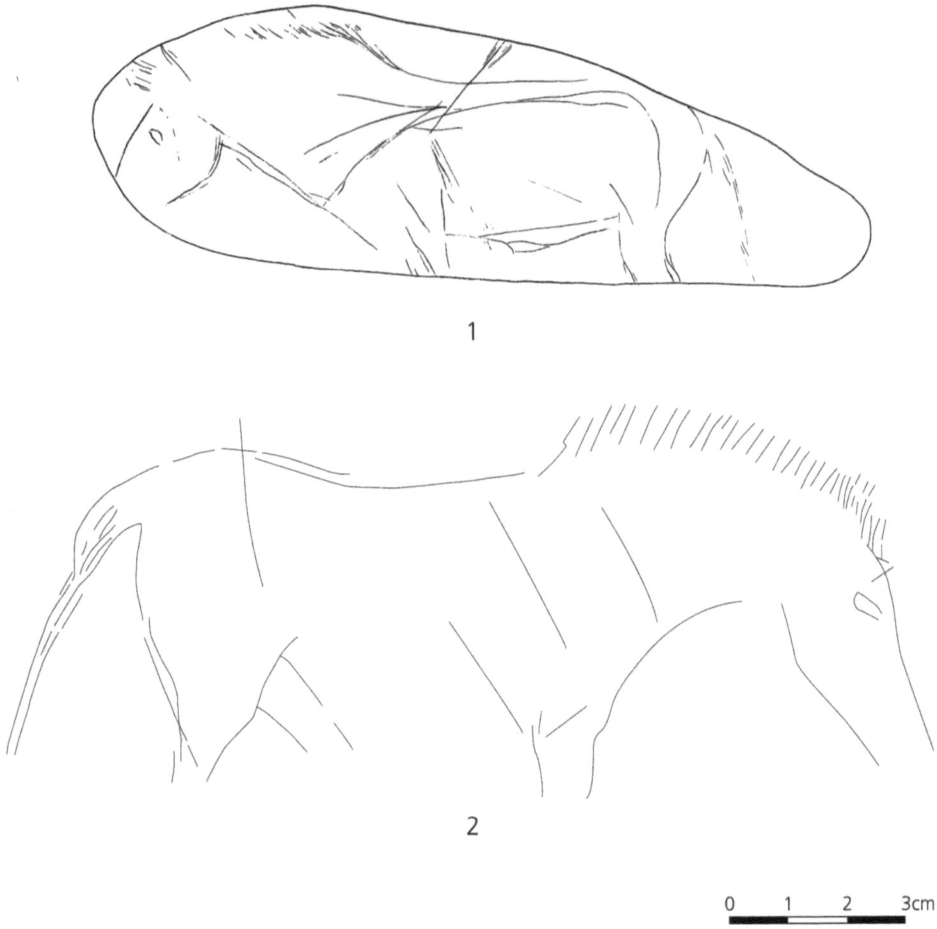

Fig. 6.6 Depictions of horses on pebbles showing speared horses from the Magdalenian site of Oelknitz (Thuringia, D). **1** Redrawn from Gaudzinski-Windheuser (2013, Fig. 272). **2** Redrawn from Feustel (1970, Fig. 1)

example have a high potential to show notches, while in compact bones perforations dominate.

Punctures are the rarest type of damage, while perforations are the most frequent.

All marks described in this context can be confused with marks left by tool use, carnivore modification and other taphonomically induced damage (Stodiek 1993; Morel 2000; Sudhues 2004; Castel 2008; Letouneux and Pétillon 2008).

According to our current state of knowledge, damage caused by osseous and lithic projectiles can sometimes be distinguished from each other as punctures and perforations can reflect the cross-section of the projectile used (Smith et al. 2007; Letouneux and Pétillon 2008). The exact differences in damage traces caused by projectiles made of either bone or flint have not yet been studied in detail.

As Leduc (2012) notes, hunting lesions produced in these experimental studies are only rarely observed in zooarchaeological assemblages, a point also stressed by Pétillon et al. (2016). A number of factors are usually invoked to explain this, among which are the nature of the experimental research, taphonomic agents that blur any existing evidence of hunting, as well as the inability of zooarchaeologists to identify the traces produced in the experimental work or at least to interpret them in terms of hunting (Leduc 2012). Other authors additionally emphasize climatic conditions and the use life of projectiles as potentially responsible factors (Pétillon et al. 2016). A far simpler explanation however would be that up until the beginning of the late Upper Palaeolithic/Final Palaeolithic projectiles tipped with flint were not in regular use, as indicated by the direct archaeological evidence for only lateral lithic hafting of weaponry.

Direct Evidence for Hunting Lesions in Late Upper Palaeolithic/Final Palaeolithic and Mesolithic Archaeological Contexts

Hunting lesions have repeatedly been reported from northern European Late Glacial and Post Glacial sites. Lesions with embedded flint fragments classified as fragments of pointed arrow projectiles are known from a number of Northern European sites dating from the Early to Final Mesolithic [e.g., Lundby Mose, DK (Møller Hansen et al. 2004); Mullerup, DK (Leduc 2012); Vig, DK (Noe-Nygaard 1973); Schwenningen, D (Ströbel 1959); Aldersro, Kongemose, Ringkloster, Hendriksholm, Svenstrup, DK (Noe-Nygaard 1974)]. Healed hunting lesions, sometimes with flint fragments still embedded in the bone, indicate that Mesolithic hunting was not always successful [e.g., Mullerup, DK (Leduc 2012); Vig, DK (Noe-Nygaard 1973); Maglelyng complex, Åmose bog, DK (Noe-Nygaard 1974), Starr Carr, GB (Noe-Nygaard 1975)].

A topic not covered here is the identification of (hunting) lesions in humans. Mesolithic cemeteries all over Europe give ample evidence here. In her seminal work on Mesolithic burials of Europe Grünberg (2000) points out that in almost all Mesolithic cemeteries and in many single graves humans carry fragments of projectiles in skulls, thoraxes or pelves. Of particular importance here is the late Mesolithic cemetery of Vasil'evka III (UA) (Telegin 1961) where almost a third of all of the individuals uncovered carried shot injuries.

The most important Pleistocene sample of genuine hunting lesions comes from the sites of Meiendorf and Stellmoor (Rust 1937, 1943). The sites are located in the Ahrensburger Tunnel valley of northern Germany. Both sites are considered to result from various activities connected to repeatedly occurring hunting of reindeer (Rust 1937, 1943; Bratlund 1990).

Hunting lesions have been reported from the Late Upper Palaeolithic Hamburgian (Meiendorf) and Final Palaeolithic Ahrensburgian levels (Meiendorf and Stellmoor). Almost 30 hunting lesions with flint inclusions have been identified by Bratlund (1990), Möller (1975) and Rust (1943) [Meiendorf, Hamburgian level, n = 5 (Bratlund 1990); Stellmoor, Hamburgian level, n = 5 (Rust 1943; Bratlund 1990); Stellmoor, Ahrensburgian level, n = 25 (Rust 1943; Möller 1975; Bratlund 1990)]. Bratlund points out that morphological difference in hunting lesions between Late Upper Palaeolithic and Final Palaeolithic contexts are not discernable.

The morphology of lesions left by direct hits is mostly predetermined by the convex bone surface and the shape of the flint projectile (Bratlund 1990). Other determining variables are presumably bone mass, impact velocity, and angle of impact, as well as bone density and structure.

These lesions are generally characterised by smooth sharp cuts which lie in an opposed angle to an irregularly shaped, chipped edge (Bratlund 1990) (Fig. 6.7). Chipping occurs due to the flint point hitting a mostly convex or concave bone surface, thus overlapping bone compacta is wedged away. Bratlund states that the sharp cuts resemble traces from blows by axes or metal knifes (Bratlund 1991).

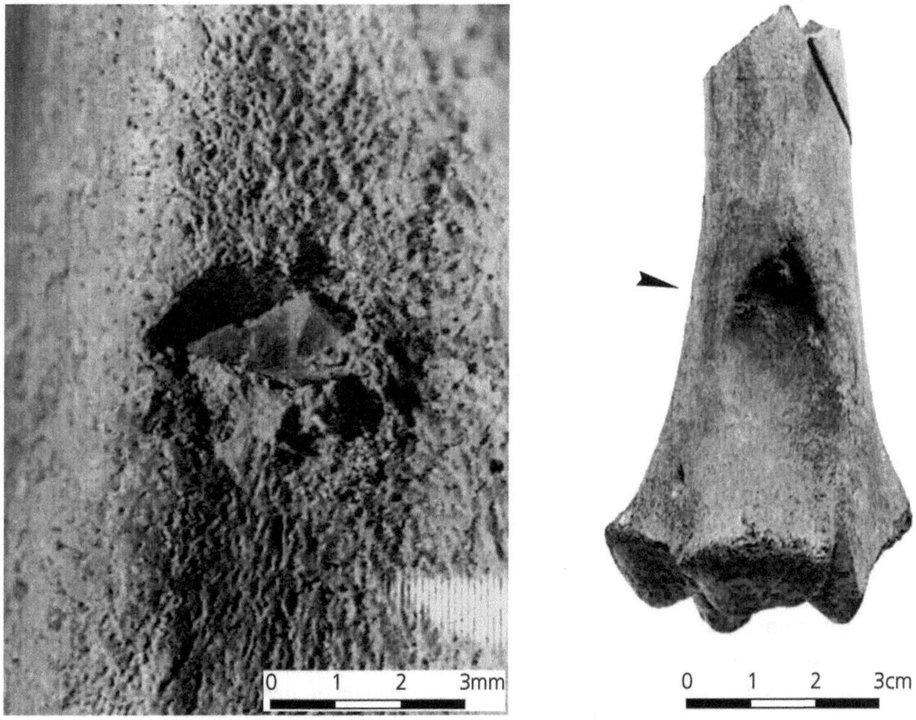

Fig. 6.7 Stellmoor (D), Ahrensburgian. Projectile shot into a right femur of a reindeer (*Rangifer tarandus*). Right: complete bone, left: shot wound in detail (after Bratlund 1990: Fig. 30) (photo by courtesy of Stiftung Schleswig Holsteinische Landesmuseen Schloss Gottorf, Schleswig)

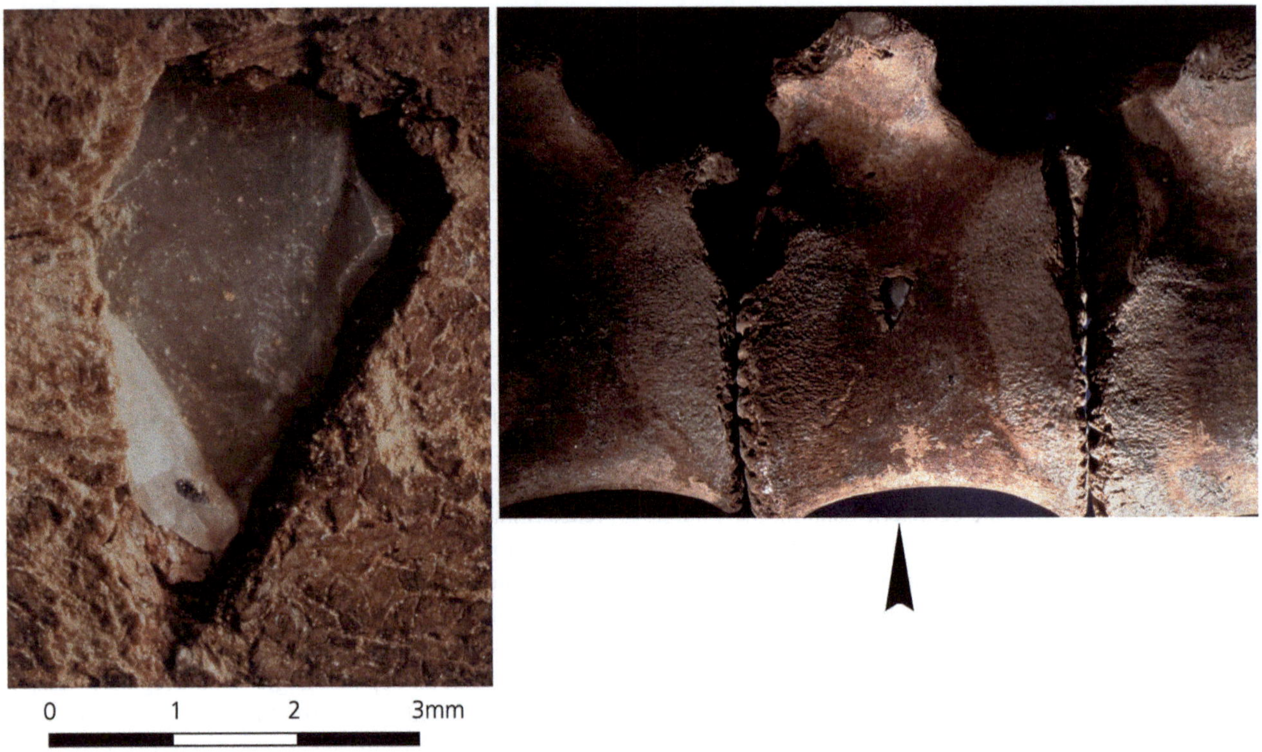

Fig. 6.8 Stellmoor (D), Ahrensburgian. Hunting lesion in thoracic vertebra of a reindeer (*Rangifer tarandus*) (after Bratlund 1990: Fig. 22) (photos MONREPOS)

In rare cases straight direct hits were documented with fractures of regular triangular or quadrangular shape with only minor edge damage, the fracture shape being predetermined by the shape of the flint projectile (Fig. 6.8). According to Bratlund it is impossible to mistake these damages for damages caused during subsequent butchering. The morphology of lesions left by differently manufactured arrows (e.g., arrows equipped with a pointed wooden tip) remained unstudied for this assemblage (Bratlund 1990).

For the Ahrensburgian at Stellmoor we witness a hunting scenario with uncontested use of bow and arrow whereas for the Hamburgian at Meiendorf the use of bow and arrow is not indisputed but considered highly plausible (Bratlund 1990). Analysis of hunting lesions by Bratlund (1990) allowed the reconstruction of the hunting strategies used. Based on the study of the projection angle, for the Ahrensburgian a mass kill scenario with bow and arrow where hunters took advantage of the individual topography of the landscape was suggested. In contrast, for the Hamburigan level at this particular site stalking and ambushing was the major component that determined hunting tactics.

Roughly contemporaneous with the evidence from the Hamburgian level at Meiendorf is evidence from the Magdalenian site of Schussenquelle (Germany) (Schuler 1994). The topographic situation is comparable to Meiendorf and Stellmoor. Due to the dominance of reindeer (MNI = 41) and evidence of activities connected to reindeer exploitation the site was interpreted as a location for repeated mass hunting of reindeer occupied mainly during the late summer/autumn period. A hunting lesion in form of an entry channel in the *Crista scapulae* carried splinters of chert (Schuler 1994: Fig. 51). The lesion most probably results from a projectile point with inserted backed bladelets. A javelin propelled by a spear thrower is thought to be the weapon (Schuler 1994).

A similar weapon was proposed to have caused a grazing wound in the left mandible of a reindeer from the Magdalénien site of Kesslerloch (CH). The shot caused severe bone damage in which the fragment of a backed bladelet got stuck (Fig. 6.9). Detailed analysis was able to demonstrate that the animal could have survived for approximately two weeks before its remains ended up in the bone debris of Kesslerloch cave (Napierala et al. 2010).

An important discovery was made at the Epigravettian site of Lugovskoe in Western Siberia, dated between 18,000–10,000 BP (Orlova et al. 2004). Lugovskoe represents a mass accumulation of mammoth bones found in a swamp that was used by the animals over a considerable amount of time as a sort of mineral lick (Leshchinsky 2006). That humans took advantage of this situation is indicated by the presence of almost 300 stone tools also discovered in the deposit. What sort of interaction between mammoths and humans we are dealing with is amply illustrated by a thoracic vertebra of a

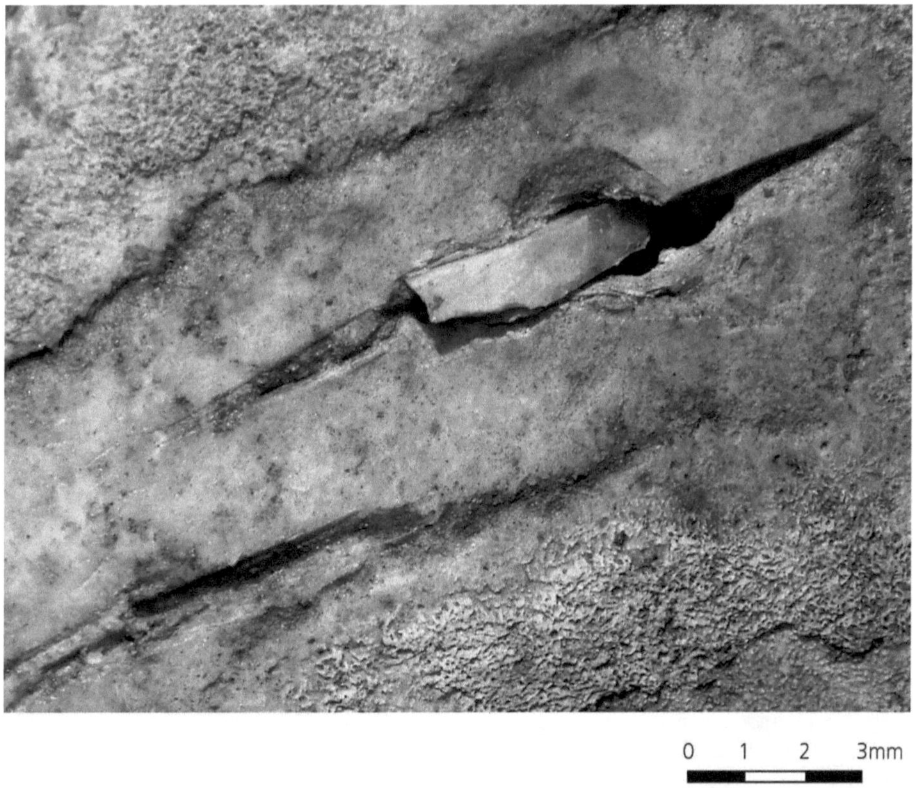

Fig. 6.9 Kesslerloch (CH), Magdalenian. Hunting lesion left by a grazing shot in the right mandible of a reindeer (*Rangifer tarandus*) (after Napierala et al. 2010) (photo by courtesy of H. Napierala)

mammoth that showed a hunting lesion in its corpus (Fig. 6.10). The shot was obviously executed with enormous force from very close distance, as it could be assumed that the weapon must have penetrated soft tissue and the scapula before it hit the vertebra. Judging from the lithic fragments that got stuck in the hunting lesion, it can be assumed that the weapon used was an organic point laterally equipped with lithic implements (Maschenko et al. 2005; Leshchinskiy 2012). Most interestingly it was recently pointed out that many of the mammoth individuals represented in the mass bone accumulation suffered from serious diseases of the bones. The vertebra was directly dated to 13,465 ± 50 uncal BP (Orlova et al. 2004; Zenin et al. 2006).

At the Archaeological Zone of the cave La Garma A (E), the mandible of a horse was uncovered in which a 4 mm long flint fragment was still embedded. The flint stuck on the buccal side of the bone, just underneath the P4. The authors suggest that this fragment was shot into the bone and probably represents the remains of a lithic projectile. The evidence dates to 14,500 cal BC (Arias Cabal et al. 2005).

In addition Lartet and Christy (1864) reported a hunting lesion in the vertebra of a reindeer from a Magdalenian context in the Grotte des Eyzies (France), although it is difficult to evaluate the character of the lesion on the basis of the existing publication (Lartet and Christy 1864).

Contemporaneous with the Ahrensburgian level at Stellmoor is an ensemble unearthed within the Final Palaeolithic site of Grotte du Bichon (CH). With this evidence one of the most impressive examples for a hunting accident survived. The record consisted of a skeleton of a female brown bear, which covered the almost complete skeleton of a human. Numerous remains of charcoal were additionally found as were a number of 18 lithics. Important for the interpretation of the archaeological record is a lithic point made from flint that was found shot in one of the bear's cervical vertebrae. Thus the archaeological record was interpreted in terms of a hunting accident. Morel (1998) reconstructs that one or several hunters attacked the animal at the entrance of Grotte du Bichon. It is not clear whether bow and arrow or a spear thrower was used to inflict the hunting lesion. The lesion however (see Morel 1998: Fig. 7) shows characteristics, described from archaeological contexts as typical for damages caused by bow and arrow reported from Stellmoor (Fig. 6.7). The injured animal withdrew into the cave and was followed by a hunter who tried to smoke the animal out of the cave as is known from ethnographic contexts. In the course of this encounter the hunter died. The series of lithics, representing an Azilian point, a blade interpreted to have been used as a knife and lamelles à dos was accordingly interpreted as an individual hunting kit.

Fig. 6.10 Lugovskoe (RUS), Epigravettian. Hunting lesion in a vertebra of a mammoth (*Mammuthus primigenius*) (photos by courtesy of S. Leshchinski)

Indirect Evidence for Hunting Lesions in Late Upper Palaeolithic/Final Palaeolithic and Mesolithic Archaeological Contexts

In addition to the indisputed evidence for hunting lesions additional archaeological source material is discussed here. In this context, the above mentioned European sites are among others that gave evidence for bone damages similar to the ones described for Meiendorf and Stellmoor but without flints being embedded (Noe-Nygaard 1974).

In addition, perforations especially on scapulae have numerously been observed within these and other assemblages and debated in the context of indirect hunting lesions. It is assumed that these damages result from failed heart shots which represent the most fatal wounds which can be inflicted on an animal (Noe-Nygaard 1974). A very illustrative example that justifies this interpretation comes from the Ahrensburgian level at Stellmoor. Here a reindeer scapula showed a perforation characterised by an incision, typical of shots with flint arrow projectiles. A flint splinter was still sticking to this bone damage (Rust 1943: plate 31; Bratlund 1990: Fig. 18a, b).

Unhealed damages observed on scapulae from a variety of species (e.g., *Sus scrofa*, *Cervus elaphus*, *Alces alces*, *Capreolus capreolus*, *Bos primigenius*) follow a uniform pattern. Length and breadth of the perforations are considered to reflect the diameter of the hunting weapon used. Noe-Nygaard reports a perforation on a scapula from *Bos primigenius*, found at the Grænge mose, DK, (Maglemose culture) which she considers to have probably been inflicted by a spear (Noe-Nygaard 1974: Fig. 3, plate I, c1). In contrast, the almost circular hole with a little distal notch in the scapula of *Cervus elaphus* from Kongemose, DK, (Kongemose culture) is assumed to result from a hunting scenario where a barbed point had been used (Noe-Nygaard 1974: Fig. 5, plate III, b3, b3).

The morphology of the damages on the rims of the perforations follows a uniform pattern. On the internal face of the bone, where the impact occurred, the damage is characterised by a clear cut rim, while on the external face numerous bone flake scars produce an irregular margin around the rim of the perforation (Noe-Nygaard 1974; Leduc 2012) (Fig. 6.11).

The exact location of unhealed perforations observed for scapulae from Mesolithic contexts was contextualised against the location of perforations observed by Rust (1943) at Stellmoor. Whereas perforations observed in Mesolithic contexts are located in the thinnest part of the scapulae, perforations scatter over the entire bone in the Final Palaeolithic Stellmoor assemblage. Noe-Nygaard sees here a reflection of differing hunting tactics employed (Noe-Nygaard 1974).

The interpretation of damage described on these scapulae as deriving from hunting is not indisputed, however. A number of taphonomic agents – among which are carnivores and humans – can produce perforations with morphologies that are indistinguishable from hunting lesions.

Bratlund (1990) reports an interesting example from the Roman period, where perforations in scapulae served the hanging off the shoulder of a prey for meat conservation. Thus, for the majority of perforation damages we face the problem of equifinality in the interpretation of these traces.

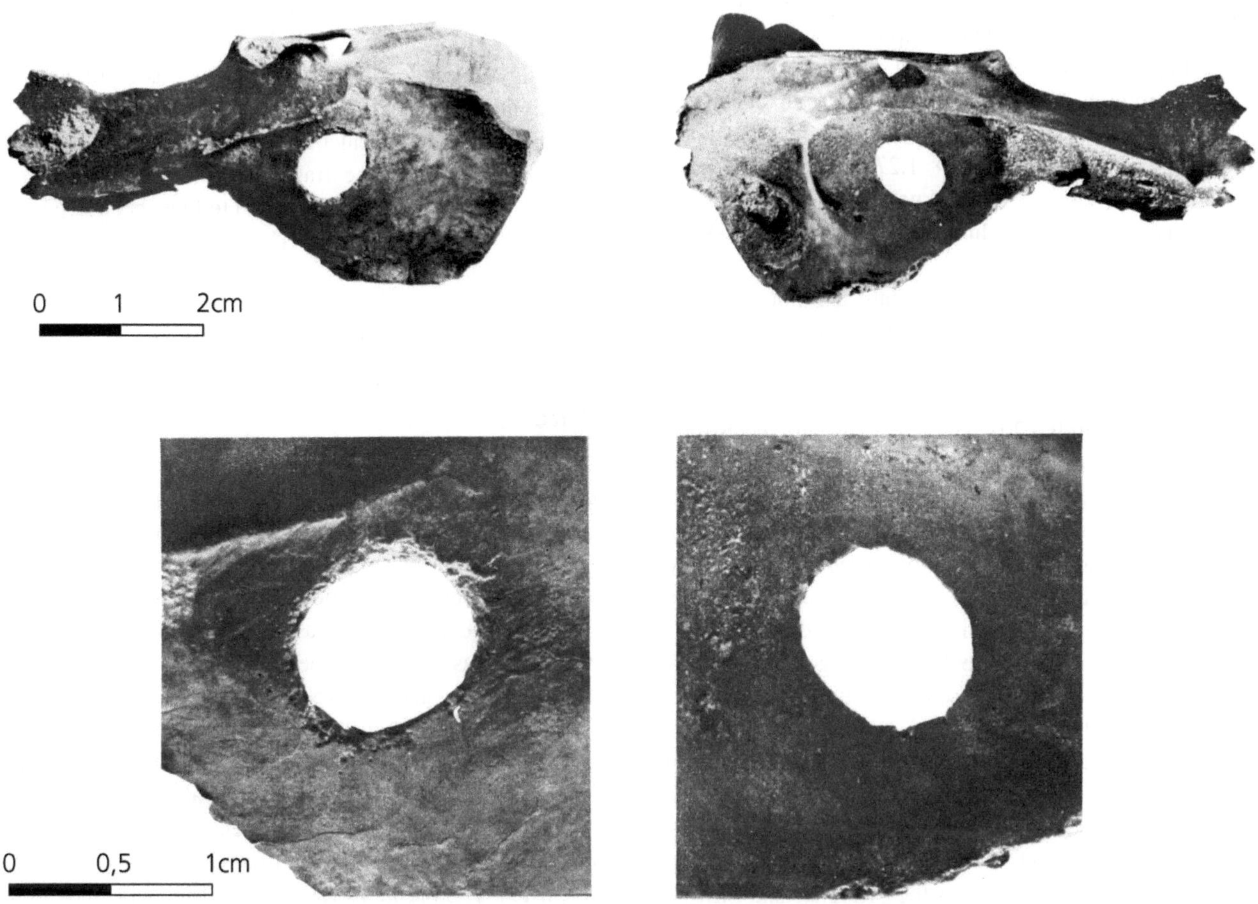

Fig. 6.11 Morphology of indirect hunting lesions on bones demonstrated by a damage on a ptarmigan (*Lagopus* sp.) pelvis from Meiendorf (D). Right: entry wound, left: exit hole (after Rust 1937: plate 53)

Direct Evidence for Hunting Lesions in Early- and Mid-Upper Palaeolithic Contexts

Compared to Late Glacial and early Holocene contexts, the number of direct indications for hunting lesions dramatically declines the further we go back in time. For the earlier phases of the Upper Palaeolithic not more than three examples have prominently been published, which will be described in the following.

The earliest examples for hunting lesions in Upper Palaeolithic contexts have been reported from Yana YMAM in Arctic Siberia dating between 29,000 and 27,000 ^{14}C BP (Nikolskij and Pitulko 2013). The site represents a mass accumulation of mammoth bones representing at least 31 individuals. The site is contemporaneous with the neighbouring archaeological site Yana RHS. For Yana YMAM mammoth hunting for ivory has been postulated. This is based on evidence for indirect hunting lesions in form of perforations on the right iliac bone of a pelvis of a young mammoth as well as on the right scapula of a juvenile mammoth. More important is the report of stone tools found embedded in mammoth scapulae. A fragment (1.5 cm × 1.1 cm × 0.15 cm) reported to represent a flat convex siltstone fragment was embedded in the right scapula of a young mammoth. Comparable damage was documented on a right mammoth scapula in form of a larger flat convex siltstone fragment (<1.1 cm × 1.27 cm × 0.5 cm) and a smaller flake, 0.45 cm in thickness. Between these stone fragments a thin ivory splinter was located (Nikolskij and Pitulko 2013: Fig. 3A, B).

Even though the lithic fragments are not particularly diagnostic the archaeological evidence indicates short projectile weapons with main shaft and long foreshaft made of ivory, probably tipped with elongated triangular microliths (Nikolskij and Pitulko 2013) which are usually reconstructed to have been laterally hafted to an organic shaft (Höck 2001; Yaroshevich et al. 2013). Moreover the use of simple ivory thrusting spears was proposed (Nikolskij and Pitulko 2013).

From the positions of the hunting lesions and an analysis of the population structure for mammoth the authors reconstruct the hunting tactics employed, that focussed on the exploitation of adolescent and young adult small animals. Hunters clearly focussed on killing by heart shots (Nikolskij and Pitulko 2013).

Indisputable evidence for a hunting lesion attesting the killing of a cave bear by humans comes from the Hohle Fels Cave in Germany. The cave looks back on a long history of research that goes back to the 19th century. Several archaeological horizons attributed to the Aurignacian, Gravettian and Magdalenian period have been excavated, rich archaeological archives have been unearthed in this cave. With the exception of the Magdalenian period *Ursus spelaeus* is a dominant element in the faunal assemblages associated with the Aurignacian and the Gravettian (cf. Münzel and Conard 2004: Table 1). In layer AH IIcf attributed to the Early Gravettian at around 20,000 BP (Münzel et al. 2001), a flint fragment of triangular shape was found embedded in the *Processus transversus* of a thoracic vertebra of an adult cave bear (Münzel and Conard 2004: Figs. 10 and 11). It is assumed that a spear was used to kill the animal for which a lying position is assumed when the attack occurred. The flint fragment is not diagnostic. The scenario was interpreted as a killing event during the period of hibernation (Münzel and Conard 2004).

A further example for a hunting lesion comes from the Siberian site of Kokorevo I, layer 3, interpreted as a home base and attributed to Mid/Late Upper Palaeolithic contexts (Abramova 1982). A proximal fragment of an antler point was discovered in a left scapula of a large adult bovid. The shot was performed with enormous force, thus the author suggests a confrontational hunting episode. The shot caused the *in situ* fragmentation of the bone during the lifetime of the individual and damaged the below lying muscle tissue. The entry wound is characterised by sharply defined edges (Fig. 6.12) whereas the exit wound (compare Canby 1979: 535) shows chipped edges (Abramova 1982).

Finally, an ambiguous hunting lesion was reported from an Aurignacian context of Combe-Buisson cave (F). A tiny fragment of a probably calcinated bone point was found in a spiral fracture of a bone fragment from a medium sized mammal. The fragment was reported to have been deposited in a hearth (Moirenc et al. 1921).

Direct Evidence for Hunting Lesions in Lower and Middle Palaeolithic Contexts

For the entire Lower and Middle Palaeolithic period only three examples for direct evidence of hunting lesions have prominently been published. The most recent discovery originates from the open-air site of Umm el Tlel in Central Syria (Boëda et al. 1999). Umm el Tlel has delivered a long stratigraphy with numerous Middle Palaeolithic layers in a lacustrine milieu. The Mousterian Level IV3b'1 represents an archaeological record where Levallois points and bones from wild ass are particularly well represented. This particular level has been dated to around 50,000 years (Boëda et al. 1999).

Within this layer a mesial fragment of a Levallois point, embedded in a cervical vertebra of a wild ass was discovered (Fig. 6.13). The mesial Levallois point fragment stuck in the vertebral foramen and showed bending fractures on both edges, indicating that the basal and distal part of the tool had

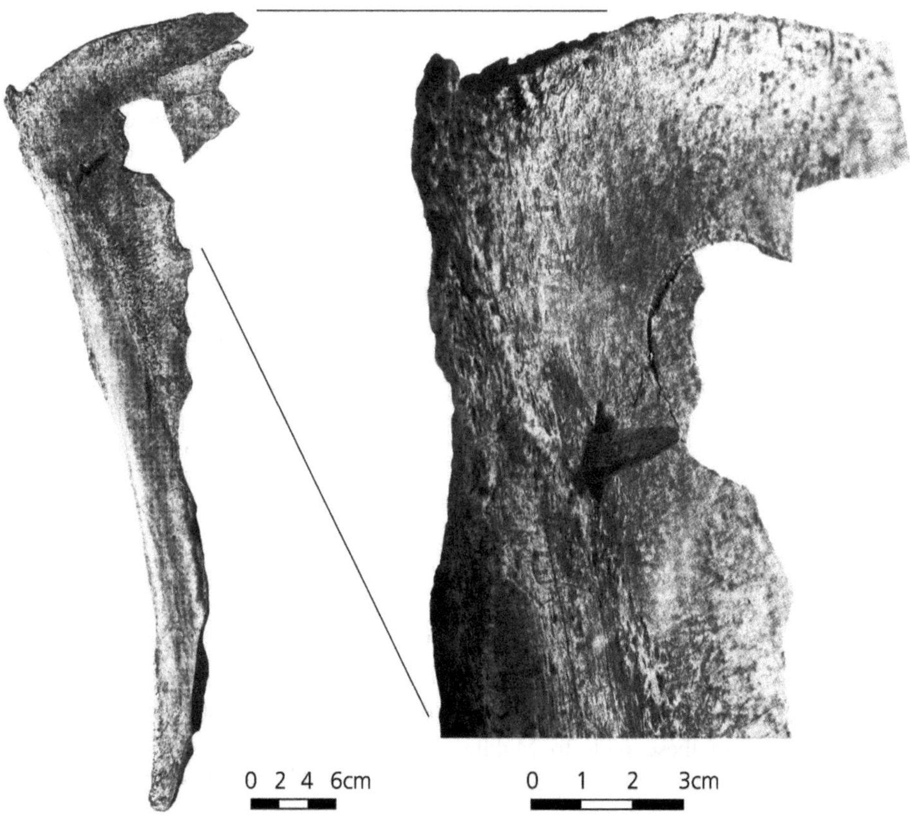

Fig. 6.12 Kokorevo I (RUS). Scapula of a large bovid with a fragment of an antler projectile (after Abramova 1982)

broken off. It has been suggested that the Levallois point was already damaged and lacking its distal part when it penetrated the wall of the vertebra with enormous force, as this part of the tool was missing from the medullary canal (Boëda et al. 1999). It needs to be mentioned that from the published pictures (Boëda et al. 1999: Fig. 2a–c) it is difficult to evaluate the exact relation between the embedded lithic fragment and the bone damage it must have caused, as this area of the vertebra shows extensive damage of which it is difficult to judge whether it had been caused by the shot itself of by other taphonomic processes.

For a final assessment of the evidence from Umm el Tlel it would be welcome if the taphonomic background from which this evidence originates would be outlined in detail. What is striking here is that the vertebra was obviously fragmented into two pieces (Boëda et al. 1999: Figs. 2a–c and 5), with the Levallois point fragment sandwiched in between. Certain taphonomic processes in lacustrine environments could be responsible for a comparable find situation, given that we would not be dealing with an *in situ* deposition and/or heavy geological overprinting of the site.

The unusual preservation conditions at Umm el Tlel are further indicated by evidence reported from the site. In level VI3d' a flint flake was uncovered embedded in a fragment of an ostrich pelvis. Level VI3d' is attributed to OIS 5a/4. The flake was located in the region of the acetabulum of the bird. According to a detailed functional analysis of the flake based on use wear studies and the analysis of bitumen traces found

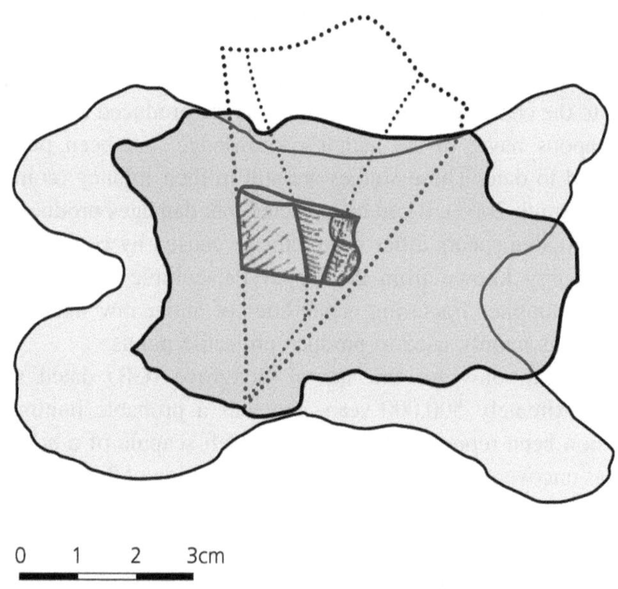

Fig. 6.13 Umm el Tlel (Syria), Middle Palaeolithic. Position of a Levallois point embedded in a vertebra of a wild ass (*Equus asinus*). Light gray = ventral, dark-grey = dorsal (after Boëda et al. 1999)

on the specimen, the evidence was interpreted in terms of a butchering scenario, where the flake accidentally remained stuck in the bird's pelvis (Bonilauri et al. 2007). Comparable butchering accidents have numerously been reported from the MSA assemblage of Klasies River Mouth (South Africa). One of these stone inclusions was argued to represent the broken tip of a stone point which was embedded in a cervical vertebra of a *Pelorovis* (Milo 1998).

It is clear that evidence from Level IV3b'1 has very little in common with damages interpreted as hunting lesions from Late Glacial and early Holocene archaeological records in that the flint was not found jammed in the bone as is obvious from all examples reported here.

The evidence from Syria however is not the only example interpreted as a direct hunting lesions reported from Middle Palaeolithic contexts. From the Mousterian of the site of La Quina (F) Henri-Martin (1907, 1934) reported on a first phalange of a bovid in which a flint fragment caused a local infection. Two further examples were mentioned in this context. One is a reindeer ulna with a silex point (Martin 1907), the other is a bone point embedded in the corpus of a reindeer vertebra (Martin 1934: Fig. X). It is not clear however whether the attribution of these finds to Middle Palaeolithic contexts is secure.

Indirect Evidence for Hunting Lesions in Lower and Middle Palaeolithic Contexts

As outlined above, the only indisputed evidence for hunting weapons during the Middle and Lower Palaeolithic are wooden spears and lances. Systematic experimental studies as to the character of lesions which can be produced by these weapons have, to the author's knowledge, not been published to date. These studies are still in their infancy (compare Smith 2003). It can be expected that damages produced by wooden spears differ from damage caused by composite weaponry known from the Upper Palaeolithic due to the much tougher fracturing capabilities of antler raw material that was mainly used to produce projectile points.

So far, only for the site of Boxgrove (GB) dated to approximately 500,000 years ago, has a probable hunting lesion been reported. At Boxgrove a left scapula of a horse was uncovered in an *in situ* context, consisting of flint debris, flint tools and bone fragments belonging to the carcass of a horse (Roberts and Parfitt 1999: Fig. 289). Due to its semicircular morphology a bone damage observed on the bone was considered to probably result from the impact of a spear (Roberts and Parfitt 1999) (Fig. 6.14).

Unfortunately for the interpretation of the traces observed at Boxgrove we are facing the problem of equifinality. As perforations can be caused by a variety of taphonomic agents, it is difficult to argue exclusively for hunting lesions in this context. Actualistic studies which might provide an interpretative frame of reference here have not been published to date.

A plausible cause for damage is illustrated by bone damage observed on a horse scapula from the German Middle Palaeolithic site of Salzgitter Lebenstedt. Two perforations next to each other were observed (Fig. 6.15). Here, it is possible to offer a straightforward interpretation as carnivore damage; two juxtaposed perforations were probably caused by the carnivore's canine teeth (Fig. 6.15).

Discussion and Conclusions

The author does not claim that the evidence for Palaeolithic and Mesolithic hunting lesions discussed above is complete. It can however be assumed that the majority of evidence is collated here.

If we evaluate the evidence for assertions on human behaviour, studies on hunting lesions can demonstrate several issues, which are often only implicitly assumed, because they have been observed in the ethnographical record and/or somehow appear "logical". In this context the study of hunting lesions could demonstrate that during the early phases of the Upper Palaeolithic confrontational hunting was practised and followed a risk minimising strategy [cf. Yana (RUS), Hohle Fels (D), Lugovskoe (RUS), Kokorevo I (RUS)], which is also evident for individual cases reported from the Middle Palaeolithic [cf. Lehringen and Gröbern (D)].

Moreover, the study of hunting lesions contextualised with the overall archaeological record demonstrates that communal hunting in the Final Upper Palaeolithic and Final Palaeolithic was regularly practiced independent from the targeted species (cf. mass hunting vs. targeting individual prey) [cf. Meiendorf and Stellmoor (D), Grotte du Bichon (CH), High Furlong (GB)].

The examination of healed hunting lesions (cf. Noe-Nygaard 1974) contextualized against wildlife census, animal ethology (i.e., birthing rates, territory size) and the overall individual archaeological record which could provide data on the temporal resolution of a particular site allows conclusions on the territory size of Mesolithic humans, which can otherwise only be reconstructed based on raw material studies of lithics.

Very rarely, is it even possible to reconstruct individual hunting strategies in detail on the basis of the study of hunting lesions [cf. Yana (RUS), Meiendorf and Stellmoor (D)].

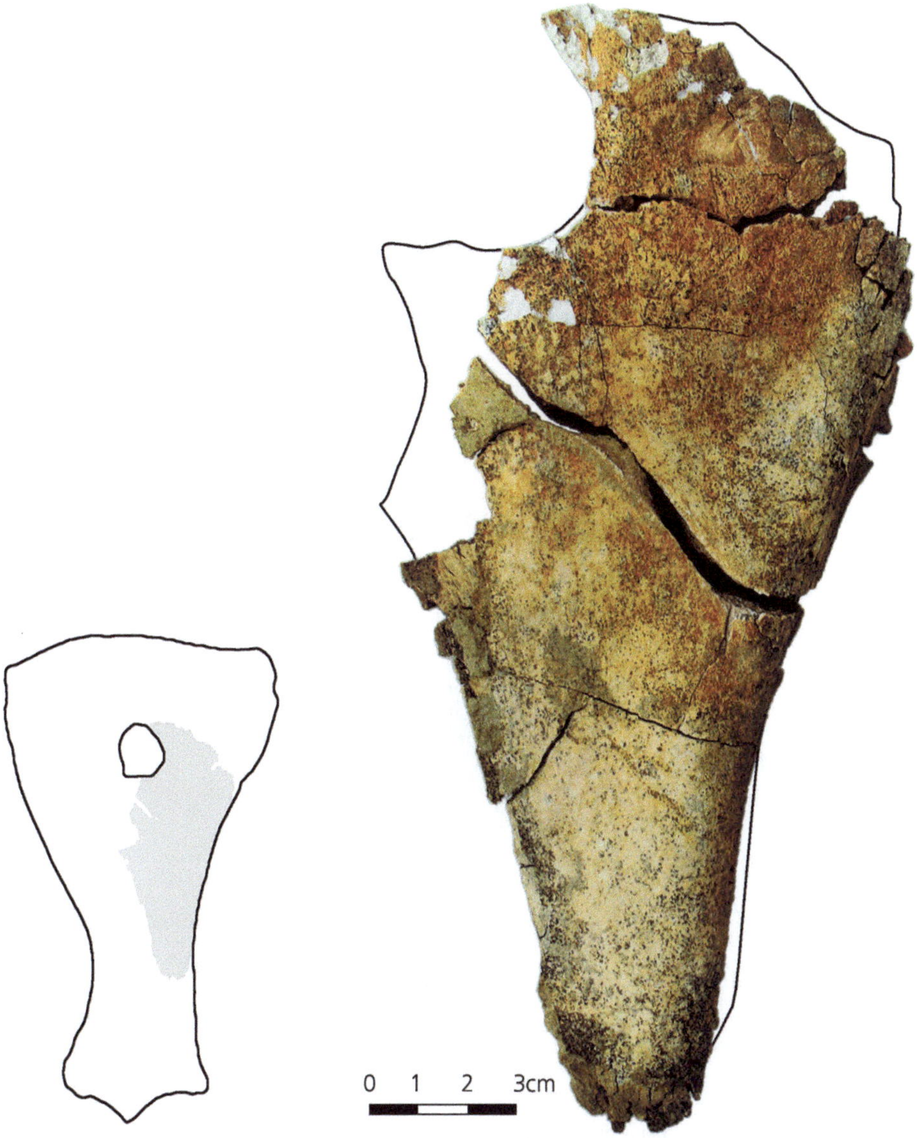

Fig. 6.14 Boxgrove (GB), Lower Palaeolithic. Fragment of a scapula of horse (*Equus ferus*) with semicircular fracture (photo by courtesy of Geoff Smith, MONREPOS)

Moreover the survey showed that we witness a considerable quantitative increase of hunting lesions by the end of the Upper Palaeolithic, where we have the evidence that, in addition to the spear thrower, bow and arrow came regularly into use. The lesions were caused by fragments of lithic projectiles which had been shot and had become embedded in animal carcasses. The fragments of flint armoury found in hunting lesions dating to the earlier phases of the Upper Palaeolithic are either undiagnostic or have been identified as triangular microliths and backed bladelets and are thus indicative only of lateral hafting.

The overall record for hunting lesions is completed by injuries caused by organic projectiles with laterally hafted lithics. These weapons are usually attributed to have been shot with a spear thrower.

In conclusion, reading the evidence considered here at face value, it would seem that hominins remained without lithic projectile technology before the Late Upper Palaeolithic or even Final Palaeolithic.

Assuming that lithic projectiles were only rarely used could explain the mismatch between results of experimental ballistic studies using shouldered lithic points which numerously produced a variety of impact traces which can only very rarely be traced in the archaeological record (Morel 2000; Smith et al. 2007; Castel 2008).

It is difficult to evaluate the exact timing as well as why, plausibly, two weaponry systems – the spear thrower/javelin and the bow and arrow – coexisted during the Late Glacial. Both javelins and arrows can be armed with the light lithic

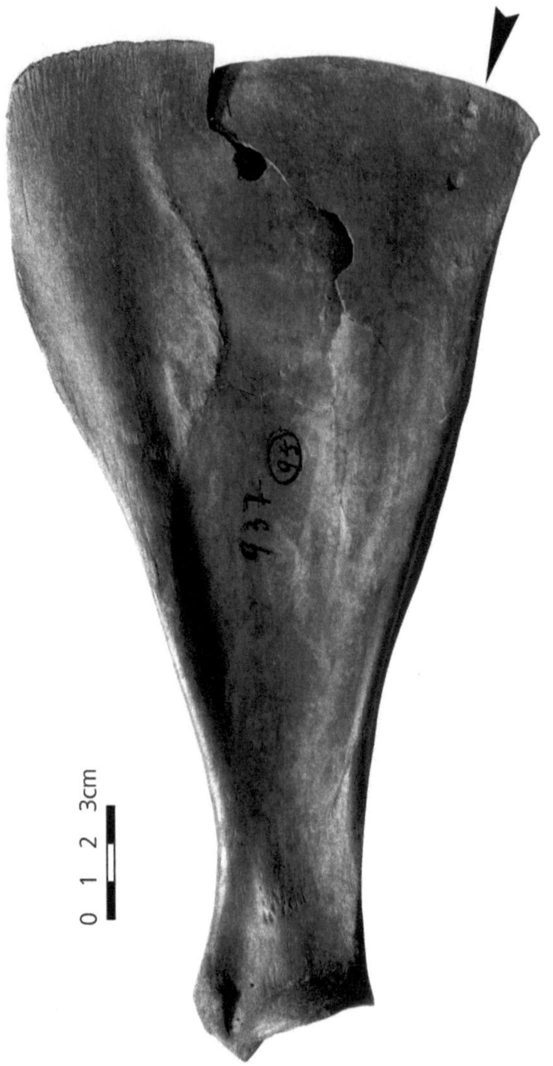

Fig. 6.15 Salzgitter-Lebenstedt (D), Middle Palaeolithic. Scapula of a horse (*Equus* sp.) with traces of carnivore damage (photo Sabine Gaudzinski-Windheuser, MONREPOS)

projectiles that are characteristic of the period (Cattelain 1997; Knecht 1997). Both systems have been associated with different hunting tactics, i.e., driving and ambushing (Rozoy 1992, but compare Cattelain 1997) as well as the exploitation of differently sized species and variations in individual hunting circumstances (Pelegrin 2000; Bignon 2008). Bratlund however points out that even though hunting tactics differed in the Final Palaeolithic/Ahrensburgian levels from tactics observed the Late Upper Palaeolithic/Hamburgian levels at Meiendorf (i.e., driving and stalking) differences in hunting lesions were not observed. An analysis of the embedded flint fragments was not part of her study and the question as to the weapon system used remains ambiguous at least for the Late Upper Palaeolithic/Hamburgian level. Only for the Final Palaeolithic Ahrensburgian the use of bow and arrow is attested here as already outlined.

The bow and arrow can be regarded as a highly flexible and due to the arrow's straight trajectory accurate weapon system. This weapon however has a lot of disadvantages and costs. The weapon is very costly to make and does not work well under particular weather conditions. Compared to a spear the arrow has a much lower impact force. Thus it can be assumed that outrunning of wounded prey was regularly practised.

Shooting with bow and arrow can be considered a low risk activity in terms of risk of injury for the hunter compared to the more confrontational use of a spear at closer range and the fact that this weapon is due to its curved trajectory not as unerring as bow and arrow. As one must regard hunting lesions to be indicative of accidents, it should be expected that these incidents are more affordable with the bow and arrow, given that in confrontational situations they attest a high risk of a lethal injury to the hunter. This might partly explain the disproportionately high number of hunting lesions and the high number of failed shots that we observe in the Mesolithic (cf. Noe-Nygaard 1974). A further variable is probably the relatively small size of territories exploited during the Mesolithic. From a more general perspective it can thus be assumed that the appearance of bow and arrow might be related to the disintegration and/or restructuring of the large social networks into smaller social units at the end of the Late Glacial in which hunting activities must have been socially embedded. Hunting in smaller hunting units might have demanded the development of a risk averting weapon to compensate for the lack of security provided by large hunting units.

A further aspect that needs to be outlined and supports the hypothesis for a late onset of lithic projectile use is the dichotomous pattern we witness when considering the evidence for the Lower/Middle Palaeolithic and the Upper Palaeolithic/Mesolithic.

For the entire Lower and Middle Palaeolithic period we have only a single possible example of a hunting lesion from Umm-el-Tlel, but as discussed above this contrasts markedly from the nature of damage reported in Upper Palaeolithic contexts. One would wish for more information on the taphonomical background from which the inserted fragment of the Levallois point fragment stem, to allow a better evaluation of the evidence. Other claims for hunting lesions remain so far ambiguous in their interpretation or assignment to a Lower/Middle Palaeolithic context.

The dichotomy in the quality and quantity of evidence between the Upper Palaeolithic/Mesolithic and earlier periods and is even more pronounced when one considers the enormous differences in the overall time spans covered by these different periods. Ironically, compared to the European Lower and Middle Palaeolithic for which we have a wealth of well-preserved bone assemblages, the well-preserved faunal record of the Upper Palaeolithic, Final Palaeolithic and Mesolithic is relatively meagre.

These discrepancies in the quantity of evidence are also underlined by the quality of *in situ* Pleistocene and early Holocene archaeological sites interpreted to represent singular hunting events as will be outlined below. Examples have been reported from the Lower/Middle Pleistocene onwards though most of these sites date to the Late Glacial or early Holocene period.

For most of the Late Glacial and early Holocene sites, the armament which was used to kill animals can incontestably be identified although even for this period we lack comprehensive knowledge as to the variety of intrinsic factors which governed the application of particular weapons. This is why remains of successful and unsuccessful hunting scenarios uncovered *in situ* e.g., at the Final Palaeolithic site of High Furlong (GB) (Hallam et al. 1973; Jacobi et al. 2009) are of utmost importance.

At High Furlong the almost complete skeleton of an adult elk (*Alces alces*) was uncovered. Found in association with the carcass two fragmented very similar bone harpoons were unearthed. Re-analysis by Paul Pettitt, Peter Rowley-Conwy and Janet Montgomery indicates that the harpoons might have been mounted together like a "leister" (Paul Pettitt, oral communication). One of them was clearly embedded in the foot of the animal whereas the other was found in the area of its chest. Re-analysis of this hunting scenario suggests that the animal was shot from behind, and that the leister broke as one harpoon was embedded in its foot and the rest continued on into the animal's chest (Paul Pettitt, oral communication). An earlier detailed discussion on the hunting scenario can be found in Pettitt and White (2012).

A further example which illustrates the above made point comes from the Mesolithic site of Prejlerup (DK), attributed to the Maglemose culture. At Prejlerup the carcass of a large aurochs bull (*Bos primigenius*) was unearthed. A number of 15 microliths were found mainly concentrated in the left hind quarter, very close to the bone. The find situation is completed by a wooden arrowshaft. The scenario is interpreted as a hunting event where the subsequent exploitation of the carcass was abandoned due to logistic problems, the bull weighing about a ton (Aaris-Sørensen and Petersen 1986).

Comparable find situations are known though very rarely e.g., from European *in situ* find situations from high resolution interglacial archives attributed to Lower/Middle Palaeolithic contexts. In contrast to the Final Palaeolithic/Early Mesolithic evidence the weapon used to kill the animals cannot be clearly identified in these sites unless we are dealing with the organic preservation of wood.

This is amply illustrated by the well known sites of Lehringen and Gröbern (D), dated to approximately 125,000 years ago. At the Eemian Interglacial site of Lehringen we have indisputable evidence for a wooden lance which was found between the ribs of a straight-tusked elephant (*Palaeoloxodon antiquus*) (Thieme and Veil 1985). The find situation uncovered at Gröbern can be characterised as a butchering scenario with a number of 27 simple flakes distributed among the bones of a complete straight-tusked elephant carcass (*Palaeoloxodon antiquus*) (Mania et al. 1990). At both sites hominins took advantage of ill and old individuals.

As the spatial organisation of Middle and Upper Palaeolithic humans differed as did their logistical use of landscapes, it is difficult to directly compare archaeological records from the Upper Palaeolithic/Mesolithic to earlier periods. This is amply illustrated when the Early Upper Palaeolithic evidence from Hohle Fels, indicating amongst other activities cave bear hunting (Münzel and Conard 2004) is considered against the Middle Palaeolithic record from Balve Cave in Germany which provides an even earlier indisputable proof for cave bear hunting and exploitation during the late Middle Palaeolithic (Kindler 2012). For the Early Upper Palaeolithic signals for organic tool production mask with its spatial organisation all earlier signals for human subsistence, whereas for the Middle Palaeolithic the overall spatial organisation can only hardly be read.

Moreover, high resolution, single period archives such as Lehringen and Gröbern are rare exceptions in the Lower and Middle Palaeolithic, most sites being palimpsests of long and/or unknown duration.

Many Lower and Middle Palaeolithic faunal assemblages, especially those deriving from cave sites show high secondary modification by large carnivores, such as large hyaenas, which are not part of Late Glacial biotopes. Carnivore modification is one variable which is responsible for highly fragmented faunal remains in these early periods. These and other natural taphonomical differences in faunal preservation have been invoked to explain the absence of hunting lesions in Lower and Middle Palaeolithic contexts and it is expected that hunting lesions might hide among the amalgam of differently marked, highly fragmented faunal remains that so regularly characterise the faunal record from these periods (cf. Morel 2000; Castel 2008; Letouneux and Pétillon 2008). This assumption however also indicates that Upper Palaeolithic faunal assemblages are generally less fragmented than faunal assemblages from the Lower/Middle Palaeolithic, which especially if we generally consider Late Glacial faunas from archaeological sites, is definitely not the case.

To conclude, concerning the timing and importance of lithic projectile technology a dichotomous pattern between data provided by studies on hunting lesions in the faunal record and current perceptions based on lithic and its actualistic data is apparent. This needs to be clarified and one way to explain the virtual absence of hunting lesions caused by tipped lithic projectiles in the Lower, Middle- and major parts of the Upper Palaeolithic is to suggest that it was simply not part of the regular weaponry system used.

The hypothesis of a late onset is in stark contrast to recent claims for composite projectile technology as early as 0.5 Ma years ago (Wilkins et al. 2012; Wilkins and Schoville 2016). It provides a perspective anti-cyclical to the current perception and research agenda on projectile technology so well illustrated in this volume. Thus, it invites the broadening of its interpretative framework and provides a new perspective.

If one accepts the hypothesis of a late onset of lithic projectile technology numerous interesting questions arise. For the majority of our past wooden weaponry prevailed. Neanderthals, considered as the "top predators" within their ecological niche, who were very effective hunters able to regularly kill prime adults of big and dangerous animals (e.g., Gaudzinski 1995, 2000; Speth and Tchernov 2007) used wooden weaponry for their confrontational encounters only. These confrontational encounters continue during the Upper Palaeolithic where they were obviously as fraught with risk as confrontational hunting encounters by Neanderthals, judging from the frequency of bone injuries recorded for Middle and Upper Palaeolithc hominins (Trinkaus 2012). Hunting large game has a much longer ancestry though, that goes way beyond the time prior to the earliest occupation of Europe (Gaudzinski-Windheuser 2005; Dominguez-Rodrigo et al. 2007).

In this context it is intriguing to ask why stone tipped projectiles were invented at all, when for the majority of our existence they played no or only a very marginal role in hunting strategies. This question becomes even more intriguing when results of experimental studies are considered which markedly illustrate the difficulty to pinpoint its immediate advantage over wood tipped projectiles (Waguespack et al. 2009; Salem and Churchill 2016). Or claims are heard outlining the brittleness of lithic points as a crucial factor "militating against use on thrusting spears" (Ellis 1997:60), or use wear studies are considered showing that even the "icons" proposed for use as lithic projectiles such as Font-Robert- and especially Levallois points had multifunctional use (Boëda et al. 1999; Milks et al. 2016).

Several suggestions have been made that socially related factors might have played a role here (Waguespack et al. 2009). This is underlined by the fact that changes in weaponry are not necessarily connected to improved hunting success or habitat adaptation by humans (see Grimm 2013). Studies on the composition of prey from faunal accumulations dating to the Mousterian and Aurignacien in South-Western France (Grayson and Delpech 2002) were not able to demonstrate significant differences, probably related to similarities in the taxonomic composition during both epochs (Grayson and Delpech 2002). This is true though lithic technology had changed and organic projectiles invented.

Apart from the social perspective, the invention of lithic projectiles could also have followed utilitarian aspects, related to the invention of the spear-thrower. Using stone tipped projectiles might e.g., increase the weight of a projectile without decreasing penetration (John Speth, oral communication).

Finally, it remains to hope that the current study provides enough food for thought to trigger new "lateral" ideas and thus new perspectives on the topic of lithic projectile technology.

Acknowledgments The author would like to thank Radu Iovita (MONREPOS Archaeological Research Centre for Human Behavioural Evolution, Germany) and Katsuhiro Sano (University of Tokyo, Japan) for the invitation to contribute a survey on hunting lesions to this volume. The author's thanks go Olaf Jöris, Paul Pettitt, John Speth and two anonymous reviewers for valuable insights, comments and inspiration and to Regina Hecht (MONREPOS Archaeological Research Centre for Human Behavioural Evolution, Germany) who was responsible for the figures of this contribution. The MONREPOS Archaeological Research Centre and Museum for Human Behavioural examines the historical processes of the evolution of human behaviour through the contextualised and interdisciplinary study of Pleistocene and early Holocene material remains. Research at MONREPOS follows a structured research strategy which constitutes the interface between chronology and chorology, adaptive strategies and social networks and provides the synergies necessary for understanding the process of becoming human. This paper is an integral part of research devoted to hominin behavioural strategies connected to the evolution of human diet.

References

Aaris-Sørensen, K., & Brinch Petersen, E. (1986). The Prejlerup aurochs – an archaeozoological discovery from Boreal Denmark. *Striae, 24*, 111–117.

Arias Cabal, P., Ontañón Peredo, R., Álvarez Fernández, E., Aparicio, M. T., Chauvin, A., Clemente Conte, I., et al. (2005). La estructura Magdaleniense de La Garma A. Aproximación a la organización espacial de un hábitat paleolitico. In N. Ferriera Bicho (Ed.), *O paleolitico* (pp. 123–141). Faro: Actas do Congresso de Arqueologia Peninsular 4.

Abramova, Z. A. (1982). Zur Jagd im Jungpaläolithikum. *Archäologisches Korespondenzblatt, 12*, 1–9.

Allain, L. (1979). L' industrie lithique et osseuse de Lascaux. In A. Leroi-Gourhan & J. Allain (Eds.), *Lascaux inconnu* (pp. 87–119). Paris: Éditions CNRS.

Bader, O., & Bader, N. (2000). *Homo sungirensis. Upper Palaeolithic man: Ecological and evolutionary aspects of the investigation*. Moscow: Nauchny Mir.

Binford, L. R. (1981). *Bones: Ancient men and modern myths*. New York: Academic Press.

Binford, L. R. (1985). Human ancestors: Changing views of their behaviour. *Journal of Anthropological Archaeology, 4*, 292–327.

Bignon, O. (2008). *Chasser les chevaux à la fin du Paléolithique dans le Bassin parisien. Stratégie cynégétique et mode de vie au Magdalénien et à l'Azilien ancien*. Oxford: BAR International Series 1747.

Boëda, E., Geneste, J. J., Griggo, C., Mercier, N., Muhesen, S., Reyss, J. L., et al. (1999). A Levallois point embedded in the vertebra of a wild ass (*Equus africanus*): Hafting, projectiles and Mousterian hunting weapons. *Antiquity, 73*, 394–402.

Bonilauri, S., Boëda, É., Griggo, C., Al-Sakhel, H., & Muhesen, S. (2007). Un éclat de silex Moustérien coincé dans un bassin d'autruche (Struthio camelus) à Umm el Tlel (Syrie Centrale). *Paléorient, 33*, 39–46.

Bosinski, G. (1969). Der magdalenien-Fundplatz Feldkirchen-Gönnersdorf, Kr. Neuwied. *Germania, 47*, 1–38.

Bosinski, G. (1987). Die große Zeit der Eiszeitjäger. *Jahrbuch des Römisch-Germanischen Zentralmuseums, 34*, 1–139.

Bratlund, B. (1990). Rentierjagd im Spätglazial. Eine Untersuchung der Jagdfrakturen an Rentierknochen von Meiendorf und Stellmoor, Kreis Stormarn. *Offa, 47*, 7–34.

Bratlund, B. (1991). A study of hunting lesions containing flint fragments on reindeer bones at Stellmoor, Schleswig-Holstein, Germany. In N. Barton, A. J. Roberts & D. Roe (Eds.), *The late glacial in north-west Europe. Human adaptations and environmental change at the end of the Pleistocene* (pp. 193–207). London: CBA Research Report 77.

Burov, G. M. (1980). Der Bogen bei den mesolithischen Stämmen Nordosteuropas. *Mesolithikum in Europa. Internationales Symposium Potsdam, April 1978* (pp. 373–388). Veröffentlichungen des Museums für Ur- und Frühgeschichte Potsdam, 14–15.

Callow, P. (1986). The flint tools. In P. Callow & J. Cornford (Eds.), *La Cotte de St. Brelade* (pp. 251–314). Norwich: Geo Books.

Canby, Th Y. (1979). The search for the first Americans. *National Geographic, 156*, 330–363.

Castel, J.-C. (2008). Identification des impacts des projectiles sur le squelette des grands ongulés. *Annales de Paléontologie, 94*, 103–118.

Cattelain, P. (1989). Un crochet de propulseur solutréen de la grotte de Combe-Saunière 1. *Bulletin de la Société préhistorique française, 86*, 213–216.

Cattelain, P. (1997). Hunting during the Upper Palaeolithic: Bow, spearthrower or both? In H. Knecht (Ed.), *Projectile technology* (pp. 213–240). New York: Plenum Press.

Churchill, S. E. (1993). Weapon technology, prey size selection, and hunting methods in modern hunter-gatherers: implications for hunting in the Palaeolithic and Mesolithic. In G. L. Peterkin, H. M. Bricker & P. A. Mellars (Eds.), *Hunting and animal exploitation in the later Palaeolithic and Mesolithic in Europe* (pp. 11–24). Washington: Archaeological Papers of the American Anthropologist Association 4.

Clottes, J. (1995). *Les cavernes de Niaux. Art préhistorique en Ariège*. Paris: Seuil.

Cluzel, J.-P., & Cleyet-Merle, J.-J. (Eds.). (2011). *Mille et une femmes de la fin des temps glaciaires*. Paris: Grand Palais.

Cziesla, E. (2001). Weitere Bemerkungen zu organischen Geschossspitzen aus Brandenburg. *Die Kunde N.F., 52*, 133–144.

Domínguez-Rodrigo, M. (2002). Hunting and scavenging by early humans: The state of the debate. *Journal of World Prehistory, 16*, 1–54.

Domínguez-Rodrigo, M., Egeland, C. P., & Barba, R. (2007). *Deconstructing Olduvai: A taphonomic study of the Bed I sites*. Dordrecht: Springer.

Ellis, C. J. (1997). Factors influencing the use of stone projectile tips. In H. Knecht (Ed.), *Projectile technology* (pp. 37–74). New York: Plenum Press.

Feustel, R. (1970). Statuettes féminines Paléolithiques de la Républic Démocratique Allemande. *Bulletin de la Société Préhistorique Française, 67*, 12–16.

Feustel, R. (1974). *Die Kniegrotte. Eine Magdalénien-Station in Thüringen*. Weimar: Veröffentlichungen des Museums für Ur- und Frühgeschichte Thüringens 5.

Gaudzinski, S. (1995). Wallertheim revisited: A re-analysis of the fauna from the Middle Palaeolithic site of Wallertheim (Rheinhessen/Germany). *Journal of Archaeological Science, 22*, 51–66.

Gaudzinski, S. (2000). On the variability of Middle Palaeolithic procurement tactics. *International Journal of Osteoarchaeology, 10*, 396–406.

Gaudzinski-Windheuser, S. (2005). *Subsistenzstrategien frühpleistozäner Hominiden in Eurasien. Taphonomische Faunenbetrachtungen der Fundstellen der 'Ubeidiya Formation (Israel)*. Mainz: Verlag des Römisch Germanischen Zentralmuseums.

Gaudzinski-Windheuser, S. (2013). *Raumnutzungsmuster des späten Jungpaläolithikums in Oelknitz (Thüringen, Deutschland)*. Mainz: Verlag des Römisch Germanischen Zentralmuseums.

Gaudzinski-Windheuser, S., & Kindler, L. (2012). Research perspectives for the study of Neandertal subsistence strategies based on the analysis of archaeozoological assemblages. *Quaternary International, 247*, 59–68.

Gaudzinski-Windheuser, S., & Roebroeks, W. (2011). On Neanderthal subsistence in Last Interglacial forested environments in Northern

Europe. In N. Conard & J. Richter (Eds.), *Neanderthal lifeways, subsistence and technology* (pp. 61–71). Dordrecht: Springer.

Geneste, J.-M., & Plisson, H. (1993). Hunting technologies and human behaviour: Lithic analysis of Solutrean shouldered points. In H. Knecht, A. Pike-Tay & R. White (Eds.), *Before Lascaux. The complex record of the Early Upper Paleolithic* (pp. 117–135). Boca Raton: CRC Press.

Ginter, B., & Połtowicz, M. (2007). Magdalenian settlement in Poland before the Bølling oscillation. In M. Korbusiewicz & J. Kabaciński (Eds.), *Studies in the Final Palaeolithic settlement of the Great European Plain* (pp. 7–19). Poznań: Institute of Archaeology and Ethnology, Polish Academy of Sciences.

Grayson, D. K., & Delpech, F. (2002). Specialized Early Upper Paleolithic hunters in Southwestern France? *Journal of Archaeological Science, 29*, 1439–1449.

Grünberg, J. M. (2000). *Mesolithische Bestattungen in Europa. Ein Beitrag zur vergleichenden Gräberkunde*. Rahden: Verlag Marie Leidorf.

Guthrie, R. D. (2005). *The nature of Paleolithic art*. Chicago: University of Chicago Press.

Hahn, J. (1969). Gravettien- Freilandstationen im Rheinland: Mainz-Linsenberg, Koblenz-Metternich und Rhens. *Bonner Jahrbuch, 169*, 44–87.

Hahn, J. (1977). *Aurignacien. Das ältere Jungpaläolithikum in Mittel- und Osteuropa*. Köln: Fundamenta A9.

Hallam, J. S., Edwards, B. J. N., Barnes, B., & Stuart, A. J. (1973). The remains of a late Glacial elk with associated barbed points from High Furlong, near Blackpool, Lancashire. *Proceedings of the Prehistoric Society, 39*, 100–128.

Höck, C. (2001). *Das Magdalénien der Kniegrotte. Ein Höhlenfundplatz bei Döbritz, Saale-Orla-Kreis*. Stuttgart: Theiss Verlag.

Jacobi, R. M., Higham, T. F. G., & Lord, T. C. (2009). Improving the chronology of the human occupation of Britain during the Late Glacial. In M. Street, N. Barton & T. Terberger (Eds.), *Humans, environment and chronology of the late glacial of the north European plain*. Proceedings of the Workshop 14 (Commission XXXII "The Final Palaeolithic of the Great European Plain") of the 15*th* UISPP Congress, Lisbon, September 2006 (pp. 7–25). Mainz: Verlag des Römisch-Germanisches Zentralmuseums.

Jungmanns, J. (2001). *Pfeil und Bogen. Herstellung und Gebrauch in der Jungsteinzeit*. Biel: Verlag Museum Schwab.

Kindler, L. (2012). *Die Rolle von Raubtieren bei der Einnischung und Subsistenz jungpleistozäner Neandertaler. Archäozoologie und Taphonomie der mittelpaläolithischen Fauna aus der Balver Höhle (Westfalen)*. Mainz: Verlag des Römisch Germanischen Zentralmuseums.

Knecht, H. (Ed.). (1997). *Projectile technology*. New York: Plenum Press.

Larsson, L., & Sjöström, A. (2011). Bog sites and wetland settlement during the Mesolithic: Research from a bog in Central Scania, Southern Sweden. *Archäologisches Korrespondenzblatt, 41*, 457–472.

Lartet, E., & Christy, H. (1864). Sur des figures d'animaux graves et sculptés et d'autres produits d'art et d'industrie rapportables aux temps primordiaux de la période humaine. *Revue archéologique, 41*, 233–267.

Leshchinskiy, S. V. (2012). Paleoecological investigation of mammoth remains from the Kraków Spadzista Street (B) site. *Quaternary International*. doi:10.1016/j.quaint.2012.05.025.

Leshchinskiy, S. V. (2006). Lugovskoye: Environment, taphonomy, and the origin of a paleofaunal site. *Archaeology, Ethnology and Anthropology of Eurasia, 25*, 33–40.

Leduc, C. (2012). New Mesolithic hunting evidence from bone injuries at Danish Maglemosian sites: Lundby Mose and Mullerus (Sjælland). *International Journal of Osteoarchaeology, 24*, 476–491.

Leroi-Gourhan, A. (1971). *Préhistoire de l'art occidental*. Paris: Éditions d'Art Lucien Mazenod.

Leroi-Gourhan, A. (1983). Une tête de sagaie à armature de lamelles de silex à Pincevent (Seine-et-Marne). *Bulletin de la Société préhistorique française, 198*, 154–155.

Leroi-Gourhan, A., & Allain, J. (1979). *Lascaux inconnu*. Paris: Éditions du CNRS.

Letourneux, C., & Pétillon, J.-M. (2008). Hunting lesions caused by osseous projectile points: Experimental results and archaeological implications. *Journal of Archaeological Science, 35*, 2849–2862.

Malmer, M. P. (1969). Die Mikrolithen in dem Pfeil-Fund von Loshult. *Meddelanden från Lunds Universitets Historiska Museum, 1966–1968*, 249–255.

Mania, D., Thomae, M., Litt, T., & Weber, T. (Eds.). (1990). *Neumark-Gröbern. Beiträge zur Jagd des mittelpaläolithischen Menschen*. Berlin: Deutscher Verlag der Wissenschaften.

Marom, A., McCullagh, J. S. O., Higham, T. F. G., Sinitsyn, A. A., & Hedges, R. E. M. (2012). Single amino acid radiocarbon dating of Upper Paleolithic modern humans. *Proceedings of the National Academy of Sciences of the United States of America, 109*, 6878–6881.

Martin, H. (1907). *Recherches sur l'evolution du Moustérien dans le gisement de La Quina. Tome I*. Paris: Schleicher.

Martin, H. (1934). Pathologie osseuse préhistorique. Blessure par arme de jet sur une vertèbre de Renne. *Association française pour l'Avancement des Sciences, 58*, 183–184.

Maschenko, E. N., Narochnyi, V. A., & Telegina, S. N. (2005). New data on the Late Pleistocene mammoth population from the Lugovskoe locality, western Siberia, Russia. In L. D. Agenbroad & R. L. Symington (Eds.), *2nd World of elephants congress. Short papers and abstracts* (pp. 109–113). Mammoth Site of Hot Springs, South Dakota.

Milks, A., Dinnis, R., & Pope, M. (2016). Morphometric Variability of Early Gravettian Tanged "Font-Robert" Points, and Functional Implications. In: R. Iovita & K. Sano (Eds.), *Multidisciplinary approaches to the study of Stone Age weaponry, Vertebrate Paleobiology and Paleoanthropology* (pp. 135–146). Dordrecht: Springer.

Milo, R. G. (1998). Evidence for hominid predation at Klasies River Mouth, South Africa, and its implications for the behaviour of early Modern Humans. *Journal of Archaeological Science, 25*, 99–133.

Moirenc, A., Cotte, J., & Cotte, C. (1921). Une inclusion remarquable dans un os paléolithique. *Revue des Études anciennes, 23*, 117–119.

Møller Hansen, K., Brinch Petersen, E., & Aaris-Sørensen, K. (2004). Filling the gap: Early Preboreal Maglemose elk deposits at Lundby, Sjælland, Denmark. In T. Terberger & B. V. Eriksen (Eds.), *Hunters in a changing world – Jäger in einer sich wandelnden Umwelt* (pp. 75–84). Rahden: Marie Leidorf Verlag.

Morel, P. (1993). Impacts de projectiles sur le gibier: quelques elements d'une approche expérimentale. In P. C. Anderson, S. Beyries, M. Otte, & H. Plisson (Eds.), *Traces et fonction: Les gestes retrouvés* (pp. 55–57). Liège: Service de Préhistoire de l'université de Liège.

Morel, P. (1998). La Grotte du Bichon (La Chaux-de-Fonds, canton de Neuchâtel, Suisse). In C. Cupillard & A. Richard (Eds.), *Les derniers chasseurs-cueilleurs du Massif Jurassien et de ses marges (13,000-5,500 avant Jésus-Christ)* (pp. 88–93). Lons-le-Saunier: Centre Jurassien du Patrimoine.

Morel, P. (2000). Impacts de chasse et archéozoologie: quelques observations expérimentales. In C. Bellier, P. Cattelain, & M. Otte (Eds.), *La chasse dans la préhistoire* (pp. 54–59). Liége: Editions ERAUL.

Münzel, S., & Conard, N. (2004). Cave bear hunting in Hohle Fels Cave in the Ach Valley of the Swabian Jura. *Revue de Paléobiologie, 23*, 877–885.

Münzel, S., Langguth, K., Conard, N., & Uerpmann, H. P. (2001). Höhlenbärenjagd auf der Schwäbischen Alb vor 30,000 Jahren. *Archäologisches Korrespondenzblatt, 31*, 317–327.

Napierala, H., Höhneisen, M., & Trautmann, M. (2010). „Stirb an einem anderen Tag". Schussverletzungen eines Rentiers vom Magdalénienfundplatz Kesslerloch bei Thayngen (Kt. Schaffhausen). *Archäologisches Korrespondenzblatt, 40*, 457–466.

Nikolskij, P., & Pitulko, V. (2013). Evidence from the Yana Palaeolithic site, Arctic Siberia, yields clues to the riddle of mammoth hunting. *Journal of Archaeological Science, 40*, 4189–4197.

Noe-Nygaard, N. (1973). The Vig bull, new information on the final hunt. *Meddelelser fra Dank Geologisk Forening, 22*, 244–248.

Noe-Nygaard, N. (1974). Mesolithic hunting in Denmark illustrated by bone injuries caused by human weapons. *Journal of Archaeological Science, 1*, 217–248.

Noe-Nygaard, N. (1975). Two shouldered blades with healed lesions from Star Carr. *Proceedings of the Prehistoric Society, 41*, 10–16.

Oakley, K. P., Andrews, A., Keeley, L. H., & Clark, J. D. (1977). A reappraisal of the Clacton spearpoint. *Proceedings of the Prehistoric Society, 43*, 13–30.

Ødum, H. (1920). *Et Elsdyrfund fra Taaderup paa Falster* (pp. 1–11). IV: Denmarks geologiske undersøgelse.

Orlova, L. A., Zenin, V. N., Stuart, T. A. J., Higham, T. F. G., Grootes, P. M., Leshchinskiy, S. V., et al. (2004). Lugovskoe, Western Siberia: A possible extra-arctic mammoth refugium at the end of the Late Glacial. *Radiocarbon, 46*, 363–368.

Otte, M., & Caspar, J. P. (1987). Les pointes de la Font-Robert: outils emmanchés? In D. Stordeur (Ed.), *La main et l'outil. Manches et emmanchements préhistoriques* (pp. 65–74). Lyon: Travaux de la Maison de l'Orient 15.

Pasda, C. (1998). *Wildbeuter im archäologischen Kontext. Das Paläolithikum in Südbaden.* Bad Bellingen: Archäologie im Südwesten 2.

Passemard, E. (1944). La caverne d'Isturitz en Pays Basque. *Préhistoire, 9*, 1–95.

Pelegrin, J. (2000). Les techniques de débitage laminaire au Tardiglaciaire: critères de diagnose et quelques réflexions. In B. Valentin, P. Bodu & M. Christensen (Eds.), *L'Europe centale et septentrionale au Tardiglaciaire. Confrontations des modèles régionaux de peuplement* (pp. 73–86). Nemours: Mémoires du Musée de Prehistoire d'Ile-de-France.

Pericot Garcia, L. (1942). *La Cueva del Parpalló (Gandía).* Madrid: Consejo Superior de Investigaciones Cientificas, Inst. Diego Velázquez.

Pétillon, J. M., Plisson, H., & Cattelain, P. (2016). Thirty years of experimental research on the breakage patterns of Stone Age Osseous points. Overview, methodological problems and current perspectives. In: R. Iovita & K. Sano (Eds.), *Multidisciplinary approaches to the study of Stone Age weaponry* (pp. 47–63). Dordrecht: Springer.

Pétillon, J. M., Bignon, O., Bodu, P., Cattelain, P., Debout, G., Langlais, M., et al. (2011). Experimental manufacture and use of Magdalenian composite projectile tips. *Journal of Archaeological Science, 38*, 1266–1283.

Pettitt, P., & White, M. (2012). *The British Palaeolithic. Human societies at the edge of the Pleistocene world.* London: Routledge.

Peyrony, D., & Peyrony, E. (1938). *Laugerie-Haute près des Eyzies (Dordogne).* Paris: Masson.

Rabinovich, R., Gaudzinski-Windheuser, S., Kindler, L., & Goren-Inbar, N. (2011). *The Acheulian site of Gesher Benot Ya'aqov. mammalian taphonomy. The assemblages of Layers V-5 and V-6.* Dordrecht: Springer.

Rieder, H. (2007). Zur Qualität der Schöninger Speere als Jagdwaffen – aus der Sicht der Sportwissenschaft. In H. Thieme (Ed.), *Die Schöninger Speere. Mensch und Jagd vor 400,000 Jahren* (pp. 159–162). Stuttgart: Theiss Verlag.

Roberts, M. B., & Parfitt, S. A. (Eds.). (1999). *Boxgrove. A Middle Pleistocene hominid site at Eartham Quarry, Boxgrove, West Sussex.* London: English Heritage Archaeological Reports 17.

Rosendahl, G., Beinhauer, K.-W., Löscher, M., Kreipl, K., Walter, R., & Rosendahl, W. (2006). Le plus vieil arc du monde? Une pièce intéressante en provenance de Mannheim, Allemagne. *L'Anthropologie, 110*, 371–382.

Rozoy, J.-G. (1992). Le propulseur et l'arc chez les chasseurs préhistoriques. Techniques et démographies compares. *Paléo, 4*, 175–193.

Rust, A. (1937). *Das altsteinzeitliche Rentierjägerlager Meiendorf.* Neumünster: Karl Wachholz Verlag.

Rust, A. (1943). *Die alt- und mittelsteinzeitlichen Funde von Stellmoor.* Neumünster: Karl Wachholtz Verlag.

Salem, P. E., & Churchill, E. (2016). Penetration, tissue damage, and lethality of wood-versus lithic-tipped points. In: R. Iovita & K. Sano (Eds.), *Multidisciplinary approaches to the study of Stone Age weaponry* (pp. 203–212). Dordrecht: Springer.

Schuler, A. (1994). *Die Schussenquelle. Eine Freilandstation des Magdalénien in Oberschwaben.* Stuttgart: Konrad Theiss Verlag.

Shea, J. J. (2006). The origin of Lithic Projectile Technology: Evidence from Africa, the Levant, and Europe. *Journal of Archaeological Science, 33*, 823–846.

Shea, J. J. (2009). The impact of projectile weaponry on Late Pleistocene Hominin Evolution. In J. J. Hublin & M. P. Richards (Eds.), *The evolution of Hominin Diets. Integrating approaches to the study of Palaeolithic subsistence* (pp. 189–200). Dordrecht: Springer.

Smith, G. M. (2003). Damage inflicted on animal bone by wooden projectiles: experimental results and archaeological implications. *Journal of Taphonomy, 1*, 105–114.

Smith, M. J., Brickley, M. B., & Leach, S. L. (2007). Experimental evidence for lithic projectile injuries. Improving identification of an under-recognised phenomenon. *Journal of Archaeological Science, 34*, 540–553.

Stodiek, U. (1993). *Zur Technologie der jungpaläolithischen Speerschleuder. Eine Studie auf der Basis archäologischer, ethologischer und experimenteller Erkenntnisse.* Tübingen: Tübinger Monographien zur Urgeschichte 9.

Ströbel, R. (1959). Tardenoisspitze in einem Bovidenknochen von Schwenningen am Neckar (Kr. Rottweil). *Fundberichte aus Schwaben N.F., 15*, 103–106.

Speth J. D., & Tchernov, E. (2007). The Middle Palaeolithic occupations at Kebara Cave: a faunal perspective. In O. Bar-Yosef & L. Meignen (Eds.), *Kebara Cave, Mt. Carmel, Israel: The Middle and Upper Palaeolithic Archaeology. Part 1* (pp. 165–260). Cambridge: Peabody Museum of Archaeology and Ethnology.

Sudhues, H. (2004). Wundballistik bei Pfeilverletzungen. PhD Dissertation, Westfälische Wilhelms-Universität Münster, Münster.

Szmidt, C., Pétillon, J.-M., Cattelain, P., Nomand, C., & Schwab, C. (2009). Premières dates radiocarbone pour le Magdalénien d'Isturitz (Pyrénées-Atlantiques). *Bulletin de la Société préhistorique française, 106*, 588–592.

Telegin, D. J. A. (1961). Vasilivs' kij tretij nekropol' v Nadporižži. *Archeologija Kiew, 13*, 3–19.

Thieme, H. (Ed.). (2007). *Die Schöninger Speere. Mensch und Jagd vor 400,000 Jahren.* Stuttgart: Theiss Verlag.

Thieme, H., & Veil, S. (1985). Neue Untersuchungen zum eemzeitlichen Elefanten-Jagdplatz Lehringen, LDKr. Verden. *Die Kunde N. F., 36*, 11–58.

Trinkaus, E. (2012). Neandertals, early modern humans, and rodeo riders. *Journal of Archaeological Science, 39*, 3691–3693.

Valensi, P., & Psathi, E. (2004). Faunal Exploitation during the Middle Palaeolithic in south-eastern France and north-western Italy. *International Journal of Osteoarchaeology, 14*, 256–272.

Villa, P., & Lenoir, M. (2009). Hunting and hunting weapons of the Lower and Middle Palaeolithic in Europe. In J. J. Hublin & M. Richards (Eds.), *The evolution of hominin diets. Integrating approaches to the study of Palaeolithic subsistence* (pp. 59–85). Dordrecht: Springer.

Voormolen, B. (2008). Ancient hunters, modern butchers. Schöningen 13II-4, a kill-butchery site dating from the northwest European Lower Palaeolithic. *Journal of Taphonomy, 6*, 71–247.

Waguespack, N. M., Surovell, T. A., Denover, A., Dallow, A., Savage, A., Hyneman, J., et al. (2009). Making a point: Wood-versus stone-tipped projectiles. *Antiquity, 83*, 786–800.

Washburn, S. L., & Lancaster, C. S. (1968). The evolution of hunting. In R. B. Lee & I. de Vore (Eds.), *Man the hunter* (pp. 293–303). Chicago: Aldine.

Weniger, G.-C. (1995). *Widerhakenspitzen des Magdaléniens Westeuropas*. Mainz: Philip von Zabern.

Wilkins, J., & Schoville, B. J. (2016). Edge Damage on 500-thousand-year-old spear tips from Kathu Pan 1, South Africa: The combined effects of sper use and taphonomic processes. In: R. Iovita & K. Sano (Eds.), *Multidisciplinary approaches to the study of Stone Age weaponry* (pp. 101–117). Dordrecht: Springer.

Wilkins, J., Schoville, B. J., Brown, K. S., & Chazan, M. (2012). Evidence for early hafted hunting technology. *Science, 338*, 942–946.

Wojtal, P. (2007). *Zooarchaeological studies of the Late Pleistocene sites in Poland*. Kraków: Polish Academy of Sciences.

Yaroshevich, A., Nadel, D., & Tsatskin, A. (2013). Composite projectiles and hafting technologies at Ohalo II (23 ka, Israel): analysis of impact fractures, morphometric characteristics and adhesive remains on microlithic tools. *Journal of Archaeological Science, 40*, 4009–4023.

Yaroshevich, A., Zaidner, Y., & Weinstein-Evron, M. (2016). Projectile damage and point morphometry at the Early Middle Paleolithic Mislya Cave, Mount Carmel (Israel): Preliminary results and interpretations. In: R. Iovita & K. Sano (Eds.), *Multidisciplinary approaches to the study of Stone Age weaponry, Vertebrate Paleobiology and Paleoanthropology* (pp. 119–134). Dordrecht: Springer.

Zenin, V. N., Leshchinskiy, S. V., Zolotarev, K. V., Grootes, P. M., & Nadeau, M.-J. (2006). Lugovskoe: Geoarchaeology and culture of a Paleolithic site. *Archaeology, Ethnology and Anthropology of Eurasia, 25*, 41–53.

Chapter 7
Edge Damage on 500-Thousand-Year-Old Spear Tips from Kathu Pan 1, South Africa: The Combined Effects of Spear Use and Taphonomic Processes

Jayne Wilkins and Benjamin J. Schoville

Abstract This paper explores the effect of taphonomic processes on 500-thousand-year-old stone points from Kathu Pan 1, South Africa by statistically comparing archaeological edge damage distributions on the points to competing models of edge damage formation. We found that both taphonomic and behavioral processes influenced edge damage formation on the KP1 points, and the KP1 edge damage distribution is best explained by a combination of taphonomic effects and use as spear tips. The edge damage distribution method employed here advances studies of Stone Age weaponry because it can be used to quantitatively assess the effect of taphonomic and behavioral processes on stone tips without relying on subjective evaluations that attribute causation to individual wear features.

Keywords Edge damage • Spear tips • Points • Taphonomy • Middle Pleistocene • Middle Stone Age • South Africa • Fauresmith

Introduction

Correctly identifying Stone Age hunting technology has enormous implications for our understanding of human evolution. Humans are unique among extant primates for relying on a skill-intensive strategy used to acquire nutrient-dense, large package food resources (Kaplan et al. 2000). Based on ethnographic evidence, the use of spears is generally directed toward large-bodied animals (Churchill 1993), and hunting large game may have co-evolved with other traits unique to the *Homo* lineage, including prolonged childhood and adolescence, grandmother and paternal investment in child-rearing, and increased cognitive capacities for information flow and storage (Kaplan et al. 2000). Changes in hunting technology through time also shed light on the development of human sociality and cooperation (e.g., Boyd and Richerson 2005; Marlowe 2005; Hill et al. 2009), as well as advances in cognitive facilities related to long-term planning and language (e.g., Wadley et al. 2009; Ambrose 2010; Wynn and Coolidge 2011; Lombard and Wadley 2016).

Unfortunately, "smoking-gun" evidence in the Paleolithic and Stone Age records are rare. What archaeologists recover are stone tools and fragments whose morphology is consistent with hunting armatures. At best, comparing morphological attributes of ethnographic weapon tips with archaeological tools provides an indication of whether tools could have feasibly functioned as weapon tips (Sisk and Shea 2011). Actually establishing past function requires wear-trace evidence. A shortcoming of some use wear approaches is that wear-trace morphologies from behavioral processes may sometimes appear similar to a variety of processes, including post-depositional ones.

The recently reported assemblage of points and point-fragments from the site of Kathu Pan 1 (KP1), South Africa (Fig. 7.1) suggests that the earliest hafted spear technology could date to at least 500,000 years-ago (Wilkins et al. 2012). Multiple lines of evidence were used to establish that the points functioned as spear tips, including morphological attributes (i.e., size and symmetry) and wear-trace morphology ("diagnostic impact fractures"). Furthermore, it was demonstrated that edge damage distributions on the KP1 points are quantitatively inconsistent with expectations for post-depositional processes and use as cutting tools. Consistent with the spear tip hypothesis, the KP1 points exhibit increased frequencies of damage at the tips and similar distributions between the left and right edges of the ventral

J. Wilkins (✉) · B.J. Schoville
Department of Archaeology, University of Cape Town, Private Bag, Rondebosch 7701, South Africa
e-mail: jayne.wilkins@uct.ac.za

B.J. Schoville
e-mail: benjamin.schoville@asu.edu

J. Wilkins · B.J. Schoville
School of Human Evolution and Social Change, Institute of Human Origins, Arizona State University, 872402 Tempe, AZ 85287-2402, USA
e-mail: jrwilki2@asu.edu

Radu Iovita and Katsuhiro Sano (eds.), *Multidisciplinary Approaches to the Study of Stone Age Weaponry*, Vertebrate Paleobiology and Paleoanthropology, DOI 10.1007/978-94-017-7602-8_7

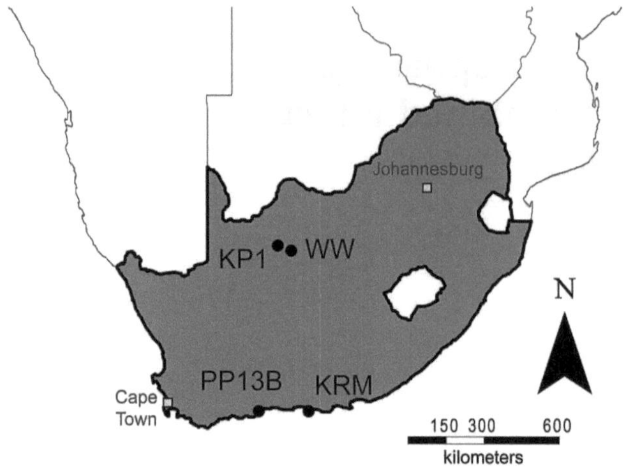

Fig. 7.1 Map of South Africa showing location of Kathu Pan 1 (KP1) and Pinnacle Point cave 13B (PP13B) and other sites mentioned in text; Wonderwerk Cave (WW) and Klasies River (KRM)

surface. In this paper, we further explore the assemblage distribution of edge damage on the KP1 points. By directly comparing KP1 point edge damage to multiple models for the effect of behavioral and taphonomic processes, we are able to assess the relative contributions of damage from multiple processes, and add further support to the spear tip hypothesis for the KP1 points.

Background

Middle Paleolithic and Middle Stone Age Points

Middle Stone Age (MSA) and Middle Paleolithic (MP) points were sometimes hafted onto spears. The most direct evidence comes from stone fragments embedded in a Pelorovis vertebra at Klasies River Mouth, South Africa (Milo 1998) and an equid vertebra at Umm el Tlel, Syria (Boëda et al. 1999; but see Gaudzinski-Windheuser 2016). Microwear and impact fractures on MP points indicate that Levantine Neandertals possessed stone tipped spears (Shea 1988; Yaroshevich et al. 2016). Experimental work utilizing a calibrated crossbow corroborates this evidence and demonstrates the effectiveness of hafted Levallois points as spear tips (Shea et al. 2001). In Europe, traces of hafting wear and mastic, as well as impact fractures on MP points also support their occasional use as hunting weapons (e.g., Hardy et al. 2001; Mazza et al. 2006; Villa et al. 2009; Rots 2013, 2016). Points are common in African MSA contexts and many studies have emphasized their roles as hunting weapons (e.g., Milo 1998; McBrearty and Brooks 2000; Donahue et al. 2004; Lombard 2004, 2005, 2007; Villa et al. 2005; Brooks et al. 2006; Van Peer et al. 2008; Rots et al. 2011).

However, not all MP and MSA points were used as hunting weapons (Kuman 1989: 241, 286; Beyries and Plisson 1998; Bird et al. 2007; Schoville 2010; Iovita 2011). Studies indicating a cutting function for points show that it is invalid to assume function based on morphology alone. As discussed above, post-depositional processes also play a role in creating edge damage. Especially relevant to interpretations of MSA tool function, Pargeter (2011) showed that trampling results in fractures that would be categorically considered "diagnostic impact fractures", which are used by many researchers to identify use as hunting weapons. Thus, functional studies of points need to rule out taphonomic explanations for observed wear characteristics and consider multiple lines of evidence.

Edge Damage Distribution Method

The edge damage distribution method for assessing tool function quantifies the location of fractures along the lateral edges of lithic pieces, using low-power microscopy and GIS software (Bird et al. 2007; Schoville 2010; Schoville and Brown 2010; Wilkins et al. 2012). There are advantages to this type of analysis over other types of use wear analyses. First, the method can be applied to a wide range of weathering conditions and raw material types. The chemical alteration of lithic surfaces (patination) can also obscure some microscopic traces, especially polishing and striations (Keeley 1980; Levi Sala 1986). Patination on most of the KP1 points makes them unsuitable for microwear analysis. Also, for some areas and time periods, such as the African Stone Age, coarse lithic raw materials are common and create challenges for more traditional microwear methods (Rots et al. 2011: 637).

Second, some tools are used ephemerally and consequently, do not develop extensive wear. When this is the case, it can be difficult to determine function confidently for an individual artifact. The edge damage distribution method employed here pools data for a large sample of artifacts, making it possible to determine regions of the tool edges that were damaged more frequently. By aggregating data for a large sample of artifacts, we are able to observe assemblage-level patterns of edge damage formation.

Furthermore, the method acknowledges that taphonomic processes sometimes result in damage that mimics use-related damage. Minor damage and scarring along the

Fig. 7.2 Examples of KP1 complete retouched points. All banded ironstone except a, g, m, o, t, w, ad (black chert), c and v (quartzite)

edges of lithic tools can result from multiple processes. Post-depositional processes such as trampling and agitation of a flake in water and/or sediments can fracture tool edges (Tringham et al. 1974; Keeley 1980; Shea and Klenck 1993; Pargeter 2011). Some use wear analysts distinguish post-depositional fractures from use related fractures, based on the assertion that post-depositional scars are generally isolated or discontinuous, elongated, of variable size and direction, and randomly distributed along the edge (e.g., Tringham et al. 1974: 192; Odell and Odell-Vereecken 1980: 96; Grace 1989). However, blind tests show that qualitative assessments of use wear are prone to error; wear on unused flakes can be misidentified as use wear even on fine-grained unweathered raw materials (e.g., Odell and Odell-Vereecken 1980: 114; Shea and Klenck 1993). Using the edge damage distribution method, it is possible to statistically assess the probability that taphonomic processes explain observed patterns of edge damage on archaeological tools.

Kathu Pan 1

Numerous retouched and non-retouched points (Fig. 7.2) were recovered from ~500 thousand-year-old sediments at KP1, an open-air doline site located in the Northern Cape, South Africa (Fig. 7.1: Beaumont 1990, 2004; Porat et al. 2010; Wilkins and Chazan 2012; Wilkins et al. 2012). Points are the most striking and one of the most abundant categories of tools in the ~500 thousand-year-old KP1 assemblage. The majority of these points are manufactured on banded ironstone. They range in maximum length from 28 to 123 mm with a mean of 70 mm (n = 127, sd = 19.7). Retouch on points is usually unifacial on the dorsal surface of both lateral edges. Retouch sometimes is distributed across the entire lateral edge, but more often it is concentrated near the distal end. If there is retouch on the ventral side, it is minimal. The majority of the points were manufactured on blades extracted from Levallois-like prepared cores (Wilkins and Chazan 2012). Typologically and technologically, these points are similar to those common in MP and MSA assemblages that date to less than ~300 thousand-years-ago.

In this study, we directly compare the observed edge damage distribution on the KP1 points to multiple distribution models. The distribution models include proxies for taphonomic processes (post-patination scars on the KP1 points), behavioral processes (MSA cutting tools, experimental spear tips), and combinations of these processes. By statistically comparing these distribution models, we are able to assess the probability that the observed edge damage distribution on the KP1 points was the result of taphonomic processes alone, behavioral processes alone, or a combination of processes.

Methods

Generating the Distribution Models

Multiple competing models of edge damage distribution were generated to compare with the distribution of damage on the KP1 points (Table 7.1). These models are based on edge damage distribution results from three lithic samples: KP1 stratum 4a points and experimental spear tips (Wilkins

Table 7.1 Summary of generated edge damage distribution models

Model		Description
Post-depositional		Distribution of post-patination scars on the KP1 points (n = 107) that necessarily have a taphonomic origin
Spear tips		Distribution of scars on experimental points (n = 32) used as spear tips
Cutting tools		Distribution of scars on PP13B MSA points (n = 238) with inferred cutting function
Spear tips/post-depositional	75/25	75% contribution from above spear tip model and 25% contribution from above post-depositional model
	50/50	50% contribution from above spear tip model and 50% contribution from above post-depositional model
	25/75	25% contribution from above spear tip model and 75% contribution from above post-depositional model
Cutting tools/post-depositional	75/25	75% contribution from above cutting tool model and 25% contribution from above post-depositional model
	50/50	50% contribution from above cutting tool model and 50% contribution from above post-depositional model
	25/75	25% contribution from above cutting tool model and 75% contribution from above post-depositional model
Cutting tools/spear tips	75/25	75% contribution from above cutting tool model and 25% contribution from above spear tip model
	50/50	50% contribution from above cutting tool model and 50% contribution from above spear tip model
	25/75	25% contribution from above cutting tool model and 75% contribution from above spear tip model

et al. 2012; Wilkins 2013), and MSA points from Pinnacle Point 13B (Schoville 2010; Schoville and Brown 2010).

Samples

The three samples used to generate the distribution models are:

1. The sample of KP1 points (n = 107) consisted of all complete retouched points (n = 69), and complete non-retouched convergent flakes and blades (n = 38) from stratum 4a in squares F21, F23, C21, C23 (Fig. 7.2).
2. A sample of 32 experimental spear tips were used create experimental patterns of edge damage resulting from behavioral processes. Experimental retouched points and convergent flakes and blades similar to those recovered from KP1 were replicated by Kyle S. Brown using banded ironstone. Each point was hafted to a wooden dowel using a combination of Acacia karoo mastic and cow tendon. A calibrated crossbow designed after Shea et al. (2001) was used to deliver and control the draw force. Two springbok (Antidorcas *marsupialis*) carcasses culled from a nearby ranch served as the targets. Each surviving point was thrust until there was visible damage, which sometimes occurred after a single trial, and up to a maximum of nine trials (Wilkins et al. 2012, supplementary material)
3. All complete convergent points from the coastal MSA site of Pinnacle Point 13B cave (PP13B) (Marean et al. 2007; Marean 2010) were analyzed (n = 238). The majority of points were manufactured on quartzite, and retouch frequency is very low (Thompson et al. 2010). The points were described by Bird et al. (2007) and Schoville (2010) who both argued that the edge damage frequency and distribution exhibited patterns consistent with use as cutting tools, and inconsistent with taphonomic damage or use as weapon tips (Schoville and Brown 2010). Therefore, we use the distribution of edge damage from PP13B as a proxy for use as cutting tools.

Mapping Edge Damage

The dorsal and ventral side of each point was photographed on a 1 cm by 1 cm grid. The digital images were georeferenced in ESRI ArcGIS 10 using the grid as landmarks for the appropriate coordinates. A shape file was created for each point and used to trace the perimeters. Tracing started at the edge of the platform at the base of each lateral edge, so that the platform was excluded from the outline. While being traced in ArcGIS, point edges were observed for fractures. Only fractures visible to the naked eye were mapped, but low-power microscopy (10–50×) was used as an aid to confirm the presence and nature of the damage.

On the KP1 points, three types of damage were coded; "potential edge damage" (PED), post-patination, and retouch.

PED is used as the descriptor for damage of unknown origin following Bird et al. (2007). These are scars or snaps visible to the naked eye that occurred before patination (i.e., they are the same color as the rest of the surface of the tool). These represent potential use damage, but could also have resulted from post-depositional processes. Causation is not attributed to individual PED scars. PED is used roughly synonymous to large "microfractures" (Shea 1992).

Post-patination is a descriptor for a scar or snap that exposes "fresh" raw material that is a different color from the patinated surface of the tool. This kind of damage certainly occurred after deposition, and after enough time had passed to patinate the surface of the tool. KP1 has a complex depositional history, the details of which are currently under further investigation. Artifacts from stratum 4a are concentrated within spring vents (Porat et al. 2010), and therefore may have been subject to a variety of taphonomic processes including turbation and compaction. After excavation, artifacts were curated together in boxes and may have incurred storage and handling damage post-excavation. The post-patination damage on KP1 points likely reflects a combination of many processes – turbation, compaction, excavation, and "drawer damage". The patination of the KP1 points provides an advantage in this case. Because post-patination scars are easily identified on many points, the KP1 points provide an opportunity to determine the actual distribution of non-use related fractures. The post-patination distribution on the KP1 points can be used to test the null hypothesis that pre-patination fractures (i.e., PED) resulted mainly from post-depositional processes rather than use. We do not consider the distribution of post-patination damage described here as indicative of one specific process, but rather a palimpsest of taphonomic, non-use related damage to artifact edges.

Continuous zones of large and invasive patinated flake scars are identified as retouch. Retouch was coded so that point edges with retouch could be separated from point edges with no retouch. PED fractures were never identified within retouched zones. Because PED scars cannot confidently be identified within zones of retouch, retouch on edges will affect the resulting distributions of PED (i.e., PED scars will appear to be absent from zones with retouch only because they could not be identified). For that reason, when distributions of PED are examined, only non-modified (i.e., not retouched) point edges are included. For example, if a point has retouch on both dorsal laterals, then only the ventral edges are used to analyze frequencies and distributions of PED. It is however, still possible to distinguish post-patination scars within these zones of retouch.

Calculating Edge Damage Distributions

Once damage was digitized in ArcGIS, line lengths for each scar and each edge between scars were calculated and exported to Excel, which was used to calculate total edge length. Total edge length was then scaled to 100 to remove the effect of size and calculate the relative location of each scar with respect to the tip and base (Schoville 2010: 383–384). The resulting data matrix consists of each point edge (i.e., specimen number 3353, ventral, right) in rows and 100 locations in columns. Each location represents one percent of the total edge length. The presence or absence of each damage type is expressed as either "1" where there is damage, or "0" where there is no damage. The resulting output expresses the relative location of each "scar" as a percent of the total edge length. Data for each edge for all points was pooled to determine assemblage-level distribution patterns.

Edge damage distribution on a given edge is calculated as the frequency of damage at each location divided by the total frequency of damage for that edge (i.e., dorsal left, ventral right, etc.). For example, assume that there are 9 occurrences of PED at location 25 on the dorsal left edge of the KP1 points and 904 total PED locations on unmodified dorsal left edges. The distribution is expressed as 9/904 (0.99%); or in other words, of all the PED scars that occur on the dorsal left edge, 0.99% occur at location 25 (25% up from the platform toward the tip). This distribution is depicted as a vertical line graph where one line represents each edge – the dorsal left, dorsal right, ventral left, and ventral right edges (e.g., Fig. 7.3). The side designations are based on the viewer's perspective of the point with the platform down and the tip up. When the dorsal side is up, the left edge is on the left, and likewise, when ventral side is up, the left edge is on the left. The y-axis shows the relative location on the point edge, with 0 representing the base (i.e., where the edge contacts the platform) and 100 representing the tip. The x-axis shows the relative frequency of damage with the left edge depicted on the left side of the graph and the right edge depicted on the right side of the graph.

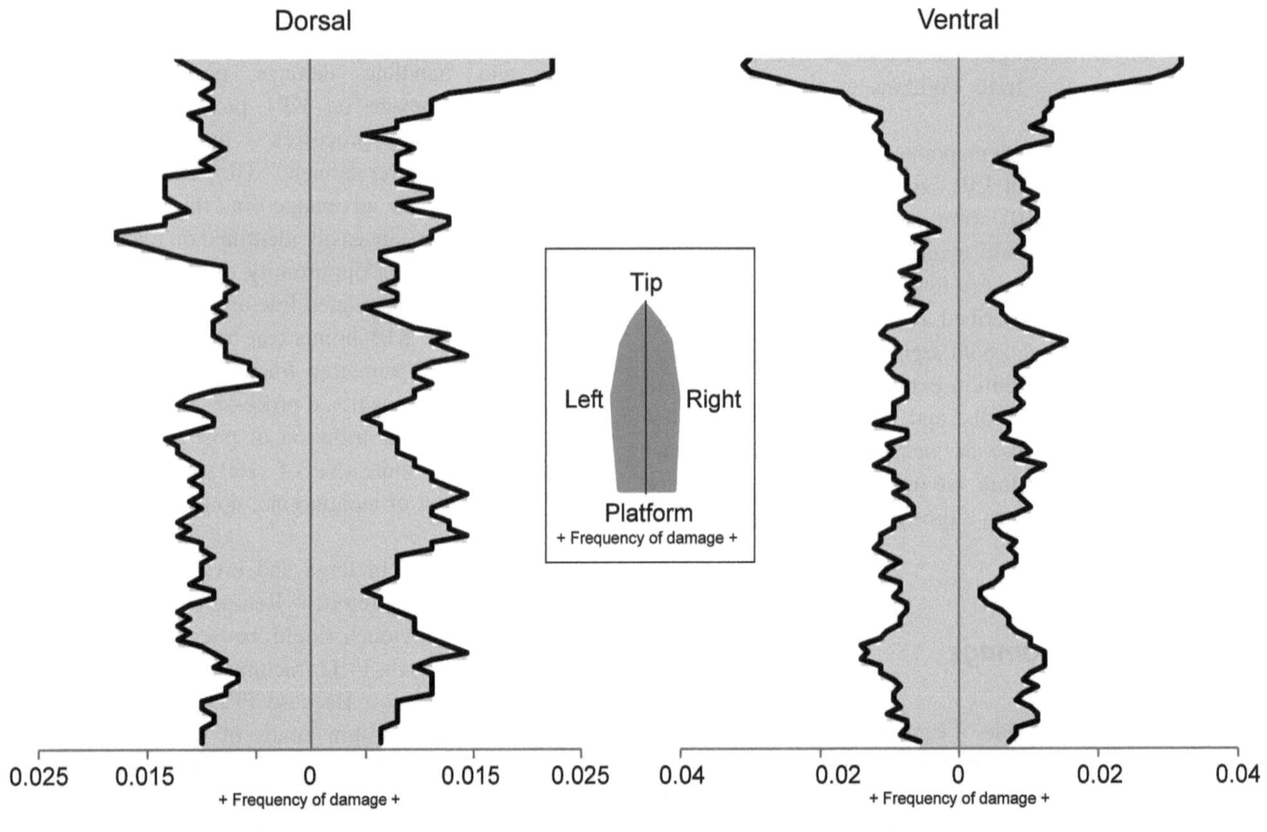

Fig. 7.3 KP1 PED distribution. The percent frequency of damage is given on the x-axis for each location along the y-axis. The bottom of the y-axis represents the base of the point and the top represents the tip. The distribution for the left edge is given on the left side of the y-axis and the distribution for the right edge is given on the right side of the y-axis

The samples we analyzed permit us to characterize and compare multiple models of edge damage distribution (Table 7.1). These models include a proxy for taphonomic processes, as evidenced by the distribution of post-patination scars on the KP1 points. The distribution of scars on the experimental spear tips provides one model for edge damage distribution resulting from behavioral processes (i.e., use as weapon armatures), and the distribution of scars on the MSA points from PP13B provide another (i.e., use as cutting tools).

Unlike the experimental spear tips, the cutting function of the PP13B tools is inferred. However, the interpretation of the PP13B points as cutting tools is robust. Taphonomic processes do not explain the observed distribution on the PP13B points (Schoville 2010). It is clear that the majority of PP13B points were not used as spear tips, because there is less damage near the tip of the points than along the edges and very low frequencies of impact fractures (Schoville 2010). There is a strong foundation based on previous experimental work for interpreting these points as cutting tools; tools used in a longitudinal motion for cutting are damaged along the utilized edge on both the dorsal and ventral face (Tringham et al. 1974), as was observed on the PP13B points (Schoville 2010). Admittedly, some percentage of the damage may be taphonomic in origin, however there does not appear to be any correlation between post-depositional disturbance and edge damage frequency that would be anticipated from taphonomic damage formation (Schoville 2014).

Combining Distributions

It is unlikely for artifacts to be subjected to a single process of edge damage formation. In order to generate models that stand as proxies for multiple processes, we combined the taphonomic and behavioral distributions from the models above. This was accomplished by taking a weighted mean of each model for the frequency of damage at each location. For example, a 75% spear tip/25% post-depositional model was generated by multiplying the frequency of damage on experimental spear tips at each location by 0.75 and the frequency of post-patination damage by 0.25 and summing the values together. The resulting distribution represents expectations for the combined effect of taphonomic and behavioral processes, with relatively more input from behavioral processes. Three types of combined distributions were generated. The combined experimental spear tip and post-patination distributions (75:25, 50:50, 25:75) were used as proxies for the combined effect of use as spear tips and taphonomic processes. The combined PP13B and post-patination distributions (75:25, 50:50, 25:75) were used as proxies for the combined effect of use as cutting tools and taphonomic processes. The combined PP13B and experimental spear tip distributions (75:25, 50:50, 25:75) were used as proxies for the combined effect of use as cutting tools and use as spear tips.

Comparing the Distribution Models

To compare the competing edge damage distribution models, the data were transformed into cumulative distributions and following Schoville (2010), subjected to the Kolmogorov-Smirnov (KS) test. The KS test is non-parametric and commonly used in archaeology to compare the cumulative distributions of two samples to determine whether they may have been drawn from populations with the same distributions (Shennan 1997).

Table 7.2 Sample sizes and frequency of damage

		KP1 PED	Post-patination	Experimental spear tips	PP13B MSA points
Number of points		107	107	32	238
Number of observable edges	Dorsal left	52	107	2	238
	Dorsal right	47	107	5	238
	Ventral right	100	107	29	238
	Ventral left	101	107	30	238
	Total	300	428	66	952
Number of scarred locations	Dorsal left	904	1185	6	1538
	Dorsal right	625	1223	70	1153
	Ventral right	971	1152	314	680
	Ventral left	1065	1028	404	767
	Total	3565	4588	794	4138
Scars per edge	Dorsal left	17.4	11.1	3.0	6.5
	Dorsal right	13.3	11.4	14.0	4.8
	Ventral right	9.7	10.8	10.8	2.9
	Ventral left	10.5	9.6	13.5	3.2
	Total	11.9	10.7	12.0	4.3

Two values are compared in the KS test. D_{max} is $1.36\sqrt{[(n_1 + n_2)/n_1 n_2]}$, and is the minimum difference between two cumulative distributions that will be significant at $p = 0.05$. D_{max} is dependent on sample size (n), which in this case, is the total number of scarred locations on the edge of interest (reported in Table 7.2). D_{obs} is the maximum observed difference between two cumulative distributions. If D_{obs} is greater than D_{max}, than there is less than 5% probability that the two distributions are drawn from the same population.

For the combined distributions, n is the smaller of the sample sizes used to generate the combined model, so that distributions based on a smaller number of scars are not artificially given more statistical power.

Results

KP1 PED Distribution

Figure 7.3 shows the distribution of PED on the KP1 points, based on a total of 3565 scarred locations on 300 unretouched edges of 107 points (Table 7.2). While the frequency of damage along the dorsal edges is relatively consistent, especially the dorsal left edge, the frequency of damage on the ventral edges is much greater near the tip. On the ventral surface, there is a roughly three-fold increase in damage frequency starting at the last 10% of each ventral edge. The dorsal right edge does exhibit a slight increase at the tip, but not to the same degree as the ventral edges (Fig. 7.3).

Based on the KS tests, the distribution of damage on the dorsal left and dorsal right edges do not significantly differ from each other, nor do the distributions between the ventral left and ventral right edges (Table 7.3).

Post-depositional Model

The first step in any archaeological analysis is to rule out taphonomic processes as the sole explanation for observed features and patterns (Schiffer 1987; Dibble et al. 2006). Our samples provide one proxy for taphonomic processes – the post-patination scars on the KP1 points – that we can characterize and compare to archaeological patterns of edge damage.

Post-patination Damage on the KP1 Points

On the KP1 points, the post-patination damage is distributed relatively evenly across the edges (Fig. 7.4). While there is a very slight increase in damage at the tip on some of edges

Table 7.3 KS results of intra-sample comparisons of edge damage distribution

Model		dorsal left vs. dorsal right	ventral right vs. ventral left
KP1 PED		D_{obs}=0.062, D_{max}=0.071	D_{obs}=0.0600, D_{max}=0.0603
Post-depositional		D_{obs}=0.056, D_{max}=0.055*	D_{obs}=0.132, D_{max}=0.058*
Spear tips		D_{obs}=0.552, D_{max}=0.579	D_{obs}=0.078, D_{max}=0.102
Cutting tools		D_{obs}=0.095, D_{max}=0.053*	D_{obs}=0.115, D_{max}=0.072*
Spear tips/post-depositional	75/25	D_{obs}=0.425, D_{max}=0.579	D_{obs}=0.088, D_{max}=0.102
	50/50	D_{obs}=0.298, D_{max}=0.579	D_{obs}=0.1015, D_{max}=0.1023
	25/75	D_{obs}=0.170, D_{max}=0.579	D_{obs}=0.117, D_{max}=0.102*
Cutting tools/post-depositional	75/25	D_{obs}=0.078, D_{max}=0.056*	D_{obs}=0.068, D_{max}=0.072
	50/50	D_{obs}=0.070, D_{max}=0.056*	D_{obs}=0.062, D_{max}=0.072
	25/75	D_{obs}=0.063, D_{max}=0.056*	D_{obs}=0.090, D_{max}=0.072*
Cutting tools/spear tips	75/25	D_{obs}=0.110, D_{max}=0.579	D_{obs}=0.077, D_{max}=0.102
	50/50	D_{obs}=0.258, D_{max}=0.579	D_{obs}=0.054, D_{max}=0.102
	25/75	D_{obs}=0.405, D_{max}=0.579	D_{obs}=0.065, D_{max}=0.102

* = significantly different ($D_{obs} > D_{max}$), $p < 0.05$

Dark grey cells = low statistical power because of small sample size
Light grey cells = similar to KP1 PED

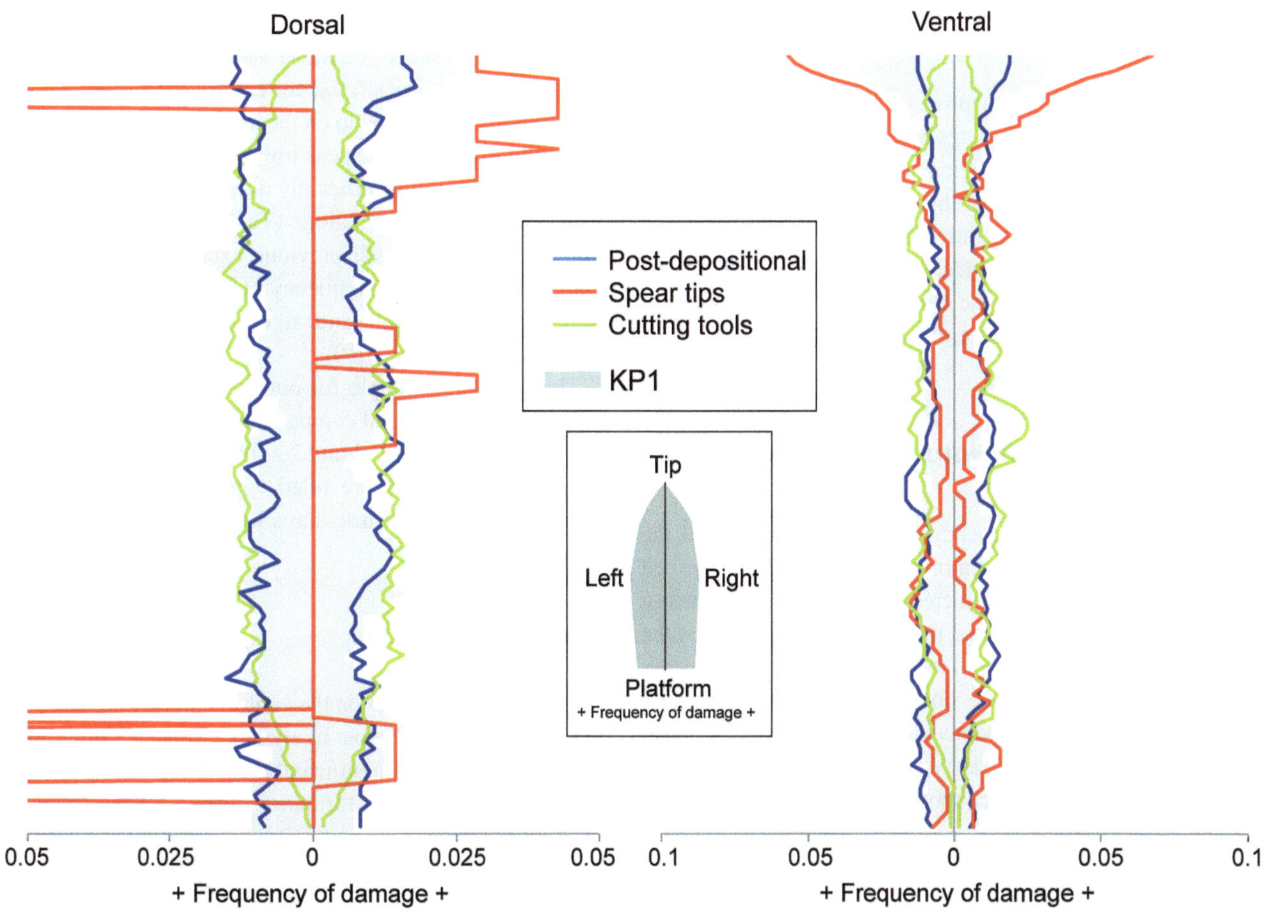

Fig. 7.4 Edge damage distribution models. The percent frequency of damage is given on the x-axis for each location along the y-axis. On the y-axis, zero represents the base of the point and 100 represents the tip. For comparison, the KP1 PED distributions for the equivalent edges are depicted with grey shading

(ventral right and dorsal right), the increase is even less pronounced on other edges (dorsal left and ventral left).

The dorsal left and dorsal right edges have significantly different distributions from each other, as do the ventral left and ventral right edges (Table 7.3).

Comparison with KP1

Visually, the dorsal KP1 PED distributions exhibit similarities with the dorsal post-patination distribution (Fig. 7.4). In contrast, the ventral KP1 PED distributions differ from the post-patination distributions, showing a much higher concentration of damage at the tip (Fig. 7.4).

Unlike the KP1 PED dorsal left and dorsal right distributions, the post-patination dorsal left and dorsal right distributions are significantly different from each other (Table 7.3). The same is true for the ventral left and ventral right distributions (Table 7.3).

To statistically evaluate the probability that the PED distribution resulted from post-depositional processes, we directly compared the PED and post-patination distributions using KS tests. The KP1 PED distributions for each edge were compared to the post-patination distribution for the equivalent edge (i.e., the PED distribution on the ventral left edge was compared to the post-patination distribution on the ventral left edge). Comparisons were done between equivalent edges, because each edge could respond slightly differently to each process.

KS tests indicate that the dorsal left and right PED distributions are not significantly different from the dorsal left and right post-patination distributions (Table 7.3). The ventral left and right PED distributions are significantly different from the ventral left and right post-patination distributions (Table 7.3). The taphonomic model could explain the dorsal KP1 PED distribution, but it does not explain the ventral PED distribution. It is unlikely that the distribution of damage on the ventral surface results solely from post-depositional processes, because the distribution of PED differs significantly from the distribution of post-patination damage.

Behavioral Models

After ruling out taphonomic processes as the sole explanation for observed patterns of damage on an archaeological sample, one can compare archaeological distributions to behavioral models. For the KP1 points, the dorsal distributions are not statistically different from our post-depositional model, but the ventral distributions are. Here we consider two behavioral models for understanding the ventral KP1 PED distribution; use as spear tips and use as cutting tools.

Damage on Experimental Spear Tips

There is an increased frequency of damage at the tip of the experimental points on the ventral surface, compared to the rest of point edge (Fig. 7.4). This increase begins at roughly the last 15–20% of the point edge and there is roughly five times the damage at the tip compared to the point edges (Fig. 7.4). Our experimental sample for the dorsal edges is small, because nearly all of the replicate points were retouched on the dorsal side and retouched edges were excluded from calculations of edge damage distribution (see above). The small sample size for the dorsal left (n = 6) results in an uninformative distribution. The dorsal right side has a larger sample of scarred locations (n = 70) than the dorsal left and one can see that the frequency of damage is greatest at the tip on the dorsal right side (Fig. 7.4).

On the experimental spear tips, the ventral left and right distributions are not significantly different based on a KS test of the cumulative distributions (Table 7.3). These observations are consistent with previous experimental spear tips, which show increased frequency at the tip and similar distributions between left and right sides (Schoville 2010; Schoville and Brown 2010).

The dorsal left sample for our experimental spear tips is too small for meaningful comparisons. The results of all tests are reported in Tables 7.3 and 7.4, but the cells involving the dorsal left distribution are filled with dark grey to indicate that in those cases the tests have low statistical power.

Comparison with KP1

The distribution of PED on the ventral unmodified edges can be used to choose the most likely model for KP1 point function, because it differs significantly from the post-depositional model for edge damage formation. While we report all the

Table 7.4 KS results of comparisons of KP1 PED to each of the edge damage distribution models (inter-sample comparisons)

		dorsal left	dorsal right	ventral right	ventral left
Post-depositional		D_{obs}=0.040, D_{max}=0.060	D_{obs}=0.040, D_{max}=0.067	D_{obs}=0.083, D_{max}=0.059*	D_{obs}=0.130, D_{max}=0.059*
Spear tips		D_{obs}=0.535, D_{max}=0.557	D_{obs}=0.388, D_{max}=0.171*	D_{obs}=0.193, D_{max}=0.088*	D_{obs}=0.198, D_{max}=0.079*
Cutting tools		D_{obs}=0.091, D_{max}=0.057*	D_{obs}=0.137, D_{max}=0.068*	D_{obs}=0.200, D_{max}=0.068*	D_{obs}=0.142, D_{max}=0.064*
Spear tips/post-depositional	75/25	D_{obs}=0.404, D_{max}=0.557	D_{obs}=0.289, D_{max}=0.171*	D_{obs}=0.128, D_{max}=0.088*	D_{obs}=0.118, D_{max}=0.079*
	50/50	D_{obs}=0.273, D_{max}=0.557	D_{obs}=0.189, D_{max}=0.171*	D_{obs}=0.064, D_{max}=0.088	D_{obs}=0.037 D_{max}=0.079
	25/75	D_{obs}=0.142, D_{max}=0.557	D_{obs}=0.096, D_{max}=0.171	D_{obs}=0.068, D_{max}=0.088	D_{obs}=0.051 D_{max}=0.079
Cutting tools/post-depositional	75/25	D_{obs}=0.067, D_{max}=0.060*	D_{obs}=0.113, D_{max}=0.068*	D_{obs}=0.169, D_{max}=0.068*	D_{obs}=0.136, D_{max}=0.064*
	50/50	D_{obs}=0.050, D_{max}=0.060	D_{obs}=0.089, D_{max}=0.068*	D_{obs}=0.139, D_{max}=0.068*	D_{obs}=0.129, D_{max}=0.064*
	25/75	D_{obs}=0.044, D_{max}=0.060	D_{obs}=0.064, D_{max}=0.068	D_{obs}=0.111, D_{max}=0.068*	D_{obs}=0.123, D_{max}=0.064*
Cutting tools/spear tips	75/25	D_{obs}=0.077, D_{max}=0.557	D_{obs}=0.058, D_{max}=0.171	D_{obs}=0.103, D_{max}=0.088*	D_{obs}=0.100, D_{max}=0.079*
	50/50	D_{obs}=0.230, D_{max}=0.557	D_{obs}=0.141, D_{max}=0.171	D_{obs}=0.055, D_{max}=0.088	D_{obs}=0.105, D_{max}=0.079*
	25/75	D_{obs}=0.382, D_{max}=0.557	D_{obs}=0.260, D_{max}=0.171*	D_{obs}=0.107, D_{max}=0.088*	D_{obs}=0.128, D_{max}=0.079*

* = significantly different ($D_{obs} > D_{max}$), $p < 0.05$

Dark grey cells = low statistical power because of small sample size

Light grey cells highlight comparisons that are NOT statistically different from KP1 PED distribution

results in Fig. 7.4; Tables 7.3 and 7.4 only the ventral distributions are discussed further to assess tool function, since we have found that the dorsal distributions can be explained by taphonomic processes.

Both the KP1 PED and spear tip ventral distributions show increases in damage at the tip. The relative increase in damage is greater for the experimental spear tips than it is for the KP1 PED. The intra-sample comparisons also give similar results. For both the KP1 PED distributions and the experimental spear tip distributions, the ventral left and right sides are not significantly different (Table 7.3).

Based on KS tests directly comparing KP1 PED and experimental spear tips (inter-sample comparisons), the ventral left distributions are significantly different (Table 7.4). Likewise, the ventral right distributions are significantly different (Table 7.4). These differences indicate that it is unlikely that the use as spear tips alone explains the distribution of PED on the ventral surface of the KP1 points, even though the similarities in the shape of the distribution are suggestive of a spear tip function.

Damage on PP13B MSA Cutting Tools

The PP13B points demonstrate higher frequencies of damage in the mid-section of the edge, with lower frequencies near the base and tip (Fig. 7.4). This pattern is true for both the dorsal and ventral edges (Fig. 7.4).

Intra-sample KS tests indicate that the dorsal left and dorsal right distributions differ significantly on the PP13B MSA cutting tools, as do the ventral left and ventral right distributions (Table 7.3).

Comparison with KP1

Based on the PP13B model for cutting tools, if the KP1 points were used mainly as cutting tools and not as spear tips, we would expect low frequencies of damage near the tip of the point and high frequencies of damage along the edges. We would also expect significantly different distributions between the ventral left and ventral right edges. Unlike the PP13B cutting tools, PED on the KP1 points shows relatively low frequencies of damage along the edges of the points and high frequencies at the tip (Fig. 7.4), and the ventral left and ventral right distributions on the KP1 points are not significantly different (Table 7.3).

Based on inter-sample KS tests, the KP1 PED and PP13B cutting tool ventral left distributions are significantly different from eachother (Table 7.4). Likewise, the ventral right distributions are significantly different (Table 7.4). These differences indicate that it is unlikely that the use as cutting tools alone explains the distribution of PED on the ventral surface of the KP1 points.

Combination Models

The KP1 PED distribution is compared to a total of nine combination models, each representing various degrees of input from behavioral and taphonomic processes. Again, the focus here will be on the ventral distributions, because it was found that dorsal distribution could be explained by post-depositional processes alone, though the results for the dorsal distributions are reported in Table 7.4 and Fig. 7.5.

Combination Spear Tip/Post-depositional Model

The combination spear tip/post-patination distribution on the ventral surface exhibits an increased frequency of damage at the tip of the points with an even distribution across the remainder of the edge (Fig. 7.5a). Comparing the three spear tip/post-depositional models to each other demonstrates that as the input from taphonomic processes increases (from 25%, to 50%, to 75%) and input from use as spear tip decreases (from 75%, to 50%, to 25%), the relative frequency of damage at the tip compared to the edge decreases. However, even with only a 25% contribution from spear tips, there is a marked increase in damage at the tip.

Intra-sample KS tests indicate that the distributions between the ventral left and ventral right edges are not significantly different for the 75/25 and 50/50 models (Table 7.3). They are significantly different for the 25/75 model (Table 7.3). This finding fits our expectations because the ventral left and ventral right distributions are significantly different for the post-depositional model but not for the spear tip model. As input from the spear tip distribution decreases and input from the post-depositional distribution increases, the relative influence of the post-depositional pattern increases and the ventral left and ventral right distributions become significantly different.

Comparison with KP1

The shape of the KP1 ventral PED distributions are similar to the spear tip/post-depositional models, with increased damage at the tip and similar frequencies of damage across the remainder of the edge. The observed frequency of PED on the KP1 points is most similar to the 50/50 and 25/75 spear

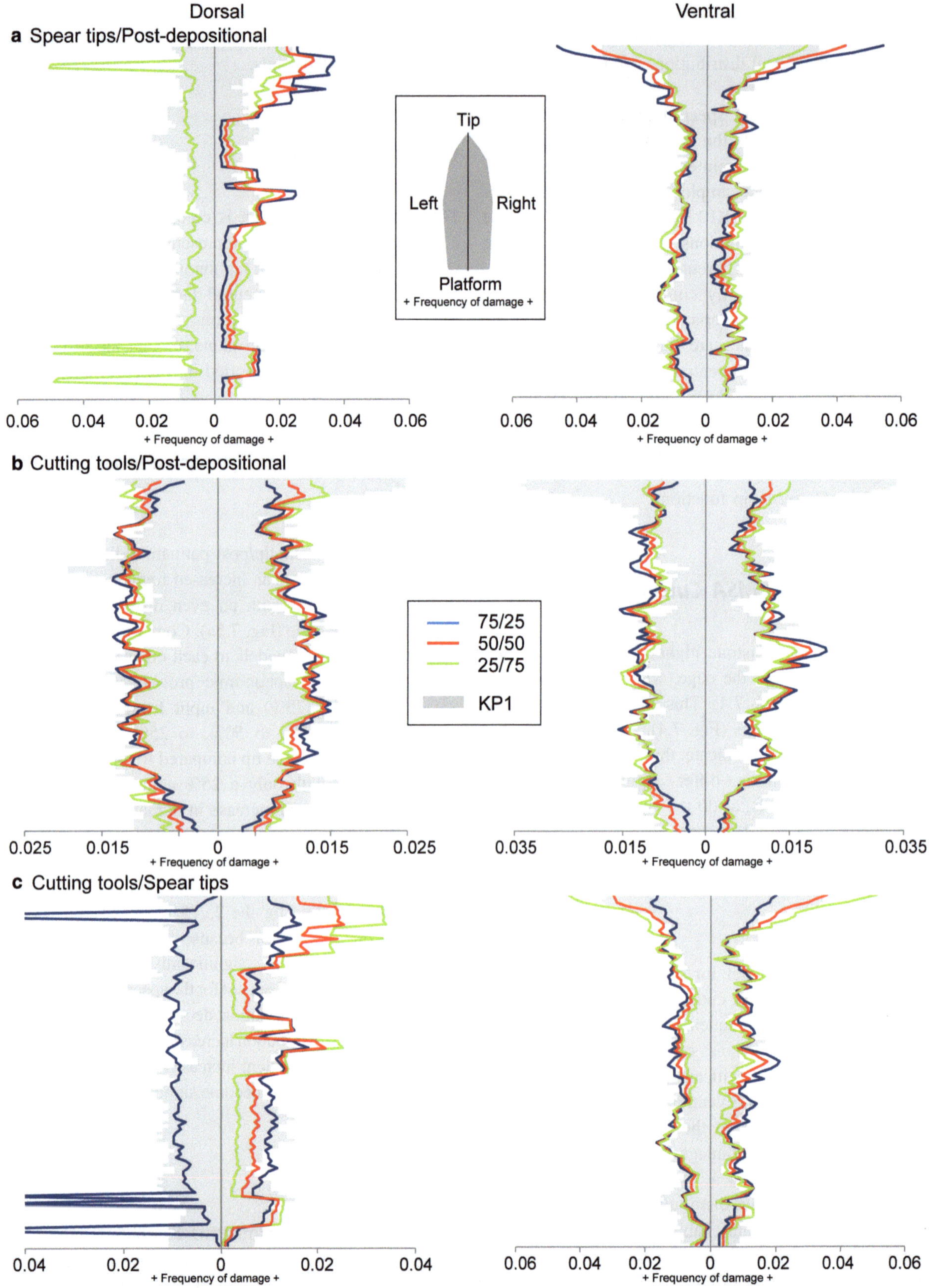

Fig. 7.5 Combination edge damage distribution models. **a** Combination spear tip/post-depositional model, **b** combination cutting tool/post-depositional model, **c** combination cutting tool/spear tip model. The distribution for the left edge is given on the left side of the y-axis and the distribution for the right edge is given on the right side of the y-axis. For comparison, the KP1 PED distributions for the equivalent edges are depicted with grey shading. Some distributions on the dorsal left side are excluded because of small sample sizes

tip/post-depositional combination model, with the frequency of PED at location 100 on the KP1 points falling between the frequencies predicted by these two models (Fig. 7.5a).

With respect to intra-sample KS test comparisons of edge damage distribution, the KP1 PED distributions are most similar to the 50/50 and 75/25 spear-tip/post-depositional models. The intra-sample ventral right and ventral left distributions are not significantly different (Table 7.3).

KS tests directly comparing the KP1 PED ventral left distribution to each of the spear tip/post-depositional models show that the 75/25 model is significantly different from KP1 PED, but the 50/50 and 25/75 models are not significantly different (Table 7.4). The same is true for the ventral right distribution (Table 7.4). In other words, the KP1 ventral PED distributions do not differ significantly from the 50/50 and 25/75 spear tip/post-depositional distributions and these models provide a reasonable explanation for the observed KP1 ventral PED distributions. The 50/50 model currently stands as the best model, because of the similarities in the intra-sample edge comparisons.

Combination Cutting Tool/Post-depositional Models

Edge damage for the combination cutting tool/post-patination distributions is not concentrated at the tips of the points, but is distributed along the entire edge, with the highest frequencies near the mid-section (Fig. 7.5b). The ventral right edge shows a slightly more pronounced increase in edge damage near the mid-section compared to the ventral left. This increase becomes less pronounced as input from the post-depositional distribution increases.

Intra-sample KS tests show that the ventral left and ventral right distributions are not significantly different for the 75/25 and 50/50 cutting tool/post-depositional combination models (Table 7.3). This finding is the opposite for the pure cutting tool and post-depositional models (Table 7.3). The maximum difference between the left and right sides occurs at different locations for the two models, so that when they are combined, the maximum difference decreases. When the input from the post-depositional distribution increases to 75%, the ventral left and right distributions are significantly different again (Table 7.3). Similar reversal patterns are observed for the dorsal and ventral sides of the left and right edges (Table 7.3).

Comparison with KP1

Visually, the KP1 ventral PED distribution differs from the cutting tool/post-depositional combination models. None of the combination models demonstrate an increased frequency of damage at the tip. Rather, the highest frequency is near the mid-section of the point edge.

Intra-sample comparisons show similarities; the KP1 ventral left and right PED distributions are not significantly different, nor are the cutting tool/post-depositional ventral left and right distributions (Table 7.3).

Statistically, the KP1 ventral PED distributions are significantly different from all the cutting tool/post-depositional models based on KS tests (Table 7.4). The cutting tool/post-depositional models do not provide a good explanation for the KP1 ventral PED distribution.

Combination Cutting Tool/Spear Tip Models

For the combination cutting tool/spear tip models, there is an increased frequency of damage at the tip of the points on the ventral surface (Fig. 7.5c). Comparing the three cutting tool/spear tip models to each other demonstrates that as the input from use as a cutting tool decreases and input from use as spear tip increases, the relative frequency of damage at the tip compared to the edges increases. In those ways, the combination cutting tool/spear tip models are similar to the combination spear tip/post-depositional models.

However, the spear tip/cutting tool models differ from the spear tip/post-depositional models with respect to how damage is distributed between the base and the mid-section. For the cutting tool/spear tip models, there is a relatively lower frequency of damage at the base, and a relatively higher frequency in the mid-section.

Intra-sample KS tests show that for all the cutting tool/spear tip models, there is no significant difference between the ventral left and ventral right edges (Table 7.3).

Comparison with KP1

Near the tip of the point, there are visual similarities between the KP1 ventral PED distributions and the cutting tool/spear tip combination models (Fig. 7.5c). The 50/50 and 25/75 models show marked increases at the tip like the KP1 ventral PED distributions. Near the base of the point, however, the KP1 PED frequencies are much higher relative to the cutting tool/spear tip models. The KP1 ventral PED does not show an increase in damage frequency at the mid-section compared to the base of the point like the cutting tool/spear tip combination models.

The intra-sample edge comparisons are similar between the KP1 ventral PED distributions and the cutting tool/spear tip combination models; there are no significant differences between the ventral left and ventral right edges (Table 7.3).

Inter-sample comparisons of edge damage show that most of the KP1 ventral PED distributions are significantly different from the cutting tool/spear tips combination models. However, the ventral right distribution for 50/50 model is not significantly different from KP1 PED (Table 7.4).

Discussion

The PED fractures on the KP1 points occurred prior to patination, but each individual scar could have resulted from taphonomic or behavioral processes. However, even if each individual fracture has unknown causation, assemblage-level distribution patterns permit, first, an evaluation of whether the cause of the distribution pattern could be related to taphonomic or behavioral processes, and second, an evaluation of which processes or combination of processes best explain the non-taphonomic distribution patterns.

The dorsal distributions of PED on the KP1 points do not differ significantly from our post-depositional model, but the ventral distributions do. Taphonomic processes appear to be a major contributor to the observed pattern on the dorsal surface, and the dorsal surface cannot be used to assess tool function. In contrast, non-taphonomic processes played some role in the observed edge damage distribution on the ventral surface of the KP1 points and can be used to assess tool function.

The KP1 ventral PED distributions were compared to multiple behavioral and combined behavioral/taphonomic models. Alone, neither the spear tip only model nor the cutting tool model adequately explains the KP1 ventral distribution. KS tests demonstrate that the KP1 ventral PED distributions differ significantly from these models. Likewise, the KP1 ventral PED distributions are significantly different from combination cutting tool/post-depositional models.

The majority of the cutting tool/spear tip combination models differ significantly from the KP1 ventral PED distribution. Only the ventral right edge in the 50/50 model is not significantly different.

There are no significant differences between the KP1 ventral distributions and the 50/50 and 25/75 spear tip/post-depositional combination models, and these models provide a robust explanation for the KP1 ventral PED distributions. The 50/50 spear tip/post-depositional model is also similar to the KP1 ventral PED distributions with respect to intra-sample edge comparisons; the ventral left and ventral right edges are not significantly different from each other. Thus, out of the tested models, the 50/50 spear tip/post-depositional model stands as the best candidate for explaining edge damage distribution patterns on the ventral surface of the KP1 points.

The results presented here best support a spear tip function for the KP1 points. KP1 points were subjected to ~500-thousand-years of taphonomic processes that obscured the behavioral signal on the dorsal surface, but the behavioral signal on the ventral surface was not completely obliterated. On the ventral surface, the behavioral signal was strong enough for functional interpretations even with taphonomic effects. Our analysis modeled expectations for the combined effect of multiple processes and found that the ventral surface of the KP1 points best fits a combined spear tip and post-depositional model.

There is a potential explanation for why the dorsal edges were affected more by post-depositional processes than the ventral edges. More damage may occur on the ventral surface of experimental spear tips (Table 7.2), especially near the tips, suggesting that the ventral spear-tip signal may be stronger to begin with. However, given the small sample size of dorsal edges on the experimental spear tips more research should be done to confirm this interpretation.

This study further demonstrates the value of the edge damage distribution method for addressing functional questions. Identifying individual tool function is the goal of most traditional functional analyses (Semenov 1964), but the statistical confidence of functional attributions is rarely addressed (c.f., Nance 1979). Appeals are often made to visual similarities between an experimental wear trace and an archaeological wear trace on individual tools, despite the widespread acknowledgment that especially with coarse grained raw materials, equifinality is a major concern and blind tests show that the probability of misinterpreting an individual tool can be quite high (Odell and Odell-Vereecken 1980; Newcomer et al. 1986; Bamforth 1988; Shea and Klenck 1993; Crowther and Haslam 2007; Wadley and Lombard 2007). The edge damage distribution method pools data so that we can assess the probability that two observed wear patterns are equivalent. Our analysis of the KP1 points identified distribution wear patterns without interpreting the cause of individual wear features, and statistically compared the archaeological patterns of wear to expected patterns given different processes and combinations of processes. We can state with statistical confidence that the edge damage distribution on the KP1 points is not solely explained by taphonomic processes, but is most likely due to a combination of taphonomic and behavioral processes.

Furthermore, because this approach pools assemblage-level data, it has the ability to draw out patterns that would otherwise be invisible on an individual tool. Our ability to confidently ascribe behavioral inferences from use wear is related to effect intensity and sample size (Sokal and Rohlf 1995). Inferences based on individual tools will be more prone to error compared to assemblage distributions of wear. Raw material durability also influences the probability of wear trace formation. Therefore, individual tool edges might not reflect the whole suite of diagnostic damage traces needed to classify it to behavioral function. An assemblage approach acknowledges these issues by expanding the unit of analysis to an entire assemblage of artifact edges. With respect to the KP1 points, the ventral surface may have been more intensely affected by behavioral processes than the dorsal surface, resulting in an observable behavioral signal. This behavioral signal is largely undetectable on individual tools, but when pooled, there is a clear pattern differentiating ventral distributions on the KP1 points from post-depositional and cutting tool models of edge damage distribution.

The edge damage distribution data supporting a spear tip function for the KP1 points does not stand alone. Several lines of evidence – "diagnostic impact fractures", basal modifications, metrics, and shape analysis – further support the hypothesis that the KP1 points were used as spear tips (Wilkins et al. 2012). A relatively high frequency of DIFs on the KP1 points also indicate that point tips were subjected to the kind of strong longitudinal forces that experimental hunting weapons experience. The frequency at KP1 is higher than expected for a taphonomic explanation (Wilkins et al. 2012: Fig. 2b). The KP1 points exhibit basal modifications suggestive of hafting. The size and shape of the KP1 points indicate that they could feasibly function as spear tips. The smaller KP1 points are as symmetrical as the larger KP1 points, which supports the spear tip hypothesis. For cutting tools, the expectation is for the smaller points to be asymmetrical compared to the larger points, because they represent later stages of resharpening (Iovita 2011). In that case, reduction of the point edges would be maximizing edge length (important for cutting) rather than symmetry (important for spear tips). While we have showed that some damage on some points was caused by post-depositional processes, the assemblage-level pattern is one that supports the use of KP1 points as spear tips. Points from MSA and MP sites were often used as weapon tips, and evidence for this behavior dates back to ~500 ka. The evidence for hafted hunting technologies ~500-thousand years ago at KP1 is currently the oldest known evidence of its kind to date, but is not unexpected, given that both Neandertals and MSA hominins appear to have used hafted hunting technology. The scarcity of chronometrically dated sites for the early Middle Pleistocene and limits imposed by the kinds of methods used to confidently assess point function may explain the lack of prior evidence.

Conclusion

The edge damage distribution method employed here does not depend on subjective evaluations of individual scars and wear features and quantitatively assesses the effect of taphonomic and behavioral processes on stone tips. It is a powerful tool for functional analyses of lithic tools, because it has the capacity to rule out taphonomic processes as the best explanation for observed wear patterns and different hypotheses about tool function can be statistically tested. We generated models representing the combined effect of different taphonomic and behavioral processes, and these models are likely to represent a more realistic scenario for wear-feature formation than any individual process alone. Our results showed that ~500-thousand year old KP1 points best fit a combined spear tip/post-depositional model and the spear tip function is supported by several additional lines of evidence. This result has important implications for understanding human evolution prior to the split in Neanderthal and modern human lineages. The edge damage distribution method represents one strategy for improving the scientific rigor of stone tool functional analysis (sensu Hutchings 2016). We think Stone Age weaponry studies can benefit from an application of this method, especially in conjunction with other types of functional analyses.

Acknowledgements We thank Kyle S. Brown and Michael Chazan. This research was supported by a Social Sciences and Humanities Research Council Joseph-Armand Bombardier Canada Graduate Scholarship and funding from the University of Toronto to J.W. The McGregor Museum and the South African Heritage Resources Agency permitted temporary export of the points for study. Fieldwork at KP1 was supported by funding from the Social Sciences and Humanities Research Council of Canada to Michael Chazan. B.J.S. was funded by grants from NSF (BCS-0524087 and BCS-1138073) to Curtis Marean, the Hyde Family Foundations, and from the Institute of Human Origins at Arizona State University. We also thank D. Morris, N. Porat, L.K. Horwitz, C. Marean, and P. Beaumont. Any errors or inconsistencies are the sole responsibility of the authors.

References

Ambrose, S. H. (2010). Coevolution of composite-tool technology, constructive memory, and language: Implications for the evolution of modern human behavior. *Current Anthropology, 51*, S135–S147.

Bamforth, D. B. (1988). Investigating microwear polishes with blind tests: The Institute results in context. *Journal of Archaeological Science, 15*, 11–23.

Beaumont, P. B. (1990). Kathu Pan. In P. B. Beaumont & D. Morris (Eds.), *Guide to archaeological sites in the Northern Cape* (pp. 75–100). Kimberley: McGregor Museum.

Beaumont, P. B. (2004). Kathu Pan and Kathu Townlands/Uitkoms. In D. Morris & P. B. Beaumont (Eds.), *Archaeology in the Northern Cape: Some key sites* (pp. 50–52). Kimberley: McGregor Museum.

Beyries, S., & Plisson, H. (1998). Pointes ou outils triangulaires? Données fonctionnelles dans le Moustérien levantin. *Paléorient, 24*(1), 5–24.

Bird, C., Minichillo, T., & Marean, C. W. (2007). Edge damage distribution at the assemblage level on Middle Stone Age lithics: An image-based GIS approach. *Journal of Archaeological Science, 34*, 771–780.

Boëda, E., Geneste, J. M., Griggo, C., Mercier, N., Muhesen, S., Reyss, J., et al. (1999). A Levallois point embedded in the vertebra of a wild ass (Equus africanus): Hafting, projectiles and mousterian hunting weapons. *Antiquity, 73*, 394–402.

Boyd, R., & Richerson, P. J. (2005). *The origin and evolution of cultures*. New York: Oxford University Press.

Brooks, A., Nevell, L., Yellen, J., & Hartman, G. (2006). Projectile technologies of the African MSA: Implications for modern human origins. In E. Hovers & S. L. Kuhn (Eds.), *Transitions before The Transition: Evolution and stability in the Middle Paleolithic and Middle Stone Age* (pp. 233–255). New York: Springer.

Churchill, S. E. (1993). Weapon technology, prey size selection, and hunting methods in modern hunter-gatherers: Implications for hunting in the Palaeolithic and Mesolithic. *Archeological Papers of the American Anthropological Association, 4*, 11–24.

Crowther, A., & Haslam, M. (2007). Blind tests in microscopic residue analysis: Comments on Wadley et al. (2004). *Journal of Archaeological Science, 34*, 997–1000.

Dibble, H., McPherron, S. P., Chase, P., Farrand, W. R., & Debenath, A. (2006). Taphonomy and the concept of Paleolithic cultures: The case of the Tayacian from Fontéchevade. *PaleoAnthropology, 1*, 1–21.

Donahue, R. E., Murphy, M. L., & Robbins, L. H. (2004). Lithic microwear analysis of Middle Stone Age artifacts from White Paintings Rock Shelter, Botswana. *Journal of Field Archaeology, 29*, 155–163.

Gaudzinski-Windheuser, S. (2016). Hunting lesions in pleistocene and early Holocene European bone assemblages and their implications for our knowledge on the use and timing of lithic projectile technology. In R. Iovita & K. Sano (Eds.), *Multidisciplinary approaches to the study of Stone Age weaponry* (pp. 273–286). Dordrecht: Springer.

Grace, R. (1989). *Interpreting the function of stone tools: The quantification and computerization of microwear analysis*. Oxford: BAR.

Hardy, B. L., Kay, M., Marks, A. E., & Monigal, K. (2001). Stone tool function at the Paleolithic sites of Starosele and Buran Kaya III, Crimea: Behavioral implications. *Proceedings of the National Academy of Sciences of the United States of America, 98*, 10972–10977.

Hill, K., Barton, M., & Hurtado, A. M. (2009). The emergence of human uniqueness: Characters underlying behavioral modernity. *Evolutionary Anthropology, 18*, 187–200.

Hutchings, K. (2016). When is a point a projectile? Morphology, impact fractures, scientific rigor, and the limits of inference. In R. Iovita & K. Sano (Eds.), *Multidisciplinary approaches to the study of stone age weaponry* (pp. 3–12). Dordrecht: Springer.

Iovita, R. (2011). Shape variation in Aterian tanged tools and the origins of projectile technology: A morphometric perspective on stone tool function. *PLoS ONE, 6*, e29029.

Kaplan, H., Hill, K., Lancaster, J., & Hurtado, A. M. (2000). A theory of human life history evolution: Diet, intelligence, and longevity. *Evolutionary Anthropology: Issues, News, and Reviews, 9*, 156–185.

Keeley, L. H. (1980). *Experimental determination of stone tool uses: A microwear analysis*. Chicago: University of Chicago Press.

Kuman, K. A. (1989). *Florisbad and ≠ Gi: The contribution of open-air sites to the study of the Middle Stone Age in Southern Africa*. Philadelphia: University of Pennsylvania Press.

Levi Sala, I. (1986). Use wear and post-depositional surface modification: A word of caution. *Journal of Archaeological Science, 13*, 229–244.

Lombard, M. (2004). Distribution patterns of organic residues on Middle Stone Age points from Sibudu Cave, Kwazulu-Natal, South Africa. *South African Archaeological Bulletin, 59*, 37–44.

Lombard, M. (2005). Evidence of hunting and hafting during the Middle Stone Age at Sibudu Cave, KwaZulu-Natal, South Africa: A multianalytical approach. *Journal of Human Evolution, 48*, 279–300.

Lombard, M. (2007). Evidence for change in Middle Stone Age hunting behaviors at Blombos Cave: Results of a macrofracture analysis. *South African Archaeological Bulletin, 62*, 62–67.

Lombard, M., & Wadley, L. (2016). Hunting technologies during the Holwiesons Poort at Sibudu Cave: What they reveal about human cognition in KwaZulu-Natal, South Africa, between ~65 and 62 ka. In R. Iovita & K. Sano (Eds.), *Multidisciplinary approaches to the study of Stone Age weaponry* (pp. 77–100). Dordrecht: Springer.

Marean, C. W. (2010). Pinnacle Point Cave 13B (Western Cape Province, South Africa) in context: The Cape floral kingdom, shellfish, and modern human origins. *Journal of Human Evolution, 59*, 425–443.

Marean, C. W., Bar-Matthews, M., Bernatchez, J., Fisher, E., Goldberg, P., Herries, A. I. R., et al. (2007). Early human use of marine resources and pigment in South Africa during the Middle Pleistocene. *Nature, 449*, 905–908.

Marlowe, F. (2005). Hunter-gatherers and human evolution. *Evolutionary Anthropology, 14*, 54–67.

Mazza, P. P. A., Martini, F., Sala, B., Magi, M., Colombini, M. P., Giachi, G., et al. (2006). A new Palaeolithic discovery: Tar-hafted stone tools in a European Mid-Pleistocene bone-bearing bed. *Journal of Archaeological Science, 33*, 1310–1318.

McBrearty, S., & Brooks, A. S. (2000). The revolution that wasn't: A new interpretation of the origin of modern human behavior. *Journal of Human Evolution, 39*, 453–563.

Milo, R. G. (1998). Evidence for hominid predation at Klasies River Mouth, South Africa, and its implications for the behaviour of early modern humans. *Journal of Archaeological Science, 25*, 99–133.

Nance, J. (1979). Fundamental statistical considerations in the study of microwear. In B. Hayden (Ed.), *Lithic use-wear analysis* (pp. 351–363). New York: Academic Press.

Newcomer, M., Grace, R., & Unger-Hamilton, R. (1986). Investigating microwear polishes with blind tests. *Journal of Archaeological Science, 13*, 203–217.

Odell, G. H., & Odell-Vereecken, F. (1980). Verifying the reliability of lithic use-wear assessments by blind tests: The low-power approach. *Journal of Field Archaeology, 7*, 87–120.

Pargeter, J. (2011). Assessing the macrofracture method for identifying Stone Age hunting weaponry. *Journal of Archaeological Science, 38*, 2882–2888.

Porat, N., Chazan, M., Grün, R., Aubert, M., Eisenmann, V., & Horwitz, L. K. (2010). New radiometric ages for the Fauresmith industry from Kathu Pan, southern Africa: Implications for the Earlier to Middle Stone Age transition. *Journal of Archaeological Science, 37*, 269–283.

Rots, V. (2013). Insights into early Middle Palaeolithic tool use and hafting in Western Europe. The functional analysis of level IIa of the early Middle Palaeolithic site of Biache-Saint-Vaast (France). *Journal of Archaeological Science, 40*, 497–506.

Rots, V. (2016). Projectiles and hafting technology. In R. Iovita & K. Sano (Eds.), *Multidisciplinary approaches to the study of Stone Age weaponry* (pp. 167–185). Dordrecht: Springer.

Rots, V., Van Peer, P., & Vermeersch, P. M. (2011). Aspects of tool production, use, and hafting in Palaeolithic assemblages from Northeast Africa. *Journal of Human Evolution, 60*, 637–664.

Schiffer, M. B. (1987). *Formation processes of the archaeological record*. Albuquerque: University of New Mexico.

Schoville, B. J. (2010). Frequency and distribution of edge damage on Middle Stone Age lithic points, Pinnacle Point 13B, South Africa. *Journal of Human Evolution, 59*, 378–391.

Schoville, B. J. (2014). Testing a taphonomic predictive model of edge damage formation with Middle Stone Age points from Pinnacle Point cave 13B and Die Kelders cave 1, South Africa. *Journal of Archaeological Science, 48*, 84–95. doi:10.1016/j.jas.2013.10.002

Schoville, B. J., & Brown, K. S. (2010). Comparing lithic assemblage edge damage distributions: Examples from the Late Pleistocene and preliminary experimental results vis-a-vis. *Explorations in Anthropology, 10*, 34–49.

Semenov, S. A. (1964). *Prehistoric technology* (M. W. Thompson, Trans.). Bath: Adams and Dart.

Shea, J. (1988). Spear points from the Middle Paleolithic of the Levant. *Journal of Field Archaeology, 15*, 441–450.

Shea, J. (1992). Lithic microwear analysis in archaeology. *Evolutionary Anthropology, 1*, 143–150.

Shea, J., Davis, Z., & Brown, K. (2001). Experimental tests of Middle Palaeolithic spear points using a calibrated crossbow. *Journal of Archaeological Science, 28*, 807–816.

Shea, J., & Klenck, J. D. (1993). An experimental investigation of the effects of trampling on the results of lithic microwear analysis. *Journal of Archaeological Science, 20*, 175–194.

Shennan, S. J. (1997). *Quantifying Archaeology* (2nd ed.). Iowa City: University of Iowa Press.

Sisk, M. L., & Shea, J. J. (2011). The African origin of complex projectile technology: An analysis using tip cross-sectional area and perimeter. *International Journal of Evolutionary Biology, 2011*. doi:10.4061/2011/968012

Sokal, R. R., & Rohlf, F. J. (1995). *Biometry* (3rd ed.). New York, NY: W.H. Freeman and Co.

Thompson, E., Williams, H. M., & Minichillo, T. (2010). Middle and late Pleistocene Middle Stone Age lithic technology from Pinnacle point 13B (Mossel Bay, Western Cape Province, South Africa). *Journal of Human Evolution, 59*, 358–377.

Tringham, R., Cooper, G., Odell, G., Voytek, B., & Whitman, A. (1974). Experimentation in the formation of edge damage: A new approach to lithic analysis. *Journal of Field Archaeology, 1*, 171–196.

Van Peer, P., Rots, V., & Vermeersch, P. M. (2008). A wasted effort at the quarry: Wear analysis and interpretation of an MSA lanceolate point from Taramsa-8, Egypt. *PaleoAnthropology, 2008*, 234–250.

Villa, P., Boscato, P., Ranaldo, F., & Ronchitelli, A. (2009). Stone tools for the hunt: Points with impact scars from a Middle Paleolithic site in southern Italy. *Journal of Archaeological Science, 36*, 850–859.

Villa, P., Delagnes, A., & Wadley, L. (2005). A late Middle Stone Age artifact assemblage from Sibudu (KwaZulu-Natal): Comparisons with the European Middle Paleolithic. *Journal of Archaeological Science, 32*, 399–422.

Wadley, L., Hodgskiss, T., & Grant, M. (2009). Implications for complex cognition from the hafting of tools with compound adhesives in the Middle Stone Age, South Africa. *Proceedings of the National Academy of Sciences, 106*, 9590–9594.

Wadley, L., & Lombard, M. (2007). Small things in perspective: The contribution of our blind tests to micro-residue studies on archaeological stone tools. *Journal of Archaeological Science, 34*, 1001–1010.

Wilkins, J. (2013). *Technological change in the Early Middle Pleistocene: The onset of the Middle Stone Age at Kathu Pan 1, Northern Cape, South Africa*. PhD dissertation, University of Toronto.

Wilkins, J., & Chazan, M. (2012). Blade production ~500 thousand years ago at Kathu Pan 1, South Africa: Support for a multiple origins hypothesis for early Middle Pleistocene blade technologies. *Journal of Archaeological Science, 39*, 1883–1900.

Wilkins, J., Schoville, B. J., Brown, K. S., & Chazan, M. (2012). Evidence for early hafted hunting technology. *Science, 338*, 942–946.

Wynn, T., & Coolidge, F. L. (2011). The implications of the working memory model for the evolution of modern cognition. *International Journal of Evolutionary Biology, 2011*. doi:10.4061/2011/741357

Yaroshevich, A., Zaidner, Y., & Weinstein-Evron, M. (2016). Projectile damage and point morphometry at the Early Middle Paleolithic Misliya Cave, Mount Carmel (Isral): Preliminary results and interpretations. In R. Iovita & K. Sano (Eds.), *Multidisciplinary approaches to the study of Stone Age weaponry* (pp. 119–134). Dordrecht: Springer.

Chapter 8
Projectile Damage and Point Morphometry at the Early Middle Paleolithic Misliya Cave, Mount Carmel (Israel): Preliminary Results and Interpretations

Alla Yaroshevich, Yossi Zaidner, and Mina Weinstein-Evron

Abstract This contribution presents analyses of projectile damage and morpho-metric characteristics of various point types from the Early Middle Paleolithic Misliya Cave, Mount Carmel, Israel. All the types present in the assemblage exhibit diagnostic impact fractures. Four types, i.e., Levallois points, Abu Sif points, Hummal points and the newly defined Misliya points appear to be the most frequently used as tips of hunting weapons. These four types differ in their morpho-metric characteristics, as well as in terms of the frequencies of diagnostic impact fractures. We suggest that the variability in points may reflect the use of different kinds of weapons, including composite projectiles – a possibility supported by the faunal evidence from Levantine MP sites and Misliya Cave, in particular. Whether the diversity in point types and sizes reflects use in different kinds of hunting weapons or variability within the same kind, the study can contribute significantly to our understanding of the techno-logical and subsistence transformations associated with the emergence of the Middle Paleolithic in the Levant.

Keywords Early Middle Paleolithic • Levant • Hunting weapons • Impact fractures • Blade technology • Mount Carmel

A. Yaroshevich (✉)
Israel Antiquities Authority, P.O. Box 586, 91004 Jerusalem, Israel
e-mail: allayaroshe@gmail.com

Y. Zaidner · M. Weinstein-Evron
Zinman Institute of Archaeology, University of Haifa, 31905 Haifa, Israel
e-mail: yzaidner@research.haifa.ac.il

M. Weinstein-Evron
e-mail: evron@research.haifa.ac.il

Introduction

The appearance of points in flint tool assemblages is one of the distinctive features characterizing the emergence of the Middle Paleolithic (MP) in the Levant. This tool class is especially dominant and diverse during the early phase of the period, the Early Middle Paleolithic (EMP). A variety of blanks obtained through different reduction methods, including prismatic blade technology (e.g., Bar-Yosef 1998), were modified into points with a broad array of forms from simple Levallois points (unmodified) to carefully and intensively retouched Abu Sif points (Copeland 1985; Gordon 1993; Wojtczak 2011; Zaidner and Weinstein-Evron 2012). The latter became the type fossil of the Early Levantine Mousterian (Copeland 1975, 1983; Neuville 1951; Meignen 1998, 2011).

The appearance of stone points in the prehistoric record implies changes in hunting related technology. Indeed, studies by Shea (1988, 1989a, b, 1991, 1993) identified the function of Levallois points from a number of Levantine Mousterian sites as tips of hunting weapons based on the presence of projectile damage (but see Plisson and Beyries 1998 for an alternative view suggesting that Levallois points were mainly used for cutting plant material). Retouched point types from an EMP context have never been studied in detail with regard to their function as projectile weapons, thus the connection between the variability of point assemblages and hunting weapons technology associated with the emergence of the MP remains poorly understood.

The global prehistoric record has provided only a few findings directly indicating use of particular kinds of weapons. Use of simple projectiles is evident from wooden spears found in several European sites (Dennel 1997; Theime 1997 and references therein). The earliest complex projectiles, i.e., spearthrowers and darts, as well as bows and arrows came from the context of the European Upper Paleolithic (Rust 1943; Garrod 1955; Rausing 1967; Stodiek 1992, 1993; Morales and Straus 2009). The function of the

Radu Iovita and Katsuhiro Sano (eds.), *Multidisciplinary Approaches to the Study of Stone Age Weaponry*,
Vertebrate Paleobiology and Paleoanthropology, DOI 10.1007/978-94-017-7602-8_8

MP/MSA (Middle Stone Age) points as tips of hunting weapons is evident from findings of points embedded into vertebrae of large ungulates (i.e., Milo 1998; Boëda et al. 1999) and the presence of fractures diagnostic of impact. These were found on Levantine (Shea 1988, 1989a, b, 1991, 1993), European (Villa and Lenoir 2006; Villa et al. 2009a; Villa and Soriano 2010) and African points (Lombard et al. 2004; Lombard 2005, 2007, 2008; Villa and Lenoir 2006; Lombard and Pargeter 2008; Villa et al. 2009b) and interpreted, in most cases, as tips of simple projectiles, i.e., spears for thrusting or throwing by hand.

Several recent works indicate that particular types of MSA points served as tips of complex projectiles. Brooks and colleagues (2006) suggested that the decrease in point length, width, thickness and weight alongside the unchanging angle of the distal tips (55°–60°) during the MSA sequences in Botswana and Ethiopia reflects adoption of a complex projectile system. Another metric characteristic, Tip Cross Sectional Perimeter (TCSP), based on the maximal width and maximal thickness (see below for calculation) of various MSA point types and compared with ethnographic North American dart tips showed a theoretical plausibility that points from Porc Epic cave in Ethiopia served as tips of darts thrown with spearthrower (Sisk and Shea 2009, 2011). Lombard and Philipson (2010) and Lombard (2011) showed that backed segments, the type fossil of the MSA Howiesons Poort culture, were used as transversal arrowheads. This interpretation is based on several kinds of evidence including the location and the direction of macro- and micro- Diagnostic Impact Fractures (DIF), residue location and the small size of the segments.

Here we present analyses of projectile damage and morpho-metric characteristics of points from the EMP Misliya Cave, Mount Carmel, Israel. The aim of this contribution is to describe the variability of the point assemblage and to provide possible interpretations for the diversity of the types and sizes in terms of their use as tips of hunting weapons. We believe that our contribution will comprise a base for further investigations of hunting-related technological transformations associated with the emergence of the MP in the Levant.

The Site and the Point Assemblage

Misliya Cave is located on the western slopes of Mount Carmel, slightly to the south of Nahal (Wadi) Sefunim, at an elevation of ca. 90 m, some 12 km south of Haifa (Fig. 8.1) and ca. 7 km north of Nahal Me'arot (Wadi el-Mughara) and the caves of Tabun, el-Wad and Skhul (Garrod and Bate 1937; McCown 1937; Jelinek 1982a, b; Jelinek et al. 1973). Excavations in 2001–2010 revealed a rich EMP layer spread over the Upper Terrace of this collapsed cave (Fig. 8.2), below a residual rock shelter or overhang (Weinstein-Evron et al. 2003). The dating of the archaeological sequence is still in process. Preliminary TL dates on burned flint artifacts from the site suggest that they are older than 200 ka (Valladas et al. 2013), thus corroborating the dates recently obtained for the same cultural phase in the nearby Tabun Cave (ca. 260–190 ka BP; Mercier and Valladas 2003, and references therein) and at Hayonim Cave, in the western Galilee (230–170 ka BP; Mercier et al. 2007) and broadly assigning the site to marine isotope stage (MIS) 7. An ongoing technological and typological analysis of the lithic industry indicates that points of various forms comprise about 40% of the tool assemblage (Zaidner and Weinstein-Evron 2012).

The typological classification of points is based on their morphological and technological features as follows: **Levallois points** (Fig. 8.3a); **Retouched Levallois points** (Fig. 8.3b); **Abu-Sif points** (elongated Mousterian points): points retouched along both edges by continuous and

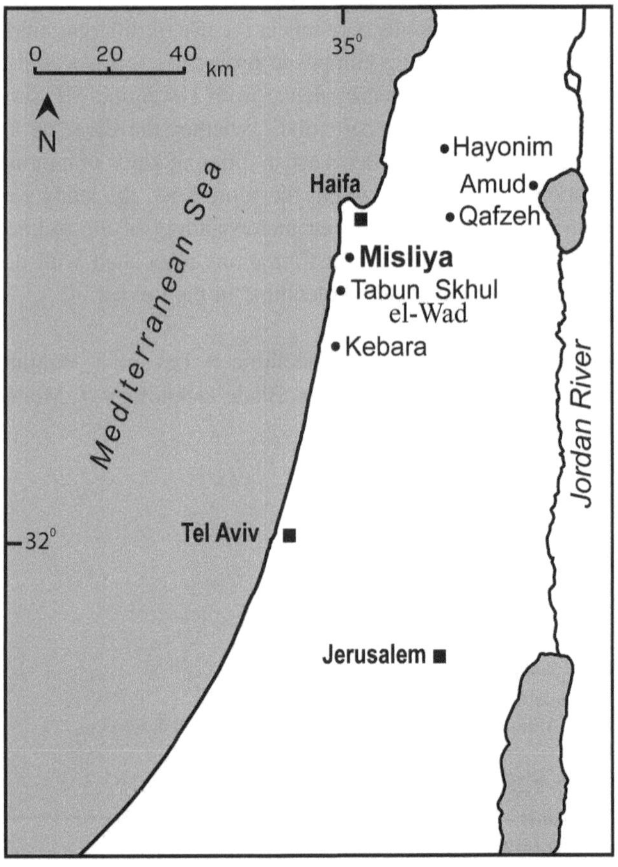

Fig. 8.1 Location map

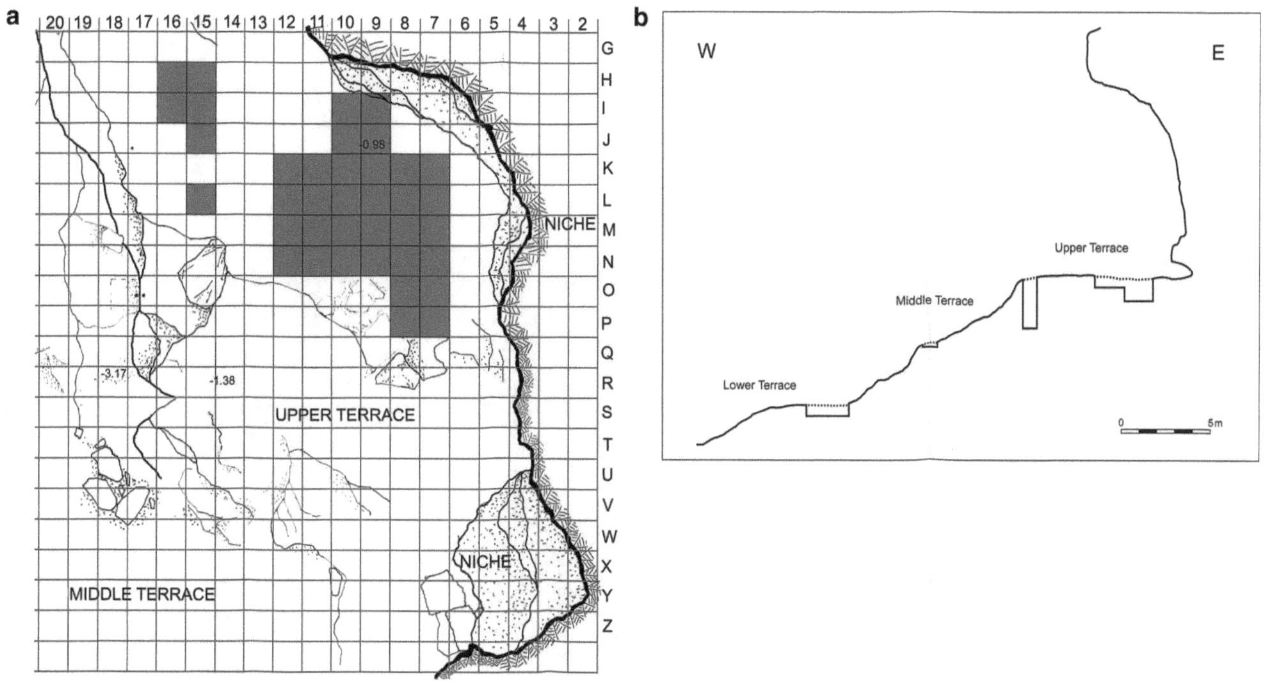

Fig. 8.2 Misliya Cave, excavated area. **a** Plan; **b** section through the three terraces

invasive or short retouch. These are made either on elongated Levallois points, elongated flakes or narrow blades (Fig. 8.3c, d); **Hummal points**: points with one fully or almost fully retouched edge opposite an edge that is either unretouched or retouched only on the tip (Fig. 8.3e). Made predominantly on blades, some are possibly made on Levallois blanks. The retouch is usually regular but not invasive and changes only slightly the original form of the blank; **Misliya points** a newly defined point type, with tip modified by abrupt retouch in the form of an oblique truncation (Fig. 8.3f). Misliya points are made on small thin blades, Levallois as well as non-Levallois, or on small Levallois points; **Points with bifacial, alternate or ventral retouch**: points made on Levallois and non-Levallois elongated blanks and modified with invasive retouch which may be either bifacial, alternating or on the ventral surface (Fig. 8.3g); **Off-set points**: points with retouch creating either an oblique truncation or an arch-like back (Fig. 8.3h). In both cases the tip of the point is offset relative to the striking axis of the blank.

For the present project we studied points from the material excavated until the 2009 season. The assemblage consists of 291 points. Levallois points (N = 90) comprise the largest group; the second largest group are Hummal (N = 46) followed by retouched Levallois points (N = 36), Abu Sif points (N = 36) and Misliya points (N = 21). Points with bifacial, alternate or ventral retouch (N = 9) and off-set points (N = 7) complete the studied assemblage. Fifty broken distal tips which could not be assigned confidently to any particular type were not included in the analysis. Figure 8.4 represents the distribution of the points within the EMP layer of Misliya Cave, showing possible contemporary use of various types.

Methods

Types and Frequencies of Diagnostic Impact Fractures

Fischer et al. (1984) delineated two types of macro-fractures, *spin-off* (Fig. 8.5a) and *step terminating bending* (Fig. 8.5b) as diagnostic of projectile impact (Hayden 1979). These two types, along with burin-like removals – another type of impact damage described in experimental studies (e.g., Barton and Bergman 1982; Bergman and Newcomer 1983) were recognized in subsequent archery experiments and analyses of archaeological points (e.g., Odell and Cowan 1986; Nuzhnyy 1989, 1990, 1999, 2008; Lombard et al. 2004; Lombard and Pargeter 2008; Yaroshevich 2010; Yaroshevich et al. 2010; Petillon et al. 2011). On a microscopic level, diagnostic impact damage appears as linear polishes and striations (Fischer et al. 1984; Crombe et al. 2001). Recent experiments

Fig. 8.3 Misliya Cave, point types. **a** Levallois point; **b** Retouched Levallois point; **c**, **d** Abu Sif points; **e** Hummal point; **f** Misliya point; **g** points with bifacial, alternate or ventral retouch; **h** off-set points

by Pargeter (2011) showed that step terminating bending fracture, spin-off fracture and burin-like fractures can occur in low frequencies (up to 3%, depending on the type of the fracture) as a result of trampling. Therefore, the frequencies of macro-fractures are important for delineating projectile function of archaeological stone points.

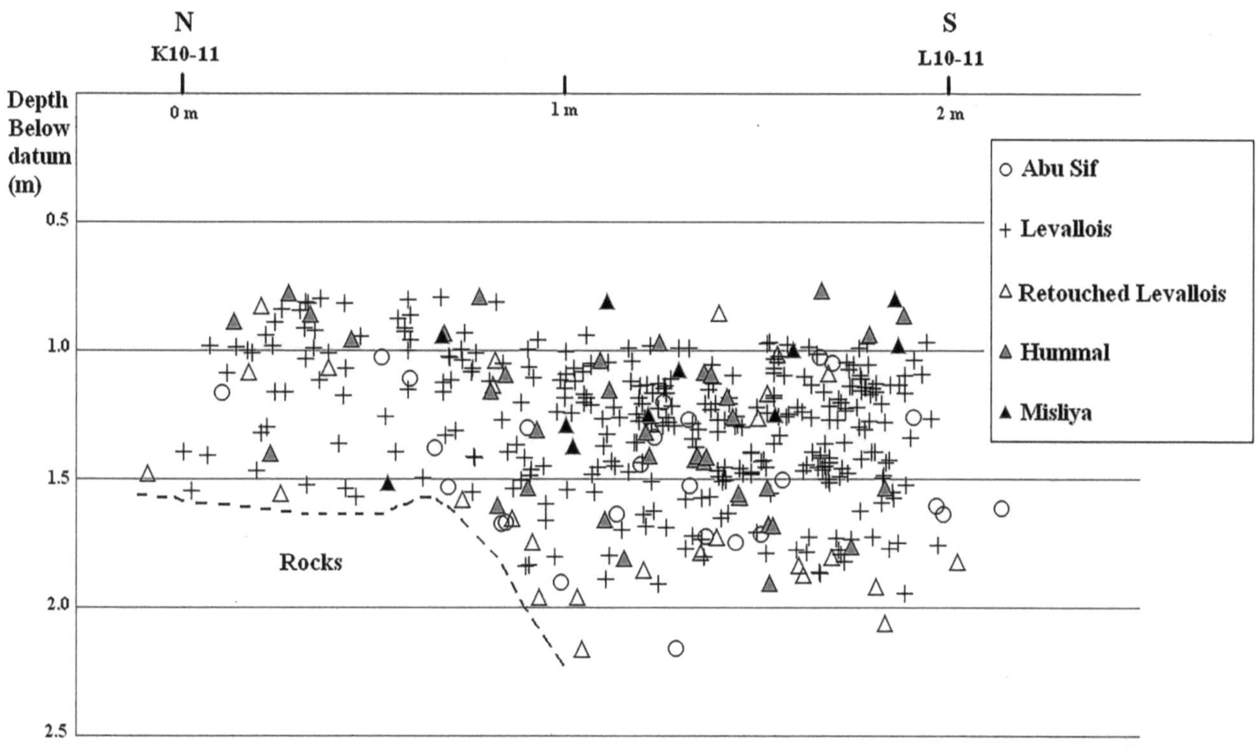

Fig. 8.4 Misliya Cave: vertical distribution of points from squares K10-11 and L10-11 in the EMP sequence, Upper Terrace (Fig. 8.2)

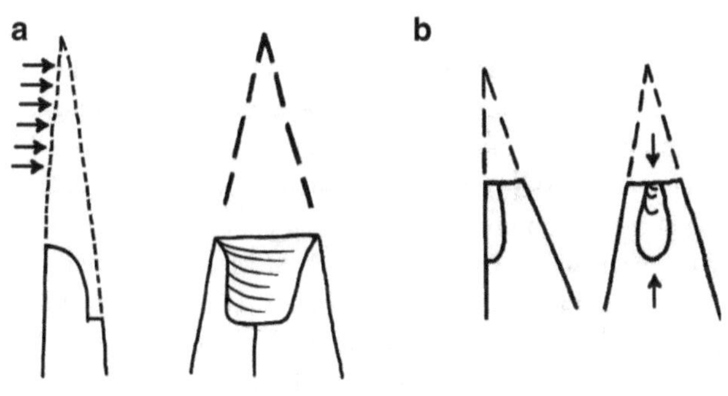

Fig. 8.5 Types of fractures diagnostic of impact. **a** Step terminating bending; **b** spin-off (after Fischer et al. 1984)

All points from the Misliya assemblage were observed for the presence of macro-DIF; their frequencies were recorded according to point type. Some points with macro-DIF were subsequently observed through Scanning Electron Microscopy (SEM) in an attempt to identify micro-DIF.

Morpho-metric Characteristics

All the points were measured in terms of their length, maximal width and maximal thickness (Fig. 8.6). Complete or nearly complete points were weighed. In order to evaluate the tip angle we outlined the distal part of the point (about 1.5–2.0 cm) and then measured the angle with a protractor. This method differs from that used in the study of Brooks et al. (2006). We believe that our approach (Fig. 8.7a, b) is more appropriate to the assemblage from Misliya Cave as many of the points at the site have either a curved lateral edge or truncation, as opposed to African points that appear to have roughly straight edges (Fig. 8.7c). Applying the method of Brooks and colleagues to points from Misliya Cave would reduce considerably the value of the tip angle and would not reliably convey the true variation in the assemblage. We also calculated the TCSP for all point types from Misliya Cave as follows: TCSP = $MaxWidth + 2\sqrt{(MaxWidth/2)^2 + MaxThickness^2}$ (Sisk and Shea 2009, 2011). For comparative purposes, data from the following assemblages were recorded: North American ethnographic dart tips, based on the collections published by Thomas (1978) and Shott (1997); archaeological points from the MSA sites of Aduma 5 and Porc Epic Cave, Ethiopia, suggested as possible tips of complex projectiles (Brooks et al. 2006; Sisk and Shea 2011). The assemblage of points from Misliya Cave and the North American ethnographic dart tips were compared through one-way analysis of variance (ANOVA) tests including Sheffé post hoc comparisons. ANOVA tests whether one or more sample means are significantly different from each other; Sheffé post hoc comparisons determine which or how many sample means are different.

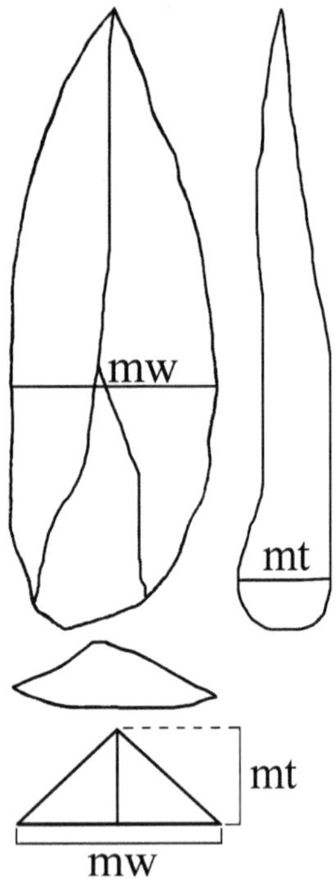

Fig. 8.6 Metric characteristics measurements based on the outline of Abu Sif points. mw Maximal width; mt Maximal thickness

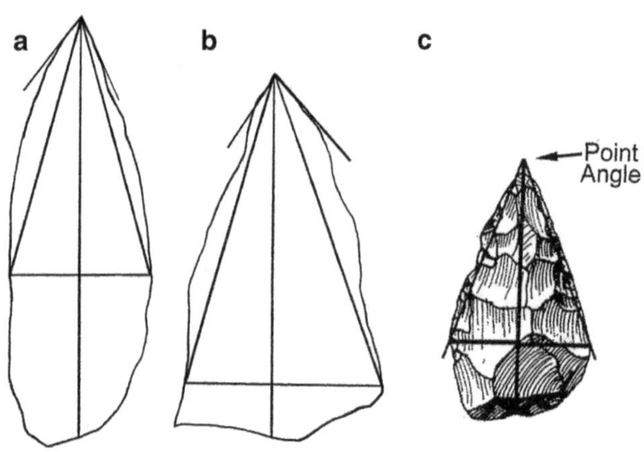

Fig. 8.7 Angle of distal tip measurements. **a, b** Applied in the present study, shown on the outlines of Abu Sif and Levallois points; **c** according to Brooks et al. (2006)

In terms of the distal tip, ANOVA analysis was applied only to points from Misliya Cave as there is no data for this characteristic for ethnographic dart tips. In addition, comparisons were made with Porc Epic points, previously suggested as possible tips of complex projectiles.

Results

Fractures Diagnostic of Projectile Impact

Table 8.1 shows the frequencies of DIF for various point types from Misliya Cave. The highest frequencies were observed among Levallois points (Figs. 8.8, 8.9 and 8.10) and Abu Sif points (Figs. 8.11, 8.12 and 8.13): 22.2 and 19.4%, respectively. Hummal points (Figs. 8.14 and 8.15) and Misliya points (Figs. 8.16 and 8.17) exhibit less than half the frequency compared to Levallois/Abu Sif points: 8.9 and 9.5%, respectively. Two retouched Levallois points with DIF comprise 6.3% of the group. Off-set points and points with bifacial or alternate retouch each have one representative with DIF, making up 11.1 and 14.3% of the group, respectively.

In the majority of the cases DIF were observed on the distal tip of the point. A few points were broken either at their proximal third or half their length with burin-like DIF (Fig. 8.13). Some exhibited DIF on both the distal tip and the breakage (Figs. 8.11 and 8.14). Fifteen points with macro-DIF on their distal tips were observed through SEM with linear striations occurring on five points (33%; e.g., Figures 8.8b, 8.11b and 8.16b). The relatively low frequency of micro-striations on the points from Misliya cave may be explained by the fact that observations were made only on the area of macro-fracture while striations may have been present on other areas of a point's surface. In previous works analyzing either experimental or archaeological assemblages of points the ratios of micro-striations vary. For example, Fischer et al. (1984) observed micro-striations on 60% of experimental points. Among eleven experimental microliths with macro-DIF observed through SEM only five exhibited micro-striations (Yaroshevich et al. 2010). For archaeological points, values of 40% (Crombe et al. 2001) and ca. 55% (16 of 29 segments, Lombard 2011) were reported.

Table 8.1 Misliya Cave: frequencies of DIF according to point type

Point type	With DIF		Total	
	N	%	N	%
Retouched Levallois	2	6.3	32	100
Levallois	20	22.2	90	100
Abu-Sif	7	19.4	36	100
Hummal	4	8.9	45	100
Misliya	2	9.5	21	100
Points with bifacial and/or alternate retouch	1	14.3	7	100
Off-set	1	11.1	9	100
Total	38	15.8	241	100

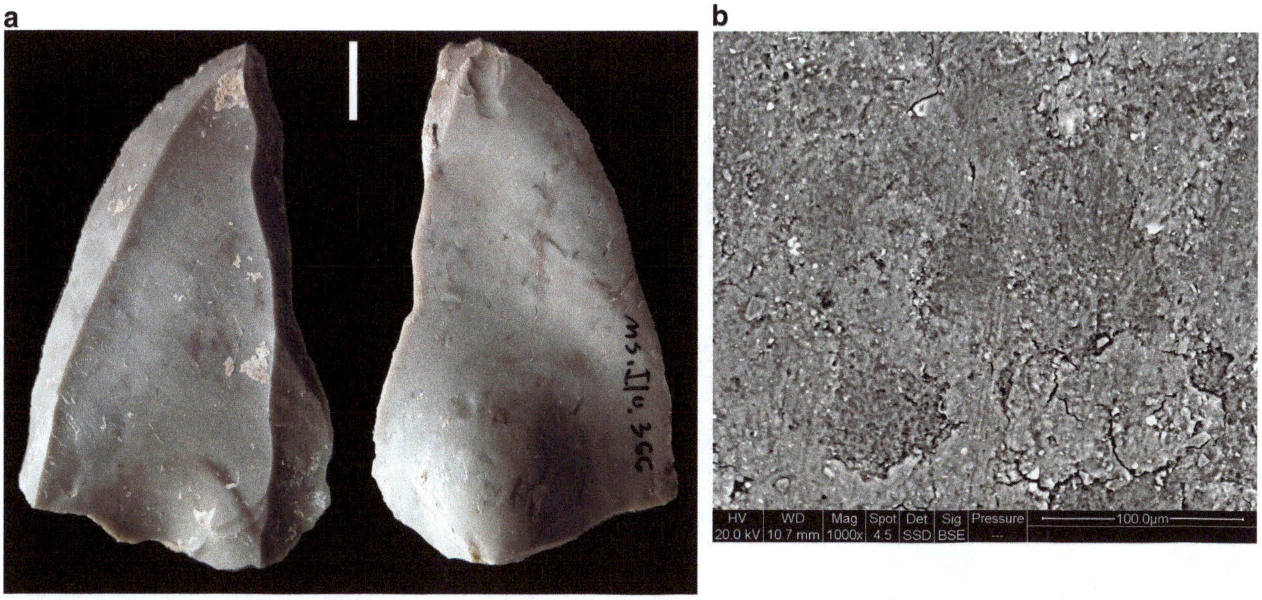

Fig. 8.8 Misliya Cave: DIF on Levallois point. **a** Macro-DIF. The scale is 5 mm; **b** micro-DIF

Fig. 8.9 Misliya Cave: Levallois point with macro-DIF. The scale is 5 mm

Fig. 8.10 Misliya Cave: Levallois point with macro-DIF. The scale is 5 mm

Morpho-metric Characteristics

The subsequent morpho-metric analyses we applied to the four types which are the most common and exhibit the highest frequencies of DIF, i.e., Levallois, Abu Sif, Hummal and Misliya points. Results of ANOVA analysis are shown for each metric characteristic separately. In addition, for each characteristic we present box plots where values for points with and without DIF are presented separately.

Maximal Width (Table 8.2, Fig. 8.18)

In terms of maximal width, the points create three distinctive groups: the first contains North American ethnographic dart tips, Misliya points and Hummal points with average values of 23.0, 21.5 and 25.7 mm, respectively. Abu Sif (30.5 mm) forms the second group whereas Levallois points (36.3 mm) belong to the third group. Hummal points bearing DIF appear at the lower end of the range for the type (Fig. 8.18) and their maximal width (19.1 mm, Table 8.3), is statistically similar

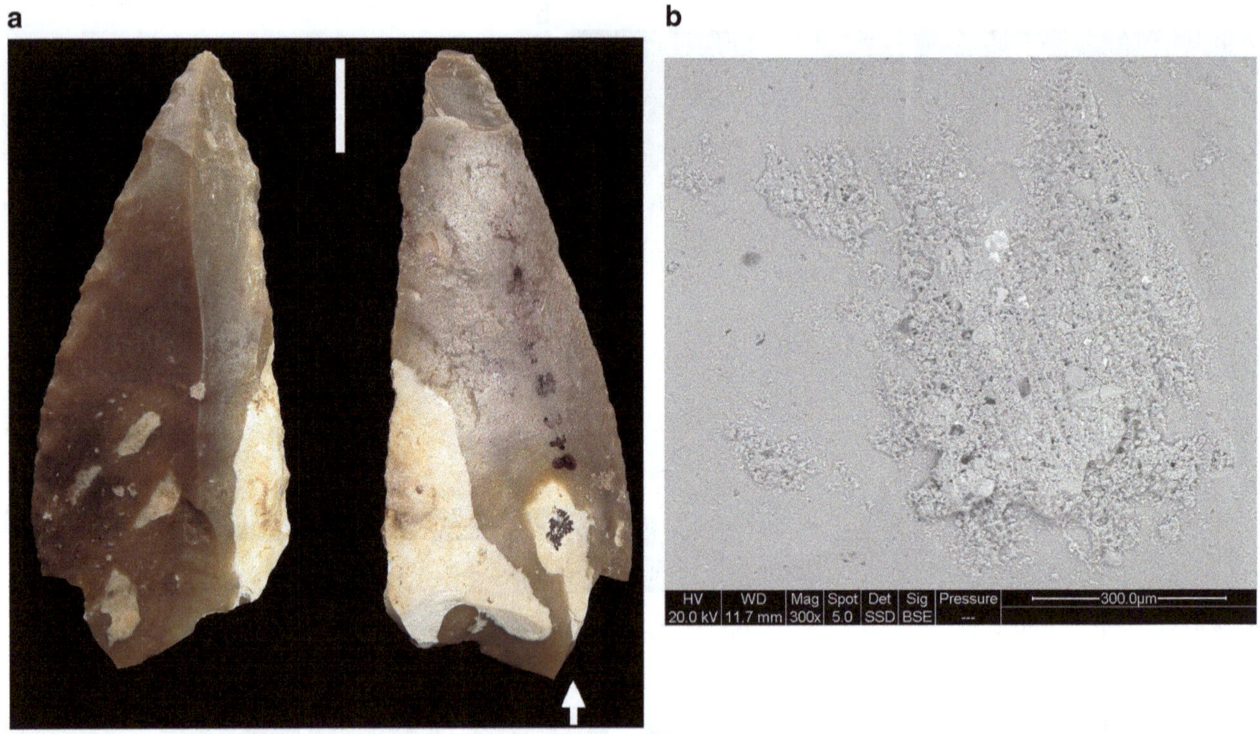

Fig. 8.11 Misliya Cave: DIF on Abu Sif point. **a** Macro-DIF. The scale is 5 mm; **b** micro-DIF

Fig. 8.12 Misliya Cave: Abu Sif point with macro-DIF. The scale is 5 mm

Fig. 8.13 Misliya Cave: Abu Sif point with macro-DIF. The scale is 5 mm

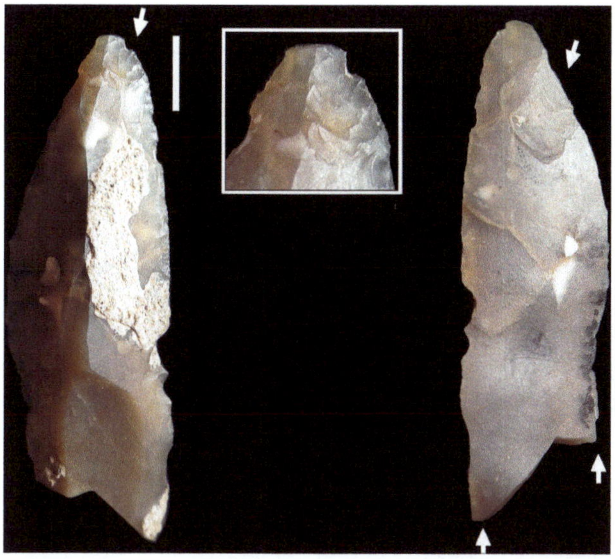

Fig. 8.14 Misliya Cave: Hummal point with macro-DIF. The scale is 5 mm

Fig. 8.15 Misliya Cave: Hummal point with macro-DIF. The scale is 5 mm

to North American ethnographic darts (Table 8.4). Also, in terms of maximal width Misliya and Hummal points with DIF are statistically similar (Table 8.4) to Porc Epic bifacial and unifacial points (23.61 and 23.15 mm, respectively, Sisk and Shea 2011) and have similar values with Aduma 5 points (about 23 mm, Brooks et al. 2006, Fig. 9).

Maximal Thickness (Table 8.5, Fig. 8.19)
In terms of mean maximal thickness North American dart tips (5.0 mm), Misliya (6.3 mm), Levallois (7.8 mm) and Abu Sif (9.3 mm) comprise four separate groups whereas Hummal (8.2 mm) belong to the third and in the fourth groups, meaning Hummal points are statistically similar to both, Levallois and Abu Sif points in terms of their maximal thickness.

Again, Hummal points with DIF have the lowest values within the type (6.2 mm). Misliya points and Hummal points with DIF are statistically similar (Table 8.4) to Porc Epic bifacial and unifacial points (8.36 and 7.45 mm, respectively, Sisk and Shea 2011) and are practically identical to Aduma 5 points (6.5 mm, Brooks et al. 2006, Fig. 9) in terms of their maximal thickness.

TCSP (Table 8.6, Fig. 8.20)
In terms of average TCSP the points create four distinct groups with Misliya (46.5 mm) and North American ethnographic dart tips (47.2 mm) comprising the first. Hummal (56.4 mm), Abu Sif (66.4 mm) and Levallois (76.0 mm) each represent separate groups. T-tests (Table 8.4) show that Hummal points with DIF are statistically similar to North American dart tips, as well as to Porc Epic bifacial and unifacial points (50.25 and 50.93 mm, respectively, Sisk and Shea 2011). Misliya points are statistically similar to Porc Epic bifacial points and even smaller than Porc Epic unifacial points in terms of TCSP (Table 8.4).

Weight (Table 8.7, Fig. 8.21)
In terms of mean weight, the points create three groups with a considerable overlap between them. North American ethnographic tips (4.4 gr.) belong to the first group; Misliya (10.6 gr.) belong to the first and to the second; Hummal (16.2 gr.) and Levallois (14.7 gr.) belong to the second and to the third; Abu Sif (22.7 gr.) belong solely to the third group. Misliya points have weights similar to Aduma 5 points (10 gr., Brooks et al. 2006, Fig. 11b).

Angle of the Distal Tip (Table 8.8, Fig. 8.22)
In terms of the average angle of the distal tip, the points create two distinctive groups with Abu Sif (58.9°), Misliya (62.0°) and Hummal (62.9°) belonging to the first and Levallois (73.1°) comprising the second. Abu Sif, Misliya and Hummal points with DIF show values lower than their type in general: 57.2°, 60° and 51°, respectively. These values are similar to Aduma points (55°–60°, Brooks et al. 2006).

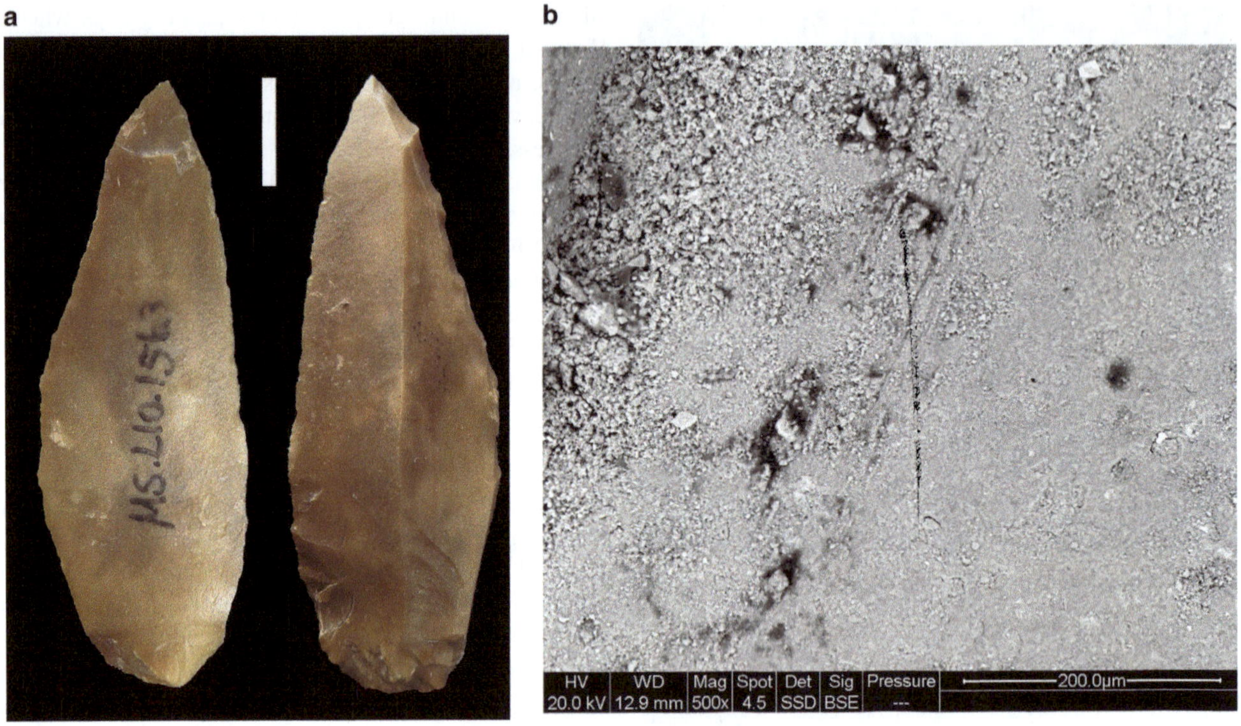

Fig. 8.16 Misliya Cave: DIF on Misliya point. **a** Macro-DIF. The scale is 5 mm; **b** micro-DIF

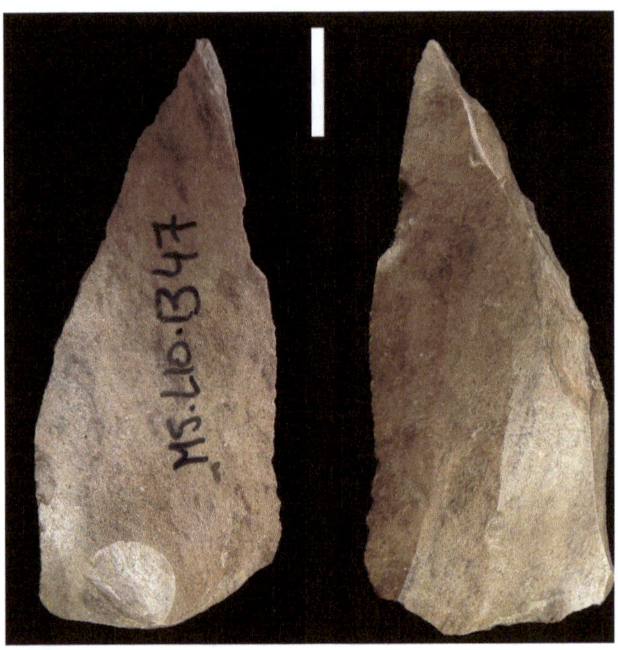

Fig. 8.17 Misliya Cave: Misliya point with macro-DIF. The scale is 5 mm

Table 8.2 Mean maximal widths for points from Misliya Cave and North American ethnographic dart tips: Scheffé homogeneous subsets based on one-way analysis of variance

Point type	N	Subset for alpha = 0.05		
		1	2	3
Misliya	18	21.5		
North American dart tips	40	23.0		
Hummal	45	25.7		
Abu-Sif	32		30.5	
Levallois	89			36.3
Significance		0.064	1.000	1.000

Discussion and Conclusions

While all point types present in the EMP layer of Misliya Cave seem to have been applied as tips of hunting weapons, there are four types, i.e., Levallois, Abu Sif, Hummal and Misliya which were most frequently used in this function. These four types differ in terms of their morpho-metric characteristics, as well as in terms of DIF ratios. Levallois

Table 8.3 Misliya Cave: morpho-metric characteristics of various types of points bearing DIF

Point type		Length	Maximal width	Maximal thickness	Angle	Weight	TCSP
Levallois	Mean	63.7	38.9	8.4	75.7	19.0	81.6
	N	14	19	19	11	12	19
	S.D.	10.6	7.0	1.8	13.0	8.0	13.9
Abu Sif	Mean	72.9	28.1	7.9	57.2	17.8	60.3
	N	2	5	5	4	4	5
	S.D.	13.4	2.5	1.5	6.9	6.2	5.5
Hummal	Mean	–	19.1	6.2	51.0	–	41.9
	N	–	4	4	4	–	4
	S.D.	–	4.4	1.5	5.2	–	9.6
Misliya	Mean	56.0	22.8	6.2	60.0	9.4	48.8
	N	2	2	2	2	2	2
	S.D.	0.9	3.2	1.4	14.14	2.0	7.5

Table 8.4 T-test probabilities comparing Misliya and Hummal points with North American ethnographic dart tips and Porc Epic bifacial and unifacial points. $p < 0.05$: the two samples differ with 95% confidence; $p > 0.05$: the two samples cannot be distinguished with 95% confidence. TCSP for Porc Epic points were compared to Misliya and Hummal points with DIF (http://in silico.net/statistics/ttest/two-sample) using standard deviation values provided by Sisk, personal communication, 2011

Point type	Versus North American dart tips			Versus Porc Epic unifacial points			Versus Porc Epic bifacial points		
	TCSP	Width	Thickness	TCSP	Width	Thickness	TCSP	Width	Thickness
Misliya	**0.79**	**0.21**	0.00	**0.12**	**0.15**	0.00	0.04	**0.10**	0.01
Hummal with DIF	**0.26**	**0.05**	0.03	**0.17**	**0.13**	**0.05**	**0.15**	**0.16**	**0.21**

and Abu Sif points, the two largest types, show relatively high frequencies of DIF, around 20%. Misliya and some Hummal points, specifically those bearing DIF are the smallest in the assemblage and statistically similar to North American ethnographic dart tips, as well as to MSA Porc Epic and Aduma 5 points in terms of metric characteristics. The frequencies of DIF for Misliya and Hummal points (ca. 10%), are only one-half of those occurring on Levallois and Abu Sif points.

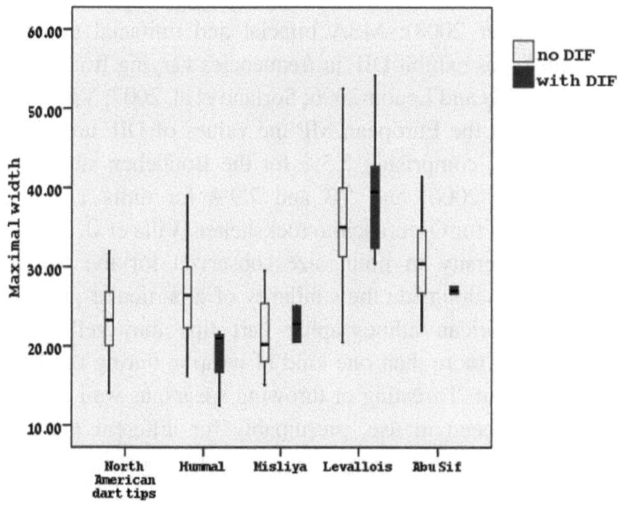

Fig. 8.18 Boxplots of maximal width values for various types of points from Misliya Cave and North American ethnographic dart tips

The largest types, Levallois and Abu Sif points differ statistically in terms of width, thickness, TCSP and the angle of the distal tip, with Levallois points being wider and thinner on average and having duller tips. Experiments with thrusting spears showed that greater width enhances penetrating ability of the point (Shea et al. 2001) while greater thickness makes the point more durable on impact, but reduces its penetrating capacity (Hughes 1998). Based on this evidence we suggest that Levallois points provided better penetration whereas Abu Sif points were designed to be more durable on impact. Abu Sif points show the lowest values of distal tip angle, a characteristic which increases

Table 8.5 Mean maximal thicknesses for points from Misliya Cave and North American ethnographic dart tips: Scheffé homogeneous subsets based on one-way analysis of variance

Point type	N	Subset for alpha = 0.05			
		1	2	3	4
North American ethnographic darts	40	5.0			
Misliya	18		6.3		
Levallois	89			7.8	
Hummal	45			8.2	8.2
Abu Sif	32				9.3
Significance		1.000	1.000	0.912	0.170

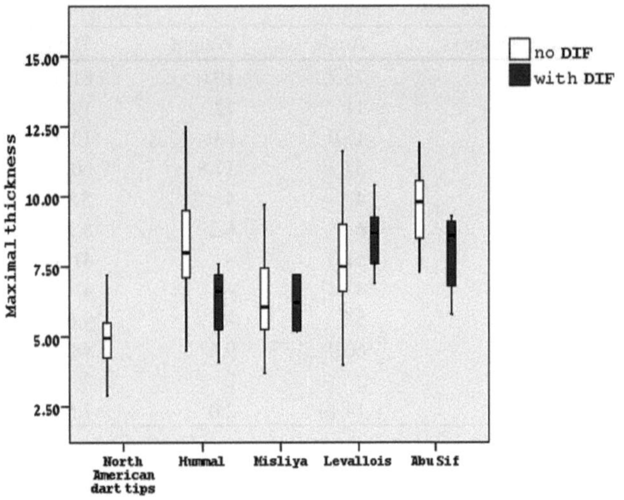

Fig. 8.19 Boxplots of maximal thickness values for various types of points from Misliya Cave and North American ethnographic dart tips

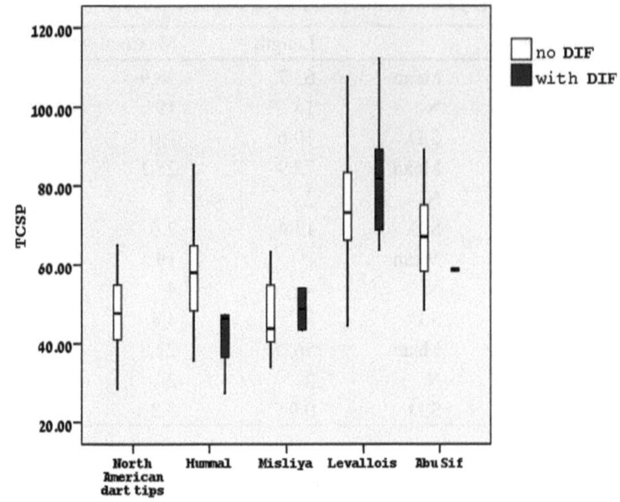

Fig. 8.20 Boxplots of TCSP values for various types of points from Misliya Cave and North American ethnographic dart tips

Table 8.6 Mean TCSPs for points from Misliya Cave and North American ethnographic dart tips: Scheffé homogeneous subsets based on one-way analysis of variance

Point type	N	Subset for alpha = 0.05			
		1	2	3	4
Misliya	18	46.5			
North American ethnographic darts	40	47.2			
Hummal	45		56.4		
Abu Sif	32			66.4	
Levallois	89				76.0
Significance		1.000	1.000	1.000	1.000

Table 8.7 Mean weights for points from Misliya Cave and North American ethnographic dart tips: Scheffé homogeneous subsets based on one-way analysis of variance

Point type	N	Subset for alpha = 0.05		
		1	2	3
North American ethnographic darts	10	4.4		
Misliya	18	10.6	10.6	
Levallois	76		14.7	14.7
Hummal	40		16.2	16.2
Abu Sif	28			22.7
Significance		0.408	0.515	0.156

penetrating abilities (Hughes 1998) thus reducing the influence of their relatively greater thickness.

There is a possibility that some DIF observed on the points from Misliya Cave occurred as a result of trampling as shown in experiments conducted by Pargeter (2011). However, the frequencies of DIF at Misliya Cave (between 6 and 22%) are considerably higher than those created in trampling experiments, up to 3%, depending on the type of the fracture (Pargeter 2011). The relatively high frequencies of DIF on Misliya Cave points precludes the possibility that these were created only as a result of trampling or post depositional processes. There are also parallels from other MP/MSA sites. For example, Shea (1988, 1993) reported relatively high frequencies of DIF for the Levantine MP, comprising about a third of all points bearing use-wear. In these analyses Shea included crushing and abrasion on tips in the criteria he used to infer projectile impact. If these are deducted, the frequency of DIF would comprise 10–20% (Shea, personal communication 2011). For the MSA Howiesons Poort segments the frequencies vary from 21 to 24% in different sites (Lombard and Pargeter 2008); MSA bifacial and unifacial point of various types exhibit DIF in frequencies varying from 5.3 to 13.4% (Villa and Lenoir 2006; Soriano et al. 2007; Villa et al. 2009a). For the European MP the values of DIF are somewhat lower, comprising 5.3% for the Bouheben site (Villa and Lenoir 2006) and 5.3 and 7.9% for units 1 and 2, respectively for Oscurusciuto rockshelter (Villa et al. 2009b).

The diversity in point sizes observed for the Misliya assemblage alongside the similarity of a particular group to North American ethnographic dart tips may reflect the presence of more than one kind of weapon during the EMP in the Levant. Thrusting or throwing spears, as well as darts may have been in use, presumably for different game or biotopes. Faunal evidence supports this possibility. The

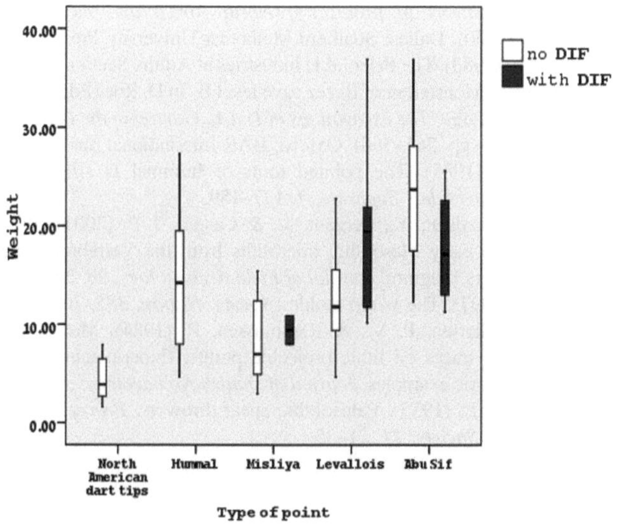

Fig. 8.21 Boxplots of weight values for various types of points from Misliya Cave and North American ethnographic dart tips

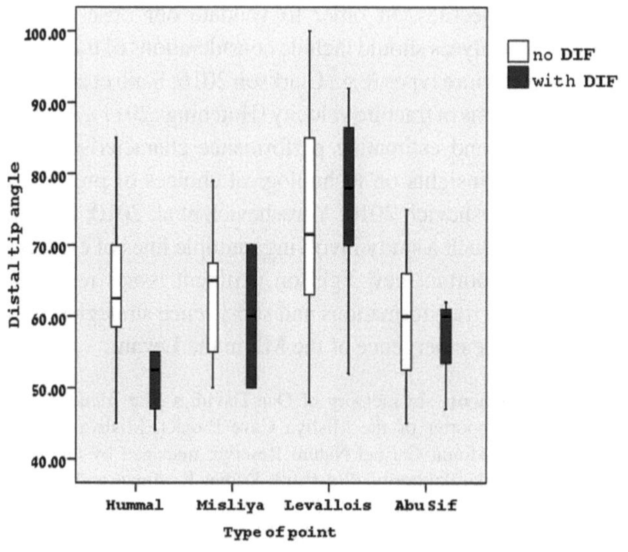

Fig. 8.22 Boxplots of the angle of the distal tip values for various types of points from Misliya Cave

Table 8.8 Mean distal angles for points from Misliya Cave: Scheffé homogeneous subsets based on one-way analysis of variance

Point type	N	Subset for alpha = 0.05	
		1	2
Abu Sif	27	58.9	
Misliya	17	62.0	
Hummal	44	62.9	
Levallois	75		73.1
Significance		0.703	1.000

emergence of the MP in the southern Levant was associated with hunting of mountain gazelle (*Gazella gazella*) in considerable numbers, a species extremely rare in the preceding Lower Paleolithic archaeofaunas. Moreover, this species, living in open terrain and hunted, according to ethnographic record (Churchill 1993) with complex projectiles was probably preferred by MP hunters in the Levant. The preference of gazelle is evident from the comparative analysis of faunal remains from a natural pitfall trap, Rantis Cave, and a number of anthropogenic cave sites (Yeshurun 2012). While in the natural trap Mesopotamian fallow deer (*Dama mesopotamica*) outnumber mountain gazelles, the anthropogenic caves, including Misliya (Yeshurun et al. 2007), show roughly equal presence of both species or an abundance of the latter. Whether this transformation in hunting behavior can be related to environmental changes i.e., prevalence of arid conditions during 285–255 or 240–230 ka BP (Vaks et al. 2010), close to the emergence of the MP, needs further research. The prevalence of aridity could have increased the population of gazelles who thrive in open, arid environments. This, in turn, could have led to the adoption of new hunting strategies and technologies, such as use of various stone tipped weapons, including long-distance projectiles.

The possibility of use of more than one kind of weapon during the EMP remains, however, theoretical. It is equally possible that the variability within Misliya Cave points actually reflects the range within one particular kind of weapon. Estimations of tip weight provided by Hughes (1998, Table IX) show that the range for Australian unfletched dart tips is 9–70 gr. Thus, dart tips seem not to be limited by weight or metric characteristics (see also Clarkson 2016) and theoretically all point types from Misliya Cave could have served as dart tips. Since there is no available data about dimensions of ethnographic (i.e., efficient) spearheads, we cannot exclude any archaeological type from being used as a tip for this kind of weapon, either. Experiments by Shea et al. (2001) showing that small and thin points are not efficient as tips of thrusting spears may be of relevance here. These provide further support for the possibility that Misliya points and Hummal points with DIF, the smallest points in our assemblage, served as tips of complex projectiles. Even so, at the present state we cannot rule out the option that these types represent the smallest efficient spearheads, probably for throwing by hand.

In sum, the diversity of Levantine EMP points in terms of their morpho-metric characteristics and the similarity of a particular group with North American dart tips support the possibility of the presence of a variety of weapons, including complex projectiles. In order to validate our observations, additional analyses should include considerations of the size of particular fracture types (e.g., Clarkson 2016; Sano et al. 2016) and calculations of fracture velocity (Hutchings 2011). Archery experiments and estimating performance characteristics can also provide insights on technological choices of prehistoric hunters (Yaroshevich 2010; Yaroshevich et al. 2010; Petillon et al. 2011). Such a study involving multiple lines of evidence will shed important new light on pertinent issues regarding technological transformations and subsistence strategies associated with the emergence of the MP in the Levant.

Acknowledgements In memory of Dan David, a dear friend and an enthusiastic supporter of the Misliya Cave Project. Misliya Cave is located in the Mount Carmel Nature Reserve, managed by the Israel Nature and Parks Authority. We thank Daniel Kaufman and Reuven Yeshurun for their comments and Reuven Kapul for preparing Fig. 8.4. Special thanks are given to Dr. Alex Berner and Larissa Popilevsky from the Electron Microscopy Centre at the Faculty of Materials Engineering, Technion, for the SEM analysis. The excavations at Misliya Cave are supported by the Dan David Foundation, the Leakey Foundation, the Irene Levi-Sala Care Archaeological Foundation and the Faculty of Humanities, University of Haifa. Israel Antiquities Authority permit numbers for the Misliya Cave excavations: G-16/2001, G-39/2002, G-14/2003, G-29/2004, G-12/2005, G-12/2006, G-4/2007, G-54/2008, G-52/2009, G-50/2010.

References

Barton, R. N. E., & Bergman, C. A. (1982). Hunters at Hengistbury: Some evidence from experimental archaeology. *World Archaeology, 14*, 236–248.

Bar-Yosef, O. (1998). The chronology of the Middle Paleolithic of the Levant. In T. Akazawa, K. Aoki & O. Bar-Yosef (Eds.), *Neandertals and modern humans in western Asia* (pp. 39–56). New York: Plenum Press.

Bergman, C. A., & Newcomer, M. H. (1983). Flint arrowhead breakage, examples from Ksar Akil. *Journal of Field Archaeology, 10*, 238–243.

Boëda, E., Geneste, J.-M., Griggo, C., Mercier, N., Muhesen, S., Reyss, et al. (1999). A Levallois point embedded in the vertebra of a wild ass (*Equus africanus*): Hafting, projectiles and Mousterian hunting weapons. *Antiquity, 73*, 394–402.

Brooks, A. S., Yellen, J. E., Nevell, L., & Hartman, G. (2006). Projectile technologies of the African MSA: Implications for modern human origins. In E. Hovers & S. L. Kuhn (Eds.), *Transitions before the Transition* (pp. 233–256). NewYork: Plenum/Kluwer.

Churchill, S. E. (1993). Weapon technology, prey size selections, and hunting methods in modern hunter-gatheres: Implications for hunting in the Paleolithic and Mesolithic. In G. L. Peterkin., H. Bricker & P. A. Mellars (Eds.), *Hunting and animal exploitation in the Later Palaeolithic and Mesolithic of Eurasia* (pp. 11–24). Washington, DC: American Anthropological Association.

Clarkson, C. (2016). Testing archaeological approaches to determining past projectile delivery systems using ethnographic and experimental data. In R. Iovita & K. Sano (Eds.), *Multidisciplinary approaches to the study of Stone Age weaponry* (pp. 189–201). Dordrecht: Springer.

Copeland, L. (1975). The Middle and Upper Paleolithic in Lebanon and Syria in the light of recent research. In F. Wendorf & A. Close (Eds.), *Problems in prehistory: North Africa and the Levant* (pp. 317–350). Dallas: Southern Methodist University Press.

Copeland L. (1983). The Paleolithic industries at Adlun. Section III: The Levalloiso-Mousterian of Bezez cave level B. In D. Roe (Ed.), *Adlun in the Stone Age: The excavations of D.A.E. Garrod in the Lebanon, 1958–1963* (pp. 261–366). Oxford: BAR International Series, 159.

Copeland, L. (1985). The pointed tools of hummal Ia (El-Kowm, Syria). *Cahiers de l'Euphrate, 4*, 177–189.

Crombe, P., Perdaen, Y., Sergant, J., & Caspar, J.-P. (2001). Wear analysis on early Mesolithic microliths from the Verrebroek site, East Flanders Belgium. *Journal of Field Archaeology, 28*, 253–269.

Dennel, R. (1997). The world's oldest spears. *Nature, 385*, 767–768.

Fischer, A., Hansen, P. V., & Rassmussen, P. (1984). Macro- and micro-wear traces on lithic projectile points. Experimental results and prehistoric examples. *Journal of Danish Archaeology, 3*, 19–46.

Garrod, D. A. E. (1955). Palaeolithic spear-throwers. *Proceedings of Prehistoric Society, 21*, 21–35.

Garrod, D. A. E., & Bate, D. M. A. (1937). *The Stone Age of Mount Carmel* (Vol. I). Excavations at the Wadi Mughara. Oxford: Clarendon Press.

Gordon, D. (1993). Mousterian tool selection, reduction, and discard at Ghar, Israel. *Journal of Field Archaeology, 20*, 205–218.

Hayden, B. (1979). Ho-Ho nomenclature committee. The Ho-Ho classification and nomenclature committee report. In B. Hayden (Ed.), *Lithic use-wear analysis* (pp. 133–135), New-York: Academic Press.

Hughes, S. S. (1998). Getting to the point: Evolutionary change in prehistoric weaponry. *Journal of Archaeological Method and Theory, 5*, 345–408.

Hutchings, K. (2011). Measuring use-related fracture velocity in lithic armatures to identify spears, javelins, darts, and arrows. *Journal of Archaeological Science, 38*, 1737–1746.

Jelinek, A. J. (1982a). The Tabun Cave and Paleolithic man in the Levant. *Science, 216*, 1369–1375.

Jelinek, A. J. (1982b). The middle Paleolithic in the Southern Levant, with comments on the appearance of modern *Homo sapiens*. In A. Ronen (Ed.), *The transition from Lower and Middle Paleolithic and the origin of modern man* (pp. 57–104). Oxford: BAR International Series, 151.

Jelinek, A. J., Farrand, W. R., Hass, G., Horowitz, A., & Goldberg, P. (1973). New excavations at the Tabun Cave, Mount Carmel, Israel: A preliminary report. *Paléorient, 1*, 151–183.

Lombard, M. (2005). Evidence of hunting and hafting during the Middle Stone Age at Sibudu Cave, KwaZulu-Natal: A multianalytical approach. *Journal of Human Evolution, 48*, 279–300.

Lombard, M. (2007). Evidence for change in Middle Stone Age hunting behaviour at Blombos Cave: Results of a macro-fracture analysis. *South African Archaeological Bulletin, 62*, 62–67.

Lombard, M. (2008). Finding resolution for the Howiesons Poort through the microscope: Micro-residue analysis of segments from Sibudu Cave, South Africa. *Journal of Archaeological Science, 35*, 26–41.

Lombard, M. (2011). Quartz-tipped arrows older than 60 ka: Further use-trace evidence from Sibudu, KwaZulu-Natal, South Africa. *Journal of Archaeological Science, 38*, 1918–1930.

Lombard, M., Parsons, I., & van der Ryst, M. M. (2004). Middle Stone Age lithic experimentation for macro-fracture and residue analyses: the process and preliminary results with reference to the Sibudu Cave points. *South African Journal of Science, 100*, 159–166.

Lombard, M., & Pargeter, J. (2008). Hunting with Howiesons Poort segments: Pilot experimental study and the functional interpretation of archaeological tools. *Journal of Archaeological Science, 35*, 2523–2531.

Lombard, M., & Philipson, L. (2010). Indications of bow and stone-tipped arrow use 64 000 years ago in KwaZulu-Natal, South Africa. *Antiquity, 84*, 635–648.

McCown, T. D. (1937). Mugharet es-Skhul: Description and Excavations. In D. A. E. Garrod & D. M. A. Bate (Eds.), *The Stone Age of Mount Carmel: Excavations at the Wady el-Mughara* (Vol. I, pp. 91–112). Chapter VI Oxford: Clarendon Press.

Meignen, L. (1998). Hayonim Cave lithic assemblage in the context of the near eastern Middle Paleolithic. In T. Akazawa, K. Aoki & O. Bar-Yosef (Eds.), *Neandertals and modern humans in Western Asia* (pp. 165–180). New York: Plenum Press.

Meignen L, (2011). The contribution of Hayonim cave assemblages to the understanding of the so-called early Levantine Mousterian. In J.-M. Le Tensorer, R. Jagher & M. Otte (Eds.), *The Lower and Middle Paleolithic in the middle east and neighbouring regions* (pp. 85–100). ERAUL 126.

Mercier, N., & Valladas, H. (2003). Reassessment of TL age estimates of burnt flints from the Palaeolithic site of Tabun Cave, Israel. *Journal of Human Evolution, 45*, 401–409.

Mercier, N., Valladas, H., Frojet, L., Joron, J.-L., Ryess, J.-L., Weiner, S., et al. (2007). Hayonim Cave: A TL-based chronology for this Levantine Mousterian sequence. *Journal of Archaeological Science, 34*, 1064–1077.

Milo, R. G. (1998). Evidence for hominid predation at Klasies River Mouth, South Africa, and its implications for the behaviour of early modern humans. *Journal of Archaeological Science, 25*, 99–133.

Morales, M. R. G., & Straus, L. G. (2009). Extraordinary early Magdalenian finds from El Mirón Cave, Cantabria (Spain). *Antiquity, 83*, 267–281.

Neuville, R. (1951). *Le Paleolithique et le Mesolithique du desert du Judee*. Archives de Musee de Paleontologie Humaine memoire 24. Paris: Masson et Cie.

Nuzhnyy, D. (1989). L'utilisation des microlithes geometriques et non geometriques comme armatures de projectiles. *Bulletin de la Societe Prehistorique Francaise, 86*, 88–96.

Nuzhnyy, D. (1990). Projectile damage on Upper Paleolithic microliths and the use of bow and arrow among Pleistocene hunters in the Ukraine. In *The interpretative possibilities of microwear studies. Proceedings of the International Conference of Lithic Use-Wear Analysis, 15–17th February 1989 in Uppsala, Sweden* (pp. 113–124). Uppsala: AUN 14.

Nuzhnyy, D. (1999). Microlithic projectile weapons of the Late Palaeolithic and Mesolithic hunters of the Crimea. *Archeologija, 1*, 5–24 (Ukrainian).

Nuzhnyy, D. (2008). *Development of microlithic technology during the Stone Age: Projectile weapons of prehistoric hunters*. Kiev: KNT. (Ukrainian).

Odell, G. H., & Cowan, F. (1986). Experiments with spears and arrows on animal targets. *Journal of Field Archaeology, 13*, 195–212.

Pargeter, J. (2011). Assessing the macrofracture method for identifying Stone Age hunting weaponry. *Journal of Archaeological Science, 38*, 2882–2888.

Pétillon, J.-M., Bignon, O., Bodu, P., Cattelain, P., Debout, G., Langlais, M., et al. (2011). Hard core and cutting edge: Experimental manufacture and use of Magdalenian composite projectile tips. *Journal of Archaeological Science, 38*, 1266–1283.

Plisson, H., & Beyries, S. (1998). Pointes ou outils triangulaires? Données fonctionnelles dans le Moustérien levantin. *Paléorient, 24*, 5–24.

Rausing, G. (1967). *The Bow: Some notes on its origin and development*. Lund: C. W. K. Gleerups.

Rust, A. (1943). *Die Alt-und Mittelsteinzeitlishe Funde von Stellmoor*. Neumünster: Washholtz.

Sano, K., Denda, Y., & Oba, M. (2016). Experiments in fracture patterns and impact velocity with replica hunting weapons from Japan. In R. Iovita & K. Sano (Eds.), *Multidisciplinary approaches to the study of Stone Age weaponry* (pp. 29–46). Dordrecht: Springer.

Shea, J. J. (1988). Spear points from the Middle Palaeolithic of the Levant. *Journal of Field Archaeology, 15*, 441–450.

Shea, J. J. (1989a). A functional study of the lithic industries associated with hominid fossils in the Kebara and Qafzeh Caves, Israel. In P. Mellars & C. Stringer (Eds.), *The human revolution* (pp. 611–625). Edinburgh, Edinburgh University Press.

Shea, J. J. (1989b). Tool use in the Levantine Mousterian of Kebara Cave, Mount Carmel. *Mitekufat HaEven* (*Journal of the Israel prehistoric society*), *22*, 15–30.

Shea, J. J. (1991). *The behavioral significance of Levantine Mousterian industrial variability*. PhD dissertation, Harvard University.

Shea, J. J. (1993). Lithic use-wear evidence for hunting by Neandertals and early modern humans from the Levantine Mousterian. In G. L. Peterkin, H. Bricker & P. A. Mellars (Eds.), *Hunting and animal exploitation in the later Paleolithic and Mesolithic of Eurasia* (pp. 11–24). Washington DC: American Anthropological Association.

Shea, J., Davis, Z., & Brown, K. (2001). Experimental tests of middle palaeolithic spear points using a calibrated crossbow. *Journal of Archaeological Science, 28*, 807–816.

Shott, M. J. (1997). Stones and shafts redux: The metric discrimination of chipped stone dart and arrow points. *American Antiquity, 62*, 86–101.

Sisk, M. L., & Shea, J. J. (2009). Experimental use and quantitative performance analysis of triangular flakes (Levallois points) used as arrowheads. *Journal of Archaeological Science, 36*, 2039–2047.

Sisk, M. L., & Shea, J. J. (2011). The African origin of complex projectile technology: An analysis using tip cross-sectional area and perimeter. *International Journal of Evolutionary Biology, 2011*, 1–8.

Soriano, S., Villa, P., & Wadley, L. (2007). Blade technology and tool forms in the middle Stone Age of South Africa: The Howiesons Poort and post-Howiesons Poort at Rose Cottage Cave. *Journal of Archaeological Science, 34*, 681–703.

Stodiek, U. (1992). A propos de l'emmanchement des propulseurs au Paléolithique Supérieur. In J. P. Rigaud, H. Laville, & B. Vandermeersch (Eds.), *Le peuplement Magdalenien: Paléogeographie physique et humaine* (pp. 317–331). Paris: Comité des Travaux Historiques et Scientifiques.

Stodiek, U. (1993). *Zur Technologie der jungpalaolithischen Speerschleuder: Eine Studie auf der Basis archaologischer, ethnologischer, und experimenteller Erkenntnisse*. Tubingen: Verlag Archaeologica Venatoria.

Theime, H. (1997). Lower Paleolithic hunting spears from Germany. *Nature, 385*, 807–810.

Thomas, D. H. (1978). Arrowheads and Atl-atl Darts – How stones got shaft. *American Antiquity, 43*, 461–472.

Vaks, A., Bar-Matthews, M., Matthews, A., Ayalon, A., & Frumkin, A. (2010). Middle-Late Quaternary paleoclimate of northern margins of the Saharan-Arabian Desert: Reconstruction from speleothems of Negev Desert, Israel. *Quaternary Science Reviews, 29*, 2647–2662.

Valladas, H., Mercier, N., Hershkovitz, I., Zaidner, I., Tsatskin, A., Yeshurun, R., et al. (2013). Dating the Lower to Middle Paleolithic transition in the Levant: A view from Misliya Cave, Mount Carmel, Israel. *Journal of Human Evolution, 65*, 585–593.

Villa, P., & Lenoir, M. (2006). Hunting weapons of the Middle Stone Age and the Middle Palaeolithic: Spear points from Sibudu, Rose Cottage and Bouheben. *South African Humanities, 18*, 89–122.

Villa, P., Boscato, P., Ranaldo, F., & Ronchitelli, A. (2009a). Stone tools for the hunt: Points with impact scars from a Middle Paleolithic site in southern Italy. *Journal of Archaeological Science, 36*, 850–859.

Villa, P., Soressi, M., Henshilwood, C. S., & Mourre, V. (2009b). The still bay points of Blombos Cave (South Africa). *Journal of Archaeological Science, 36*, 441–460.

Villa, P., & Soriano, S. (2010). Hunting weapons of Neandertals and early modern humans in South Africa. *Journal of Anthropological Research, 66*, 5–39.

Weinstein-Evron, M., Bar-Oz, G., Zaidner, Y., Tsatskin, A., Druck, D., Porat, N., & Hershkovitz, I. (2003). Introducing Misliya cave, Mount Carmel, Israel: A new continuous lower/middle Paleolithic sequence in the Levant. *Eurasian Prehistory, 1*, 31–55.

Wojtczak, D. (2011). Hummal (Central Syria) and its eponymous industry. In J.-M. Le Tensorer, R. Jagher & M. Otte (Eds.), *The Lower and Middle Paleolithic in the Middle East and neighboring regions* (pp. 289–308). ERAUL 126.

Yaroshevich, A. (2010). *Microlithic variability and design and performance of projectile weapons during the Levantine Epipaleolithic: Experimental and archaeological evidence*. PhD dissertation, University of Haifa.

Yaroshevich, A., Kaufman, D., Nuzhnyy, D., Bar-Yosef, O., & Weinstein-Evron, M. (2010). Design and performance of microlith implemented projectiles during the middle and late epipalaeolithic of the Levant: Experimental and archaeological evidence. *Journal of Archaeological Science, 37*, 368–388.

Yeshurun, R. (2012). Middle Paleolithic prey-choice inferred from a natural pitfall trap: Rantis Cave, Israel. In J. L. Clark & J. D. Speth (Eds.), *Zooarchaeology and modern human origins: Human hunting behavior during the later pleistocene* (pp. 45–75). Dordrecht: Springer.

Yeshurun, R., Bar-Oz, G., & Weinstein-Evron, M. (2007). Modern hunting behavior in the early middle Paleolithic: Faunal remains from Misliya Cave, Mount Carmel, Israel. *Journal of Human Evolution, 53*, 656–677.

Zaidner, Y., & Weinstein-Evron, M. (2012). Making a point: The Early Middle Palaeolithic tool assemblage of Misliya Cave, Mount Carmel, Israel. *Before Farming, 4*, 1–23.

Chapter 9
Morpho-Metric Variability of Early Gravettian Tanged "Font-Robert" Points, and Functional Implications

Annemieke Milks, Rob Dinnis, and Matthew Pope

Abstract Early Gravettian Font-Robert points – tanged tools created on blades – were initially defined as weapon armatures, and this is frequently referred to as their function. However, Font-Robert points have been described as a morphologically variable type, with suggestions that this morphological variability represents a functional variability. Here we discuss this issue with reference to a sample of Early Gravettian tanged artifacts (including Font-Robert points) from Maisières-Canal in Belgium, as well as two similar artifacts from Britain. Although many of the artifacts studied have a morphology and size commensurate with their function as lithic armatures, the majority are apparently unlikely to have functioned within a "complex" projectile technology, which contrasts with measurement data published on Font-Robert points from France. Instead, Font-Robert points from Maisières-Canal and Britain display a notable level of morpho-metric variability. By extension, this suggests a functional variability, a possibility that needs confirmation with use-wear analysis. These Font-Robert points may have served as technologically simpler throwing or thrusting spears, as knives, or as versatile, multi-function tools. Overall, we stress that morpho-metric data complements use-wear studies, when assessing potential projectile function, and can help make an assessment of which artifacts to target for such research techniques.

A. Milks (✉)
Institute of Archaeology, University College London,
31–34 Gordon Square, London, WC1H 0PY, UK
e-mail: a.milks@ucl.ac.uk

M. Pope
Institute of Archaeology, University College London,
31–34 Gordon Square, London, WC1H 0PY, UK
e-mail: m.pope@ucl.ac.uk

R. Dinnis
School of History, Classics and Archaeology, University of Edinburgh, William Robertson Wing Old Medical School, Teviot Place, Edinburgh, EH8 9AG, UK
e-mail: rdinnis@thebritishmuseum.ac.uk

Keywords Impact fractures • Tip cross-sectional area (TCSA) • Tool typologies • Upper Paleolithic • Weapon armatures

Introduction

"Font-Robert" points are one of several characteristic lithic point types found within western European Gravettian lithic assemblages. They are of variable dimensions with length measurements ranging from ca. 30 to 137 mm (Lansac 2002; Pesesse and Flas 2012; AM pers. obs.). Crafted on blades, they are unified by their steeply retouched and well-crafted tang. Retouch of tangs and the distal part varies from direct, bifacial or, more rarely, inverse retouch; it includes abrupt, semi-abrupt, flat and shallow retouch, and ranges from being minimally invasive to covering the entire artifact (De Sonneville-Bordes 1960; De Heinzelin 1973; Demars and Laurent 1992; Pesesse and Flas 2012, 2013). Demars and Laurent's definition (1992) specifies a pointed shape for the body of the artifacts: Font-Robert points are thus a group of tanged tools unified by their characteristic tang morphology and their pointed shape. Returned to below, it can also be noted here that Early Gravettian assemblages containing Font-Robert points also contain other tanged tool types with the same characteristic tang morphology, including those with rounded bodies, burin-like terminations and various truncations (e.g., De Sonneville-Bordes 1960; De Heinzelin 1973; Delporte and Tuffreau 1984; Otte and Caspar 1987; Jacobi and Higham 2011).

Due to their characteristic tang morphology Font-Robert points are considered by many to be an index fossil of the early Western European Gravettian. They have been found in abundance in Belgian and central and southern French assemblages, with a smaller number also found in northern

Radu Iovita and Katsuhiro Sano (eds.), *Multidisciplinary Approaches to the Study of Stone Age Weaponry*, Vertebrate Paleobiology and Paleoanthropology, DOI 10.1007/978-94-017-7602-8_9

France, northern Spain, Britain, Germany and Italy (Otte 1979; Campbell 1980; Demars and Laurent 1992; Palma di Cesnola 1993; Mussi 2001; Conard and Moreau 2004; Jacobi 2007).

The earliest archaeologically secure radiocarbon dates for contexts containing Font-Robert points are ca. 28.0 ^{14}C kBP from the open-air site of Maisières-Canal in Belgium (Haesaerts and Damblon 2004; Jacobi et al. 2010). Dates for the Early Gravettian of other regions potentially match this early date (Conard and Moreau 2004; Pesesse and Flas 2012). With specific reference to Northwest Europe, it has been proposed that Gravettian facies containing Font-Robert points appear later in southern France than at Maisières-Canal, perhaps as a result of population movement relating to the onset of a particularly cold climatic event, and that these facies subsequently persist in both Belgium and southern France for several millennia (Desbrosse and Kozlowski 1988; Djindjian and Bosselin 1994; Djindjian et al. 1999). This model references radiocarbon dates from both regions as well as techno-typological differences between the lithic assemblage at Maisières-Canal and those of other Belgian and southern French sites which also contain Font-Robert points. In fact, the chronology of many of these sites is imprecise. In particular, chrono-stratigraphic problems at the key Font-Robert point site of La Ferrassie (see Bertran et al. 2008) and also potential methodological problems with radiocarbon dating of this period (see Higham 2011) make a direct reading of the radiocarbon data potentially misleading.

Irrespective of its precise chronology, lithic material from Maisières-Canal and from British findspots is commonly considered as particularly technologically and morphologically similar (e.g., Otte 1974; Campbell 1980; Jacobi 1980; Jacobi and Higham 2011). Often referred to are the use of flat distal retouch to shape the tanged tools and the shared presence of small and delicate transverse removals at the base of the tang of Font-Robert points and other tanged tools. In addition, there is a notable scarcity of backed bladelets in the Maisières-Canal assemblage, and no Gravettian backed artifacts are known from any British assemblage (Otte 1974, 1979; Campbell 1980; Jacobi 1980, 2007; Jacobi et al. 2010; Jacobi and Higham 2011; Pesesse and Flas 2013). Campbell (1980, 1986) referred to British material as "Maisièrian" to stress these similarities, a term which continues to be used to describe the Maisières-Canal assemblage (e.g., Pesesse and Flas 2012). By logical extension, this implies that British artifacts and those from Maisières-Canal are distinct from those in assemblages elsewhere. One point of difference stressed by Campbell (1986) is the larger size of "Maisièrian" tanged tools in comparison to the French tanged tools.

As a result of these perceived cultural differences, and probably also due to the absence of a detailed type definition, nomenclature used to describe Font-Robert points within all of these western European assemblages is markedly inconsistent. Some prefer to reserve the term "Font-Robert point" for those non-Maisièrian examples closer to southern French types (e.g., Otte and Noiret 2007). Gravettian tanged artifacts from Maisières-Canal and Britain are variously referred to as "tanged points", "stemmed points", "tanged tools", "*pointe à soie*", "*pointe à pédoncule*" or "*outils pédonculées*" (e.g., De Heinzelin 1973; Rots 2002; Otte and Noiret 2007; Dinnis 2009; Flas 2009; Jacobi et al. 2010).

Despite this, most *pointed* artifacts from all of these contexts fit the morphological criteria for Font-Robert points as outlined by Demars and Laurent (1992), and the broad similarity of all of these examples has led to some preferring the term "Font-Robert points" for Maisièrian and non-Maisièrian examples (e.g., Jacobi 2007; Pettitt and White 2012). Furthermore, in some cases the term "Font-Robert point" has been used to describe fragmentary tanged artifacts where little more than the tang is present, and therefore the morphology for the body of the artifact is completely unknown (e.g., Jacobi 2007; Jacobi et al. 2010; Pesesse and Flas 2013). Here, for the sake of simplicity, we use "Font-Robert point" to refer to those tanged artifacts which fulfill basic typological criteria as outlined by Demars and Laurent (1992) (i.e., steeply backed tangs opposed by a "point"), and we use "tanged artifact" as an inclusive term of all Gravettian tanged artifacts, irrespective of their tip morphology. Thus, Font-Robert points are a subset of Gravettian tanged artifacts. This study examined a sample of Gravettian tanged artifacts from Maisières-Canal (n = 52) and one example each from two British find-spots. The study focuses on the morpho-metrics of these tools, and in particular the Font-Robert points (n = 27).

It is important to clarify the terms used in this study from the outset. In order to avoid confusion, we accept the definition that "complex" projectile technology refers to higher velocity projectiles such as the bow and arrow or spearthrower and dart, though "projectile" will still be used as a term inclusive of hand-thrown spears. Therefore "dart" refers to a spear thrown with a spearthrower, while "spear points" refers to spears, whether lower velocity hand-thrown spears or thrusting spears (Shea and Sisk 2010; Villa and Soriano 2010). The term "point" refers to the entire pointed lithic artifact, and "tip" to the part of the artifact at the opposing end of the tang.[1] "Weapon armature" is used when the points discussed could be darts, hand-thrown spear points or thrusting spear points, making an assignment to specific projectile technology difficult.

[1]Hughes (1998) uses "tip" to indicate the entire artifact, thus "tip cross-sectional area" refers to the maximum area of the entire artifact. We follow the terminology TCSA for "tip cross-sectional area", but use "tip" elsewhere to indicate the opposing end of the tang.

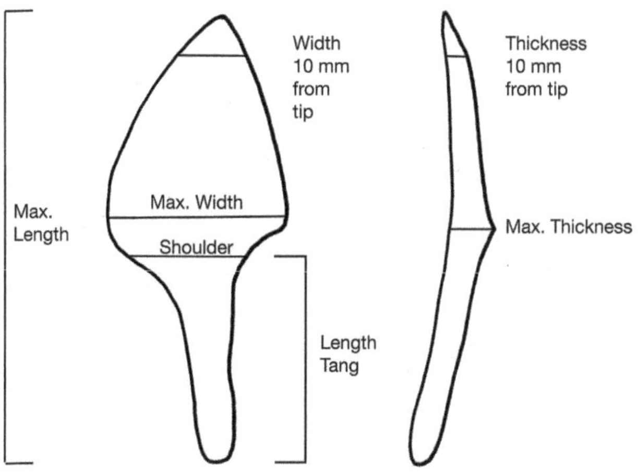

Fig. 9.1 Schematic drawing of measurements taken

Materials and Methods

A sample of tanged artifacts ($n = 54$) from Maisières-Canal and British collections was examined with the objective of determining their potential function as weapon armatures. Standard morpho-metric data were collected to allow overall metric consideration of the sample itself, and also comparison with similar data on Gravettian tanged tools from elsewhere, and specifically that of Shea (2006). Artifacts were deemed "complete" or "nearly complete" if they had a recognizable tip and tang, with only a small percentage of either missing. Only these complete and nearly complete artifacts were studied. Metric data were not collected for tangs that were incomplete. Measurement data collected can be viewed in Fig. 9.1. Mass measurements (grams) were also taken.

Sites and Material Studied

The *Champ de Fouilles* area at Maisières-Canal yielded a rich collection of nearly 35,000 artifacts in exceptionally fresh condition (Fig. 9.2) (De Heinzelin 1971; Haesaerts and De Heinzelin 1979). Clear stratigraphy, associated faunal material and radiocarbon dating have added to making the collection particularly useful for study (Haesaerts and De Heinzelin 1979; Haesaerts and Damblon 2004). Recent radiocarbon dates have confirmed earlier dating of the site, with the most secure individual date on human activity being from a reindeer bone bearing cut-marks: 27,950 ± 170 ^{14}C BP (OxA–18007) (Jacobi et al. 2010). The 130 tanged tools, of which ~36% are broken pieces, are among the most distinctive in the collection (De Heinzelin 1973; Otte and Caspar 1987; Pesesse and Flas 2012). A sample of 52 of these tanged tools was made available for study.

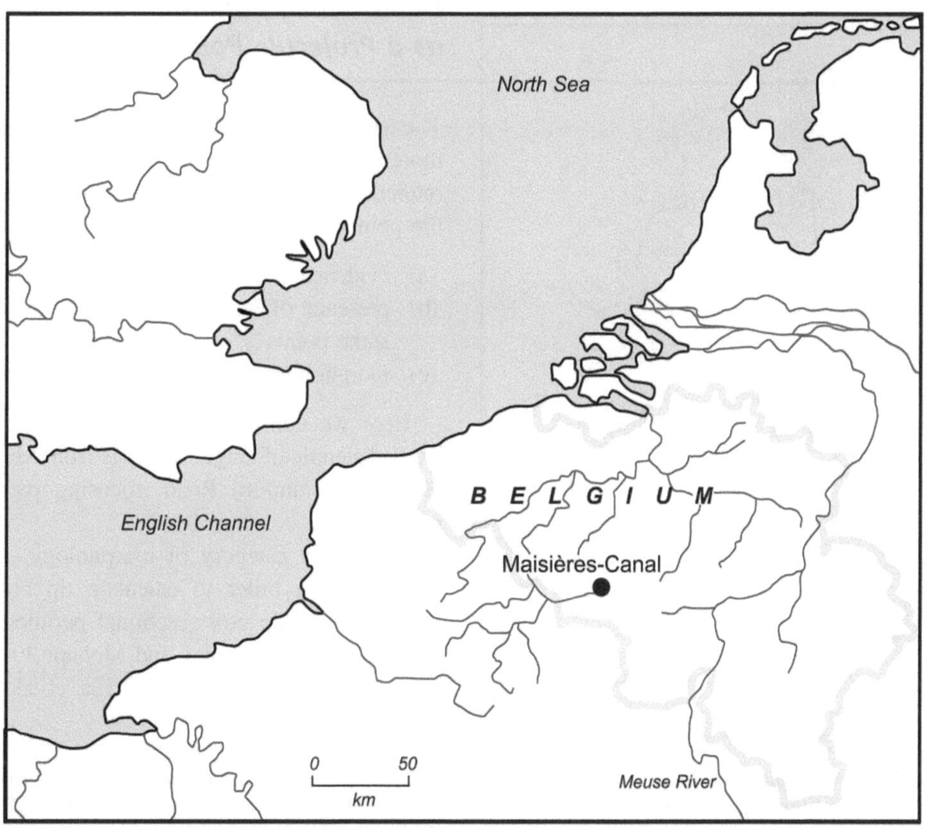

Fig. 9.2 Location of Maisières-Canal (courtesy of Sylvia Bello)

Tanged artifacts from seven British find-spots (Jacobi et al. 2010) can be attributed to the Early Gravettian largely based upon their typology (Fig. 9.3). At none of these sites have these artifacts come from a well-stratified Gravettian assemblage, and only at Kent's Cavern can a pre-Last Glacial Maximum age be soundly inferred (Jacobi and Higham 2011). However, without a precise typo-morphological parallel elsewhere in British prehistory, and as a result of the similarity to material from Maisières-Canal as described above, these artifacts are generally accepted as Gravettian (Campbell 1980; Jacobi 1980, 2007; Bahn and Pettitt 2009; Dinnis 2009).

The Bramford Road tanged tool (Fig. 9.6a) was discovered in terrace gravels of the River Gipping in Ipswich (Suffolk, England) along with a mix of artifacts from various periods (Jacobi 2007; Jacobi and Higham 2011). A complete Font-Robert point (Fig. 9.5) from Pin Hole, Creswell Crags (Derbyshire, England) is the most impressive and well-preserved of the British Gravettian tanged tools, and was one of two recovered from demonstrably Pleistocene deposits in the cave (Jacobi and Higham 2011). These two complete tanged artifacts from Bramford Road Pit (n = 1) and Pin Hole (n = 1) were studied. For curatorial reasons no other complete or near-complete British examples were available for study.

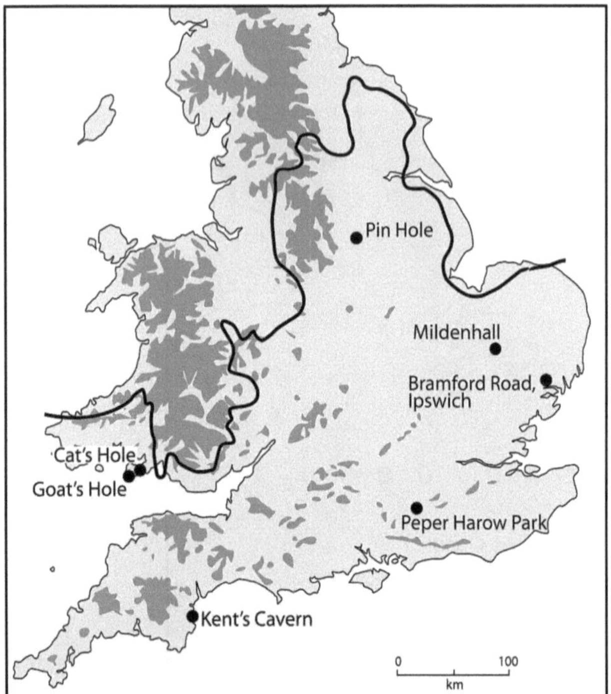

Fig. 9.3 British find-spots containing Gravettian tanged tools (From Jacobi et al. 2010: 36, reprinted with permission of Sylvia Bello and Tom Higham)

The Function of Gravettian Tanged Tools

The initial description of Gravettian Font-Robert points from the southern French site of Grotte de Font-Robert (Corrèze) defined them as weapon armatures (Bardon et al. 1908). This functional supposition continues (e.g., Jacobi 1980; Peterkin 1993; Pike-Tay and Bricker 1993; Shea 2006; Dinnis 2009), in spite of the oft-noted variation already alluded to. Two experimental studies have examined the function of Font-Robert points, reaching different conclusions. The first, by Otte and Caspar (1987), examined a small sample of tanged tools from the Maisières-Canal assemblage using combined macro- and microscopy, comparing observed use traces and breaks against an experimental replica sample. For Otte and Caspar, these artifacts are likely to have been hafted domestic tools, but unfortunately their microwear study did not include looking for traces of projectile use. The more recent study of Lansac (2002) tested replicas of Font-Robert points from the French sites of La Font-Robert, Pré-Aubert and La Grotte des Morts. Lansac's experiments produced results consistent with Font-Robert points functioning as weapon armatures, and more specifically as darts.

Determining Potential Function as a Projectile Point

The potential projectile function of tanged artifacts, and more specifically of Font-Robert points, was assessed with reference to three criteria (Villa and Soriano 2010). Projectile point use should be distinguishable via:

(a) evidence of hafting
(b) presence of impact fractures or microwear on at least some points
(c) morphology: a sharp tip and thin cross-section

Here we consider these criteria with reference to the studied sample of tanged artifacts from Maisières-Canal, Pin Hole and Bramford Road, focusing particularly on their morphology.

Within the category of morphology, metric analysis is often used in order to calculate tip cross-sectional area (TCSA),[2] and tip cross-sectional perimeter (TCSP) values (e.g., Shea 2006; Wadley and Mohapi 2008; Sisk and Shea 2009; Shea and Sisk 2010; Villa et al. 2009; Villa and

[2]TCSA values are calculated using the maximum widths and thicknesses in mm of the tools: **Area (in mm^2) = (0.5 × width) × thickness**.

Soriano 2010). Smaller TCSA improves projectile penetration, as the higher the TCSA value, the more the projectile is slowed in flight, reducing effectiveness (Hughes 1998). We can thus use TCSA values from archaeological samples as a means of testing hypotheses about weapon armature and/or projectile point usage. Alongside using TCSA values of lithic points to assess potential projectile function, it was first shown by Shea (2006) that values could be used, with caution, to assign a delivery system to potential projectiles (e.g., Shea 2006; Villa and Soriano 2010; Sisk and Shea 2011). This is useful, as the means of using macroscopic and microscopic analytical techniques to distinguish between delivery systems are only just beginning to be explored (Iovita et al. 2016; Sano et al. 2016).

TCSA can be used to sort potential projectile points into possible delivery systems, based upon correlation of sizes with different delivery systems using collated data from ethnographic and recent prehistoric examples along with experimental data (Shea 2006 and references therein, but see Clarkson 2016). Using this reference sample, points which are too large to be included in spear point categories are frequently viewed as suggestive of non-projectile functions such as knives (e.g., Harrold 1993; Shea 2006). The values are approximate, for in reality calculating the area of an object is complicated, and point cross-sections, which vary as a result of flake scars and retouch, are highly individualized. In addition, some of the sample reference collections for comparison are small, especially for spear points, and Shea (2006) cautions that comparisons should not be made without this in mind. Here we use TCSA rather than TCSP to directly compare our measurements with published TCSA values for southern French examples (Shea 2006).

Recently, use of TCSA has been questioned, with critics maintaining that it is only able to assess the *potential* but not the *probability* of lithic points to function as projectile points (e.g., see Lombard and Phillipson 2010). Sisk and Shea (2011) have themselves stressed that such studies utilizing TCSA are only intended as a first step towards understanding the *potential* of a lithic point to function as a projectile. This caveat is stated here, and we are mindful to use TCSA only to identify the potential for projectile technologies amongst the sample studied, along with a discussion on possible delivery systems. Nevertheless, TCSA remains a useful additional source of data for retaining and rejecting hypotheses about projectile function of lithic artifacts, of discussing potential delivery systems, and a way of making inter-site and inter-regional comparisons of artifacts within a typological category.

Results

Evidence of Hafting

The retouched tangs of all Gravettian tanged artifacts suggest hafting. Rots (2002) has provided direct evidence, using macro- and microscopic wear analysis on the tanged burins from Maisières-Canal, that these tangs were indeed a component of a complex haft. Given the similarity of British examples to those from Maisières-Canal, these too are assumed to have been hafted. As noted, it is the presence of a retouched tang which unifies many typological components of the Maisières-Canal assemblage. Tangs are almost exclusively created on the proximal end of wide blades, and are all dorsally retouched, with both abrupt and semi-abrupt retouch, occasionally covering the entire dorsal surface. Ventral retouch of the tang is rare, and was seen on only nine tools from the sample.

The size of these tangs is clearly deliberately standardized (Table 9.1). Standard deviations (SD) for thicknesses and widths of the tangs are small suggesting that tang widths and thicknesses were tightly controlled to fit into the hafts. In 28 tools, the maximum thickness of the tang, was equal to or exceeded the maximum thickness of the body. Lengths of the tangs are related to the size of the overall tool, but are still much smaller than the SD of the lengths of the overall points. Figure 9.4 is a scatterplot showing the relationship between overall artifact length and tang length. The r^2 value (0.55) shows that while there is a relationship between size of the artifact and size of the tang, the length of the artifact is not a good predictor for length of the tang. Tang sizes are much more standardized than overall tool size. This suggests potential rejuvenation of tools, with tangs remaining relatively constant in size while the body of the artifact is reworked or resharpened and thus reduced in size. Whatever their function, rejuvenation of hafted tools is logical, as the greatest investment in the manufacture of the tool would have

Table 9.1 Descriptive statistics for artifact length and tang measurements from sample studied

Measurement	Mean	Standard deviation	Minimum	Maximum	*n*
Tang width	13.3	3.3	8	21	49*
Tang thickness	8.9	2.7	4	15	49*
Tang length	34.8	7.5	19	51	49*
Artifact length	85.8	18.5	54	137	54

*Tangs which were partially broken were not included in measurements

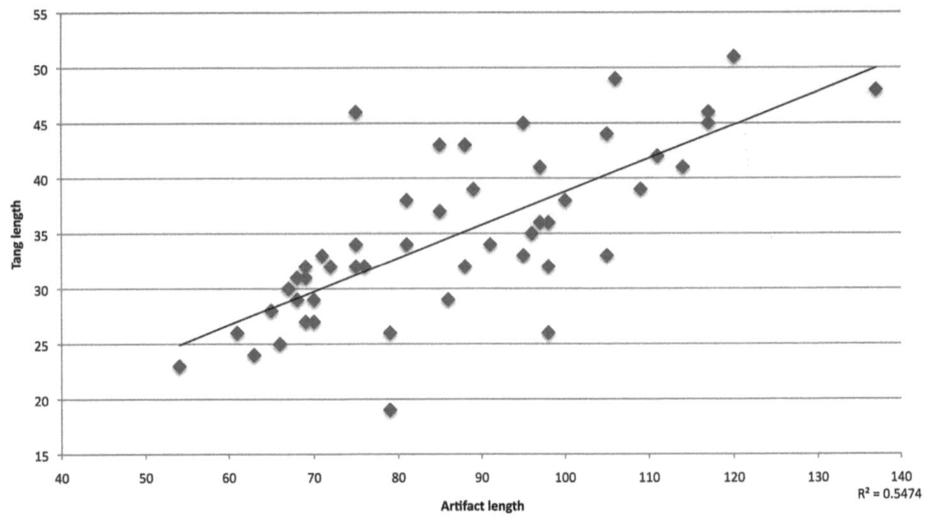

Fig. 9.4 Scatterplot showing length measurements (mm) for tanged tools' overall length to tangs' lengths (n = 49). Data for artifacts were not plotted if tangs were incomplete

been the creation of the tang and its attachment to the haft. This standardization also accords with Rots' (2002) suggestion of that they may be designed to fit directly into antler.

Impact Fractures

Pesesse and Flas (2012) recently published a macrofracture analysis of 121 tanged tools from the Maisières-Canal collection (excluding tanged burins and rough-outs), suggesting that there is good evidence that the tanged tools were used as weapon armatures. However, their analysis included crushing as well as step, feather and hinge terminating fractures, and recent experimental studies have shown that these fracture types can occur from pre- and post-depositional processes. Crushing can occur as a result of other uses for tools such as butchery, woodworking, engraving, chiseling and trampling (Shea et al. 2002), and hinge, feather and step terminating fractures can occur as a result of both manufacture and trampling (Sano 2009; Pargeter 2013 and references therein). Moreover, they do not report any bifacial spin-off fractures, or unifacial spin-off fractures >6 mm, considered by many to be reasonably reliable impact fractures (e.g., Sano 2009; Pargeter 2011). This makes the assessment of frequencies of impact fractures in the Maisières-Canal collection difficult, but there are still some clues that suggest that a projectile function for at least some of the tanged tools is possible. One line of evidence is the percentage of the tanged tools (9%) with multiple impact fractures, a percentage that is significantly higher than those produced in knapping experiments (0.6%) but still lower than those from hunting experiments (30%) (Pargeter 2013). Maisières-Canal has been interpreted as a short-term residential site (Haesaerts and De Heinzelin 1979; and see Roebroeks 2000). Frequencies wouldn't be as high at a residential site as at a hunting site (Villa and Lenoir 2009) and recycling, which has been observed on the Maisièr-es-Canal tanged tools (discussed below), would obscure the fracture signature. The other line of evidence is the location of impact fractures in relation to retouch on at least one of the tanged tools. Both Sano (2009) and Pargeter (2013) stress that the relationship of impact fractures to retouch can help to infer use as a weapon, and Pesesse and Flas (2012: Figs. 3, 4) clearly illustrate an example of a tanged tool with multiple impact fractures overlapping previous retouch. The percentages and location in relation to retouch of multiple impact fractures suggest that a proportion of the tanged tools may have served as weapon armatures.

The Pin Hole point (Fig. 9.5) also displays a possible impact fracture, with a burin-like fracture extending from the tip along the right lateral edge (30 mm). Intriguingly, Clarkson's experiments (2016) only produced burin scars >15 mm when fired as mechanically-aided projectiles, and not as hand-delivered spears. While this burin scar may conceivably relate to its use as a weapon armature, it could also have occurred from it being dropped, during knapping or possibly from some other use (Sano 2009; Pargeter 2013). As the Pin Hole point is not part of a large assemblage of tanged tools, and the burin-like fracture does not overlap previous retouch, little can be inferred with confidence about the cause of this burin-like fracture. Because of these issues with equifinality, as Shea et al. (2002) point out, it is important to incorporate data from metric analysis, as well as data from microwear analysis, where possible. We present a morpho-metric analysis of a sample of the tools below, but

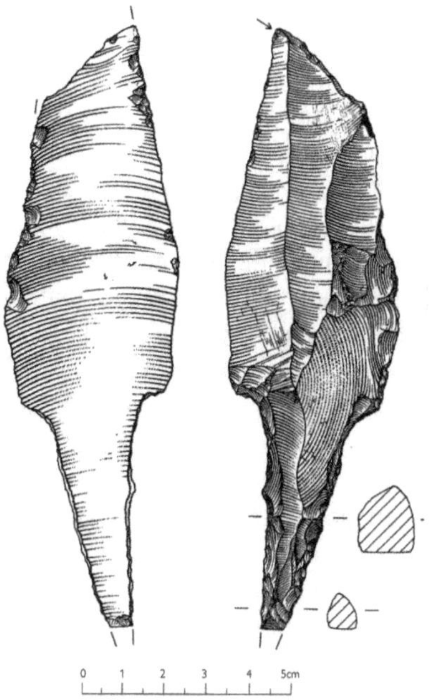

Fig. 9.5 Pin Hole tanged tool. (Drawing by Joanna Richards, from Pettitt and Jacobi 2009: 23, courtesy of Paul Pettitt and Joanna Richards)

microwear analysis to look for evidence of weapon use has yet to be undertaken on the distal body of Gravettian tanged tools.

Morphology

Blade production at Maisières-Canal is typically bi-polar, with blade detachment from opposed platform cores producing straight-profiled, relatively sturdy blades that taper distally to a point (Otte 1979; Jacobi et al. 2010; Pesesse and Flas 2012). These blanks are ideal for the production of Font-Robert points and other tanged artifacts, although it should be stressed that the blanks selected for tanged artifacts included a few with a pronounced curvature. The British artifacts are made on similar wide blade blanks. Many of the bodies and tangs of tanged tools from Maisières-Canal, as well as the complete example from Pin Hole are asymmetrical, with tangs and bodies often curving to one side (e.g., Figs. 9.6b, d, 9.7e and 9.5).

As already described, tanged artifacts at Maisières-Canal are typologically variable. 25 could not reasonably be described typologically as a "point", but, as has been noted by others, are rounded into scraper-like tips, terminate in burins, or end in a truncation (Table 9.2) (De Heinzelin

Table 9.2 Morphology of tanged tools in sample (n = 54)

Morphology	n
Pointed	28
Scraper-like/rounded	12
Burin-like	9
Truncation	2
Broken*	3
Total	54

*A few tools' tips were damaged in such a way that it was not possible to reliably infer their morphology prior to breakage

1973; Otte and Caspar 1987; Rots 2002). Furthermore, of the 27 remaining Maisières-Canal Font-Robert points, one displayed a pronounced ventral curvature. A straight profile is an important feature for lithic weapon armatures as an excessively curved profile will bend and break on impact (e.g., Bergman and Newcomer 1983; Jacobi 2007). This curved-profile point is thus excluded from consideration of TCSA values on the basis that it could not have functioned as a weapon armature. Three other points with a slight curvature have remained in the analysis. Of the two British artifacts that from Pin Hole can reasonably be described as a Font-Robert point, whereas that from Bramford Road (Fig. 9.6a) is more difficult to assign to a typological category, largely due to its heavy edge damage, though the morphology at point of discard is best described as a scraper-like tip. In total, 27 of the 54 tanged artifacts examined had the basic morphological criteria to have potentially functioned as weapon armatures.

Retouch of tanged tools is highly variable. A few pieces bear no retouch on the body itself (Table 9.3), while others have retouch completely covering the dorsal face of the tool. Type of retouch on the distal part varies from abrupt and semi-abrupt retouch to flat retouch. Ventral retouch of the body is rare (Table 9.3) and relates in every case to the tip of the tool. Of the seven tools bearing ventral retouch on the body, six have pointed tips, suggesting a concern for the pointiness and thinness of the tip of the tool itself. Six tools have dorsal and/or ventral retouch confined to 10 mm from the tip of the tool, further highlighting that within the collection there are tools for which retouch is shaping the tip of the tool. The thinness and pointiness of the tips of tools is considered to be a significant factor for performance of weapon armatures' penetration and durability (Hughes 1998; Shea et al. 2002). None of the tools' bodies have backed edges, and backed pieces occur in only 0.4% of the entire *Champ de Fouilles* lithic assemblage (Pesesse and Flas 2012).

The Pin Hole point (Fig. 9.5) is unusual amongst the British Gravettian tanged tools in that it has an absence of dorsal retouch applied to the distal part of the tool (Jacobi

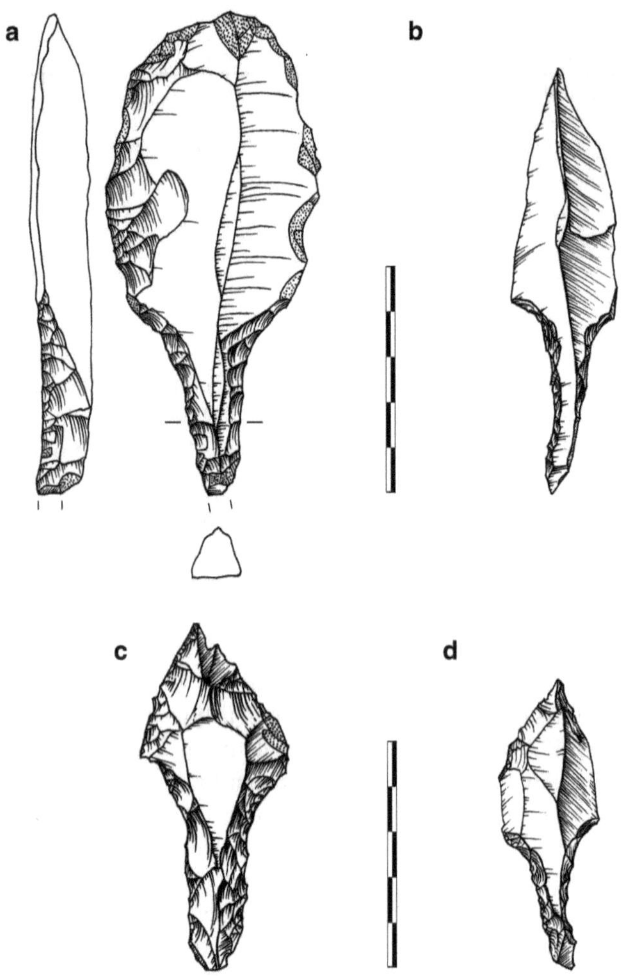

Fig. 9.6 **a** Bramford road tanged tool; **b–d** Maisières-Canal tanged tools (Drawings by M. Terrade modified after Otte 1985 [**a**] and De Heinzelin 1973 [**b–d**])

1980). There are some minimally invasive removals from the ventral side of the tool (Jacobi and Higham 2011), as well as edge damage and a burin-like removal from the tip of the tool (30 mm long), that could be use-related. Otherwise the body of the tool is largely unmodified. In contrast, the Bramford Road tool (Fig. 9.6a), with a rounded tip, has flat retouch applied to the dorsal surface, with heavy post-depositional edge damage.

Table 9.3 Retouch of distal part (body) of tanged tools (n = 54)

Retouch location	All tanged tools (n = 54)	Pointed tools (n = 27)
No retouch on body	4	2
Ventral retouch on body	7	6
Retouch confined to 10 mm from tip	6	5
Edge retouch extending below 10 mm from tip*	44	20

*Includes burin removals

TCSA and Mass

Descriptive statistics for the 27 tools classified as Font-Robert points can be found in Table 9.4. In this group, TCSA values range from 60 to 279 mm^2 (Table 9.5), all of which fall within the range suggested for spear points. Additionally, only one of the tools in the whole sample has a TCSA value (413 mm^2) that extends beyond Shea's (2006) spear point reference sample's range (but see Clarkson 2016). 19 of the 27 points (70%) are too large to fall into the reference sample's dart category, with the remaining eight falling into both dart and spear categories. Thus, on the basis of TCSA values, a classification of them as darts seems unwarranted. This is in contrast to published TCSA values for Font-Robert points from France, which appear to be better characterized as darts, albeit on the large side of that category (Shea 2006), and confirms Campbell's (1986) observation that they are on the whole larger than examples from France. To illustrate with an example, one candidate for a weapon armature in the Maisières-Canal assemblage

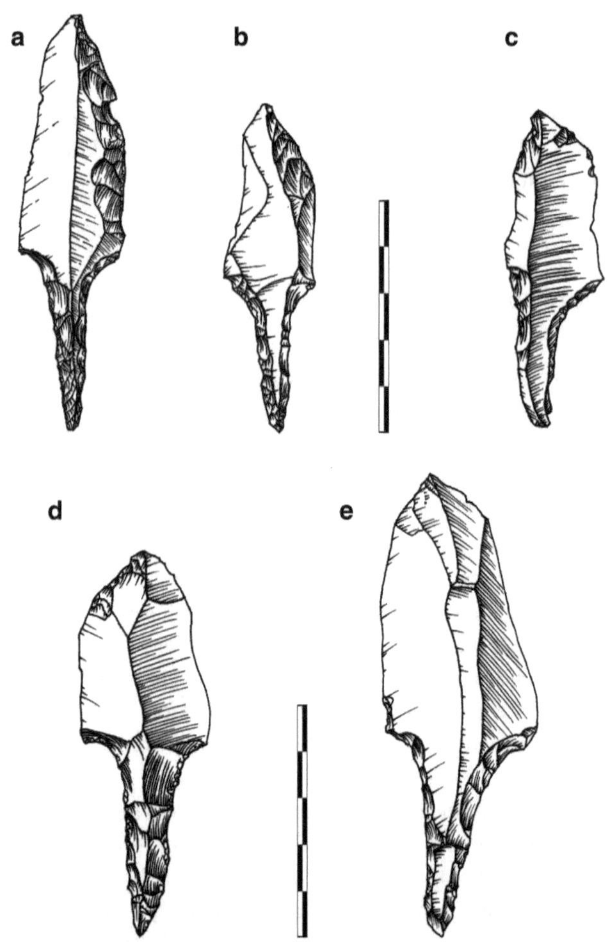

Fig. 9.7 **a–e** Maisières-Canal tanged tools (Drawings by M. Terrade modified after De Heinzelin 1973)

Table 9.4 Descriptive statistics for Font-Robert point measurements from Maisières-Canal and Pin Hole (distances in mm, mass in grams)

Measurement	Mean	Standard deviation	Minimum	Maximum	*n*
Length	87	18.1	87	137	27
Width	18.7	6.8	9	34	27
Thickness	5.5	1.9	2	9	27
Mass	19.5	11.5	6	54	27

Table 9.5 TCSA values (mm^2) of Font-Robert points from Belgian, British and French sites, compared with examples of ethnographic and recent archaeological darts and thrusting spear points

Samples	Mean	SD	Min	Max	*n*	Source[a]
Ethnographic/recent archaeological sample of darts	58	18	20	94	40	1
Experimental thrusting spear points	168	89	50	392	28	1, 2
Font-Robert points from La Ferrassie, Flageolet I, Les Vachons[b]	61	26	20	140	34	1
Font-Robert points from Maisières-Canal, Pin Hole	142	66	60	279	27	3

[a]*Sources 1* Shea 2006 using data from Thomas 1978 and Shott 1997; *2* Shea et al. 2001 cited in Shea 2006; 3, points measured by Milks
[b]Points with TCSA >120 mm^2 (n = 4) were removed from Shea's analysis. Without removing these larger points, the average TCSA is 73 mm^2 (SD = 47)

has a thin, pointed tip, with shaping of its tip by very fine, regular retouch along the left ventral edge, extending 15 mm back from the tip (Fig. 9.6b). The distal part lacks any retouch apart from that confined to the tip. This point has a TCSA of 96 mm^2, falling outside the known dart range. Based upon its TCSA value, therefore, it is unlikely to have functioned as a component of "complex" projectile technology, but it could well have been intended as a spear point.

Otte and Caspar (1987: Fig. 2: 1, 2) proposed that pointed tools from Maisières-Canal could be hafted knives. They suggest that the smallest and lightest points in the collection could be considered as projectiles, while the majority of tanged tools are better understood as domestic tools. However, as stated above, none of the Font-Robert points included in the analysis falls outside the reference sample's range for hand-delivered spears. Therefore a weapon armature function for these artifacts cannot be excluded based on size alone. Furthermore, evidence for knife-use does not rule out use as weapon armatures. We can thus retain the hypothesis that some of the Font-Robert points from Maisières-Canal, along with the Pin Hole point, could have functioned as a weapon armature on the basis that they fit the criteria discussed above, including evidence of hafting, impact damage, tip morphology and thin cross-section. Whether these potential spear points could have been hafted as hand-thrown or thrusting spears remains difficult to assess due to small or nonexistent reference samples available for comparison. Shea's (2006) data on spear point TCSA are based upon experimental points for thrusting spears only. Thus we have no metric data – apart from mass and diameter (see Noetling 1911 cited in Cundy 1989; Palter 1977; Villa and Soriano 2010) – for hand-thrown spears, which presumably would be different from those for thrusting spears. On the basis of size, the larger examples within the sample of Font-Robert points studied, if seen as components of weapons at all, would most probably have functioned as spear points. However, consideration of individual point morphology – be it the obvious differences in their shape or differences in their retouch – leaves one in little doubt that assigning function to artifacts based upon their size alone is unsatisfactory.

Discussion

Based on their typo-morphology, 50% of the 54 artifacts studied look unlikely to have functioned, at least immediately prior to the point of discard, as weapon armatures. However, there are certainly artifacts in the Maisières-Canal collection, as well as from Britain, that qualify as Font-Robert points following Demars and Laurent's (1992) definition. Once those examples that cannot be considered as weapon armatures on morphological grounds are excluded, TCSA values suggest their potential to have functioned as spear points. Clearly though, even within the homogeneous Maisières-Canal assemblage, macroscopic considerations alone suggest that Font-Robert points are not unified by a single function. This presumed functional versatility supports in part both Otte and Caspar's (1987) study suggesting their use as hand-held tools as well as Pesesse and Flas's (2012) work suggesting the presence of projectile points amongst the tanged tools.

Sizes of Font-Robert points studied are highly variable, and intriguingly have a mean TCSA value that is over twice as high as those from southern France (Table 9.5). These differences may support a chrono-cultural distinction between Maisières-Canal and other sites in Belgium and France, but their implication certainly extends to functional considerations. TCSA values, for example, may allow Shea's (2006) sample from southern France to be seen as darts and therefore a component of "complex" projectile technology. The same measurements put into question whether this would have been the function of the vast majority of Font-Robert points studied from Maisières-Canal and Britain.

In light of this consideration of artifact taxonomy, morphology and function, it is important to note that evidence for the recycling of tools at Maisières-Canal is easy to find. For example, a non-tanged "Maisières point" from Maisières-Canal housed at the British Museum has been reworked into a dihedal burin prior to its discard (RD pers. obs.), and Pesesse and Flas (2012) report that several tanged burins in the assemblage bear retouch beneath their burin facets consistent with their former life as other tool types. Recycling of artifacts certainly confuses both observable tip-shape and microscopic use-traces in terms of primary mode of use, and it is logical that where an artifact is discarded is the site where it had been recycled (Hays and Surmely 2005). Therefore evidence of recycling amongst the collection of tanged tools from Maisières-Canal supports the possibility that weapon armatures were re-worked into burins, scrapers and other domestic tools. Shott (2016) also highlights the significance of curating projectile points to extend their longevity for their original function, which would also have increased morpho-metric variability of discarded tools

It is easy to conceive how this technological versatility may have extended to a functional versatility. The morphological and, by logical extension, functional variability of the tanged tools studied may alternatively be explained by viewing them as versatile tools. They could perhaps have had multiple functions, acting as knives or spear points simultaneously, or potentially, one subsequent to the other, a practice known ethnographically, particularly for projectile points amongst highly mobile hunting groups (Ahler 1978; Greaves 1997; Nelson 1997). Gravettian people appear to have lived highly mobile lives, as indicated by movements of raw materials and shells across vast distances, and experienced downturns and rapid shifts in climate (Roebroeks 2000). There is evidence from microwear and morpho-metric studies on other Gravettian tool types which support both the use of multipurpose tools and the practice of recycling of projectile points (Harrold 1993 and references therein). However, it must be stressed that there are clear examples that cannot be satisfactorily viewed to have functioned as a weapon armature at any point in their life history. The pronounced curvature of one tanged tool, and the tanged tool with an extremely large TCSA are two such examples. Therefore the tanged tools as a group cannot be viewed either as versatile tools or recycled weapon armatures as there are examples that would not fit into either category.

As for most archaeological taxa, tool typologies are clearly necessary, but their interpretative usefulness is inevitably limited. Irrespective of the complexity and standardisation of any particular artifact type, to understand its life history and functional use(s) ideally requires investigation using many different methods (e.g., see papers in Bracco et al. 2006; also Hardy et al. 2008; Dinnis et al. 2009). In many cases a combination of technological considerations, use-wear, residue and macro-fracture analysis and experimental replication is required before the functional history of artifacts can be soundly inferred. Here, we have highlighted that basic morphological data in conjunction with metric analysis is still useful when discussing potential projectile function.

For Font-Robert points, a systematic controlled experimental program of artifacts that represent the morphological variability seen in the archaeological record – using them as darts, hand-delivered spears and knives – would contribute towards a better understanding of this complex type. Combining this with a comprehensive microwear analysis of the Maisières-Canal collection would be favorable and, given its fresh condition, possible. In the meantime, we caution against using Gravettian tanged tools as a proxy for the presence of projectile technology, in spite of some evidence supporting that possibility. Indeed, this caution extends to using Gravettian tanged tools as a proxy for any single functional activity at all.

Acknowledgement We are pleased to thank Ivan Jadin at the IRSNB, as well as the Manchester, Creswell Crags and Ipswich Museums, who granted access to their collections, to Damien Pesesse and Damien Flas for allowing us to reference their "in press" work, and to Tom Higham, Sylvia Bello, Joanna Richards, Paul Pettitt, Marcel Otte and the IRSNB (on behalf of Jean De Heinzelin) for permission to reproduce or modify images. We thank Sander-Martijn Milks and Melanie Terrade for drawing, scanning and preparing images and Mark Wigglesworth for proof reading. Finally, three anonymous reviewers gave comments on drafts of this paper that helped improve our work.

References

Ahler, S. A. (1978). *Projectile Point form and function at Rodgers Shelter*. Missouri, Columbia, MO: Missouri Archaeological Society.

Bahn, P., & Pettitt, P. (2009). *Britain's oldest art: The ice age cave art of Creswell Crags*, with contributions by Andrew Chamberlain et al. Swindon: English Heritage.

Bardon, L., Bouyssonnie, A., & Bouyssonnie, J. (1908). La grotte de La Font-Robert. *Congrès d'anthropologie et d'archéologie préhistoriques, compte rendu de la treizième session, Monaco 1906, tome 2*, 172–184.

Bergman, C. A., & Newcomer, M. H. (1983). Flint arrowhead breakage: Examples from Ksar Akil, Lebanon. *Journal of Field Archaeology, 10*, 231–243.

Bertran, P., Caner, L., Langohr, R., Lemée, L., & d'Errico, F. (2008). Continental palaeoenvironments during MIS 2 and 3 in southwestern France: The La Ferrassie rockshelter record. *Quaternary Science Reviews, 27*, 2048–2063.

Bracco, J.-P., De Araujo Igreja, M., & Le Brun-Ricalens, F. (Eds.). (2006). *Burins Préhistoriques: Formes, fonctionnements, fonctions*. Luxembourg: Archéologiques 2.

Campbell, J. B. (1980). Le problème des subdivisions du Paléolithique supérieur britannique dans son cadre européen. *Bulletin de la Société royale belge d'Anthropologie et de Préhistoire, 91*, 39–77.

Campbell, J. B. (1986). Hiatus and continuity in the British Upper Palaeolithic: A view from the Antipodes. In D. A. Roe (Ed.), *Studies in the Upper Palaeolithic of Britain and Northwest Europe* (pp. 7–42). Oxford: BAR International Series 296.

Clarkson, C. (2016). Testing archaeological approaches to determining past projectile delivery systems using ethnographic and experimental data. In R. Iovita & K. Sano (Eds.), *Multidisciplinary approaches to the study of Stone Age weaponry* (pp. 189–201). Dordrecht: Springer.

Conard, N. J., & Moreau, L. (2004). Current research on the Gravettian of the Swabian Jura. *Mitteilungen der Gesellschaft für Urgeschichte, 13*, 29–59.

Cundy, B. J. (1989). *Formal variation in Australian spear and spearthrower technology*. Oxford: BAR International Series 546.

De Heinzelin, J. (1971). Le gisement périgordien de Maisières-Canal (1966). *Bulletin de la Société royale belge d'Anthropologie et de Préhistoire, 82*, 63–76.

De Heinzelin, J. (1973). *L'industrie du site paléolithique de Maisières-Canal* (Mémoires del'Institut royal des Sciences naturelles de Belgique 171). Brussels: Institut royal des Sciences naturelles de Belgique.

De Sonneville-Bordes, D. (1960). *Le Paleolithique supérieur en Périgord*. Bordeaux: Imprimeries Delmas.

Delporte, H., & Tuffreau, A. (1984). Les industries du Périgordien V de La Ferrassie. Le grand abri de La Ferrassie. Fouilles 1968–1973. Sous la direction de H. Delporte. *Études quaternaires. Mémoire n° 7*, 235–247.

Demars, P. Y., & Laurent, P. (1992). *Types d'outils lithiques du paléolithique supérieur en Europe*. Paris: Centre national de la recherche scientifique.

Desbrosse, R., & Kozlowski, J. K. (1988). *Hommes et climats à l'âge du mammouth. Le Paléolithique supérieur d'Eurasie centrale*. Paris: Masson.

Dinnis, R. (2009). *Understanding the British Aurignacian*. PhD dissertation, University of Sheffield.

Dinnis, R., Pawlik, A., & Gaillard, C. (2009). Bladelet cores as weapon tips? Hafting residue identification and micro-wear analysis of three carinated burins from the late Aurignacian of Les Vachons, France. *Journal of Archaeological Science, 36*, 1922–1934.

Djindjian, F., & Bosselin, B. (1994). Périgordien et Gravettien: l'épilogue d'une contradiction? *Préhistoire Européenne, 6*, 117–131.

Djindjian, F., Koslowski, J., & Otte, M. (1999). *Le Paléolithique supérieur en Europe*. Paris: Armand Colin.

Flas, D. (2009). The Lincombian-Ranisian-Jermanowician and the limit of the Aurignacian spreading on the northern European plain. In F. Djindjian, J. K. Kozlowski & N. Bicho (Eds.), *Le concept de territoires dans le Paléolithique supérieur européen, UISPP, Actes du XVe congrès mondial (Lisbonne, 4–9 septembre 2006)—Session C16* (pp. 135–142). Oxford: BAR International Series 938.

Greaves, R. D. (1997). Hunting and multifunctional use of bows and arrows: Ethnoarchaeology of technological organization among Pumé hunters of Venezuela. In H. Knecht (Ed.), *Projectile technology* (pp. 287–320). New York: Plenum.

Haesaerts, P., & Damblon, F. (2004). Les dates radiocarbones de Maisières-Canal. In R. Miller, P. Haesaerts & M. Otte (Eds.), *L'atelier de taille aurignacien de Maisières-Canal (Belgique)* (pp. 27–28). Liège: ERAUL 110.

Haesaerts, P., & De Heinzelin, J. (1979). *Le Site paléolithique de Maisières-Canal*. Bruges: Dissertationes Archaeologicae Gandenses 19.

Hardy, B. L., Bolus, M., & Conard, N. J. (2008). Hammer or crescent wrench? Stone-tool form and function in the Aurignacian of southwest Germany. *Journal of Human Evolution, 54*, 648–662.

Harrold, F. B. (1993). Variability and function among Gravette points from Southwestern France. In G. L. Peterkin, H. Bricker & P. A. Mellars (Eds.), *Hunting and animal exploitation in the later Paleolithic and Mesolithic of Eurasia* (pp. 69–81). Washington DC: American Anthropological Association.

Hays, M., & Surmely, F. (2005). Réflexions sur la fonction des microgravettes et la question de l'utilisation de l'arc au Gravettien ancien. *PALEO, 17*, 145–155.

Higham, T. (2011). European middle and Upper Palaeolithic radiocarbon dates are often older than they look: Problems with previous dates and some remedies. *Antiquity, 85*, 235–249.

Hughes, S. S. (1998). Getting to the point: Evolutionary change in prehistoric weaponry. *Journal of Archaeological Method and Theory, 5*, 345–408.

Iovita, R., Schönekeß, H., Gaudzinski-Windheuser, S., & Jäger, F. (2016). Identifying weapon delivery systems using macrofracture analysis and fracture propagation velocity: A controlled experiment. In R. Iovita & K. Sano (Eds.), *Multidisciplinary approaches to the study of Stone Age weaponry* (pp. 13–27). Dordrecht: Springer.

Jacobi, R. M. (1980). The Upper Palaeolithic of Britain with special reference to Wales. In J. A. Taylor (Ed.), *Culture and environment in prehistoric wales. Selected essays* (pp. 15–100). Oxford: BAR British Series 76.

Jacobi, R. M. (2007). A collection of early Upper Palaeolithic artefacts from Beedings, near Pulborough, West Sussex, and the context of similar finds from the British Isles. *Proceedings of the Prehistoric Society, 73*, 229–325.

Jacobi, R. M., & Higham, T. F. G. (2011). The British earlier Upper Palaeolithic: Settlement and chronology. In N. Ashton, S. G. Lewis & C. Stringer (Eds.), *The ancient human occupation of Britain* (pp. 181–222). Amsterdam: Elsevier.

Jacobi, R. M., Higham, T. F. G., Haesaerts, P., Jadin, I., & Basell, L. S. (2010). Radiocarbon chronology for the early Gravettian of northern Europe: New AMS determinations for Maisières-Canal, Belgium. *Antiquity, 84*, 26–40.

Lansac, P. (2002). Fonction des pointes de la Font-Robert: Étude archéologique et expérimentale appliquée aux collections de la Font-Robert, Pré-Aubert et la Grotte des Morts (Corrèze) 1. *Préhistoire du Sud-Ouest, juin, 2002*, 1–24.

Lombard, M., & Phillipson, L. (2010). Indications of bow and stone-tipped arrow use 64,000 years ago in KwaZulu-Natal, South Africa. *Antiquity, 84*, 635–648.

Mussi, M. (2001). *Earliest Italy: An overview of the Italian Paleolithic and Mesolithic*. New York: Kluwer Academic/Plenum Publishers.

Nelson, M. C. (1997). Projectile points: Form, function, and design. In H. Knecht (Ed.), *Projectile technology* (pp. 371–384). New York: Plenum.

Noetling, F. (1911). Notes on the hunting sticks (Lughrana), spears (Perenna), and baskets (Tughbrana) of the Tasmanian Aborigines. *Papers of the Royal Society of Tasmania*, 38–61.

Otte, M. (1974). Les pointes à retouches plates du paléolithique supérieur initial de Belgique. *Études et Recherches Archéologiques de l'Université de Liège (ERAUL), Serie A, 2*.

Otte, M. (1979). Documentation archéologique. In P. Haesaerts & J. De Heinzelin (Eds.), *Le Site paléolithique de Maisières-Canal* (pp. 69–89). Bruges: Dissertationes Archaeologicae Gandenses 19.

Otte, M. (1985). *Les industries à pointes foliacees et à pointes pédonculées dans le nord-ouest européen*. Editions du C.E.D.A.

Otte, M., & Caspar, J.-P. (1987). Les pointes de la Font-Robert: Outils emmanchés? In D. Stordeur (Ed.), *La main et l'outil, Manches et emmanchements préhistoriques* (pp. 65–74). Lyon: Travaux de la Maison de l'Orient 15.

Otte, M., & Noiret, P. (2007). Le Gravettien du nord-ouest de l'Europe. *Paléo, 19*, 243–255.

Palma di Cesnola, A. (1993). *Il Paleolitico superiore in Italia*. Firenze: Garlatti e RazzaiEditori.

Palter, J. L. (1977). Design and construction of Australian spear-thrower projectiles and hand-thrown spears. *Archaeology and Physical Anthropology in Oceania, 12*, 161–172.

Pargeter, J. (2011). Assessing the macrofracture method for identifying Stone Age hunting weaponry. *Journal of Archaeological Science, 38*, 2882–2888.

Pargeter, J. (2013). Rock type variability and impact fracture formation: Working towards a more robust macrofracture method. *Journal of Archaeological Science, 40*, 4056–4065.

Pesesse, D., & Flas, D. (2012). The Maisierian, at the edge of the Gravettian. *Proceedings of the Prehistoric Society, 78*, 95–109.

Pesesse, D., & Flas, D. (2013). Which Gravettians at Spy? In H. Rougier & P. Semal (Eds.), *Spy Cave: 125 years of multidisciplinary research at the Betche-aux-Rotches (Jemeppe-sur-Sambre, Province of Namur, Belgium)*, Vol. 1. Brussels, Royal Belgian Institute of Natural Sciences, Royal Belgian Society of Anthropology and Prehistory and NESPOS Society, *Anthropologica et Praehistorica 123/2012*, 257–268.

Peterkin, G. L. (1993). Lithic and organic hunting technology. In G. L. Peterkin, H. Bricker & P. A. Mellars (Eds.), *Hunting and animal exploitation in the later Paleolithic and Mesolithic of Eurasia* (pp. 49–68). Washington DC: American Anthropological Association.

Pettitt, P., & Jacobi, R. (2009). The Palaeolithic archaeology of Creswell Crags. In P. Bahn & P. Pettitt (Eds.), *Britain's oldest art: The ice age cave art of Creswell Crags with contributions by Andrew Chamberlain et al.* (pp. 16–35). Swindon: English Heritage.

Pettitt, P., & White, M. (2012). *The British Palaeolithic: Human societies at the edge of the Pleistocene world*. London: Routledge.

Pike-Tay, A., & Bricker, H. M. (1993). Hunting in the Gravettian: An examination of evidence from Southwestern France. In G. L. Peterkin, H. Bricker & P. A. Mellars (Eds.), *Hunting and animal exploitation in the later Paleolithic and Mesolithic of Eurasia* (pp. 127–143). Washington DC: American Anthropological Association.

Roebroeks, W. (2000). A marginal matter: The human occupation of northwestern Europe-30,000-20,000 bp. In W. Roebroeks, M. Mussi, J. Svoboda & K. Fennema (Eds.), *Hunters of the Golden Age: The Mid Upper Palaeolithic of Eurasia 30000–20000 BP* (pp. 299–312). Leiden: University of Leiden.

Rots, V. (2002). Are tangs morphological adaptations in view of hafting? Macro-and microscopic wear analysis on a selection of tanged burins from Maisières-Canal. *Notae Praehistoricae, 22*, 61–69.

Sano, K. (2009). Hunting evidence from stone artefacts from the Magdalenian cave site Bois Laiterie, Belgium: A fracture analysis. *Quartär, 56*, 67–86.

Sano, K., Denda, Y., & Oba, M. (2016). Experiments in fracture patterns and impact velocity with replica hunting weapons from Japan. In R. Iovita & K. Sano (Eds.), *Multidisciplinary approaches to the study of Stone Age weaponry* (pp. 29–46). Dordrecht: Springer.

Shea, J. J. (2006). The origins of lithic projectile point technology: Evidence from Africa, the Levant, and Europe. *Journal of Archaeological Science, 33*, 823–846.

Shea, J. J., & Sisk, M. L. (2010). Complex projectile technology and *Homo sapiens* dispersal into western Eurasia. *PaleoAnthropology, 2010*, 100–122.

Shea, J. J., Davis, Z., & Brown, K. (2001). Experimental tests of Middle Palaeolithic spear points using a calibrated crossbow. *Journal of Archaeological Science, 28*, 807–816.

Shea, J. J., Brown, K., & Davis, Z. (2002). Controlled experiments with Middle Paleolithic spear points: Levallois points. In J. R. Mathieu (Ed.), *Experimental archaeology: Replicating past objects, behaviors and processes* (pp. 55–72). Oxford: BAR International Series 1035.

Shott, M. J. (2016). Survivorship distribution in experimental spear points: Implications for tool design and assemblage formation. In R. Iovita & K. Sano (Eds.), *Multidisciplinary approaches to the study of Stone Age weaponry* (pp. 245–258). Dordrecht: Springer.

Sisk, M., & Shea, J. (2009). Experimental use and quantitative performance analysis of triangular flakes (Levallois points) used as arrowheads. *Journal of Archaeological Science, 36*, 2039–2047.

Sisk, M., & Shea, J. (2011). The African origin of complex projectile technology: An analysis using tip cross-sectional area and perimeter. *International Journal of Evolutionary Biology, 2011*, 1–8.

Villa, P., & Lenoir, M. (2009). Hunting and hunting weapons of the Lower and Middle Paleolithic of Europe. In J.-J. Hublin & M. P. Richards (Eds.), *The evolution of hominin diets: Integrating approaches to the study of Palaeolithic subsistence* (pp. 59–88). Dordrecht: Springer.

Villa, P., & Soriano, S. (2010). Hunting weapons of Neanderthals and early modern humans in South Africa: Similarities and differences. *Journal of Anthropological Research, 66*, 5–38.

Villa, P., Boscato, P., Ranaldo, F., & Ronchitelli, A. (2009). Stone tools for the hunt: Points with impact scars from a Middle Paleolithic site in southern Italy. *Journal of Archaeological Science, 36*, 850–859.

Wadley, L., & Mohapi, M. (2008). A segment is not a monolith: Evidence from the Howiesons Poort of Sibudu, South Africa. *Journal of Archaeological Science, 35*, 2594–2605.

Chapter 10
Early Gravettian Projectile Technology in Southwestern Iberian Peninsula: The Double Backed and Bipointed Bladelets of Vale Boi (Portugal)

João Marreiros, Nuno Bicho, Juan Gibaja, João Cascalheira, and Telmo Pereira

Abstract Unlike other Gravettian contexts in Southern Iberian Peninsula, the Early Gravettian lithic assemblage from the archaeological site of Vale Boi (SW Portugal) is characterized by the absence of typical backed points, such as Gravettian and Microgravette points. Instead, backed technology is present in the unusual form of bipointed double backed bladelets. The presence of these backed tools in other Gravettian contexts is very rare, and their strong presence in the lithic assemblages from Vale Boi has no parallel in Southern Iberia, representing a novelty for the Gravettian record in the region. Given their morphology, this type of backed tool has been associated, in other industries, with perforation activities. In this paper, however, we present the results on technological, macro and micro-wear analyses showing the presence of fatigue traces (diagnostic impact fractures and hafting traces) commonly associated to projectile tips. These data represent a novelty in lithic projectile technology from Southwestern Iberia, and may reflect improving hunting techniques related to diet diversification and intensification and/or stylistic variation among Gravettian population.

Keywords Iberian Peninsula • Early Gravettian • Backed technology • Use-wear analysis

J. Marreiros (✉) · N. Bicho · J. Cascalheira · T. Pereira
ICArEHB. FCHS, Universidade do Algarve, Campus Gambelas, 8005-139 Faro, Portugal
e-mail: jmmarreiros@ualg.pt

N. Bicho
e-mail: nbicho@ualg.pt

J. Cascalheira
e-mail: jmcascalheira@ualg.pt

T. Pereira
e-mail: telmojrpereira@gmail.com

J. Gibaja
Consejo Superior de Investigacións Cientificias, Institución Milá i Fontanales, C/Egipcíaques, 15, 08001 Barcelona, Spain
e-mail: jfgibaja@imf.csic.es

Gravettian Backed Technology

During the Gravettian, backed technology is predominant, diverse and an important component within lithic assemblages in Central and Western Europe. Different morphotypes, such as fléchettes, Microgravette, Gravette and Font Robert points appeared and have been used elsewhere as diagnostic *fossile directeurs* in this technocomplex (Otte 1983; Bosselin and Djindjian 1994; Djindjian and Bosselin 1994; Bosselin 1996; Otte and Noiret 2007; Moreau 2012). From a technological perspective, Gravettian backed tools can be organized into two groups: (1) those with laminar dimensions (e.g., Gravette and Font Robert points), and (2) microlithic elements (e.g., Microgravettes, unilateral backed and double backed bladelets). Backed tools variability among and within Gravettian assemblages is likely related to two main aspects: (1) stylistic variation, argued as reflecting of socio-cultural and ethnic diversity among Gravettian groups in a specific or different geographical territories, and (2) function, as considerable differences in morphometric and technological attributes may suggest different functionality and, therefore, variability within and among assemblages suggest different site function (Harrold 1993).

During the last decades functional analysis focused on macro and micro wear traces contributing to one of most debated topics about Gravettian technology: function and variability among Gravettian backed lithic tools (e.g., Donahue 1988; Soriano 1998; Kimball 1992; O'Farrell 2004; Hays and Surmely 2005). Due to its morphometric aspect, laminar tools have been intuitively and commonly associated with two types of function: as knives and as projectiles. Recently, use-wear analysis confirmed this idea; comparative studies between experimental and archaeological samples revealed three types of use for these tools: (1) projectile points, by the presence of diagnostic fatigue traces associated with such activities, (2) knives, evidenced by the presence of butchering micro-traces and (3) both activities, butchering after impact damage or both functions simultaneous.

Radu Iovita and Katsuhiro Sano (eds.), *Multidisciplinary Approaches to the Study of Stone Age Weaponry*, Vertebrate Paleobiology and Paleoanthropology, DOI 10.1007/978-94-017-7602-8_10

From the Gravettian, microlithic backed elements become predominant in lithic assemblages in SW Iberia archaeological record and, from the functional perspective, reveal a different and more complex scenario than in previous times. Due to this morphology, these micro elements suggest hafting technology. In this case the lithic must be attached to another component (i.e., bone, antler or wood). Thus, component weaponry tends to be more complex and may suggest different types of hafting.

In South-southwestern Iberia, Roman and Villaverde's study (2006) suggested that there is a clear tendency for microlithization during the Gravettian, where microgravettes are more frequent than La Gravette points. Such tools may be related into two aspects: (1) different tasks, therefore indicating a functional specialization among sites in that region, and (2) cultural and stylistic differences between territories.

In this paper, we present the results of technological and use-wear analysis of bipointed double backed bladelets from the Early Gravettian of Vale Boi (Southern Portugal). Given their morphology, we initially thought that they had been used as perforators for beads and other body ornaments. During our analysis, however, using macro and microscopic approaches, we identified diagnostic marks, such as DIF (diagnostic impact fractures) and adherent residues, generally associated with projectile lithic tools. These data, thus, suggest a novel technological adaptation for the Southwestern Iberian Gravettian by the early Anatomically Modern Humans in the region.

Macro and Micro Wear Analysis on Lithic Projectiles

Use-wear analysis is the most important tool to recognize evidence of use in lithic and bone tools (Semenov 1964; Hayden 1979; Plisson and Geneste 1989; Anderson 1990). During the last decade experimental tests and macro and micro wear analyses have been used to find diagnostic evidence for the use of lithic tools as projectiles in hunting activities: the DIF and hafting traces (Frison and Bradley 1980; Bradley 1982; Bergmann and Newcommer 1983; Fischer et al. 1984; Odell and Cowan 1986; Bradley and Frison 1987; Geneste and Plisson 1993; O'Farrell 1996, 2004; Shea et al. 2002; Lombard et al. 2004; Lombard 2005; Villa and Lenoir 2006). According to these experimental tests, elements in lithic tools associated with projectiles activities are categorized in two groups and can be analyzed by two different use-wear approaches: impact traces, including DIF and striations; and hafting traces, such as polish and organic residues (adhesive gum). DIF, fractures and flaking marks, are usually observed using a macro and stereomicroscope. These kinds of marks are associated with the impact of the lithic tools with the target. There are four different types of DIF: (1) burin fracture, normally present on the lateral edge of the tip, (2) stepped, hinged and feather terminations on the dorsal or ventral face of the lithic, (3) scaled or crashed tips, and (4) fluted scars. Striations are normally linear grooves that follow the longitudinal axis associated with the direction of the impact between the tool and the target.

Hafting traces are mostly recognized by clusters of small fractures and micro-polish on the hafted edge (Rots 2003, 2011), as well as by adhering residue used to haft the lithic to its handle (Haslam 2006; Langejans 2011; Rots et al. 2011).

Regarding the DIF, several experimental studies, however, have shown that all these stigma are mainly the consequence of the contact between the tool and the animal hard material (i.e., bone or antler). The absence of these evidences is not, however, a reliable proof that specific tools were not used as projectiles. No re-use or the contact with soft material (i.e., hide and flesh), reveal that such diagnostic marks are only present in a small percentage of cases.

Our study includes 12 bipointed double backed bladelets made on chert. These tools were recovered from the Early Gravettian levels of the archaeological site of Vale Boi; 3 are from the Terrace and 9 from the Rockshelter excavation areas. Use-Wear analysis was carried out using a stereomicrosope (Leica LED5000 SLI) and a transmitted light microscope (Leica DM2500M, 50-600x).

The Site of Vale Boi

Vale Boi is a multicomponent site, located in Southwestern Portugal, with a long Upper Paleolithic chronostratigraphic record (Bicho et al. 2003, 2010; Manne et al. 2012). The sequence is accompanied by a set of more than twenty absolute dates, which makes this archaeological site an important reference for the reconstruction of the Upper Paleolithic record in Southwestern Iberia.

Vale Boi is situated at the extreme south-southwestern Atlantic coast of Iberian Peninsula (Cape St. Vicente, Algarve). From the geological perspective the valley is bordered to the north by a landscape of schist and greywacke from the Carbonic and in the South by limestone and dolomite formations from the Triassic and Jurassic periods. Some 15–20 km distant there are a couple of chert quarries near the Cape St. Vicente (Bicho et al. 2010). In this case, a small river in the valley follows North-to-South for 2 km until reaching the Atlantic Ocean. The site is limited by a 10 m high limestone outcrop (Fig. 10.1) on top of a slope, marked by a sequence of geological platforms, resulting from the erosion of the limestone bedrock.

The site was found in 1998 and tested in 2000. The team identified an area of around 10,000 m^2 of scattered artifacts

Fig. 10.1 Archaeological site of Vale Boi, *1* Southwestern Iberia geographical location; *2* Limestone outcrop, *3* Panoramic photo from the excavation areas

including flakes, cores, faunal remains and anvils (Bicho et al. 2012). There are three areas in the site: the Rockshelter, the Slope, and the Terrace.

The Stratigraphy and Chronometric Sequence

The Rockshelter has been excavated since 2006 and at the moment there are two excavation areas. One has 20 m^2 by 8 m deep, while the other (started in 2010) is 4 m^2 by 0.7 m deep. The first has one Magdalenian horizon (Z), three layers attributed to the Solutrean (A, B and C) overlaying the Gravettian layer D. Big limestone boulders resulting from the roof collapse cover the Solutrean occupations and are present in smaller number throughout the sequence (Bicho et al. 2012) (Fig. 10.2). All archaeological layers include lithic artifacts and faunal remains but the Magdalenian and the Early Gravettian assemblages are poor when compared to the Solutrean contexts (Cascalheira 2010).

The slope is marked by a stepped sequence of flattish platforms. The archaeology fills large cavities in the limestone bedrock, (Bicho et al. 2003, 2010), and likely correspond to midden deposits (two excavated areas in a total of 15 m^2). The archaeological sequence is similar to the Terrace, with Magdalenian, Solutrean, Proto-Solutrean and Gravettian assemblages.

On the Terrace, the excavation area is 5 × 5 m^2 by 2 m deep. The deposit presents a long sequence of human occupations dated to the Early Neolithic, Solutrean, Proto-Solutrean, Late Gravettian, Early Gravettian and Mousterian (Marreiros et al. 2012). The excavation has yet to reach bedrock. All Upper Paleolithic layers have a high density of remains. There are three Gravettian levels (c. 27, 29, and 32.5 cal kBP) (Table 10.1), whose lithic assemblages are associated with backed technology. Although double backed points are only present in the lower occupation, from a technological perspective, all assemblages are similar.

Results

The Early Gravettian lithic assemblage of Vale Boi includes chert, quartz and greywacke debitage, although reflecting different kinds of reduction strategies. Chert is the most complex: nodules were reduced through simple reduction strategies with very little shaping or cortex removal. Prismatic cores were predominantly used for flake and bladelet

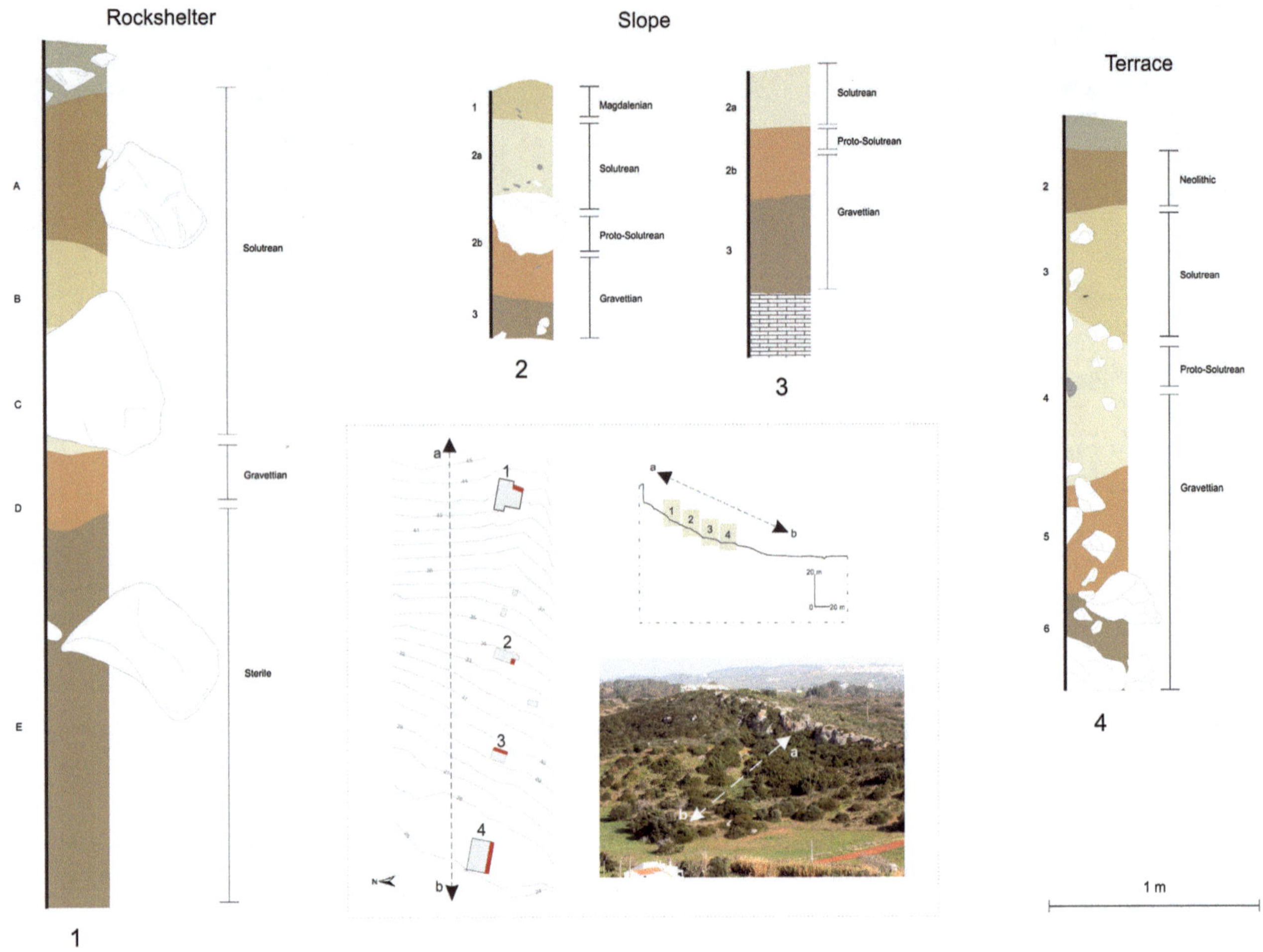

Fig. 10.2 Vale Boi, chrono-stratigraphic sequence from Terrace and Rockshelter

production. Bladelet reduction was also made by burin and carinated endscrapers exploitation (Marreiros et al. 2012) (Fig. 10.3). The functional analysis made on diverse typological elements from the lithic assemblage show that different materials were worked at the site (i.e., hide, wood, antler and butchering), which may reflect long settlement occupations (Gibaja and Bicho 2006; Bicho et al. 2010).

The assemblage we present in this paper is composed of bipointed double backed bladelets all made exclusively on local chert, and even though it is a small-sized sample, the study reflects interesting results (Fig. 10.4).

These backed tools are characterized by slightly twisted or curved sections in the long axis, with no cortex, except for a single piece from the Terrace that has a very small cortical area. The butt was frequently removed and the cross-section is mainly quadrangular. Typologically these tools are defined as a point made on a rectilinear shaped bladelet, backed retouch on both edges, and pointed at both the distal and proximal tips.

Morphometric analysis shows that the projectiles from the Rockshelter are clearly smaller with different means and maximum dimensions – 23.89 × 4.32 × 3.5 mm (length × width × thickness), while the examples from the Terrace are approximately 18.95 × 3.75 × 3.09 mm (Table 10.2), although this assemblage is too small to pursuit further interpretations.

Macroscopic observations show that 4 of the 12 tools are fractured by impact at their distal ends. Despite the low number of macroscopic breaks, the terminal fractures exhibit diagnostic microscopic impact fractures, such as burin shape

Table 10.1 Radiometric ^{14}C dates for the Upper Pleistocene levels at Vale Boi

Area	Level	Phase	Lab.	Date	Material	Date CAL BP[a]	Notes
Terrace	2	Early Neolithic	Wk-17030	6036 ± 39	Bone	6990–6785	
Terrace	2	Early Neolithic	OxA-13445	6042 ± 34	Bone	6982–6791	
Terrace	2	Early Neolithic	Wk-17842	6095 ± 40	Bone	7157–6807	
Terrace	2	Early Neolithic	Wk-13865	6018 ± 34	Bone	6950–6752	
Terrace	2	Mesolithic	TO-12197	7500 ± 90	Tooth, *H. sapiens*	8514–8056	
Shelter	Z1	Magdalenian	Wk-31088	15,660 ± 86	Tooth	19,250–18,606	
Slope	2	Solutrean	AA-63307	11,840 ± 280	Charcoal	14,821–13,131	
Slope	2	Solutrean	AA-63308	15,710 ± 320	Charcoal	19,548–18,115	
Terrace	3	Solutrean	Wk-13685	8749 ± 58	Charcoal	[b]	
Terrace	3	Solutrean	Wk-24761	8886 ± 30	Charcoal	[b]	
Terrace	3	Solutrean	AA-63305	8825 ± 57	Charcoal	[b]	
Terrace	3	Solutrean	AA-63310	8696 ± 54	Charcoal	[b]	
Terrace	3	Solutrean	Wk-36255	8664 ± 25	Olea	[b]	
Terrace	3	Solutrean	Wk-36256	8737 ± 25	Olea	[b]	
Shelter	B1	Solutrean	Wk-17840	20,340 ± 160	*Patella* sp.	24,305–23,380	Calcite
Shelter	B6	Solutrean	Wk-24765	18,859 ± 90	Charcoal	23,233–22,191	
Shelter	C1	Solutrean	Wk-24763	19,533 ± 92	Charcoal	23,720–22,684	
Shelter	C4	Solutrean	Wk-26800	20,620 ± 160	Charcoal	25,045–24,196	
Shelter	D2	Solutrean	Wk-26802	20,570 ± 158	Charcoal	25,020–24,119	
Slope	2	Solutrean	Wk-12131	17,634 ± 110	Bone	21,405–20,518	
Slope	2	Solutrean	Wk-12130	18,410 ± 165	Bone	22,357–21,505	Minimum age
Shelter	D4	Gravettian?	Wk-26803	21,859 ± 186	*Patella* sp.	[b]	Calcite
Terrace	4	Gravettian	Wk-24762	24,769 ± 180	Charcoal	30,211–29,287	–
Terrace	4	Gravettian	Wk-31090	24,549 ± 165	Bone	29,825–28,608	Minimum age – small sample with low collagen yield
Terrace	4	Gravettian	Wk-32144	24,381 ± 258	*Patella* sp.	29,307–27,981	Calcite
				23,613 ± 240	*Patella* sp.	28,440–26,919	Aragonite
Slope	3	Gravettian	Wk-13686	22,470 ± 235	Bone	27,844–26,288	–
Slope	3	Gravettian	Wk-16414	23,995 ± 230	*Patella* sp.	28,741–27,650	Calcite
Slope	3	Gravettian	Wk-12132	24300 ± 205	Charcoal	29,522–28,539	–
Slope	3	Gravettian	Wk-17841	24,560 ± 570	*Patella* sp.	30,211–27,743	Calcite
Terrace	5	Early Gravettian	Wk-31089	24,183 ± 161	Bone	[b]	Minimum age – small sample with low collagen yield
Terrace	5	Early Gravettian	OxA-25710	25,050 ± 100	*Patella* sp.	29,565–28,636	Calcite
Terrace	5	Early Gravettian	Wk-30677	25,196 ± 103	*Patella* sp.	29,906–28,620	Calcite
				22,235 ± 173		[b]	Aragonite
Terrace	5	Early Gravettian	Wk-32145	25,181 ± 293	*Pecten* sp.	30,200–28,600	Minimum age—burnt sample
Terrace	5	Early Gravettian	Wk-30679	25,317 ± 99	*Patella* sp.	30,141–29,246	Calcite

(continued)

Table 10.1 (continued)

Area	Level	Phase	Lab.	Date	Material	Date CAL BP[a]	Notes
				25,390 ± 255		30,331–28,970	Aragonite
Terrace	5	Early Gravettian	Wk-26801	27,720 ± 370	Charcoal	[b]	–
Terrace	6	Early Gravettian	Wk-30678	25,579 ± 98	*Patella* sp.	30,232–29,487	Calcite
Terrace	6	Early Gravettian	Wk-35713	25,930 ± 122	*Pecten* sp.	30,482–29,599	–
Terrace	6	Early Gravettian	Wk-35714	25,964 ± 110	*Pecten* sp.	30,570–29,585	–
Terrace	6	Early Gravettian	Wk-35712	26,026 ± 114	*Nassarius* sp.	30,590–29,645	–
Terrace	6	Early Gravettian	Wk-30676	24,318 ± 90	*Patella* sp.	[b]31,096–29,740	Calcite
				26,353 ± 284			Aragonite
Terrace	6	Early Gravettian	Wk-32147	27,141 ± 365	*Acanthocardia* sp.	31,502–30,474	Aragonite
Terrace	6	Early Gravettian	Wk-32146	28,321 ± 422	*Pecten* sp.	33,070–31,240	Calcite
Terrace	6	Early Gravettian	Wk-35717	28,012 ± 192	*Arbutus* sp.	32,875–31,566	–
Shelter	D4	Early Gravettian	Wk-31087	28,140 ± 195	*Littorina obtusata*	32,324–31,253	Aragonite

[a]Calibration with OxCal version 4.2 (Bronk Ramsey 1995) with the IntCal09 curve (Reimer et al. 2009). Marine data (Delta-R 209 ± 102) from Reimer et al. (2009)
[b]Non-calibrated results due to inversion, contamination or recrystallization of samples

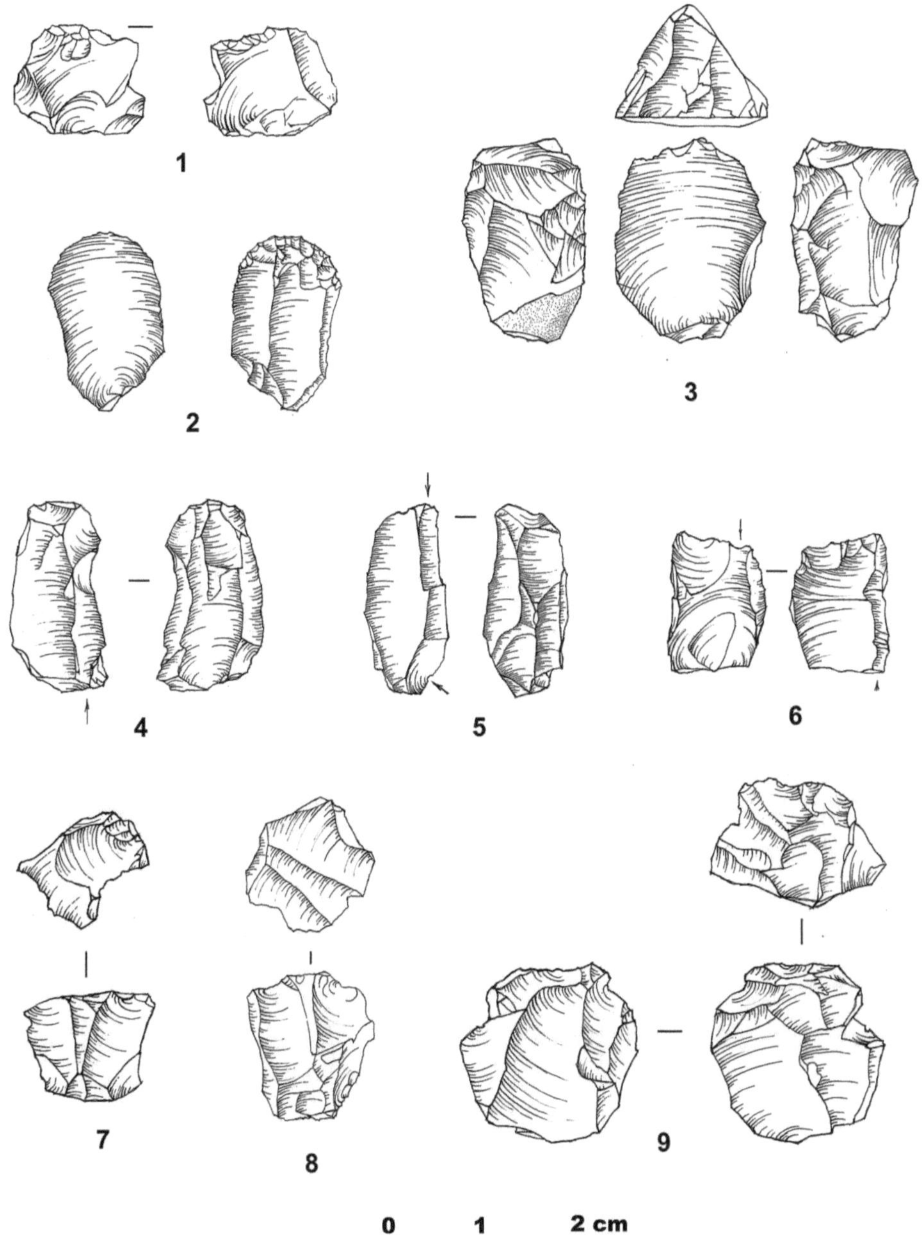

Fig. 10.3 Gravettian lithic assemblage: *1* Splintered piece; *2, 3* Carinated endscrapers; *4–6* Burins; *7–9* Flake cores. (Drawings by Júlia Madeira)

or small sized grooves (spin-off) (<1 mm) (Fig. 10.5a, b, e and f). The presence of small fractures, usually at 90°, in some of the examples, is not diagnostic of their use as projectiles. Instead, 90° fractures have been associated to different causes such as: knapping or trampling and abandonment. In any case, these types of pieces show a set of modifications that are possibly the result of contact with antler and other materials used during their production.

The chert from Vale Boi is medium and coarse grained and usually of relatively poor quality. This kind of raw material difficult the preservation of certain micro traces, such as micropolish or longitudinal groove marks. On

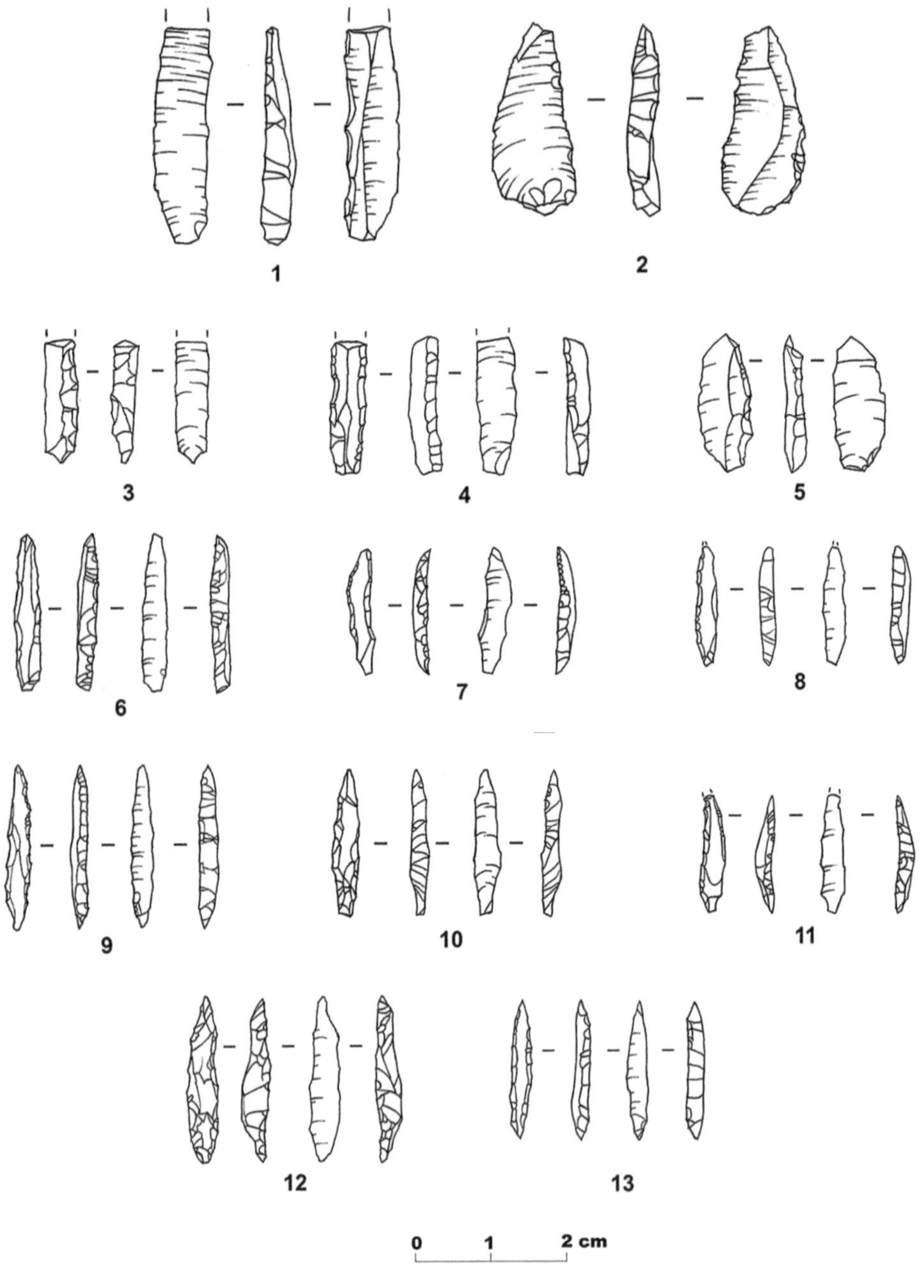

Fig. 10.4 The Gravettian double backed and bipointed tools: *1* and *3* Backed bladelet; *2* Chalterperronian point; *4–13* Double backed and bipointed bladelets. (Drawings by Júlia Madeira)

irregular surfaces, such as these, micropolish development tends to be slower and groove marks are often difficult to observe. However on three tools we did observe the presence of polishing known as "mirror" or micropolish "G" (Fig. 10.5b). The micropolish on projectiles is usually explained by friction, generated in the hafting area, by small chips coming from the pressure and the tool itself. Also, on the retouched edges of two of the backed tools micropolish similar to traces produced by wood were observed (Fig. 10.5b, d), typically linked to the contact with the

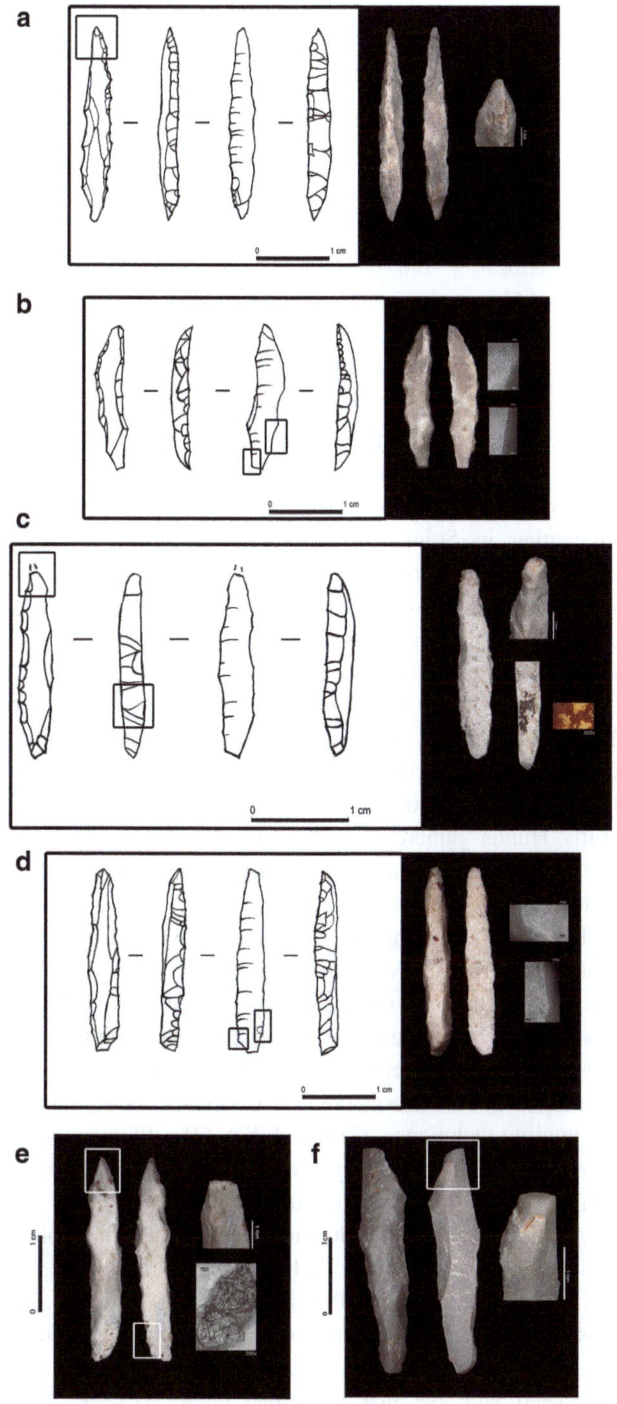

Fig. 10.5 Diagnostic impact fractures. Burin-like impact scars (**c**, **e**, **f**), stepped or tongue-shaped towards the dorsal face (**a**). Use-wear traces from micropolish (**b**, **d**) and adhesive residue (**c**, **e**)

handle (Rots 2003; Rots et al. 2011). Three points exhibit the possible presence of an organic and/or adhesive material on their lateral edge (Fig. 10.5c, e); however, this should be tested in the future.

Discussion

Due to the morphology of these tools (bi-pointed and abrupt retouch on both edges) it was assumed *a priori* that these tools could have been used as perforators, following other authors (Pesesse 2006). Typically, the drilling of hard materials, semi-hard or abrasive as skin, wood, bone, shell, etc., tend to generate diagnostic polish after a few minutes of use, a very pronounced rounding on the contact area and hard breaks as dents of different morphology and size – none of these elements are seen in the Vale Boi backed tool assemblage and, given these data, we discarded this possibility since the tips have not the slightest evidence related to this type of work.

As mentioned previously, these tools were likely hafted and the presence of organic adhesive and specific polish traces confirm this idea. We found no evidence of butchering or any other scrape or cutting traces. In fact, these tools are too small to be used as knives, and even when attached to a handle and used as side knives they would not need to be either bipointed nor double backed.

Many studies have been focusing on the evidence of the propulsion techniques in hunting activities by prehistoric hunter-gatherers (Odell and Cowan 1986; Cattelain 1994, 1997; Hays and Surmely 2005; Shea 2006; Shea and Sisk 2010). These studies showed the possibility of using two propulsion techniques: arrow and bow, and spear, launched by hand or by a propellant (i.e., atlatl). Experimental and ethnographic studies showed that these types of microlithic backed elements may have been used in different propulsion projectile systems, revealing significant shooting techniques variability, which may be related with hunting strategies and prey targets, and/or stylistic variation among hunter-gatherer groups.

In Vale Boi the low number of double backed tools with fractures of impact may be related with their morphology and/or hafting method. The relation among length/width/thickness makes them strong, sturdy and able to handle strong direct impacts. Thus, when surface is in contact with the animal's skeleton the impact was minimal and did not suffer fractures, more so if the contact was tangential. Fractures most likely occurred only in those cases where the apex of the projectile hit directly the animal bone. Fragmentation has not caused a total loss of their effectiveness, but still, both tips could still be used, retouched and reused. In fact, it is likely significant that the pieces were abandoned without most of them being broken.

The faunal assemblage associated with these tools contains well-preserved marine and terrestrial remains. The assemblage is composed by three main dietary species: rabbit (*Oryctolagus cuniculus*), red deer (*Cervus elaphus*) and horse (*Equus caballus*). Although in low quantities other mammallian species are present: aurochs (Bos primigenius),

Table 10.2 Double backed points morphological attributes (mm)

		N	Minimum	Maximum	Mean	St. deviation
Rockshelter						
	Length		22.03	26.19	23.89	2.115
	Width		3.85	4.89	4.32	0.527
	Thickness		2.97	3.87	3.553	0.505
	Valid N	3				
Terrace						
	Length		15.21	25.26	18.95	3.12
	Width		2.77	4.51	3.75	0.57
	Thickness		2.57	4.01	3.09	0.64
	Valid N	7				

ibex (*Capra pyrenaica*), wild boar (*Sus scrofa*), and few skeleton remains of carnivores, voles, and medium and small large birds. Evidence suggests that carnivores were hunted and processed by humans (Manne et al. 2012). The intensive diversification suggested by the huge exploitation of rabbits and marrow acquisition through grease rendering suggests some subsistence pressure that may have lead hunter-gatherers to improve hunting techniques.

The presence of ornaments and portable art, hearths, lithic technology, bone points, and sizeable quantities of marine and terrestrial faunal remains, suggest that the site of Vale Boi was used as a residential camp. Lithic raw materials, including chert, were exploited in a simple way, likely related to the characteristics of those local rocks, suggesting local knapping. The proximity to local resources (i.e., lithic and prey) corroborates the use of Vale Boi as a residential settlement complemented by foraging expeditions to resources acquisition (Bicho et al. 2010).

As mentioned before, the bipointed doubled backed tools are different from the typical Gravettian points and very unusual among Gravettian assemblages in Southern Iberia. In both the Spanish and Portuguese Gravettian assemblages (e.g., Bajondillo, Zafarraya, Cabeço do Porto Marinho) (Zilhão 1997; Barroso 2003; Cortés 2007) there are a series of doubled backed bladelets, either projectile or not, but they are not bipointed. It is possible that such lack of parallels could be related to the misclassification of these artifacts merely as backed bladelets, probably due to its reduced number in each lithic assemblage or the fact that they are just broken fragments. Such singularity may reflect high investment certainly connected to a functional specialization related to development of effective hunting techniques.

Conclusions

The archaeological site of Vale Boi shows new data for the Gravettian lithic projectile technology in South-southwestern Iberia. Despite a small number of specimens forming the studied assemblage, in this paper we present data that clearly show a significant contribution for the knowledge of the weaponry system used for hunting activities during the Gravettian in Southern Iberia. Experimental and ethnographic studies show that similar types, such as Microgravettes and backed bladelets, may have been used in different propulsion techniques, suggesting that hunting techniques, such as the bow and arrow, were possibly known and used during the Early Gravettian in Central and Western Europe (Hays and Surmely 2005). The improvement of different hunting skills may be a result of resource pressure, brought above by demographic, climatic and/or landscape shifts. Hunter-gatherers improved their ecological dynamics to the new system, and these adaptations are likely reflected in techno-cultural changes.

Projectile technology is seen as a strategic adaptation of hunter-gatherer behavior. Innovation and diversity among projectile technology inevitably means significant ecological advantage. Differences between simple and complex technology suggest different weapons systems and has been seen as a reflex of human adaptation to ecological niche changes: (1) hunting techniques, (2) broad diet, (3) settlement strategies and (4) ethnicity language among hunter-gatherers in contiguous territory (Bicho 2009).

At this moment, the predominance of this type of bipointed doubled backed tools in the Early Gravettian contexts of Vale Boi has no published parallels in the Iberian Peninsula. The closest and better-documented case is Southwestern France (e.g., the site of Vigne Brun), where these artifacts are also associated with the first Gravettian phase (Pesesse 2006). Lithic projectiles as well as lithic technological strategies from Vale Boi, show singular aspects when compared to their contiguous areas, and double backed and bipointed weaponry has been argued to be a distinctive mark of the Early Gravettian in Portugal (Marreios et al. 2013, in press; Marreiros and Bicho 2013). This contradicts the possibility of homogeneity during this period in the Iberian Gravettian (Villaverde et al. 1998) and reinforces the idea of considerable regional variety. In this cultural mosaic, the territories were marked by socio-cultural boundaries, possibly reflecting some demographic pressure (Djindjian and Bosselin 1994; Villaverde 2001; Bazile 2007; Fullola et al. 2007).

Ethnoarchaeological studies show that projectiles are one of the most stylistic and symbolic meaningful elements for social identity and personal style (Wiessner 1983; Binford 1984; Sacket 1985, 1986). Living hunter-gatherers use different kinds of points as a distinctive marker to individualize their community from others, especially from contiguous regions. This phenomenon was also inferred for past populations in Western Iberia (Zilhão 1997; Roman and Villaverde 2006; Bicho 2009). Regarding this idea, some authors recently used typometric, technological and functional analysis of the lithic projectiles of these communities to define cultural and territorial patterns for the Iberian Gravettian (Klaric et al. 2009). Klaric and colleagues argued that such cultural unity would be highly improbable. In fact, by opposition, they suggested the idea of a cultural mosaic, most probably characterized by extensive networks that allowed the exchange of technological solutions, in order to respond to possible environmental crisis (Bradtmöller et al. 2012; Haws 2012; Schmidt et al. 2012).

References

Anderson, P. (1990). Aspects of behavior in the Middle Paleolithic: Functional analysis of stone tools from southwest France. In P. Mellars (Ed.), *The emergence of modern humans: An archaeological perspective* (pp. 389–418). Edinburgh: Edinburgh University Press.

Barroso, C. (2003). *El Pleistoceno Superior de la Cueva del Boquete de Zafarraya*. Junta de Andalucia: Arqueología Monografías.

Bazile, F. (2007). Le Gravettien: entités régionales d'une paléoculture européenne. Paléo, Spécial table ronde (1ère partie). Les Eyzies, juillet 2004. *Revue d'archéologie préhistorique, 19*, 89–103.

Bergmann, C., & Newcomer, M. (1983). Flint arrowhead breakage: Examples from Ksar Akil, Lebanon. *Journal of Field Archaeology, 10*, 239–243.

Bicho, N. (2009). Fashion and glamour: Weaponry and beads as territorial markers in Southern Iberia. In F. Djindjian, J. Kozlowski & N. Bicho (Eds.), *Paleolithic hunter-gatherers concept of territory* (pp. 243–251). Proceedings of the XV Congress of UISPP. British Archaeological Reports (BAR).

Bicho, N., Cascalheira, J., & Marreiros, J. (2012). On the (l)edge: The case of Vale Boi rockshelter (Algarve, Southern Portugal). In K. Bergsvik & R. Skeates (Eds.), *Caves in context. The economical, social and ritual importance of caves and rockshelters* (pp. 65–81). Oxford: Oxbow Books.

Bicho, N., Gibaja, J., Stiner, M., & Manne, T. (2010). Le paléolithique supérieur au sud du Portugal: le site de Vale Boi. *L'Anthropologie, 114*, 48–67.

Bicho, N., Stiner, M., Lindly, J., Ferring, C., & Correia, J. (2003). Preliminary results from the Upper Paleolithic site of Vale Boi, southwestern Portugal. *Journal of Iberian Archaeology, 5*, 51–66.

Binford, L. (1984). An Alyawara day: Flour, spinifex gum, and shifting perspectives. *Journal of Anthropological Research, 40*, 157–182.

Bosselin, B. (1996). Contribution de l'Abri Pataud à la chronologie du Gravettien français. *Bulletin de la Société Préhistorique Française, 93*, 183–194.

Bosselin, B., & Djindjian, F. (1994). La chronologie du Gravettien français. *Préhistoire Européenne, 6*, 77–115.

Bradley, B. (1982). Flaked stone technology and typology. In G. C. Frison & D. J. Stanford (Eds.), *The Agate Basin site: A record of the Paleoindian occupation of the northwestern high plains* (pp. 181–208). New York: Academic Press.

Bradley, B., & Frison, G. (1987). Projectile points and specialized bifaces from the Horner site. In G. Frison & L. Todd (Eds.), *The Horner site. The type site of the Cody cultural complex* (pp. 199–232). New York: Academic Press.

Bradtmöller, M., Pastoors, A., Weninger, B., & Weniger, G.-C. (2012). The repeated replacement model – rapid climate change and population dynamics in Late Pleistocene Europe. *Quaternary International, 247*, 38–49.

Bronk Ramsey, C. (1995). Radiocarbon calibration and analysis of stratigraphy: The OxCal program. *Radiocarbon, 37*, 425–430.

Cascalheira, J. (2010). *Tecnologia lítia Solutrense do Abrigo de Vale Boi (Vila do Bispo)*. Cadernos da Uniarq, 5. Lisboa: Centro de Arqueologia da Universidade de Lisboa.

Cattelain, P. (1994). La chasse au paléolithique supérieur. Arc ou propulseur, ou les deux? *Archéo-Situla, 21–24*, 5–26.

Cattelain, P. (1997). Hunting during the Upper Paleolithic: Bow, spearthrower or both? In H. Knecht (Ed.), *Projectile technology*. New York: Plenum press Corporation.

Cortés, M. (2007). *El Paleolítico Medio y superior en el sector central de Andalucía (Córdoba y Málaga)*. Monografía del Museo Nacional y Centro de Investigación de Altamira 22, Madrid: Ministerio de Cultura.

Djindjian, F., & Bosselin, B. (1994). Périgordien et Gravettien: l'épilogue d'une contradiction? *Préhistoire européenne, 6*, 117–131.

Donahue, R. (1988). Microwear analysis and site function of Paglicci Cave, Level 4A. *World Archaeology, 19*, 357–375.

Fischer, A., Vemming Hansen, P., & Rasmussen, P. (1984). Macro and micro wear traces on lithic projectile points. Experimental results and prehistoric examples. *Journal of Danish Archaeology, 3*, 19–46.

Frison, G., & Bradley, B. (1980). *Folsom tools and technology at the Hanson site, Wyoming*. Albuquerque: University of New Mexico Press.

Fullola, J., Roman, D., Soler, N., & Villaverde, V. (2007). Le Gravettien de la côte méditerranéenne ibérique. Paléo. Le Gravettien: entités régionales d'une paléoculture européenne, Les Eyzies, juillet 2004. *Revue d'archéologie préhistorique, 17*, 73–88.

Geneste, J.-M., & Plisson, H. (1993). Hunting technologies and human behavior: Lithic analysis of Solutrean shouldered points. In H. Knecht, A. Pike-Tay & R. White (Eds.), *Before Lascaux: The complex record of the Early Upper Paleolithic* (pp. 117–135). Boca Raton: CRC Press.

Gibaja, J., & Bicho, N. (2006). La function de los instrumentos líticos en el asentamiento de Vale Boi (Algarve, Portugal). Estudio del utillaje gravetiense y solutrense. *Saguntum, 38*, 7–21.

Harrold, F. (1993). Variability and function among gravette points from Southwestern France. In G. Peterkin, H. Bricker & P. Mellars (Eds.), *Hunting and animal exploitation in the later Paleolithic and Mesolithic of Eurasia* (pp. 69–79). Archaeological Papers of the American Anthropological Association.

Haslam, M. (2006). Potential misidentification of *in situ* archaeological tool-residues: Starch and conidia. *Journal of Archaeological Science, 33*, 114–121.

Haws, J. (2012). Paleolithic socionatural relationships during MIS 3 and 2 in Central Portugal. *Quaternary International, 264*, 61–77.

Hayden, B. (1979). *Paleolithic reflections: Lithic technology and ethnographic excavation among the Australian Aborigines*. Canberra: Australian Institute of Aboriginal Studies.

Hays, M., & Surmely, F. (2005). Réflexions sur la function des microgravettes et la question de l'utilisation de l'arc au gravettien ancien. *Paléo, 17*, 145–156.

Kimball, L. (1992). Microwear analysis of the functions of projectile points and unifacial tools. In C. Bergman & J. Doershuk (Eds.), *Archaeological data recovery for transcontinental gas pipe line corporations 6.79-mile Leidy natural gas pipeline expansion, Padula Site (36-Nm-15), Northhampton County Pennsylvania* (pp. 238–288). Houston, TX: Technical Report Submitted to Transcontinental Gas Pipe Line Corporation.

Klaric, L., Guillermin, P., & Aubry, T. (2009). Des armatures variées et des modes de productions variables Réflexions à partir de quelques exemples issus du Gravettien d'Europe occidentale (France, Portugal, Allemagne). *Gallia préhistoire, 51*, 113–154.

Langejans, G. (2011). Discerning use-related micro-residues on tools: Testing the multi-stranded approach for archaeological studies. *Journal of Archaeological Science, 38*, 985–1000.

Lombard, M. (2005). A method for identifying Stone Age hunting tools. *South African Archaeological Bulletin, 60*, 115–120.

Lombard, M., Parsons, I., & van der Ryst, M. (2004). Middle Stone Age lithic experimentation for macro-fracture and residue analyses: The process and preliminary results with reference to the Sibudu Cave points. *South African Journal of Science, 100*, 159–166.

Manne, T., Cascalheira, J., Évora, M., Marreiros, J., & Bicho, N. (2012). Intensification during the Upper Paleolithic at Vale Boi, southwestern Portugal. *Quaternary International, 264*, 83–99.

Marreiros, J., & Bicho, N. (2013). Lithic technology variability and human ecodynamics during the Early Gravettian of Southern Iberian Peninsula. *Quaternary International, 318*, 90–101.

Marreiros, J., Cascalheira, J., & Bicho, N. (2012). Flakes technology from Early Gravettian of Vale Boi (Southwestern Iberian Peninsula). In A. Pastoors & M. Peresani (Eds.), *Flakes not blades: The role of flake production at the onset of the Upper Paleolithic in Europe* (pp. 11–23). Mettmann: Wissenschaftliche Schriften des Neanderhal Museums 5.

Marreiros, J., Gibaja, J., Cascalheira, J., Evora, M., Regala, F., Pereira, T., et al. (2013). Nuevas evidencias sobre el Paleolítico superior inicial del Suroeste Peninsular: el Gravetiense Vicentino de Vale Boi (sur de Portugal). In C. de las Heras, J. A. Lasheras, A. Arrizabalaga & M. de la Rasilla (Eds.), *Pensando El Gravetiense: Nuevos datos para la región cantábrica en sur context peninsular y pirenaico*. Monografía del Museo Nacional y Centro de Investigación de Altamira 23 (pp. 86–103). Madrid: Ministerio de Cultura.

Moreau, L. (2012). Le Gravettien ancien d'Europe centrale revisité: mise au point et perspectives. *L'Anthropologie, 116*, 609–638.

O'Farrell, M. (1996). *Approche technologique et fonctionnelle des pointes de La Gravette*. Dipomes d'Etudes Approfondies, Universite´ Bordeaux 1.

O'Farrell, M. (2004). Les pointes de la Gravette de Corbiac (Dordogne) et considérations sur la chasse au Paléolithique supérieur ancien. In P. Bodu & C. Constantin (Eds.), *Approches fonctionnelles en préhistoire* (pp. 121–138). Congrès Préhistorique de France.

Odell, G., & Cowan, F. (1986). Experiments with spears and arrows on animal targets. *Journal of Field Archaeology, 13*, 195–212.

Otte, M. (1983). Les groupes gravettiens en Europe centrale. In L. Bñesz & J. Kozlowski (Eds.), *L'Aurignacian et le Gravettien (Périgordien) dans leur Cadre écologique* (pp. 465–478). IXe Congrès UISPP, Colloque XV, 1976. ERAUL 13.

Otte, M., & Noiret, P. (2007). Le Gravettien du nord-ouest de l'Europe. Paléo. Le Gravettien: entités régionales d'une paléoculture européenne, Les Eyzies, juillet 2004. *Revue d'archéologie préhistorique, 17*, 243–255.

Pesesse, D. (2006). La «pointe à dos alternes», un nouveau fossile directeur du Gravettien? *Bulletin de la Société préhistorique française, 103*, 465–478.

Plisson, H., & Geneste, J.-M. (1989). Analyse technologique des pointes à cran solutréennes du Placard (Charente), du Founeau du Diable, du Pech de la Boissière et Combe-Saunière (Dordogne). *Paleo, 1*, 65–106.

Reimer, P.J., Baillie, M.G.L., Bard, E., Bayliss, A., Beck, J.W., Blackwell, P.G., et al. (2009). INTCal09 and Marine09 Radiocarbon age calibration curves, 0–50,000 years Cal BP. *Radiocarbon, 51*, 1111–1150.

Roman, D., & Villaverde, V. (2006). Las puntas de la gravette y las microgravettes de los yacimientos gravetienses del país valenciano: caracterización morfológica y tipométrica y análisis de sus fracturas. *Miscelánea en homenaje a Victoria Cabrera Zona Arqueológica, 1*, 440–451.

Rots, V. (2003). Towards an understanding of hafting: The macro and microscopic evidence. *Antiquity, 77*, 805–815.

Rots, V. (2011). *Prehension and hafting traces on flint tools: Methodology*. Leuven: Leuven University Press.

Rots, V., Van Peer, P., & Vermeersch, P. (2011). Aspects of tool production, use, and hafting in Palaeolithic assemblages from Northeast Africa. *Journal of Human Evolution, 66*, 637–664.

Sackett, J. (1985). Style and ethnicity in the Kalahari: A reply to Wiessner. *American Antiquity, 50*, 154–159.

Sackett, J. (1986). Style, function, and assemblage variability: A reply to Binford. *American Antiquity, 51*, 628–634.

Schmidt, I., Bradtmöller, M., Kehl, M., Pastoors, A., Tafelmaier, Y., Weninger, B., et al. (2012). Rapid climate change and variability of settlement patterns in Iberia during the Late Pleistocene. *Quaternary International, 274*, 179–204.

Semenov, S. (1964). *Prehistoric technology: An experimental study of the oldest tools and artifacts from traces of manufacture and wear*. London: Cory, Adams and Mackay.

Shea, J. (2006). The origins of lithic projectile point technology: Evidence from Africa, the Levant, and Europe. *Journal of Archaeological Science, 33*, 823–846.

Shea, J., Brown, K., & Davis, Z. (2002). Controlled experiments with Middle Palaeolithic spear points: Levallois points. In J. Mathieu (Ed.), *Experimental archaeology: Replicating past objects, behaviours and processes* (pp. 55–72). BAR International Series 1035. Oxford: Archaeopress.

Shea, J., & Sisk, M. (2010). Complex projectile technology and *Homo sapiens* dispersal into Western Eurasia. *Paleoanthropology*, 100–122.

Soriano, S. (1998). Les microgravettes du Périgordien de Rabier à Lanquais (Dordogne). Analyse technologique fonctionnelle. *Gallia Préhistoire, 40*, 75–94.

Villa, P., & Lenoir, M. (2006). Hunting weapons of the Middle Stone Age and the Middle Palaeolithic: Spear points from Sibudu, Rose Cottage and Bouheben. *South African Humanities, 18*, 89–122.

Villaverde, V. (2001). El Paleolítico superior: El tiempo de los cromañones. Periodización y características. In V. Villaverde (Ed.), *De Neandertales a Cromañones. Los Inicios del Poblamiento Humano en el País Valenciano* (pp. 177–218). València: Universitat de València.

Villaverde, V., Aura, J., & Barton, M. (1998). The Upper Paleolithic in Mediterranean Spain: A review of current evidence. *Journal of World Prehistory, 12*, 121–198.

Wiessner, P. (1983). Style and social information in Kalahari San projectile points. *American Antiquity, 48*, 253–276.

Zilhão, J. (1997). *O Paleolítico Superior da Estremadura Portuguesa*, 2 Volumes. Colibri: Lisboa.

Chapter 11
Uncertain Evidence for Weapons and Craft Tools: Functional Investigations of Australian Microliths

Richard Fullagar

Abstract At least two general hypotheses have been proposed to explain microlith function in Australia. Recent residue studies of Australian microliths, commonly called backed microliths, suggest that these small stone tools were hafted and used in a variety of tasks but lack compelling evidence of use as spear tips or barbs (Hiscock et al. 2011). In contrast, earlier studies have supported Johan Kamminga's conclusion that, on the balance of evidence, Australian microliths were "primarily the penetrating or lacerating elements of composite spears" (Kamminga 1980: 11). I argue that it is premature to reject either of these hypotheses, and argue that current evidence for microlith function is consistent with a limited range of composite tool forms including elements in spears and multi-purpose knives.

Keywords Usewear • Residues • Spears • Stone tools • Backed artifacts

Introduction

Debate about microlith functions in Australia is constrained not so much by available techniques including usewear and residue analysis, but by their limited application to a few stone artifact assemblages. These standardised tools, often called backed microliths, were made from a wide variety of stone types, are found archaeologically across most of mainland Australia (the exceptions being zones in the far north), and they first appeared in the terminal Pleistocene (Slack et al. 2004; Hiscock et al. 2011). Backed microliths did not become abundant and widespread until after the mid Holocene. Studies of usewear, including breakage, and residues on Australian backed microliths suggest that these small implements were used for a range of tasks including craft activities, multi-purpose knives, hunting spears and deadly weapons (see Case Studies below). While hafting traces have not been extensively studied in Australia, it is often presumed that Australian backed microliths were indeed hafted, largely on the basis of plant resin residues (cf. Rots 2016). Elsewhere in the world, backed microliths have been primarily identified as projectile armatures for arrows as well as spears (see Hiscock et al. 2011; Lombard and Wadley 2016; Marreiros et al. 2016). A characteristic of recent arguments about tool function has been reliance on diverse lines of evidence: usewear (including breakage patterns and impact damage), hafting traces and residues from use (Rots 2016). However, these various lines of evidence have rarely if ever been deployed together in an Australian context.

The question can be asked: what makes the Australian evidence of microlith function different from the evidence obtained in other places in the world? One response is to consider diversity of backed microlith functions in other parts of the world (Hiscock et al. 2011: 306). Here, I suggest that despite recent work indicating that Australian microliths were used on a wide range of contact materials, several details are lacking, and questions remain unanswered. For instance, what form(s) did the composite tool (composed of backed microlith elements) possess? Could one or two primary functions (e.g., spear armatures and/or multi-purpose knives) and extensive recycling account for the (apparently anomalous) variation in modes of use and contact materials observed for Australian backed microliths?

R. Fullagar (✉)
Centre for Archaeological Science, School of Earth and Environmental Sciences, University of Wollongong, Northfields Ave, Wollongong, NSW 2522, Australia
e-mail: fullagar@uow.edu.au

Radu Iovita and Katsuhiro Sano (eds.), *Multidisciplinary Approaches to the Study of Stone Age Weaponry*,
Vertebrate Paleobiology and Paleoanthropology, DOI 10.1007/978-94-017-7602-8_11

Background

The first systematic usewear study of Australian microliths was based on examination of thousands of specimens in antiquarian museum collections and professionally excavated archaeological collections (Kamminga 1978, 1980). Kamminga identified three forms of fracture damage that could be interpreted as usewear (although not necessarily in all instances): edge fracturing, tip snapping and transverse snapping. He argued that edge fracturing (bending-initiated and feather-terminated scars) on low edge angles and typically below 1 mm in size are not in any way diagnostic on their own. Kamminga suggested that tip fracturing was also of little diagnostic value – in part because this kind of damage was observed on experimental stone tools used for other activities (e.g., tip snapping on 50% of experimental stone awls used to pierce kangaroo skin). The third type of damage he observed was transverse snapping, which he also argued was not diagnostic of function because it occurred on many experimental tools in the course of manufacture and also on tools used for quite different tasks. Kamminga did not study use-polishes under vertical incident light but used a stereoscopic microscope with oblique reflected light mostly at low magnification.

Although distinctive projectile usewear (e.g., burin-like impact scars) observed under the oblique light microscope was rare, he argued that the low incidence of usewear in conjunction with other available evidence supported his interpretation that the primary function of microliths examined was to serve as penetrating or lacerating elements of composite spears. In his conclusion, Kamminga explicitly did not exclude the possibility of other backed microlith functions for assemblages or specimens in particular areas of prehistoric Australia. Burin-like impact damage or "longitudinal macrofracture" (Dockall 1997), a potentially significant form of usewear, is rare but has been found occasionally (e.g., Clark 1979). Although not commonly reported in Australia, burin-like impact is a useful indicator of head-on impact.

Boot (2005) further explored the potential of transverse snaps via experiments that included manufacturing, backing, spear throwing and woodworking. His experiments included two spears each armed with seven backed silcrete microliths, each thrown twice. Four of the barbs on the first spear, including a barb that contacted bone, were undamaged and three sustained one to four fractures (apparently after contact with the ground). Four barbs on the second spear did not have contact with the target (gel block with bone inserts) and three sustained edge fractures after contact with the ground. The three barbs that entered the target sustained tip snapping. Key conclusions were that usewear was sometimes absent but that the proportion of transverse snapping may be indicative, but not necessarily diagnostic of projectile armatures.

Recent unpublished studies by Chris Clarkson and his students at the University of Queensland have further explored the nature of damage and potentially diagnostic impact fractures on projectile tips and barbs. Preliminary results suggest a range of fracture types, scar sizes and breakage location although the incidence of tip impact fracture was often low, only about 10% (Chris Clarkson, personal communication; Clarkson 2016). A recent unpublished report (Fullagar 2011) for the Pilbara in northwestern Australia indicates that 14/74 (19%) of all backed microliths had barb or tip impact damage; and 3/8 (37%) Bondi points (asymmetrical backed microliths) have tip damage *consistent with* (rather than *diagnostic of*) impact.

As is the case for stone knives and projectile points (e.g., Akerman et al. 2002), few if any backed microlith studies in Australia have integrated all lines of evidence (e.g., usewear, manufacturing damage, hafting traces, breakage patterns, diagnostic impact marks and residues). To evaluate how reliable and convincing our current methodology is, I address evidence from two key studies that provide quite different interpretations of function. I do not question the actual results of usewear/residues in these recent studies; nor do I question the significance of context or that different site settings are indeed likely to reveal different data sets and interpretations. I simply consider the possible interpretations and ask two questions. Are the backed microlith data in Case Study 1 consistent with a different hypothesis; that the primary function was spear armatures? And are the backed microlith data in Case Study 2 consistent with an alternative hypothesis: that a couple of backed microliths were used as spear armatures but most were unrelated to the cause of death, and merely the remnants of the victim's toolkit of multi-purpose, multi-functional implements?

Case Study 1

Robertson (2005) undertook a usewear/residue study at three sites in the Mangrove Creek catchment just to the north of Sydney in southeastern Australia. Publication of the microliths in three rock shelter assemblages (Deep Creek, Emu Tracks and Mussel) revealed traces of six classes of contact material (plant, wood, bone, skin, feather, flesh) and five modes of use (cutting, drilling, incising, projectile/thrusting and scraping) in various combinations (Robertson et al. 2009). The apparent projectile/thrusting traces were associated with wood and other plant working, and consequently the interpretation identified no unequivocal evidence for hafted microliths on spears and or projectiles. Robertson et al. (2009: 305) infer that "...backed artifacts were used on multiple occasions and/or were often multi-purpose and multi-functional." Other studies in the Hunter Valley, further

Table 11.1 Frequency (%) of task association and function/mode of use for backed artefacts analysed by Robertson and colleagues. Note that the percentages refer to proportions of used specimens and multi-functional tools are counted more than once. Note also that percentages of unknown function and unknown task association are not included. (See Robertson et al. 2009 for details)

	Site		
	Deep creek n = 41 all specimens	Emu tracks n = 65 all specimens	Mussel n = 93 all specimens
Task association identified	n = 39 specimens	n = 49 specimens	n = 26 specimens
Plant (incl. wood)	24.3	43.8	34.8
Animal	81.2	66.8	6
Function/Mode of use identified	n = 37 specimens	n = 49 specimens	n = 39 specimens
Parallel to long axis	59.5	60.4	34.8
Transverse	54.1	97.9	37.8
Incising	37.8	33.3	13.6

to the north, suggest a similar range of functions but with more evidence of spear armature function (Fullagar et al. 1994).

The task associations identified included a high proportion of animal contact materials (bone, skin, feather and flesh) at Deep Creek and Emu Tracks, and a high proportion of use traces associated with directionality aligned parallel with the long axis of the backed microlith, and incising compared with transverse motion (e.g., scraping). Tip use associated with incising is also indicated in Table 11.1.

The absence of diagnostic impact traces on specimens with animal traces might be explained by robust artifact morphology, particular hafting configurations, the experimental evidence that such traces are rarely observed or the tool stone (e.g., silcrete) which usually lacks the micro-polish traces more often observed on fine-grained flint. Projectile/thrusting traces were often observed in association with plant and woodworking traces at Mussel.

Without further data and experimental testing of hafting configurations, interpretation of the residues remains uncertain. Although there is little doubt about the range of contact materials demonstrated by Robertson et al. (2009), the plant/wood residues might also be consistent with specimens hafted on wooden shafts or associated with other plant materials (e.g., as bindings). The percentage of specimens with more than one function is interesting: Deep Creek (60%), Emu Tracks (9.2%) and Mussel (41.7%). Multi-functionality in conjunction with hafting, which seems to be generally inferred for all specimens, suggests a multi-purpose, composite knife with a sharp tip, but is consistent also with a detachable spear fore-shaft.

There may be good counter arguments to these suggestions, but the scarcity of impact traces may not be conclusive evidence for the absence of backed microliths functioning as spear armatures at these sites and without more detailed study of hafting traces and configurations it remains uncertain whether backed microliths are primarily associated with more than one class of composite tool (e.g., knives, spears, drills, etc.).

Case Study 2

In the Sydney region, Fullagar (2009) and McDonald et al. (2007) examined usewear/residues and apparently diagnostic impact fractures on microliths associated with the violent death of a human victim (Table 11.2). The evidence suggested various possible weapons, which most likely included a spear (thrown or thrusted). Given the likely weapon entry orientations, it was concluded that a spear was used in at least one body penetration. Barb and tip fractures on the microliths suggested possible microlith orientations in a haft. The only surviving residue detected was bone tissue attached to microlith tips that were embedded in the human bones. Nevertheless, some of the backed artifacts displayed usewear suggestive of other functions (e.g., an awl). However, it was suggested that any microlith might serve equally well as a barb or lacerating element in a composite spear.

At least six specimens had traces of use with no definite functional assignation. And of six specimens likely to be associated with hard contact (probably from a thrusted or thrown spear) the use traces on their own do not provide unequivocal or diagnostic evidence; some uncertainty remains and an experimental testing program is needed to assess hafting arrangements (see Fig. 11.1) and the inferred impact damage. The conclusive evidence for spears (thrown or thrusted) is contextual, and provided by several specimens buried and oriented in particular skeletal remains, one with bone impacted at the tip. Although similar usewear is found on some other specimens, it is uncertain whether they are all elements of the deadly weapons used.

At least one specimen had clear micro-polish indicating skin working, most likely repeated hide penetrations. I interpreted this implement to be an awl, and not a projectile tip, since it lacked diagnostic indications of impact damage, despite the fact that the lack of impact damage is not uncommon in stone-tipped spear experiments. It is possible that this "awl" could have been subsequently hafted and used as a spear tip, but had simply avoided contact with a hard surface. Alternatively it could have had served more than one purpose, originally as part of an implement used as an awl (e.g., the tip of a composite knife) and later recycled as a spear armature.

Most specimens, which lack apparently diagnostic impact fractures, may in fact have been part of the victim's tool kit, and the remains of a few multi-functional backed microlith implements not dissimilar to the findings of Robertson et al. (2009) (Table 11.2).

Table 11.2 Summary of use-wear and residues on stone artifacts from the Narrabeen site. Reproduced from *Antiquity* (McDonald et al. 2007, Table 1 at http://www.antiquity.ac.uk/projgall/mcdonald). [a]Type refers to presence of backing and technological classes. [b]Tip cross sectional area after Shea 2006. [c]Observations are from incident light only. Residues have not yet been removed for study under transmitted light microscopy. [d]Use refers confidence of interpretations, as follows: 0 (no traces of use), 1 (possible), 2 (probable) and 3 (definite). [e]Function refers to the likely cause of dominant usewear: [f]protruding oblique impact, [g]head-on impact, [h]dominant function not from hard impact. Bold emphasis indicates conjoin set

OON No.	Type[a]	Location	Refit with no.	Stone material	Length (mm)	Width (mm)	Thick. (mm)	TCSA[b]	Retouch	Damage	Usewear	Residues[c]	Hafting	Use[d]	Function[e]
51	Backed flake	Backed blade, vertebral column		Pink-red silcrete	17.4	10	8	40	Bi-directional backing	Broken tips	Rounding on backed edge; none on chord, use scar at proximal tip; cf. barb	Grey residue on backed edge	Probably hafted. Grey residue is possibly resin	3	[f]Impact, probably projectile barb
2	Backed flake	Around skull		Grey quartzite	21	11	5	27.5	Bi-directional backing	No breakage	Rounding, polish, striae	Dark smears cellulose, starch on backed edge	Probably hafted. Dark smears are possibly resin	3	[h]Piercing and slicing skin. Not from a projectile.
3	**Backed broken flake (tip)**	**Around skull**	**4**	**Pink silcrete**	**10**	**6.3**	**4.2**	**13.2**	**Bi-directional backing**	**Crushed tip**	**Crushing at tip**		**Complex fracture, probably from hafting configuration**	**3**	[g]Likely damage from projectile tip
4	**Backed flake**		**3**	**Pink silcrete**	**16**	**10**	**11**	**55**	**Bi-directional backing**	**Steps from break**	**Scaring on chord**	**Plant tissue, charcoal, carbonate**		**3**	
5	Backed flake	West side vert column		Red silcrete	17.4	7.7	5.8	22	Uni-directional backing initiated on ventral		Impact scar on tip		Probable	3	[g]Likely damage from projectile tip
6	**Backed fragment**	**Underneath skull**	**8**	**Red silcrete**	**13.3**	**8.4**	**5.3**	**22.3**	**Bi-directional backing but rare**	**Break is probably along 'old' fracture caused by backing**	**Probable impact scar**	**Impacted yellow tissue same colour as bone fragments**	**Probable**	**2**	[g]Likely damage from projectile tip
8	**Fragment (tip)**	**Underneath skull**	**6**	**Red silcrete**	**6**	**4.6**	**2.4**	**na**						**2**	
7	Backed flake	Excavated around skull		Grey quartzite	18.8	8.6	4.6	19.8	Bi-directional backing near tip		Rounding and step scar on tip	Dark residues on backed edge	Dark residue is possibly hafting resin	3	Uncertain, tip used
9	Backed flake			Quartzite	14.2	6.5	3.6	11.7	Bi-directional backing		Possible impact scar			1	Uncertain, tip used
10	Bipolar piece	Around skull		Quartz	15.7	7	4.4	15.4	Backing not clear	Bipolar crushing and scars	Uncertain	Carbonised plaques – probably not from use		1	Uncertain, possible use of tip

(continued)

Table 11.2 (continued)

OON No.	Type[a]	Location	Refit with no.	Stone material	Length (mm)	Width (mm)	Thick. (mm)	TCSA[b]	Retouch	Damage	Usewear	Residues[c]	Hafting	Use[d]	Function[e]
11	Backed flake	Vertebral column		Quartzite	15.1	10.4	4	20.8	Bi-directional backing, not very steep	Tip broken	Scars associated with broken tip		Probable	3	[f]Hard impact, possibly from projectile
12	Backed flake			Red silcrete	15	7.5	5.6	21	Uni-directional backing, initiated on ventral	Tip broken	Rounding and longitudinal striations near tip			3	[h]Uncertain, probably not hard impact, awl or projectile?
13	Backed fragment	Dry sieved		Quartzite	19.7	10	4.2	21	Bi-directional backing	Tip crushed or broken?	Scarring on chord, tip crushed	Black residue on backed edge	Uncertain	2	Uncertain, possible use of tip, awl or projectile?
14	**Backed fragment**	**Dry sieved**	**15**	**Red silcrete**	**11**	**7.6**	**3.8**	**14.4**	**Bi-directional backing**	**Tips broken**	**Rounding, bending scars along chord**	**Sediment, unidentified particles**	**Probably hafted**	**3**	[f]Hard impact possibly from projectile; also considerable damage along chord
15	**Backed fragment (tip)**	**Inside vertebral canal. Dry sieved**	**14**	**Red silcrete**	**4.5**	**3.6**	**2**	**na**	**Bi-directional backing**	**Missing fragments.**	**Impact scar cf. barb**			**3**	
16	Backed flake	Between L1 and L2		Red silcrete	18.8	8.8	3.8	na	Bi-directional backing	Tip broken	Impact scar cf. barb, scarring		Probably hafted	3	[f]Hard impact possibly from projectile
17	Fragment			Grey quartzite	7.5	5.6	4.1	na	Possible backing, uni-directional	Broken	Impact damage unclear			1	Uncertain, possible use of tip

Fig. 11.1 Tip break of refitted specimens OON4 (left) + 3 (right), showing a long narrow impact fracture (initiated at the tip) with a step termination that initiates a spin off fracture (with step termination). The maximum length of OON3 is 10 mm

Discussion and Conclusion

As stated above, I do not doubt the range of contact materials or modes of use recently proposed for Australian backed microliths. The essential question is: are there alternative explanations that limit the kind of composite tool to which microliths were hafted? Second, did most microliths found in Australia serve one primary function? The case studies above suggest uncertainties that imply a need for more experimental and archaeological data on hafting configurations, and there is a need to further reconstruct the types of prehistoric composite implement(s) on which microliths were fixed. Moreover the archaeological context raises a key issue. The Narrabeen microliths with compelling evidence for use as hafted spear tips and barbs are found at the likely kill site. In contrast, the Mangrove Creek microliths are found at what appear to be dwelling locations where tools, even those with a dominant primary function, might be repaired, removed from hafts and used incidentally for a range of incidental tasks.

Could one or two composite tool forms account for the variation observed in backed microlith function? I suggest that without detailed study of hafting arrangements and further projectile damage experiments, it is premature to conclude that microliths were not commonly utilised elements i.e., armatures on thrown spears in Australia. Robertson et al. (2009) raise another key issue worthy of further study: "… that backed artifacts might sometimes have been modified by further retouching, perhaps in association with re-hafting events". This latter issue of further modification suggests that implement shape and extent of retouch may be linked with reduction stages.

The traces found on Australian backed microliths are consistent with two main tool forms: composite spears and multi-purpose knives with sharp tips (see Fig. 11.2). The haft configuration and variation of spear armatures has not been securely reconstructed, but evidence at the Narrabeen site suggests a series of hafted elements serving as tips and barbs. The suggested haft configuration of multi-purpose knives has not been tested experimentally, but evidence from several sites suggests that such an implement would have fixed elements (for cutting and scraping) with a protruding tip (used for awling and piercing, drilling and incising).

While study of usewear and residues has made considerable advances, future studies should target hafting traces, impact scars and breakage patterns on experimental and archaeological specimens. White (2011; see also the comments that follow his article) reviewed "utilitarian explanations" (e.g., backed microliths as standardized, portable reliable tools) and has argued that "social explanations" need to be given more weight (e.g., stylistic phenomena and symbolic associations). One way to investigate this would be via a firmer reconstruction of the complete implement(s) to which backed microliths were hafted.

White (2011) also notes previously postulated links between climate change, faunal remains, hunting, backed microliths and the need for more efficient tools. He asserts that links are based on that assumption that backed artifacts were primarily made for spear armatures, which, he goes on to say "…we now know was almost certainly not generally the case". If this is the current consensus, I cannot agree. I do not think that any study has yet demonstrated that Australian backed microliths are *generally* not projectile armatures. Robertson et al. (2009) may well be correct in their interpretations that seem to eliminate a projectile function at the analysed sites. However, the archaeological context (e.g., at habitation vs. kill sites) of microlith occurrence needs further theorising; and the diagnostic indicators of microlithic armatures requires further experimental testing with Australian tool stones. Moreover, the argument that Australian backed microliths are *generally* not elements of projectile weapons remains a proposition that needs to be tested by integration of key multiple lines of evidence: hafting traces, usewear and breakage patterns, contact residues and archaeological context.

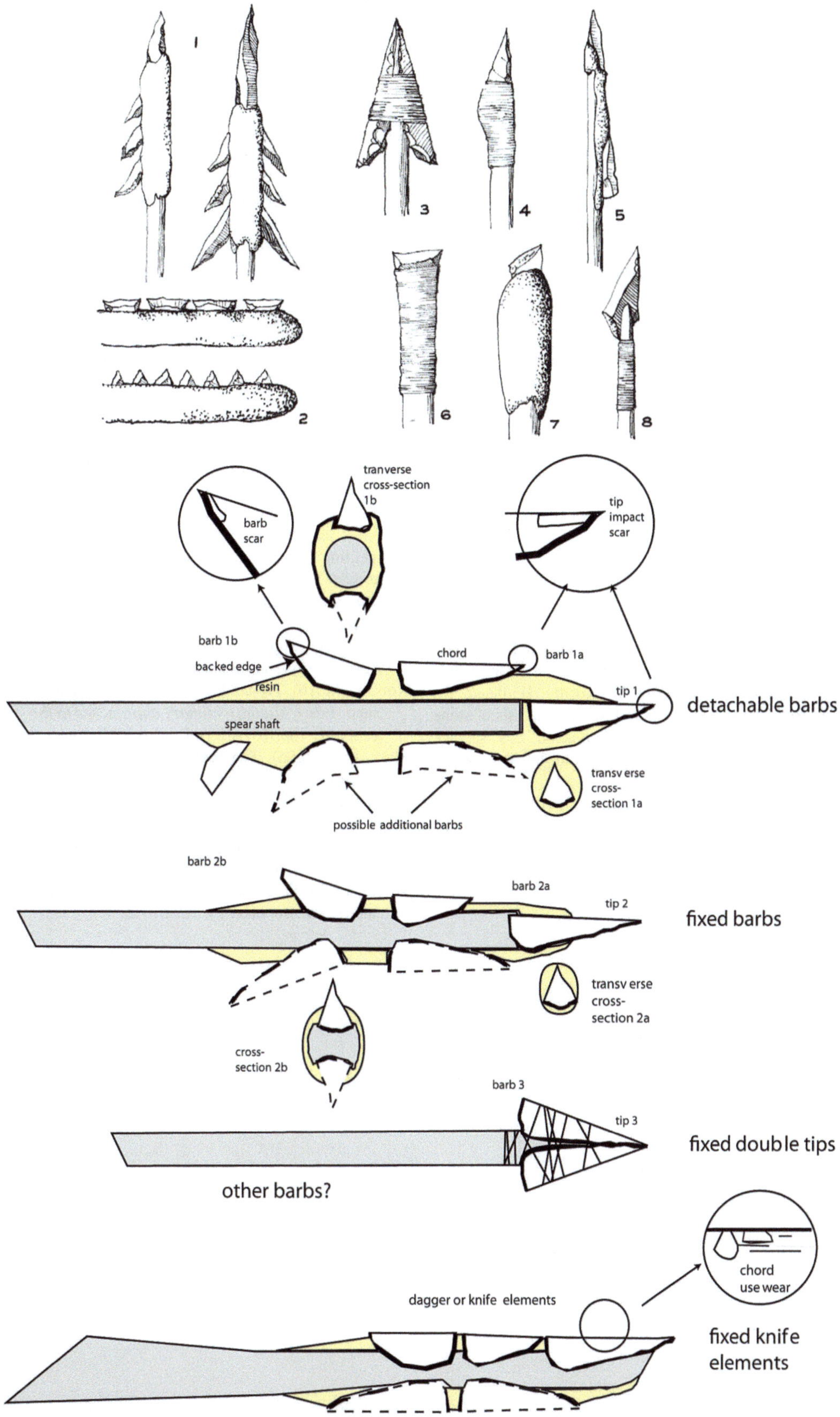

Fig. 11.2 Possible hafting arrangements of backed artifacts. McCarthy's (1976, p. 51) suggested hafting arrangements (top, nos. 1–8) reproduced with permission from The Australian Museum. [Reproduced from Fullagar et al. (2009)]

Acknowledgements I am grateful for comments from referees and particularly discussion with Chris Clarkson, Johan Kamminga, Gail Robertson and Veerle Rots. I thank also the organisers and participants in the Mainz symposium/workshop titled "Multidisciplinary scientific approaches to the study of Stone Age weaponry", September 2011.

References

Akerman, K., Fullagar, R., & van Gijn, A. (2002). Weapons and Wunan: Production, function and exchange of spear points from the Kimberley, northwestern Australia. *Australian Aboriginal Studies, 1*, 13–42.

Boot, P. (2005). Transverse snapping on stone artefacts. In I. McFarlane, M. J. Roberts & R. Paton (Eds.), *Many exchanges: Archaeology, history, community and the work of Isobel McBryde* (pp. 343–366). Canberra: Aboriginal History Monograph.

Clark, D. (1979). An impact burin on a Bondi point from Hoods Drift, South Australia. *The Artefact, 4*, 68–72.

Clarkson, C. (2016). Testing archaeological approaches to determining past projectile delivery systems using ethnographic and experimental data. In R. Iovita & K. Sano (Eds.), *Multidisciplinary approaches to the study of Stone Age weaponry* (pp. 189–201). Dordrecht: Springer.

Dockall, J. E. (1997). Wear traces and projectile impact: A review of the experimental and archaeological evidence. *Journal of Field Archaeology, 24*, 321–331.

Fullagar, R. (2011). Usewear and residues on stone tools from MIB25 PAD3, Pilbara, WA. Unpublished Report to ACHM and Rio Tinto.

Fullagar, R., Furby, J., & Brass, L. (1994). Stone artefacts from Bulga, Hunter valley. Unpublished report for M. Koettig. [Fullagar, R., J. Furby and L. Brass 1994. Use-wear and residue analysis of stone tools from Bulga. In Bulga Lease Authorisation 219 salvage excavations. A report to Saxonvale Coal Pty Ltd, by M. Koettig, Vol. 5, pp. 26–105].

Fullagar, R., McDonald, J., Field, J., & Donlon, D. (2009). Deadly weapons: Backed microliths from Narrabeen. In M. Haslam, G. Robertson, A. Crowther, S. Nugent & L. Kirkwood (Eds.), *Archaeological science under a microscope: Studies in residue and ancient DNA analysis in honour of Thomas H. Loy* (pp. 248–260). Terra Australis 30, Canberra: ANU e-Press.

Hiscock, P., Clarkson, C., & Mackay, A. (2011). Big debates over little tools: Ongoing disputes over microliths on three continents. *World Archaeology, 43*, 653–664.

Kamminga, J. (1978). *Journey into the microcosms. A functional study of prehistoric Australian stone tools*. PhD dissertation, University of Sydney.

Kamminga, J. (1980). A functional investigation of Australian microliths. *The Artefact, 5*, 1–18.

Lombard, M., & Wadley, L. (2016). Hunting technologies during the Holwiesons Poort at Sibudu Cave: What they reveal about human cognition in KwaZulu-Natal, South Africa, between ~65 and 62 ka. In R. Iovita & K. Sano (Eds.), *Multidisciplinary approaches to the study of Stone Age weaponry* (pp. 273–286). Dordrecht: Springer.

Marreiros, J., Bicho, N., Gibaja, J., Cascalheira, J., & Pereira, T. (2016). Early Gravettian projectile technology in Southwestern Iberian Peninsula: The double backed and bipointed bladelets of Vale Boi (Portugal). In R. Iovita & K. Sano (Eds.), *Multidisciplinary approaches to the study of Stone Age weaponry* (pp. 147–158). Dordrecht: Springer.

McCarthy, F. (1976). *Australian Aboriginal stone implements, including shell, bone and tooth implements* (2nd ed.). Sydney: Australian Museum Trust.

McDonald, J., Donlon, D., Field, J., Fullagar, R., Coltrain, J., Mitchell, P., et al. (2007). The first archaeological evidence for death by spearing in Australia. *Antiquity, 81*, 877–885.

Robertson, G. (2005). *Backed artefact use in eastern Australia: A residue and use-wear analysis*. PhD dissertation, University of Queensland.

Robertson, G., Attenbrow, V., & Hiscock, P. (2009). Multiple uses for Australian backed artefacts. *Antiquity, 83*, 296–308.

Rots, V. (2016). Projectiles and hafting technology. In R. Iovita & K. Sano (Eds.), *Multidisciplinary approaches to the study of Stone Age weaponry* (pp. 167–185). Dordrecht: Springer.

Shea, J. J. (2006). The origins of lithic projectile point technology: Evidence from Africa, the Levant, and Europe. *Journal of Archaeological Science, 33*, 823–846.

Slack, M., Fullagar, R., Field, J., & Border A. (2004). New Pleistocene ages for backed artefact technology in Australia. *Archaeology in Oceania, 39*, 131–137.

White, J. (2011). Backed artefacts: Useful socially and operationally. *Australian Archaeology, 72*, 67–75.

Chapter 12
Projectiles and Hafting Technology

Veerle Rots

Abstract Stone tool hafting has always been considered important, but its interpretative potential has not yet been sufficiently recognized. While wear studies have recently demonstrated the possibility of deriving hafting data from the stone tools themselves, it is essential that these kinds of data are now also integrated with regard to armature identifications. New experiments with spears and arrows show that armature identifications are complex and that no single feature on its own is diagnostic of projectile impact. Also the distinction between different projecting modes is still seriously hampered by the lack of a reliable reference. It is argued that hafting wear is essential for more adequate identifications of armatures and their projecting mode. The analysis of a number of archaeological Middle Palaeolithic and Late Palaeolithic assemblages in North West Europe allowed identifying the existence of hafted spear points for the Middle Palaeolithic sites and arrows armed with tips and barbs for the Late Palaeolithic sites.

Keywords Hafting • Armature • Projectile • Breakage • Experiments • Wear traces • Impact traces • Middle Palaeolithic

Introduction

Knowing whether and how stone tools were hafted improves our understanding of past human behaviour (Keeley 1982; Ambrose 2001, 2010; Rots 2003, 2010a; Barham 2013). It provides insight into the organic tool component that is rarely preserved, and it allows understanding the complete life cycle of stone tools, including discard patterns (Rots 2003). The choice to haft a stone tool depends on various factors, amongst which expertise with working organic materials to produce hafts and fixation agents (bindings, glues) is a necessary first step.

While hafting has often been dealt with as an inseparable category, recent functional data indicate that different degrees of hafting may play a role on a behavioral level (Rots 2015). Aside from the development of hafted tools, also the elaboration of hafting towards different tool functions and the development of differing articulations between stone tool and haft are crucial. Therefore, it seems valid to distinguish between tool uses that necessarily require a haft – if the task has to be performed with stone tools – and tools for which the addition of a haft "only" improves a tool's efficiency. Armatures are obviously examples of the former, next to hafted stone axes. Stone points cannot be used as armature if they are not hafted. This implies that any stone point that was used as armature should evidently show remains of this former hafting. Consequently, a reliable identification of armatures not only depends on knowledge regarding what use-wear evidence could be considered as diagnostic, it also requires insight into hafting wear.

When reflecting on which tools use might have stimulated the development of hafting techniques, it appears likely that it may first have concerned tools for which hafting was a necessity. These tools would first have consisted of organic material only (i.e., no hafting), like the wooden spears that were in use from about 400–300 ka onwards [e.g., *Schöningen* (Germany) (Thieme 1997; Behre 2012), *Clacton-on-Sea* (UK) (Oakley et al. 1977)]. Adding a stone element to a spear in order to produce a hafted spear point demands expertise on how it can be fixed. One may assume that the incentive to be able to use a stone tip on wooden spears or a stone blank for percussion implements is higher than for any other stone tool that can perform well without being hafted. In that case, the first attempts to haft stone tools may have concerned armatures and percussion implements, and only applied to other stone tools later on. Current archaeological evidence seems to support such a scenario (Rots and Van Peer 2006; Rots et al. 2011; Rots 2015) (see also below).

V. Rots (✉)
Chercheur Qualifié du FNRS, TraceoLab/Préhistoire, University of Liège, Quai Roosevelt 1B, 4000 Liège, Belgium
e-mail: veerle.rots@ulg.ac.be

Radu Iovita and Katsuhiro Sano (eds.), *Multidisciplinary Approaches to the Study of Stone Age Weaponry*, Vertebrate Paleobiology and Paleoanthropology, DOI 10.1007/978-94-017-7602-8_12

Hafting Evidence

Both direct and indirect evidence have been used to identify hafting, independent of tool use. *Preserved hafts* are the most direct and reliable evidence of hafting. Most examples however date to the Neolithic period. The earliest evidence of *adhesive* use dates to the late Middle Pleistocene site of Campitello (Italy) (Mazza et al. 2006). Aside from this early evidence, most current direct evidence for the use of adhesives broadly dates to around 70 ka in the Old World (Boëda et al. 1996; Hedges et al. 1998; Boëda 2008; Wadley et al. 2009; Rots et al. 2011). Recently, it was established that hafting is also identifiable based on *microscopic wear patterns* (including polish, scarring, striations, rounding) and that also the hafting arrangement can be inferred when the preservation state of the material is sufficient (Rots 2002a, 2010a). The method proves to be a reliable means to identify the existence of stone tool hafting based on the stone tools themselves, which allows for an improved understanding of both the timing and nature of hafted stone tools, independent from the preservation of organic material. The identification of the hafting arrangement is appreciably more difficult than the identification of hafting itself, in particular for older assemblages, but it is nevertheless possible. In general, *use-wear traces* never provide direct evidence of hafting; they can at most provide indirect evidence. For instance, for armatures (i.e., arrow/spear tips, barbs), a haft is a necessity and the use-wear evidence thus indirectly indicates hafting. As a result, the identification of diagnostic impact wear on a stone tool (e.g., Fischer et al. 1984) necessarily implies that the stone tool was used while hafted and that hafting evidence should be present too.

Aside from these direct arguments, several **indirect arguments** have been used over the years to argue for the existence of hafting. *Morphological adjustments* such as the removal of bulbs, proximal thinning, proximal width reduction (Rots 2005), notches, tangs (Rots 2002b), etc. have predominated. Tangs in particular have been a source of much discussion, especially with regard to Aterian points (e.g., Clark 1970). However, it is not just the choice to haft a stone tool but the chosen hafting arrangement that determines the relevance of morphological adaptations. While certain hafting arrangements set low demands on a stone tool's morphology, allowing the hafting of various morphologies and sizes; other hafting arrangements may gain significantly from specific morphological features. These morphological features thus potentially indicate the existence of a particular hafting method. By contrast, they have no value for identifying the timing of hafting as sufficient hafting modes exist that have no truly detectable requirements on the level of a stone tool's morphology.

Standardization has also been used as an argument in favor of hafting (Marks et al. 2001). However, if one wants to argue for a potential link with hafting, one first needs to differentiate between the active and non-active tool part. While hafting may have necessitated the production of more morphologically similar pieces (Bar-Yosef and Kuhn 1999), this morphological similarity essentially concerns the non-active part of a stone tool, and "standardization" in view of hafting – if it exists – may not be so easily visible in the archaeological record. Characteristics referring to the complete stone tool may create a visual perception of "standardization" without being necessarily relevant for hafting purposes (e.g., blank length, morphology of used edge, location of shaping retouch). It is clear that only a functional study can establish a potential relation between standardization and hafting.

For *small tools*, assumed problems in easy manual manipulation are generally used as arguments to advocate hafting (Bar-Yosef and Kuhn 1999). In the case of *microliths* used as projectiles, hafting can be inferred based on the presence of diagnostic impact damage from use (Fischer et al. 1984). Microliths (or bladelets) frequently proved to have been used hafted for European Late Palaeolithic and Mesolithic assemblages, but one still needs to be careful. Often, microliths are too easily assumed to represent parts of a projectile technology leading to potential interpretative errors. After all, various functions have been identified for microliths (independent of the region and period) including projectiles (tips, barbs), knives and drills (Donahue 1988; Kimball 1989; Caspar and De Bie 1996).

Some researchers have proposed that the presence of *ochre* on stone tools could be an indication of hafting (Beyries and Inizan 1982; Wadley et al. 2004). However, when no resin is found, ochre is an argument for hafting that is equally indirect as morphological adjustments are. While ochre may indeed form an ingredient of resin and potentially remain on a stone tool surface after the resin has degraded, it may have had various other functions as well and it can only be used as a valid argument for hafting in association with resin residues and/or hafting wear (Wadley et al. 2009; Rots et al. 2011).

Breakage is not frequently used as an indirect argument for hafting, but experimental studies have demonstrated that hafted use results in breakage more frequently than hand-held use (Rots 2002a, 2010a). Most hafting fractures occur at the haft limit, usually about one or two millimeters inside the haft. It is the point where the stone tool is most vulnerable when pressure is exerted, in particular in the case of thin tools. The majority of hafting fractures occurs on tools with a medial thickness of maximum 7 mm, in particular when used in high-pressure motions. The most distinctive trait for hafting fractures is abundant scarring in direct relation with the fracture (Fig. 12.1). While fractures

Fig. 12.1 High-impact related hafting fracture: experimental tool used for adzing wood

are indeed suggestive of hafting, they do not provide conclusive evidence on their own.

Diagnostic Evidence of Hunting Weapons: Wear Features and Residues

A number of armature experiments have been performed over the years, the majority concerning Late Palaeolithic and Mesolithic projectiles (Fischer et al. 1984; Odell and Cowan 1986; Bergman et al. 1988; Caspar and De Bie 1996), but some were performed on Upper Palaeolithic points (Plisson and Geneste 1989), on spear points (Odell and Cowan 1986; Plisson and Beyries 1998) or Middle Stone Age segments (Lombard and Pargeter 2008; see Rots and Plisson 2014 for an overview). Tool samples vary, but relevant data concerning potentially diagnostic wear patterns were generally obtained.

Unfortunately, armature identifications have recently suffered from a loss of rigour, both with respect to methods applied and the criteria considered as diagnostic (see Rots and Plisson 2014 for a discussion). Therefore, I will formulate some personal ideas on how a reliable armature analysis should minimally be performed and what wear features are potentially diagnostic. A macroscopic examination of scarring or fractures on potential armatures (even with the aid of a hand lens) without training and an *available* and *relevant* experimental reference collection is difficult and is not expected to significantly contribute to insights into past hunting technologies.

In my opinion, five aspects are essential on a methodological level for studies that have the intention to try and identify armatures:

- A microscopic analysis: the use of a stereoscopic binocular microscope with magnifications up to at least 50× is a minimum, and the additional use of a metallurgical microscope for high magnifications is preferable.
- One wear feature is not sufficient for a reliable identification of an armature, the wear pattern as a whole has to support the interpretation.
- An available experimental reference collection that includes reproductions of the archaeological stone tools under study or comparable examples, used for various uses, amongst which armatures but also perforating and cutting tools, for instance. If claims are made regarding the projecting mode of the armature, the collection should include experimental armatures used with different projecting modes. The experimental reference is preferrably continuously available to the analyst.
- Skill is an important element for the production of an experimental reference collection, both with regard to stone tool manufacture, hafting, ballistics and use (e.g., experienced spear-throwers and/or archers.
- The analyst requires relevant expertise regarding different wear features, not only those linked with armature use, but also those linked with other tool uses in order to adequately assess the expected and observed variability.
- The above in a sense implies that only trained microwear analysts are well-placed to perform a reliable armature analysis. This is true. On the other hand, the lack of sufficient microwear analysts and the eagerness to understand past hunting technologies have forced many researchers into using less appropriate methods, which is understandable. Nevertheless, it remains essential that every method is first rigorously tested (e.g., including blind testing) before results can be considered reliable.

Many authors have published details on what features are diagnostic to identify armatures. I particularly want to stress the importance of observing different forms of diagnostic evidence in order to produce incontestable results: not only specific wear features, but the wear pattern as a whole is crucial. One isolated tip fracture or scarring patch should *never* be considered as sufficient or reliable evidence. Aside from tip damage, also the lateral edges of armatures may suffer a lot of damage; it may perhaps not always be diagnostic on its own, but its presence is nevertheless quite characteristic. In addition, also the hafted portion may show diagnostic features that resulted from the counter-pressure against the haft or within the animal.

Step-terminating bending fractures, spin-offs and burination have frequently been cited as the most diagnostic evidence of armature use (Fig. 12.2). Far less cited are the

Fig. 12.2 Diagnostic impact use damage: **a** burination on the ventral right tip of tip 108 (12.5×); **b** spin-off on the dorsal tip of barb 29 (16×); **c** MLIT's on the ventral distal tip of tip 85 in association with tip damage (100×)

microscopic linear impact traces, abbreviated as MLIT's, (Moss 1983; Fischer et al. 1984) that are formed in direct association with tip damage (Fig. 12.2c). The reason is of course that their observation requires a metallurgical microscope, which is rarely used in current studies on armatures, next to a sufficiently good preservation of the material. MLIT's are formed by the scar flake that detaches upon impact and shortly scratches the stone surface during this process. As a result, they start at the termination of the impact scar or fracture and they are always oriented (broadly) parallel to the use axis. They should not be confused with other striations that can form as a result of knapping, use, hafting, or other processes, nor should they be confused with smears or other residual features. MLIT's can only be observed on pieces that were appropriately cleaned with chemicals (e.g., ethanol) in order to remove adhering residues. MLIT's are not always equally explicit, sometimes it is simply a faint, narrow bright line starting from the scar negative, in other cases multiple, parallel and explicit striations are observed (see examples below).

While step-terminating bending fractures, spin-offs and burinations may indeed form as a result of weapon use, these features should preferably not occur isolated. Even though experiments have demonstrated that diagnostic wear features do not form at each impact, it is nevertheless essential for archaeological pieces to show more than one wear feature in order to support their identification as armature. This implies that an ideal diagnostic wear pattern consists of explicit tip damage (step-terminating bending fractures, spin-offs, burinations, or a combination of these), associated with MLIT's, lateral impact-related scarring and impact-related damage on the hafted portion, preferably also in association with MLIT's witnessing the counter-pressure.

Residues alone are not sufficient evidence to provide a reliable identification. After all, butchering knives may show exactly the same set of residues and residue distributions (both on the level of use and hafting). They often also show explicit tip damage. The danger is real because independent of tool size, pointed stone tools (or bladelets) initially assumed to have been part of an armature arrangement instead often proved to have been used as butchering knife based on a microscopic wear analysis (e.g., Plisson and Beyries 1998; Caspar and De Bie 1996; Rots 2015). Therefore, a residue analysis should preferrably be combined with an analysis of other wear features that are more diagnostic.

While resin residues may witness the fact that a stone tool may have been used hafted – on the condition that the wear pattern confirms the distribution – resin is in itself not a diagnostic indication of a hunting weapon. Resin may be used to haft a various set of stone tools and there is significant overlap what the hafted area concerns between different kinds of tool uses. In addition, resin residues are not always reliable to delimit the hafted portion of the stone tool as resin tends to get all over the stone tool during hafting or de-hafting (see experiments).

Hunting Experiments

Over the years, I performed different experiments related to the use of hunting weapons in collaboration with the *Chercheurs de la Wallonie* at the *Préhistosite de Ramioul* (Liège). Two sets of experiments are dealt with here: an exploratory experiment regarding thrusting and throwing spear points, and a more elaborate experiment on arrows equipped with tips and barbs. Levallois points were used in the former experiment, while diverse microlithic points (retouched base, backed, obliquely truncated, crescents) were manufactured for the second experiment. Both use and hafting wear were examined.

Spear Point Experiment

The spear point experiment was performed in the framework of an analysis of different Middle Palaeolithic assemblages. The goal was to evaluate whether thrusting and throwing spear points could potentially be distinguished based on microscopic evidence, one aspect of which was testing whether lateral use damage from a rotating action upon insertion formed on thrusting spears only, a hypothesis that was put forward earlier (Rots 2009; Rots et al. 2011). In addition, the efficiency of different hafting methods was examined. The experiment was exploratory only and larger-scale follow-up experiments are currently in progress.

Eleven Levallois points were used for this experiment; five were used as thrusting spear tips, six as throwing spear tips (Table 12.1). All pieces were mounted on a wooden spear and fixed with the aid of bindings and/or resin (Fig. 12.3). One point was fixed against a straight wooden haft (i.e., no insertion) with a ball of resin, similar to

Table 12.1 Details of spear point experiment

ID	Sequence in exp.	Haft type	Haft material	Bindings	Fixation	Activity	Attempts	Result	Flint grain size
Exp.43/1	7	Male split	Wood	Leather	–	Thrusting spear	5	Usable	Medium
Exp.43/2	9	Male split	Wood	Intestines	–	Thrusting spear	5	Usable	Fine
Exp.43/3	1	Male split	Wood	–	Resin	Throwing spear	4	De-hafted	Fine
Exp.43/4	6	Juxtaposed	Wood	Leather	–	Throwing spear	5	De-hafted	Fine
Exp.43/5	2	Straight	Wood	–	Resin	Throwing spear	1	De-hafted	Fine
Exp.43/6	5	Juxtaposed	Wood	Leather	–	Throwing spear	2	Tip damage	Medium
Exp.43/9	10	Male split	Wood	Leather	–	Thrusting spear	5	No penetration	Fine
Exp.43/10	11	Juxtaposed	Wood	Intestines	–	Thrusting spear	5	Usable	Fine
Exp.43/11	8	Male split	Wood	–	Resin	Thrusting spear	5	No penetration	Fine
Exp.43/12	3	Male split	Wood	Tendons	–	Throwing spear	11	Point out of axe	Fine
Exp.43/13	4	Juxtaposed	Wood	Intestines	–	Throwing spear	15	Usable	Fine

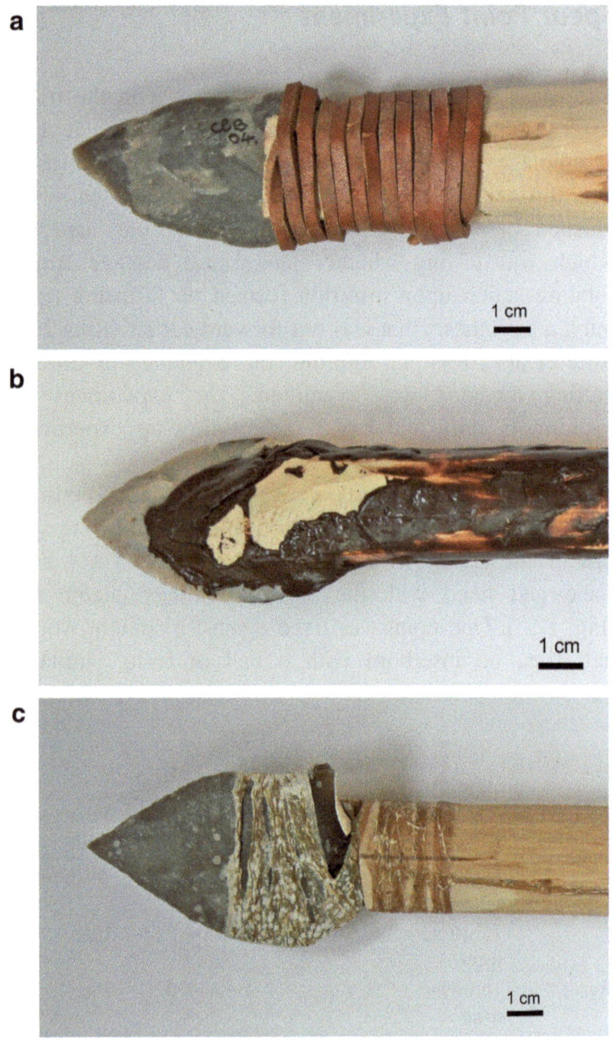

Fig. 12.3 Experimental hafted spear points, fixations with: **a** leather bindings (exp. 43/1); **b** resin (exp. 43/3); **c** intestines (exp. 43/10)

Australian Aborigines hafting modes (Hayden 1979). Throwing spear points were thrown from a distance of 6–8 m. All spears were thrown or thrusted by one and the same person, Christian Lepers, an experienced spear thrower and an overall experienced experimenter (Fig. 12.4). A freshly killed deer was used. All spears were used in 5 successful attempts, unless the point detached from the haft earlier on.

Results

Generally speaking, points proved to detach more frequently from thrown arrangements in comparison to thrusted ones. The most successful fixations proved to be resin or intestines. Wear features are most prominently present on thrown spear points, but this is also because the size of the animal and the way it was fixed as target (i.e., hung and fixed with ropes) did not allow a high pressure to be exerted with the whole body during thrusting. Less damage is formed on retouched edges in comparison to unretouched ones. The standard impact wear features were observed on the points (Table 12.2; Figs. 12.5 and 12.6). Tip fractures diagnostic of impact were nevertheless rare, in spite of the presence of other impact-related features. This stresses the importance of examining the whole wear pattern on these points instead of focussing too much on the tip only.

Fig. 12.4 Experimental setting spear point experiment: throwing spear

A diagnostic wear pattern could be observed on about half of the spear points (3 thrusting, 3 throwing). For three of these (2 thrusting, 1 throwing), the use-wear evidence alone would not be sufficient to consider the evidence as diagnostic, while it can be considered diagnostic in combination with the evidence on the hafted portion. For three thrusting spear points and one throwing spear point, the wear evidence may be suggestive for a use as spear point, but it cannot be considered as diagnostic. At least one throwing spear point detached after one attempt without the formation of diagnostic wear features (Exp. 43/5).

Discussion

While this experiment was only exploratory in nature, interesting observations were nevertheless possible. Distinct clues with regard to the distinction between thrusting and throwing spear points were not yet obtained even though the throwing spear points were on average more intensely damaged than thrusting spear points (but see earlier comments with regard to exerted pressure) and more often show diagnostic wear features (Table 12.2; Figs. 12.5 and 12.6). This counts for both the use and the hafting evidence. More abundant and more typical hafting scarring forms on throwing spears, while

Table 12.2 Wear evidence on spear points

Wear location	Wear features	Thrusting spears (5)	Throwing spears (6)
Tip	Step-terminating fracture	0	1
	Step-terminating scarring	1	3
	Other fracture	1	1
	Spin-off	1	0
	Burination	0	0
	Crushing	1	1
	MLIT	1	0
Lateral edges	Fracture	0	1
	Step-terminating scarring	2	4
	Sliced scarring	0	1
	Burination	0	0
	Crushing	1	2
	MLIT	1	0
Hafted area (impact-related features)	Step-terminating scarring	3	3
	Sliced scarring	1	2
	Burination	0	1
	MLIT	0	0

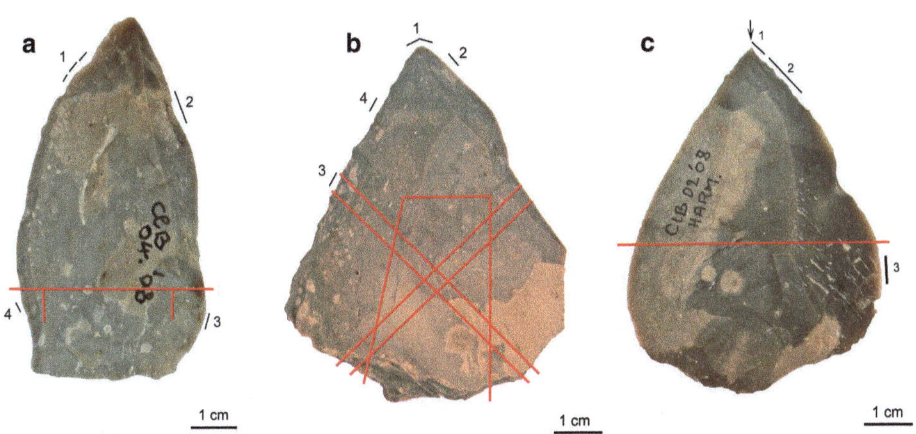

Fig. 12.5 Wear distribution on experimental thrusting spear points: **a** Exp. 43/1: *1* Impact-related step-terminating scars with curved initiation on the ventral edge, associated with MLIT's at the termination and similar scars on the dorsal edge, *2* Concentration of step-terminating scars, laterally initiated, on the ventral face, associated at the proximal side with wider and deeper scars with curved initiation, *3* Crushed scar patch on the dorsal edge, hinge- and step-terminations, *4* Scalar scars with curved initiation and feather termination (ventral edge); **b** Exp. 43/9: step-terminating scalar scar with dorsal initiation on the ventral tip, associated with spin-off on the dorsal tip, *2* Intrusive scar patch with feather- and step-terminations on the ventral edge, *3* Wide hinge-terminating hafting scar on the dorsal edge, *4* Deep scalar impact scar, partially feather- and partially hinge-terminating; **c** Exp. 43/2: *1* Faint MLIT on the ventral tip, associated with polish formation, *2* Band of bright spots, striations and polish, associated with retouch, due to knapping, *3* Sliced and sliced-into-scalar scars on the ventral edge, partially alternating, due to the contact with bindings

thrusting spear elements show few typical scars and scarring is mainly concentrated around haft boundaries. This confirms the general observation that the exerted pressure is an important factor in hafting trace formation (Rots 2002a, 2010a). It implies that hafting wear may provide relevant data for evaluating the relative amount of pressure that is exerted upon impact and thus the projecting mode.

Arrow Experiment

The goal of the arrow experiment was to examine whether a reliable distinction between tips and barbs was possible based on a microscopic analysis. In addition, the efficiency of using bindings instead of resin for hafting the pieces was tested. Two sets of experiments were performed, including 100

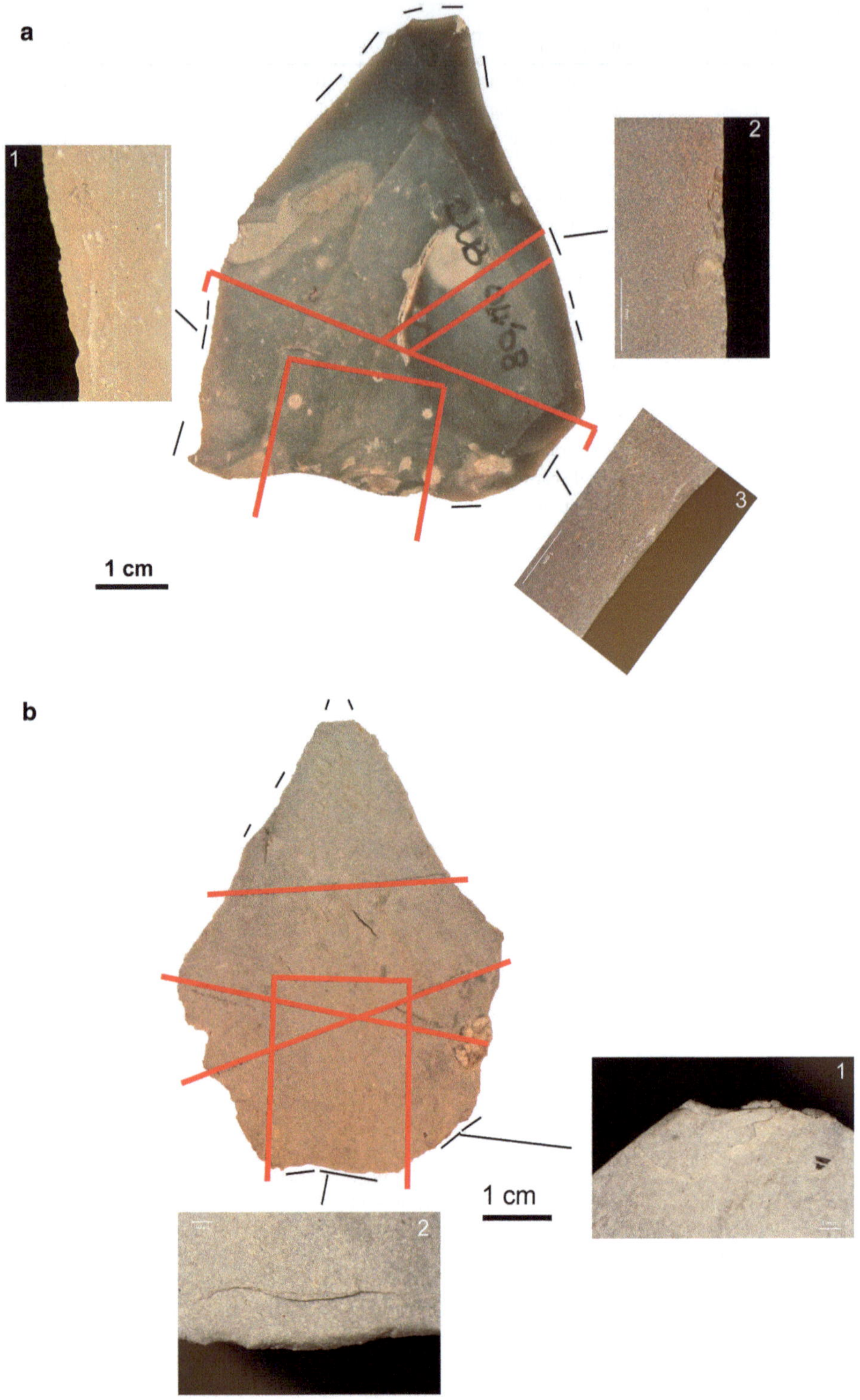

Fig. 12.6 Wear distribution on experimental throwing spear points: **a** Exp. 43/4: *1* Wide step-terminating hafting scarring on the dorsal proximal left edge (40×), *2* Hafting scar patch due to bindings on the dorsal medial right edge consisting of a large step-terminating scalar scar and smaller step-terminating scars with curved initiation (32×), *3* Sliced step-terminating non-intrusive hafting scars on the dorsal proximal right edge (32×); **b** Exp. 43/6: *1* Large scalar step-terminating scar which caused a small burination due to counter-pressure, located on the dorsal proximal right edge (10×), *2* Wide scalar scar with strongly curved initiation and step termination on the ventral base due to counter-pressure against the haft (10×)

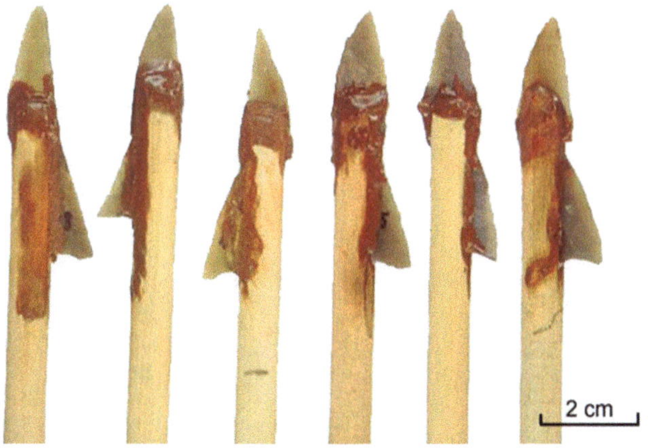

Fig. 12.7 Experimental hafted arrowheads and barbs

Fig. 12.8 Experimental setting arrow experiment

arrows in total. In total, 100 stone tips and 104 barbs were used (Fig. 12.7). This implies that four arrows in total were equipped with two barbs. Two simple plain wooden bows were used: one of 35 pounds and another one of 60 pounds. Arrows were shot by two experienced archers (Louis Baumans, Didier Cocchi), at a distance of about 18–20 m. For each arrow experiment, a freshly killed sheep was used (Fig. 12.8). Most arrowheads were fixed with resin, with bindings securing the arrow underneath the tip, but some were fixed with bindings only. Straw was placed behind the sheep in order to protect arrows that missed their target. All arrows were shot up to minimally one successful hit, unless the tip or barb detached earlier.

Only the first experiment is included in more detail here. It consists of 49 tips, 45 of which were recovered, and 51 barbs, 37 of which were recovered.

Results

Again, points appeared to detach most frequently in the case of bindings. Most pieces actually proved too small to allow a secure hafting with bindings. A combination of bindings and resin was successful. Eight tips detached as a result of impact, five of which were recovered, two of which remained stuck in a piece of wood. Twenty-four barbs detached, 10 of which were recovered. Additional fragments were found during the butchering of the sheep, but only fragments that could be recognized as a tip or barb of a specific arrow were included.

The experiment resulted in the formation of distinctive impact damage on the majority of the tips and on a good number of the barbs (Table 12.3; Figs. 12.9 and 12.10). In nearly all cases, a combination of different wear features was observed. When different types of fractures or damage were recorded on one individual point, they were separately inventoried, with a maximum of one feature type per point. The same counts for MLIT's: when several concentrations were observed, they were only counted once per point.

Tip fractures occurred on about half of the points (47% of the tips, on 57% of the barbs), but the tip fractures on the barbs were rarely diagnostic. Step-terminating scarring did not occur on tips of barbs, while it was frequent on tips. Spin-offs and burination occured on both tips and barbs. Overall, barbs showed less diagnostic damage types than tips. Lateral scarring was frequent on both tips and barbs, but sliced scarring – typical of the cutting motion upon impact – were clearly more frequent on barbs.

MLIT's were frequent and they predominated on the tips, mainly in association with tip fractures (Fig. 12.11). They also occurred frequently on barbs where they were predominantly associated with lateral damage (Fig. 12.12a). The MLIT's differed significantly in explicitness; many were narrow and faint. While striations occurred in the hafted area of barbs, they never took the form of actual MLIT's. Resin friction striations by contrast were rather frequent on the hafted portion of barbs; they resulted from the pressure and friction upon detach under impact.

While distinctions between tips and barbs have been proposed based on the distribution of wear features and their axis (e.g., Rots et al. 2003, 2005), the experiment proves that such a distinction is possible, but not straightforward. Resin distribution, for instance, is not a reliable feature as it is also influenced by the de-hafting procedure during which resin may get dispersed in non-hafted areas. The latter particularly happens when resin is heated to allow extraction. Nor is there one type of diagnostic feature that allows a distinction

Table 12.3 Wear evidence of first arrow experiment

Total number analyzed		Tips		Barbs	
		45/49	%	37/51	%
Tip	Step-terminating fracture	14	31.1	3	8.1
	Other tip fracture	7	15.6	18	48.6
	Scarring associated with tip fracture	7	15.6	3	8.1
	Crushed tip	5	11.1	0	0.0
	Step-terminating scarring on tip	16	35.6	0	0.0
	Spin-off	7	15.6	3	8.1
	Burination	9	20.0	5	13.5
	MLIT (low power)	5	11.1	1	2.7
	MLIT (high power)	26	57.8	4	10.8
Lateral edge(s) (not hafted)	Step-terminating scarring	4	8.9	5	13.5
	Spin-off	1	2.2	3	8.1
	Burination	1	2.2	2	5.4
	Sliced scar patches	5	11.1	11	29.7
	Other lateral scarring	26	57.8	21	56.8
	Alternating scar patches	7	15.6	4	10.8
	MLIT (high power)	12	26.7	17	45.9
Hafted area	Sliced scar patches	4	8.9	0	0.0
(impact related features)	Other scarring	7	15.6	2	5.4
	Notch/explicit scarring at boundary	8	17.8	8	21.6
	MLIT (high power)	7	15.6	0	0.0
Base (counter-pressure)	Step-terminating fracture	2	4.4	1	2.7
	Step-terminating scarring	7	15.6	0	0.0
	Spin-off	2	4.4	0	0.0
	Crushing	3	6.7	1	2.7
Corners of base	Burination	6	13.3	5	13.5
	Step-terminating scarring	5	11.1	6	16.2
	Fracture	9	20.0	22	59.5

between tips and barbs. It is the combination of different features and their distribution over the piece that can be diagnostic (Table 12.4).

For instance, step-terminating tip fractures proved to be more abundant on tips, while barbs generally show a small non-diagnostic fracture at the tip, but a very high number of small fractures on one corner of the base. The frequent occurrence of tip damage on barbs is perhaps unexpected, as this part is hafted in or against the shaft, but it needs to be stressed that the fractures are generally small and rarely step-terminating. In contrast to the frequent occurrence of damage on one of the proximal corners in the case of barbs, proximal damage on tips is generally located on both proximal corners, if at all present. The latter depends on the amount of protrusion of the base from the shaft. In addition, the proximal damage on barbs witnesses a twisted motion far more frequently than the one on tips. Also sliced scarring on the lateral edge is far more common on barbs. Under high magnification, the distinction between tips and barbs is generally rather explicit with MLIT's hardly occurring on the tips of barbs, but being clearly more abundant in association with damage on the lateral edges. Also bright spots are frequently associated with lateral damage on barbs.

There may however be one type of fracture that could be typical of barbs: on a number of barbs (from the second arrow experiment), a specific type of compression fracture occurs on the tips of barbs located inside the haft (Fig. 12.13). This type of fracture was only observed on barbs and can be

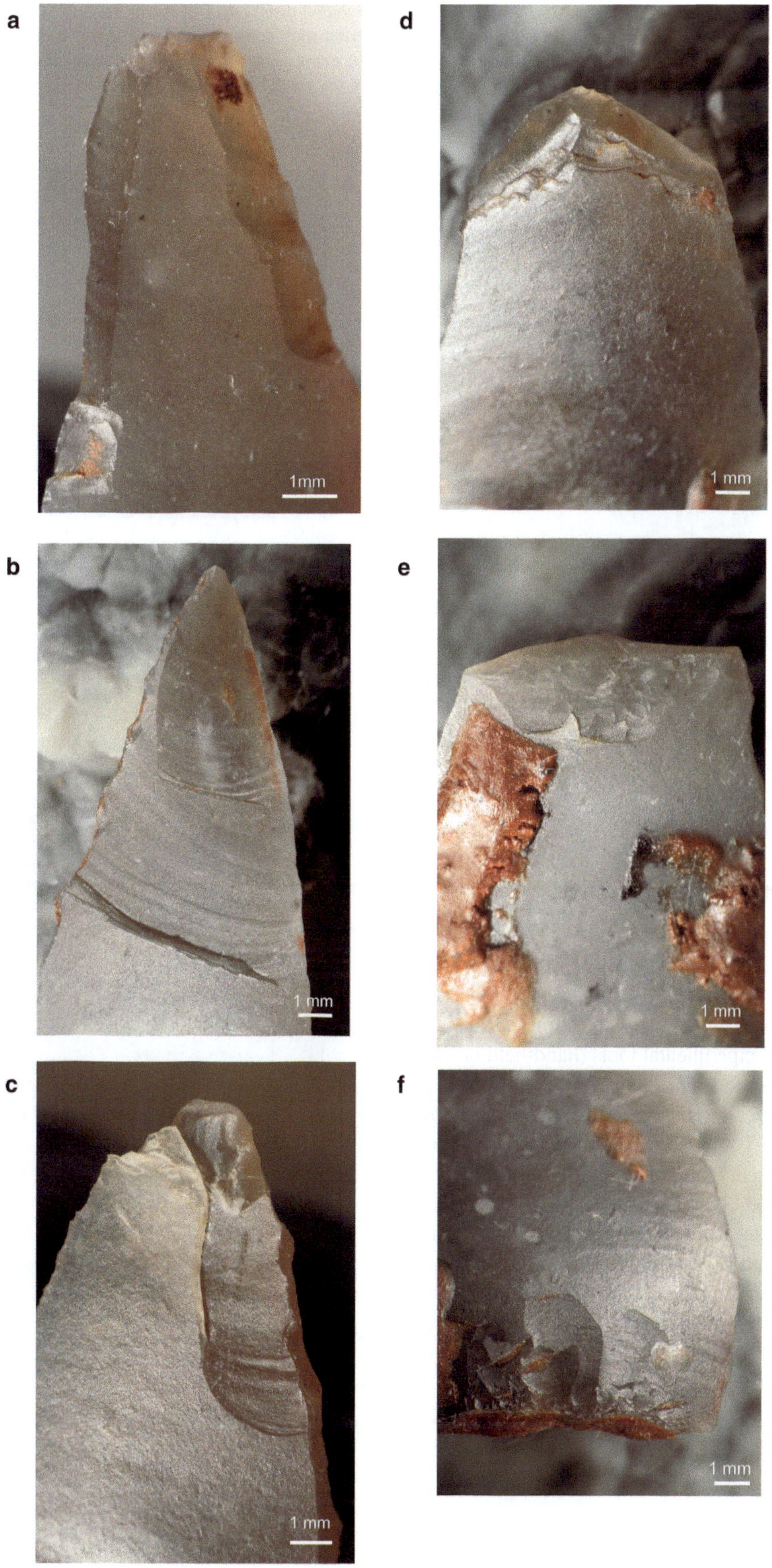
a
b
c
d
e
f
1mm
1 mm
1 mm
1 mm
1 mm
1 mm

◀ **Fig. 12.9** Low magnification wear evidence on tips: **a** burination on both ventral distal edges of the tip (tip 108) (12.5×); **b** double superposed step-terminating spin-off's on the ventral distal tip (tip 39) (8×); **c** step-terminating spin-off on ventral tip of a tip (arrow 4) (10×); **d** double step-terminating bending fracture with dorsal initiation on tip (tip114) (8×); **e** transversal fracture with associated step-terminating scarring (tip 118) (8×); **f** feather- and step-terminating scarring from counter-pressure on the ventral base of tip 19 (8×)

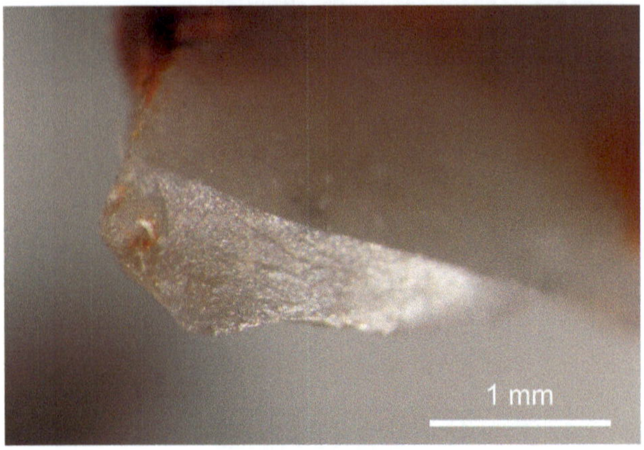

Fig. 12.10 Low magnification wear evidence on barbs: oblique burination on the ventral tip of barb 29 (25×)

attributed to a compression pressure within the haft, possibly due to a contact between the tip and the barb upon impact (i.e., tip detaching and moving backwards). This will need to be explored in more detail.

Hafting and Other Experiments

Aside from specific hunting experiments, the interpretation of hafting wear on armatures also relies on a much more elaborate experimental reference collection consisting of more than 400 used experimental tools (hand-held or hafted) (Rots 2002a, 2010a) and more than 500 experimental artifacts for technological wear patterns (knapping, retouch, etc.) (Rots 2010b). Tools were hafted in various arrangements (i.e., juxtaposed, male, male split) with different haft materials (i.e., wood, bone, antler, leather) and different fixation aids (i.e., adhesives, bindings). For more details on this experimental and methodological work, I refer to the above publications and references therein.

I only reiterate some evidence which appears relevant in this context. Resin fixation proved to result in typical resin friction wear, aside from the residues it left behind. Fixations with bindings proved to result in characteristic scarring and scar patterns. Generally speaking, resin resulted in less traces

a

b

c

Fig. 12.11 High magnification wear evidence on tips: **a** MLIT's on ventral distal tip (tip 85) in association with tip damage (100×); **b** MLIT's on ventral distal tip (tip 114) associated with tip damage (100×); **c** faint MLIT on ventral distal tip (tip 22) associated with tip damage (100×)

a

b

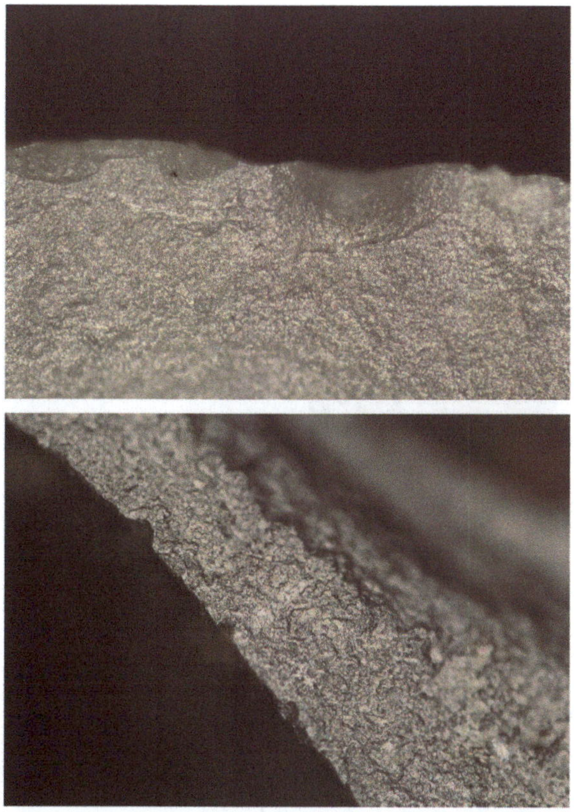

Fig. 12.12 High magnification wear evidence on barbs: **a** MLIT parallel to edge and associated with edge scarring on the dorsal medial right edge of a barb (barb 88) (100×); **b** MLIT's at the termination of a large spin-off that nearly reaches up to the other ventral edge (barb 45), the short MLIT's connect the termination with the opposite ventral edge (200×)

than bindings applied wet, which in turn caused less trace formation than bindings applied dry. Juxtaposed handles proved to result in a different wear pattern between the dorsal and ventral face, while a male handle resulted in a similar wear pattern on both faces and an explicit impact on the lateral edges. Male split handles result in a wear pattern that differs between the centre of the tool and the lateral edges.

Archaeological Case Studies

The experimental work described above has been used as a basis for the identification of armatures on different Palaeolithic sites in Europe and Northeast-Africa. It appears relevant to briefly explore the current state of knowledge on hunting weapons in the Palaeolithic based on these new functional results.

The existence of hunting weapons in the Middle Palaeolithic has been a heavily debated topic. In the past, the capacity to hunt effectively was denied for Neanderthals and they were mainly portrayed as scavengers. Due to new discoveries (Thieme 1997; Boëda et al. 1999) and results from faunal analyses (Gaudzinski and Roebroeks 2000), functional analyses (Shea 1988a; but see Plisson and Beyries 1998) and isotope studies (Richards et al. 2000), Neanderthals were considered to be expert hunters relying mainly on animal foods for their subsistence. At the same time, this expert hunting was assumed to have been undertaken with simple weapons, such as thrusting or throwing spears, while more complex weapons (e.g., spear-thrower, bow) were by definition reserved for anatomically modern humans only, with an assumed earliest introduction in Africa (Shea and Sisk 2010). Independent of the existence of supportive evidence, Neanderthals were thus once again portrayed as incapable of complex technology, in sharp contrast to behaviourally modern humans.

Such interpretations are fine if supported by actual evidence, but overall the argumentation used is rather poor. For instance, TCSA (tip cross-sectional area) values are in themselves insufficient to indicate a use as weapon and they are thus only relevant for points for which a use as armature was first demonstrated. Nor is there any support yet for the reliability of such values to infer a particular projecting mode. Similarly, the existence of a bow-and-arrow technology in South Africa around 70 ka is based largely on the small size of the segments, and on a range of indirect arguments (e.g., the assumed existence of snares – no organic remains; Lombard and Phillipson 2010).

Table 12.4 Results of the wear analysis on the microliths of a number of Dutch Late Palaeolithic sites

	Sample	Used as point	*Tips*	*Barbs*	*Combined*	Used as drill
Zeijen	35	31	*18*	*8*	*1*	2
Siegerswoude II	21	18	*15*	*3*	*0*	0
Emmerhout	13	10	*7*	*2*	*0*	possibly 1
Luttenberg	17	13	*10*	*3*	*0*	3
Total	86	72	*50*	*16*	*1*	**6**

Numbers indicate the counts of pieces identified as point or drill, for points a position and orientation is also inferred based on the observed wear patterns

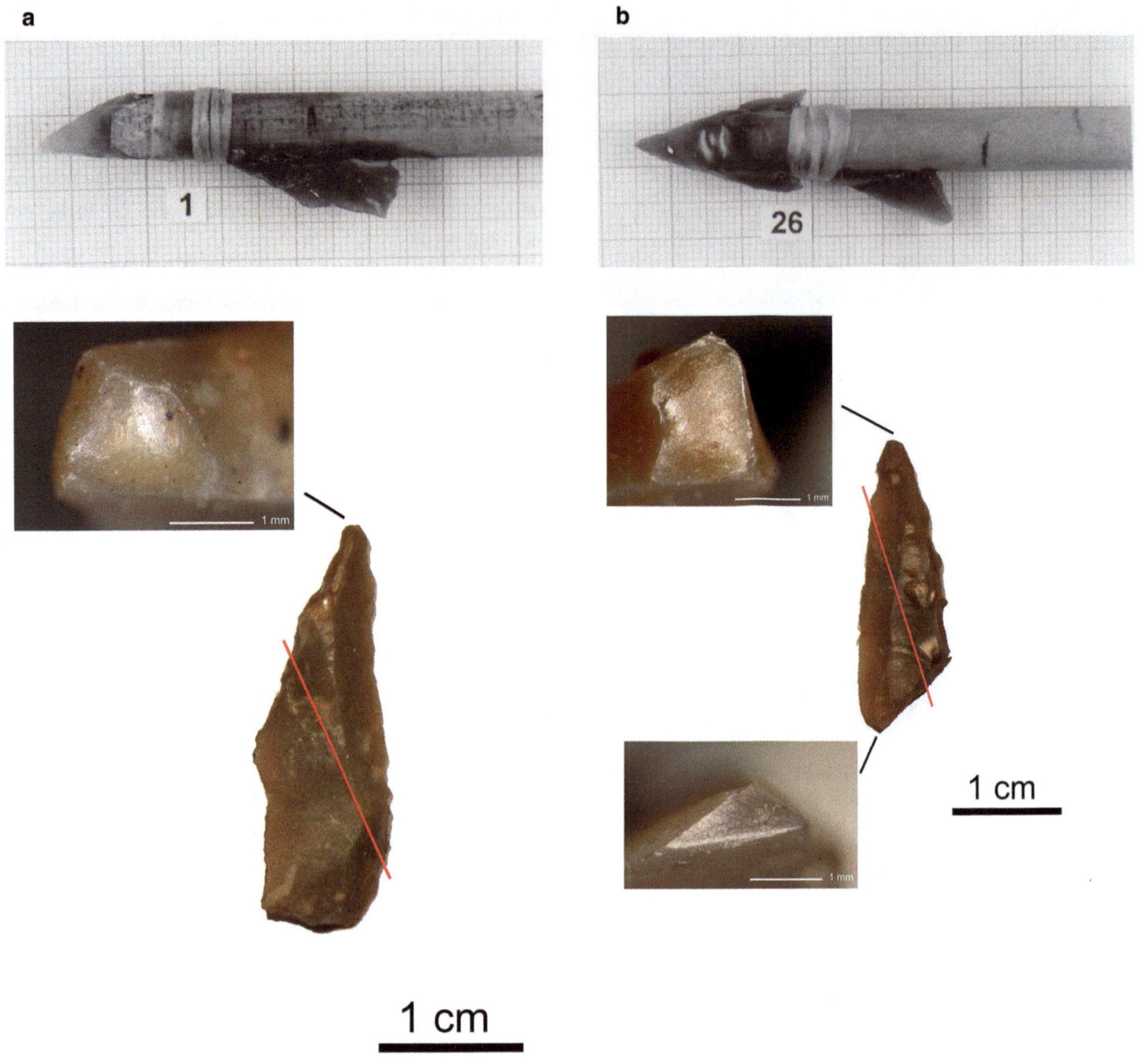

Fig. 12.13 Specific kind of tip fracture on tips of barbs located inside the haft due to a compression within the haft (arrow experiment 2): **a** compression fracture on the tip of barb 208; **b** compression fracture on the tip of barb 172, also a small oblique fracture with dorsal initiation and minor feather termination on the left proximal base

Up to now, the projecting mode of armatures has never been inferred based on a large-scale experimentation that actually supports the existence of specific diagnostic criteria that would allow such interpretations. While it is tempting to use more straightforward and more easily available arguments to advocate a certain projecting mode, such interpretations risk to be misused. While the existence of wooden spears is supported from about 400–300 ka (Movius 1950; Oakley et al. 1977; Thieme and Veil 1985; Thieme 1997), the question remains whether and when stone points

Fig. 12.14 Spear points at Biache-St-Vaast: **a** Elongated Moustier point (E8-513): *1* Burination on dorsal tip (16×), *2* Striation associated with scar on the ventral medial left edge (haft boundary) (100×), *3* Hafting scarring around the haft boundary (16×), *4* MLIT due to counter-pressure on the ventral proximal surface, initiated from the termination of the large proximal fracture (100×), *5* Large proximal fracture due to counter-pressure against the haft upon impact (8×), *6* Hafting scarring around the haft boundary on the dorsal medial right edge (8×), *7* Hafting scarring around the haft boundary on the dorsal medial right edge (8×); **b** Elongated Moustier point (I8-507): *1* Scar on the ventral tip initiated from the distal extremity, it continues into a burination on the ventral distal left edge (12.5×), *2* Hafting scarring with oblique orientation on the ventral medial left edge (16×), *3* Burination on the ventral proximal left base, initiated from the left (counter-pressure within the haft) (16×)

a
1
1 mm
2
3
4
5
6
7
1 cm

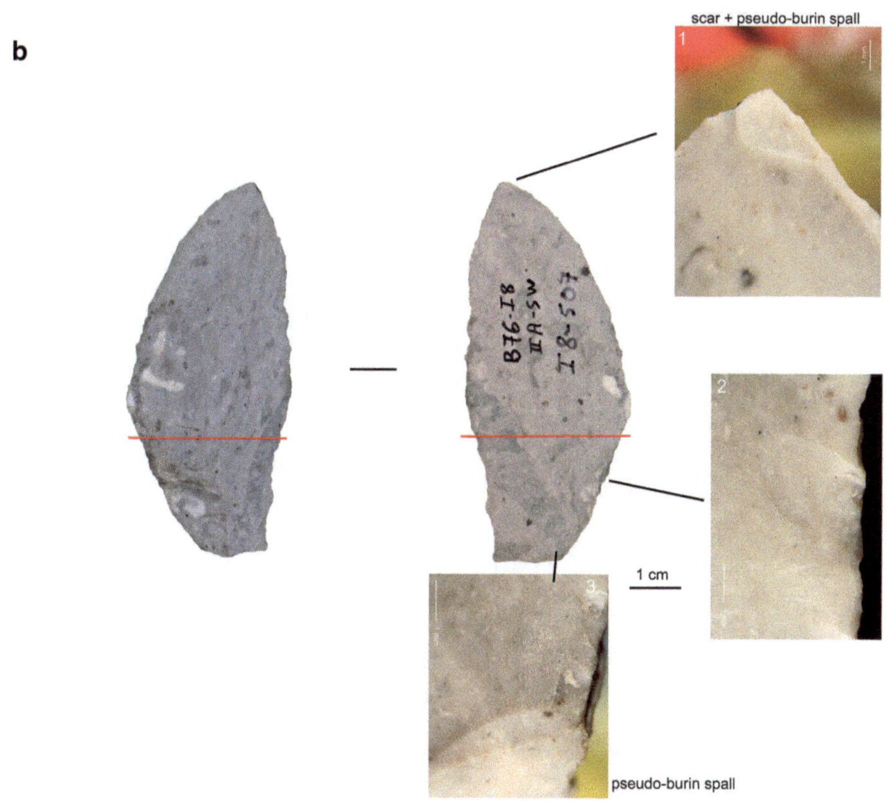
b
scar + pseudo-burin spall
1
2
3
1 cm
pseudo-burin spall

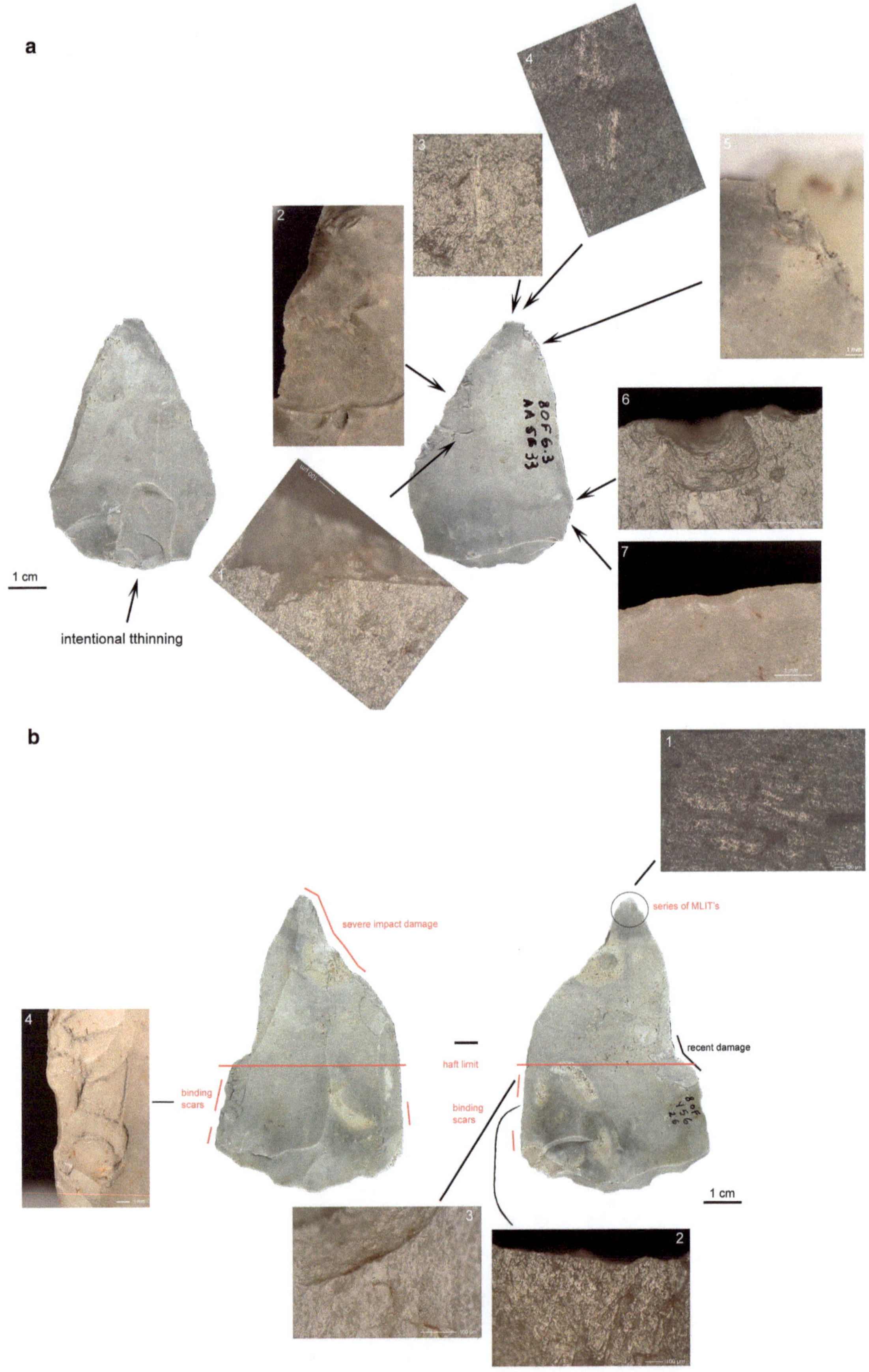
a
1 cm
intentional tthinning
b
severe impact damage
series of MLIT's
haft limit
binding scars
binding scars
recent damage
1 cm

◀ **Fig. 12.15** Spear points at Bettencourt: **a** Levallois point (AA 5a33): *1* MLIT's associated with the large ventral impact scarring on the right edge (100×), *2* Large ventral step-terminating impact scar with curved initiation on the ventral right edge (8×), *3* MLIT on the ventral distal tip (100×), *4* MLIT's on ventral distal tip (50×), *5* Use scarring on the ventral distal left edge, *6* Bright spot associated with hafting scarring on the ventral medial left edge (around haft boundary) (200×), *7* Sliced scars due to a contact with bindings on the ventral proximal left edge (20×); **b** Levallois point (Y56/26): *1* Series of MLIT's on the ventral distal tip (50×), *2* Bright spot zone due to friction within the haft on the ventral proximal right edge (100×), *3* Bright spot associated with hafting scar on the ventral medial right edge (haft boundary) (200×), *4* Hafting scar concentration on the dorsal proximal left edge consisting out of step-terminating scalar scars with curved initiation (8×)

were mounted on wooden spears. This necessitates sufficient expertise with regard to hafting and an acknowledgement of the advantages it may offer. Direct evidence for the existence of hafted stone tips was provided by the Levallois point embedded in a vertebra (Boëda et al. 1999). Given the unique nature of such finds, a reliable and broader insight in the issue is only possible based on detailed functional studies.

Based on new results from the functional analysis of *Biache-St-Vaast* (Tuffreau and Sommé 1988; Rots 2013), it is clear that hafted spear points are in use from about 200 ka. Explicit diagnostic wear patterns were observed on 16 pieces on an examined assemblage of 157 pieces (Fig. 12.14). Aside from thrusting spear points, the slender and light nature of some of the points in combination with explicit use and hafting damage, suggests that at least part of these points were also used in thrown arrangements (Fig. 12.14a). However, the typical distinction between spear points with or without damage from a rotating motion on the distal lateral edge is not observable at *Biache-St-Vaast*, even though it was observed at *Sesselfelsgrotte* (Rots 2009) and at *Sodmein Cave* (Vermeersch et al. 1994; Van Peer et al. 1996; Rots et al. 2011). While the evidence observed at *Biache-St-Vaast* is the oldest one that is currently observed, spear points were also identified at later Middle Palaeolithic sites. At *Bettencourt* (75–85 ka BP), at least 6 spear points were identified in a set of 27 examined Levallois points (Rots, In prep.) (Fig. 12.15). At *Sesselfelsgrotte* (40–46 ka BP), 17 spear points and 11 spear point fragments were identified in a total examined assemblage of 292 pieces (Rots 2009). While this only provides a very sketchy, anecdotic insight into Middle Palaeolithic hunting technology, it supports nevertheless that spear point evidence exists. It was observed on each of the examined sites, in varying numbers, which was determined by the site's function (Rots 2015). It is to be expected that more spear points will be identified in future functional studies, which will hopefully provide a more complete and balanced picture.

While my personal examination of Upper Palaeolithic sites is still on-going, I also want to draw attention to the danger of considering any microlithic point as an arrowhead or barb, and the feasibility of distinguishing arrow tips and barbs in a Late Palaeolithic context. A set of 35 tools classified as points by the excavators were examined from the Creswellian site of *Zeijen* (Rots et al. 2003), next to 21 points from the Creswellian site of *Siegerswoude II*, 13 points from the Creswellian site of *Emmerhout* and 17 points from the Hamburgian site of *Luttenberg* (Rots et al. 2005) (Table 12.4). Aside from the identification of drills among the pieces classified as points (7%), the majority showed diagnostic evidence of projectile use. Of the pieces used as projectiles, 69% proved to have been mounted as tip, against 22% as barb. Given the high rate of detachment of barbs in experimental use conditions, it is likely that a large part of the archaeological barbs was never recovered. No inferences could however be made regarding the combined or separated use of tips and barbs.

Discussion

While tip damage is a crucial aspect that is often visible on used armatures, it is important to stress that armature identifications should rely on the damage pattern visible over the whole piece. One wear feature is never sufficient for a reliable identification. Above all, a macroscopic identification of armatures is generally not reliable, as it tends to rely on fracture types only, for which criteria on what to call diagnostic are often applied insufficiently strict.

Aside from tip damage, such as step-terminating fractures, burination, and spin-off's, also lateral damage is important on the used portion. Sliced scars, for instance, witness the cutting motion upon insertion and are thus frequent. MLIT's have unfortunately been neglected recently due to the focus on what is visible under low magnification (or with the aid of a hand lens). It has been stressed that this is a regrettable evolution. MLIT's are generally only observable under high magnification, but they are actually the most reliable proof of the impact-related nature of the damage features they are associated with. Only when assemblages are heavily alterated or patinated may MLIT's no longer be visible. The hafted portion should not be neglected either because several impact-related wear features occur there as a result of the counter-pressure against the haft or within the animal. In addition, it allows determining the haft boundaries and the fixation mode used. The combination of the wear features on the used and hafted portion often allows a far more secure identification of armatures.

Also the position of the element in the shaft can only be determined based on a combination of use and hafting wear evidence, and specific wear patterns were proposed. It is clear that the occurrence of a tip fracture is not sufficient to consider an implement as a tip instead of a barb, and also the resin distribution is not reliable on its own.

With regard to the distinction between different projecting modes of hunting weapons, no reliable diagnostic identification criteria are yet available, in spite of some suggestive elements that still need to be tested on their value. There is a high need for more elaborate, large-scale experimentation in order to provide further insight and to determine the potential of wear traces for making such distinctions. TCSA values do not provide a reliable alternative and while Wallner lines (if confirmed through blind testing) may provide a solution (Hutchings 2011), it unfortunately concerns some raw materials (i.e., obsidian) only.

Conclusion

Experimental results that have been produced over the years, including the ones presented here, have allowed the proposition of a set of diagnostic microscopic wear features and patterns that allow a reliable identification of armatures in archaeological assemblages. However, these criteria have recently been used far less rigorously and analytical procedures have gradually been moving away from microscopic approaches. Here, the importance of microscopic examinations for a reliable identification of armatures is stressed and new experimental results were discussed. It is stressed that examinations of armatures should not rely on one wear feature only. Attention needs to be devoted to the association between wear features in the used and hafted portion, and to the damage pattern as a whole.

While functional results remain overall too infrequent for an adequate insight into past hunting technology, it was nevertheless demonstrated based on a microscopic functional study that hafted spear points occur from at least about 200 ka years ago in Europe. This appears to concern both thrusting and throwing spear points. The identification of the earliest weapons that were projected with a spear-thrower or bow is currently still dependent on the recovery of organic finds: no reliable diagnostic identification criteria are yet available. More elaborate and systematic experimental work seems essential if progression in this matter is to be made.

Acknowledgements I am indebted to the Fund for Scientific Research (FNRS and FWO) and to the research fund of KU Leuven for their financial support of this research. My sincerest thanks are due to the *Chercheurs de la Wallonie* and the *Préhistosite de Ramioul* who were involved in the experimental program. I especially want to thank Louis Baumans, Claude Bawin, Didier Cocchi, Fernand Collin, Christian Lepers, Louis Pirnay, and Philippe Pirson.

References

Ambrose, S. H. (2001). Paleolithic technology and human evolution. *Science, 291*, 1748–1753.

Ambrose, S. H. (2010). Coevolution of composite-tool technology, constructive memory, and language. *Implications for the evolution of modern human behavior, Current Anthropology, 51*, S135–S147.

Barham, L. (2013). *From hand to handle: The first industrial revolution*. Oxford University Press.

Bar-Yosef, O., & Kuhn, S. L. (1999). The big deal about blades: Laminar technologies and human evolution. *American Anthropologist, 101*, 322–338.

Behre, K.-E. (Ed.). (2012). *Die chronologische Einordnung der paläolithischen Fundstellen von Schöningen*. Mainz: Verlag des Römisch-Germanischen Zentralmuseums.

Bergman, C. A., McEwen, E., & Miller, R. (1988). Experimental archery: Projectile velocities and comparison of bow performances. *Antiquity, 62*, 658–670.

Beyries, S., & Inizan, M.-L. (1982). Typologie, ocre, fonction. In D. Cahen (Ed.), *Tailler! pour quoi faire: préhistoire et technologie lithique II. Recent progress in microwear studies* (pp. 313–322). Tervuren: Musée royal de l'Afrique centrale.

Boëda, E. (2008). Middle Palaeolithic bitumen use at Umm el Tlel around 70,000 BP. *Antiquity, 82*, 853–861.

Boëda, E., Connan, J., Dessort, D., Muhesen, S., Mercier, N., Valladas, H., et al. (1996). Bitumen as a hafting material on Middle Palaeolithic artefacts. *Nature, 380*, 336–338.

Boëda, E., Geneste, J. M., Griggo, C., Mercier, N., Muhesen, S., Reyss, J. L., et al. (1999). A Levallois point embedded in the vertebra of a wild ass (Equus africanus): Hafting, projectiles and Mousterian hunting weapons. *Antiquity, 73*, 394–402.

Caspar, J. P., & De Bie, M. (1996). Preparing for the hunt in the Late Paleolithic camp at Rekem, Belgium. *Journal of Field Archaeology, 23*, 437–460.

Clark, J. D. (1970). *The prehistory of Africa*. London: Thames and Hudson.

Donahue, R. E. (1988). Microwear analysis and site function of Paglicci Cave, level 4A. *World Archaeology, 19*, 357–375.

Fischer, A., Hansen, P. V., & Rasmussen, P. (1984). Macro and micro wear traces on lithic projectile points. *Journal of Danish Archaeology, 3*, 19–46.

Gaudzinski, S., & Roebroeks, W. (2000). Adults only. Reindeer hunting at the Middle Palaeolithic site Salzgitter Lebenstedt, Northern Germany. *Journal of Human Evolution, 38*, 497–521.

Hayden, B. (1979). *Palaeolithic reflections*. Lithic technology and ethnographic excavation among Australian Aborigines, Canberra: Australian Institute of Aboriginal Studies.

Hedges, R. E. M., Pettitt, P. B., Bronk Ramsey, C., & Van Klinken, G. J. (1998). Radiocarbon dates from the Oxford AMS system: Archaeometry datelist 25. *Archaeometry, 40*, 227–239.

Hutchings, W. K. (2011). Measuring use-related fracture velocity in lithic armatures to identify spears, javelins, darts, and arrows. *Journal of Archaeological Science, 38*, 1737–1746.

Keeley, L. H. (1982). Hafting and retooling—Effects on the archaeological record. *American Antiquity, 47*, 798–809.

Kimball, L. R. (1989). *Planning and functional variability in the Upper Palaeolithic: Microwear analysis of Upper Perigordian tools from the Flageolet I (Dordogne)*. Unpublished PhD dissertation. Northwestern University.

Lombard, M., & Pargeter, J. (2008). Hunting with Howiesons Poort segments: Pilot experimental study and the functional interpretation of archaeological tools. *Journal of Archaeological Science, 35*, 2523–2531.

Lombard, M., & Phillipson, L. (2010). Indications of bow and stone-tipped arrow use 64,000 years ago in KwaZulu-Natal. *South Africa. Antiquity, 84*, 1–14.

Marks, A. E., Hietala, H. J., & Williams, J. (2001). Tool standardization in the Middle and Upper Palaeolithic: A closer look. *Cambridge Archaeological Journal, 11*, 17–44.

Mazza, P. P. A., Martini, F., Sala, B., Magi, M., Colombini, M. P., Giachi, G., et al. (2006). A new Palaeolithic discovery: Tar-hafted stone tools in a European Mid-Pleistocene bone-bearing bed. *Journal of Archaeological Science, 33*, 1310–1318.

Moss, E. H. (1983). *The functional analysis of flint implements: Pincevent and Pont D'Ambon: Two case studies from the French Final Palaeolithic*. Oxford: BAR.

Movius, H. L. (1950). A wooden spear of third interglacial age from Lower Saxony. *Southwestern Journal of Anthropology, 6*, 139–142.

Oakley, K. P., Andrews, P., Keeley, L. H., & Clark, D. (1977). A reappraisal of the Clacton spearpoint. *Proceedings of the Prehistoric Society, 43*, 13–30.

Odell, G. H., & Cowan, F. (1986). Experiments with spears and arrows on animal targets. *Journal of Field Archaeology, 13*, 195–212.

Plisson, H., & Beyries, S. (1998). Pointes ou outils triangulaires? Données fonctionnelles dans le Moustérien Levantin. *Paléorient, 24*, 5–16.

Plisson, H., & Geneste, J.-M. (1989). Analyse technologique des pointes à cran solutréennes du Placard (Charente), du Fourneau du Diable, du Pech de la Boissière et de Combe Saunière (Dordogne). *Paléo, 1*, 65–105.

Richards, M. P., Pettitt, P. B., Trinkaus, E., Smith, F. H., Paunovic, M., & Karavanic, I. (2000). Neanderthal diet at Vindija and Neanderthal predation: The evidence from stable isotopes. *Proceedings of the National Academy of Sciences, 97*, 7663–7666.

Rots, V. (2002a). *Hafting traces on flint tools: Possibilities and limitations of macro- and microscopic approaches* (Unpublished Ph. D. dissertation. Belgium: Katholieke Universiteit Leuven).

Rots, V. (2002b). Are tangs morphological adaptations in view of hafting? Macro- and microscopic wear analysis on a selection of tanged burins from Maisières-Canal. *Notae Praehistoricae, 114*, 61–69.

Rots, V. (2003). Towards an understanding of hafting: The macro- and microscopic evidence. *Antiquity, 77*, 805–815.

Rots, V. (2005). Wear traces and the interpretation of stone tools. *Journal of Field Archaeology, 30*, 61–73.

Rots, V. (2009). The functional analysis of the Mousterian and Micoquian assemblages of Sesselfelsgrotte, Germany. Tool use and hafting in the European Late Middle Paleolithic. *Quartär, 56*, 37–66.

Rots, V. (2010a). *Prehension and hafting wear on flint tools. A methodology*. Leuven: Leuven University Press.

Rots, V. (2010b). Un tailleur et ses traces. Traces microscopiques de production: programme expérimental et potentiel interprétatif. *Bulletin de la Société Royale Belge d'Etudes Géologiques et Archéologiques, Les Chercheurs de la Wallonie hors-série, 2*, 51–67.

Rots, V. (2013). Insights into early Middle Palaeolithic tool use and hafting in Western Europe. The functional analysis of Level IIa of the early Middle Palaeolithic site of Biache-Saint-Vaast (France). *Journal of Archaeological Science, 40*, 497–506.

Rots, V. (2015). Hafting and site function in the European Middle Paleolithic. In N. J. Conard & A. Delagnes (Eds.), *Settlement Dynamics of the Middle Palaeolithic and Middle Stone Age* (Vol. 4, pp. 383–410). Tübingen: Kerns Verlag.

Rots, V. (In prep). The functional analysis of the Levallois points of Bettencourt-Saint-Ouen.

Rots, V., & Plisson, H. (2014). Projectiles and the abuse of the use-wear method in a search for impact. *Journal of Archaeological Science, 48*, 154–165.

Rots, V., Stapert, D., & Johanssen, L. (2003). De Cheddar- en Creswell-spitsen van Zeijen (Dr.): 'projectielen' of 'messen'? *Paleo-Aktueel, 13*, 24–29.

Rots, V., Stapert, D., & Johanssen, L. (2005). Spitsen van Siegerswoude (Fr.), Emmerhout (Dr.) en Luttenberg (OV.): gebruiksssporenonderzoek. *Paleo-Aktueel, 14–15*, 11–15.

Rots, V., & Van Peer, P. (2006). Early evidence of complexity in lithic economy: core-axe production, hafting and use at Late Middle Pleistocene site 8-B-11, Sai Island (Sudan). *Journal of Archaeological Science, 33*, 360–371.

Rots, V., Van Peer, P., & Vermeersch, P. M. (2011). Aspects of tool production, use and hafting in Palaeolithic industries from Northeast Africa. *Journal of Human Evolution, 60*, 637–664.

Shea, J. J. (1988). Spear points from the Middle Paleolithic of the Levant. *Journal of Field Archaeology, 15*, 441–456.

Shea, J., & Sisk, M. (2010). Complex projectile technology and *Homo sapiens* dispersal into Western Eurasia. *PaleoAnthropology, 2010*, 100–122.

Thieme, H. (1997). Lower Palaeolithic hunting spears from Germany. *Nature, 385*, 807–810.

Thieme, H., & Veil, S. (1985). Neue Untersuchungen zum eemzeitlichen Elefanten-Jagdplatz Laehringen, LDKR, *Die Kunde N.F., 36*, 11–58.

Tuffreau, A., & Sommé, J. (1988). Le gisement paléolithique moyen de Biache-Saint-Vaast (Pas-de-Calais). Volume I. Stratigraphie, environnement, etudes archéologiques (1ère partie). *Mémoires de la Société Préhistorique Française, 21*, 338p.

Van Peer, P., Vermeersch, P., & Moeyersons, J. (1996). Palaeolithic stratigraphy of Sodmein Cave (Red Sea Mountains, Egypt). *Geo-Eco-Trop, 20*, 61–71.

Vermeersch, P. M., Van Peer, P., Moeyersons, J., & Van Neer, W. (1994). Sodmein cave site (Red Sea Mountains, Egypt). *Sahara, 6*, 31–40.

Wadley, L., Hodgskiss, T., & Grant, M. (2009). Implications for complex cognition from the hafting of tools with compound adhesives in the Middle Stone Age (South Africa). *Proceedings of the National Academy of Sciences, 106*, 9590–9594.

Wadley, L., Williamson, B., & Lombard, M. (2004). Ochre in hafting in Middle Stone Age southern Africa: A practical role. *Antiquity, 78*, 661–675.

Part III

Measures of Weapon Performance

Chapter 13
Testing Archaeological Approaches to Determining Past Projectile Delivery Systems Using Ethnographic and Experimental Data

C. Clarkson

Abstract TCSA and TCSP are often considered valuable measures of projectile performance, particularly in terms of penetration and overall design. Proponents of this view have also argued that TCSA/TCSP may also be useful for identifying the origins and spread of more complex projectile technologies such as the spear thrower and bow. The strength of these arguments will be tested against ethnographic data and new experiments. The results suggest that TCSA/TCSP statistics are not robust measures of projectile performance, or reliable proxies for inferring delivery systems. An alternative approach is developed using experimental data that compares impact fracture size for three different diagnostic impact fracture types. This approach, while found to be valuable, also presents problems for archaeological identification of projectile technologies.

Keywords Projectile technology • Human evolution • Tip cross-sectional area • Tip-cross sectional perimeter • Experimental archaeology • Impact fracture size

Introduction

A number of recent studies have built on Hughes' (1998) observation that Tip Cross-Sectional Area (TCSA) and Tip Cross-Sectional Perimeter (TCSP) are useful ballistic measures of relevance for inferring past projectile design and use (Hughes 1998; Pargeter 2007; Wadley and Mohapi 2008; Villa and Lenoir 2009). TCSA and TCSP are calculated from maximum point width and thickness (Fig. 13.1), the rationale being that a small tip-cross sectional area or perimeter is vital for ensuring deep penetration of skin and tissue for low velocity weapons, effectively concentrating the kinetic energy of the projectile on a small area allowing the projectile to tear a hole in the skin (Hughes 1998). A common notion is that hominins might refine the manufacture of points to decrease cross-sectional dimensions and improve their killing power as they became more reliant on projectile technology, perhaps resulting in changes in prey choice, expansion into new environments and other forms of cultural change.

Shea and Sisk in particular have pursued the notion that Tip Cross-Sectional Area (TCSA) and Tip Cross-Sectional Perimeter (TCSP) as useful in determining projectile performance and the evolution of projectile systems (Shea 2006; Sisk and Shea 2009, 2011; Shea and Sisk 2010). They employ ethnographic, experimental and archaeological data to extend this proposition to propose that TCSA/TCSP may also be useful in differentiating the mode of delivery from archaeological point assemblages, effectively allowing points delivered by hand in thrusting and throwing spears to be differentiated from those launched using more complex devices such as spear throwers and bows.

Shea and Sisk argue that ethnographic and archaeological collections of hafted stone projectile points show that low TCSA and TCSP scores on stone points are only associated with mechanically projected weaponry such as bows and spear throwers, and that only a small amount of overlap exists between these two systems. Points thought to be associated with simple spear systems such as thrusting spears and javelins are thought to be much larger, although no ethnographic or archaeological evidence is presented to support this proposition. Experiments conducted by Shea and Sisk are advanced to support the notion that larger tips were effective as thrusting weapons, although the performance of such points as projectiles was not explored.

Having collected TCSA and TCSP data on a large number of points from sites in Africa, the Levant and Europe, they argue the first archaeological signs of the use of complex projectiles, as inferred from the first appearance of points with low TCSA/TCSP values, appear with modern humans around the time of exit from Africa c.50 ka, and that

C. Clarkson (✉)
School of Social Science, The University of Queensland, Brisbane, QLD 4072, Australia
e-mail: c.clarkson@uq.edu.au

Radu Iovita and Katsuhiro Sano (eds.), *Multidisciplinary Approaches to the Study of Stone Age Weaponry*, Vertebrate Paleobiology and Paleoanthropology, DOI 10.1007/978-94-017-7602-8_13

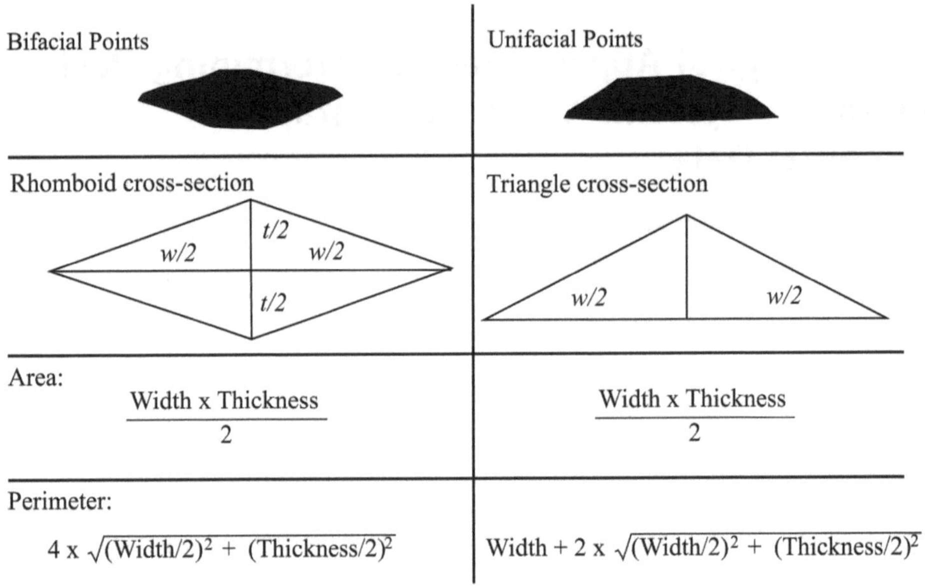

Fig. 13.1 Method of calculating TCSA and TCSP (From Sisk and Shea 2011)

this likely aided their colonization of new environments. Earlier modern humans and archaic species such as Neanderthals used much larger points that they argue represent the use of hand thrown or thrusting spears. As a result of Shea and Sisk's influential studies, TCSA/TCSP now regularly feature in discussions and comparisons of stone points and in discussions of the origins and type of projectile technology (Pargeter 2007; Costa 2012; cf. Wadley and Mohapi 2008; Moncel et al. 2009; Villa and Lenoir 2009; Villa et al. 2009; Lombard and Phillipson 2010; Lombard et al. 2010).

We raise a number of points of contention with TCSA and TCSP as useful measures of projectile performance and as valid indicators of specific types of projectile technology. First, ethnographic and archaeological examples of arrows and darts from the last 10,000 years may not be suitable models for comparison with points from much older periods of human evolution. The record of changing size and form of stone points throughout the late Pleistocene indicates that weapon tips have undergone a long process of development and it may be a mistake to think that early projectile technologies mirrored all aspects of more recent systems. Different elements of projectile systems may have changed at different times or rates. For instance, it may be that some developments in projection system, such as the introduction of the spear thrower or bow, came before reductions in point size.

Indeed, long-term changes in tipped-weapon systems exist in some regions that point to a complex and multidirectional sequence of developments in projectile technology. In the South African MSA (Lombard and Clark 2008), for example, weapon tips varied dramatically in size, hafting arrangement and even raw material type, and this was likely related to the types of prey being captured (Lombard and Clark 2008) as well as the systems of mobility and landuse employed (McCall 2007; Mackay 2010). Similar cases for non-linear change in projectile technology and the retention of the atlatl exist for the New World (Blitz 1988; Hughes 1998), while Buchanan et al. (2011) have demonstrated that no clear relationship exists between prey size and point size in Paleoindian assemblages.

Not all late prehistoric ethnographic and archaeological examples support a simple correlation between point size and projectile system. While known prehistoric arrows in Europe and the New World do appear to have had small stone tips, no such strong correlation is seen between point size and spear thrower technology in Friis-Hansen's (1990) ethnographic data, for instance, suggesting that point size is not always an accurate discriminator of projection systems.

The Australian ethnographic and archaeological evidence also indicates a huge range in TCSA values from the smallest backed artifacts, to unifacial and bifacial points, right up to the use of huge stone "leilira" blades (Newman and Moore 2013). This huge range of point types were often attached to spears thrown with a spear thrower (Fig. 13.2). Thus, in Australia a very wide range of TCSA values are associated with a single projection system, with points attached to darts from the late Holocene varying hugely in size. Some leilira blades (large stone pointed blades from northern and central Australia) have TCSA values vastly greater than anything in Shea and Sisk's database (Fig. 13.2) (contra, Shea 1997). The case of the leilira indicates that very large spear points were thrown with a spear thrower

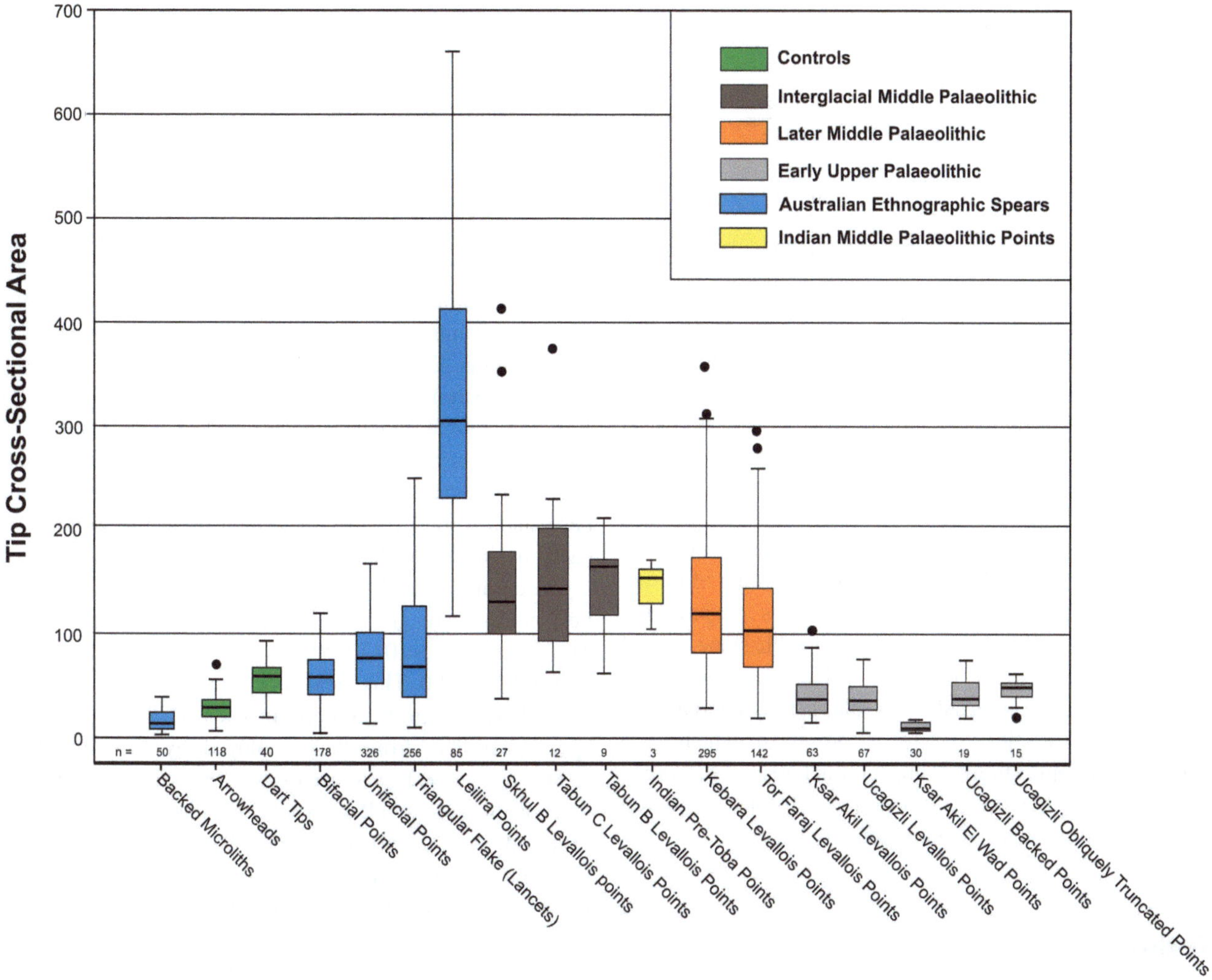

Fig. 13.2 Comparison of Australian ethnographic and archaeological stone projectile tips with Shea's (2006) global archaeological and ethnographic dataset

(Fig. 13.2), suggesting that common notions about mechanical limits on point size projected in this way are exaggerated. The existence of these spears also implies that Mousterian, Levallois and the larger bifacial points from MSA, Mousterian and transitional assemblages in Europe could all have been effectively projected as weapon tips from spear throwers, depending on spear design and hafting. This observation compels us to continue the search for early spear thrower technology.

Observations were made on thirty spears in the Northern Territory Museum and Art Gallery Collections that are tipped with large stone retouched or unretouched leilira blades from northern Australia (Thomson 1949). These spears are typically made from light and flexible wood, weigh 274 ± 45 g and 238 ± 13 cm in length, and taper in diameter by one third over their length, from 18 ± 2.5 mm below the hafting to 12 ± 1.8 mm at the butt end. The point of balance for such spears was located at 33 ± 4% of total length back from the tip. The stone tips were inserted directly into the split end of the spear shaft, without fore-shaft, and were glued and tied with beeswax and bark twine. The spears were thrown using a distinctively shaped Arnhem Land spear thrower with an average length of 86.6 ± 5.9 cm.

Made to similar specifications, prehistoric artisans would have been capable of mounting almost any large Middle Paleolithic foliate, Levallois or Mousterian point to produce highly effective, long-range, accurate spears like those used in northern Australia without need for fletching or other complex elements (Cundy 1989: 12–13). Archaeological finds of resin and pitch confirm that Neanderthals and perhaps other hominins possessed knowledge of the essential adhesive technologies required to firmly attach Levallois, Mousterian or foliate points to a spear shaft (Grünberg 2002; Boëda et al. 2008), while the Shöningen spears indicate that aerodynamic designs was likely in use from an early period (Thieme 1997).

Table 13.1 Product moment correlation coefficients for TCSA/TCSP vs penetration for experimental arrows and darts fired into gelatin blocks

Variable	R^2 arrows	P	R^2 darts	P
TCSA	0.191	0.001	0.065	0.065
TCSP	0.159	0.004	0.026	0.128
Hafted TCSA	0.264	0.006	0.005	<0.09
Tip weight	0.426	0.001	0.103	0.09
Total weight	0.395	0.001	0.152	0.03

Finally, while Hughes, Shea and Sisk assert that TCSA/TCSP are valuable indicators of projectile performance, no study has yet provided a convincing test of the strength of association between TCSA/TCSP and projectile penetration for stone-tipped weapons. Shea and Sisk's own experiments examined thrusting spears and arrows with triangular stone tips, but obtained only very low product moment correlation coefficients for their experiments (e.g., $r^2 = 0.084$ for TCSP). Here we present the results of further testing of TCSA/TCSP as proxy measures of projectile penetration for darts and arrows armed with stone tips of widely varying size and type (see Table 13.1). Following presentation of these results, an alternative approach to determining projection system is presented which uses experimental replication of impact fractures to determine whether impact fracture size measured on stone points can be used to discriminate weapon delivery systems.

Methods

The experiment set out to compare the penetration of a light, high velocity projectile (arrow, N = 51) with a heavier, lower velocity projectile (dart, N = 54) fired from a crossbow with tips of greatly varying TCSA and TCSP. The crossbow consisted of a compound bow clamped to a purpose-built frame to create a stable and accurate firing platform. A total of 105 stone points of widely varying size and form were employed in the projectile tests. The compound bow had a draw weight of 45 pounds and was positioned at a distance of 5 m from the target. With some initial practice, the crossbow was capable of launching both projectiles with sufficient accuracy to consistently hit the target. The compound bow's cam system ensured that projectiles were launched with the same force each time irrespective of small variations in draw length. While 5 m is likely too close to form a likely analogue for prehistoric hunting, the purpose was to control the launch distance while maintaining accuracy. The results are not intended to reflect real impact depths in prehistoric hunting situations. It should be noted, however, that using a lower powered delivery at close range should simulate a higher powered delivery at greater distances. Since hunting bows are typically in the order of 55–65 lbs draw weight, our lower poundage bow probably simulates the drop in impact force quite well when fired at greater ranges of 15–30 m as typically recorded in hunting situations (Catellan 1997).

Each projectile was tipped with stone points of differing TCSA/TCSP values and representing a range of formal types (see Appendix for individual point data). Points were mounted onto the ends of dowel fore-shafts of equal length and of two different diameters – 8 mm for arrows and 12 mm for darts. Points were attached using commercial adhesive putty (*Selleys Kneed It Multipurpose Epoxy Putty*) that created a very strong but relatively unobtrusive joint (Fig. 13.3). Fore-shafts were attached to the fletched main shaft using short pieces of brass tubing of appropriate diameter to create a tight but detachable join. Arrow shafts weighed 51 g and dart shafts weighed 156 g. The total projectile weights for each specimen after hafting are provided in the Appendix.

To ensure a roughly equal representation of TCSA values, points were grouped into six categories representing TCSA size-ranges: 0–50, 51–100, 101–150, 151–200, 201–250 and 251–300. This yielded between eight and ten points of each type (arrow or dart) in each TCSA group. This approach differs from that employed by Sisk and Shea (2009) where the TCSA size ranges were positively skewed, with many more small TCSA values than larger ones (see Sisk and Shea (2009), Fig. 13.2). The largest TCSA measurement employed in this study was 292.5 and the smallest was 19.5, with a mean and standard deviation of 152 ± 83.

Each point was fired into a gelatin block of dimensions 30 × 25 × 20 cm, set on a straw bale in front of a backdrop of thick carpet positioned to catch stray shots. The target was positioned 5 m from the front edge of the bow. The penetration depth in centimeters was recorded for each shot and the data entered in a Lotus Approach database. Penetration ratio as measured by Shea and Sisk was not

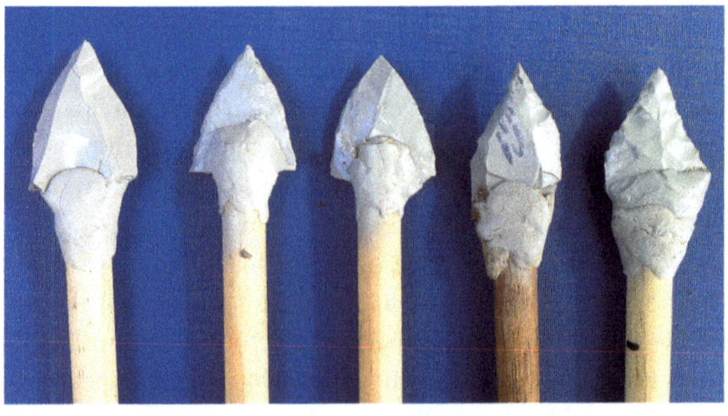

Fig. 13.3 Examples of the commercial hafting putty used to attach the tips to the shafts, as well as a range of points types used in the experiment. From left to right: Levallois Point, Unifacial Point, Levallois Point, Mousterian Point and Bifacial Point. All points were painted grey prior to use

examined here as this has no actual bearing on the lethalness of a wound. Lethalness should be understood as the likelihood of damaging vital organs by deeply penetrating the body whereas the length of the projectile itself is unlikely to be meaningful.

Each shot was aimed at an undamaged section of the block and shots intersecting an existing entry hole were discounted and the points refired. Gelatin blocks were discarded once entry holes became too numerous to consistently hit an undamaged section, or when cracks began to form in the block.

Results

The results of the experimental testing of TCSA and TCSP indicate that a very poor correlation exists between these two statistics and penetration depth (Figs. 13.4 and 13.5). Both arrows and darts return very low product moment correlation coefficients (r^2 values) for the relationship between TCSA/TCSP and penetration (Table 13.1). At first glance this appears to indicate that TCSA and TCSP are very poor predictors of penetration depth and hence are poor proxies for ballistic performance. This result cannot be explained by hafting joints or variations in point weight as hafting joints made only small and consistent differences to TCSA scores and the weights of the fore-shaft and main-shaft were kept constant.

The second major finding is that penetration depth overlaps extensively for arrows and darts within any part of the TCSA/TCSP range. However, as would be expected, arrows penetrate more deeply than darts ($t = 6.854$, $df = 102$, $p = <0.0005$), indicating that velocity and mass are more important determinants of penetration depth, given that the range of TCSA and TCSP values was identical for both arrows and darts. For each projection system, total projectile weight indeed explains much more of the variation in penetration depth for each projectile type than does TCSA or TCSP (arrow mass: $r^2 = 0.395$, $p = < 0.0005$; dart mass: $r^2 = 0.152$). Unfortunately, velocity was not measured in this experiment, but based on data presented in Hughes' (1998), arrows tend to be twice as fast as darts (46.9 vs. 23.6 m/s), although Hutchings and Brüchert (2007) found that in some cases darts can be as fast as arrows. In this experiment, however, launch force was kept constant and only mass will have affected projectile velocity.

Hughes (1998) gives the equation for penetration depth as (mass * velocity)/(tcsa * shape constant). Therefore to explore the contribution of mass and velocity, a linear regression was performed using mass * velocity (as borrowed from Hughes of 46.9 for arrows and 23.6 m/s for darts) and penetration depth. Inserting mass * velocity increases the r^2 from 0.395 (for mass alone) up to 0.478. Adding TCSP as an additional independent variables reveals that while mass * velocity is highly significant ($p = 0.0005$), TCSP is not ($p = 0.919$) and the r^2 value remains unchanged. We find therefore that TCSP, the preferred measure of Sisk and Shea, makes little difference to the penetration of arrows and darts at least within the size limits tested here, at least within the size range tested here.

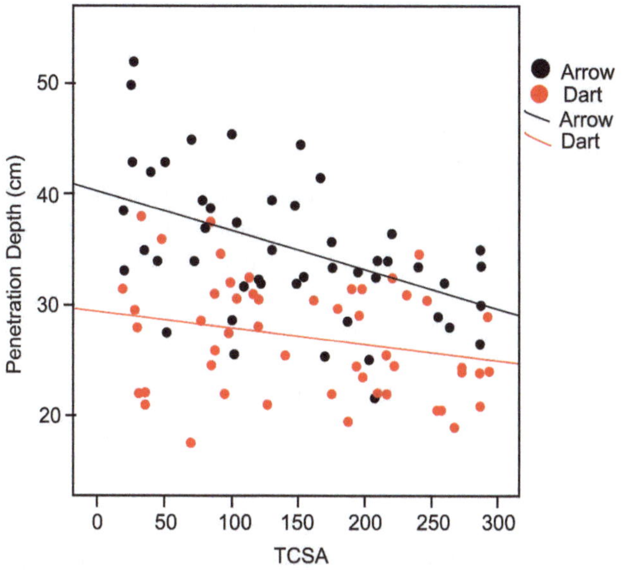

Fig. 13.4 Penetration depth for experimental arrows and darts of varying TCSA when fired into gelatin blocks. $R^2_{darts} = 0.065 (p = 0.065)$, $R^2_{arrows} = 0.191 (p = 0.001)$. For a summary of all statistics see Table 13.1

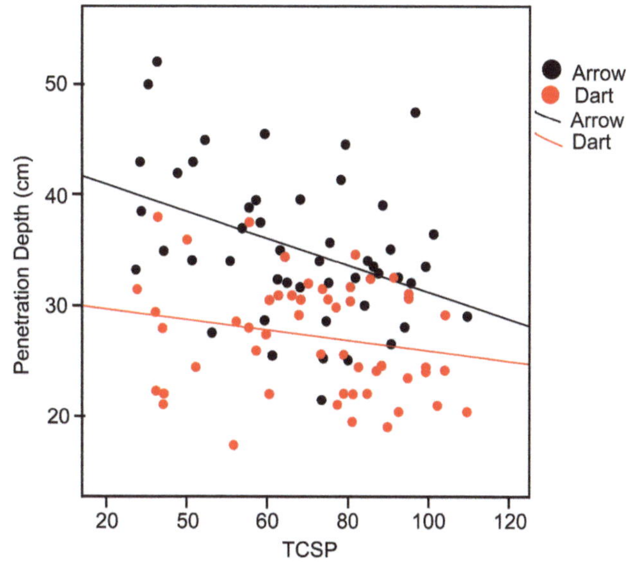

Fig. 13.5 Penetration depth for experimental arrows and darts of varying TCSP when fired into gelatin blocks. $R^2_{darts} = 0.026 (p = 0.128)$, $R^2_{arrows} = 0.159 (p = 0.004)$. For a summary of all statistics see Table 13.1

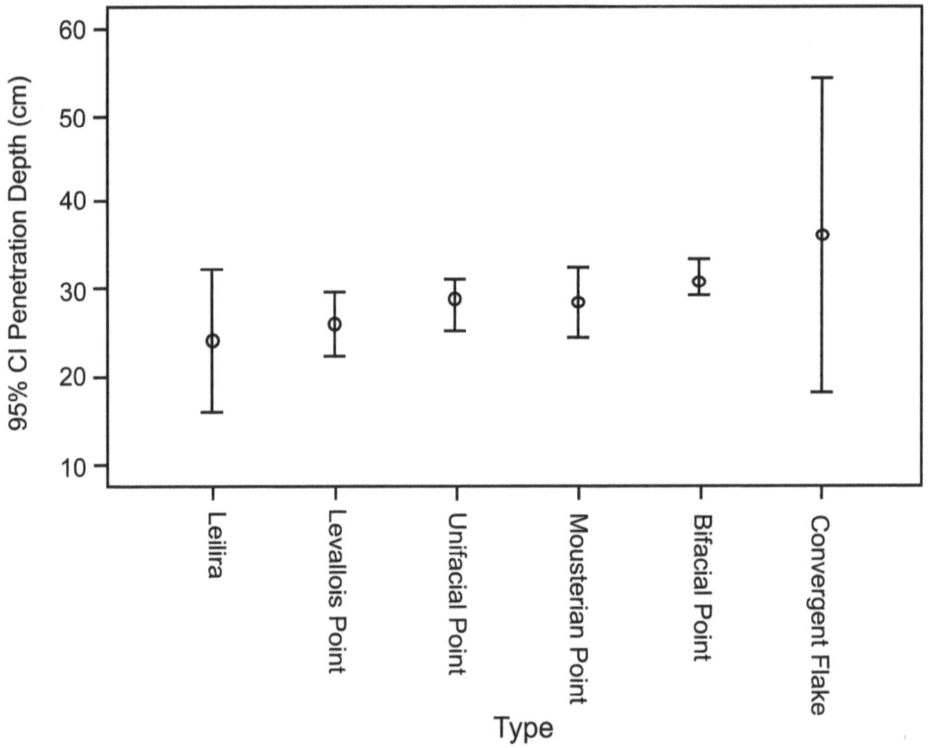

Fig. 13.6 95% confidence intervals for penetration depths for different points types when fired into gelatin blocks. Results are for darts only

Finally, no apparent differences are present in penetration depth when viewed by point type (Fig. 13.6).

The results generated from experimental testing of arrows and darts with tips of widely ranging TCSA/TCSP values have shown that these statistics are inadequate proxies for projectile performance, at least in terms of penetration depth. This suggests they are also likely to be inadequate for determining the types of projectile delivery system used in the past, as our experiments indicate that points with a TCSA of close to 300 can generate lethal wounds with penetration depths of ≥30 cm into ballistics gel when projected from either bow or spear thrower. In other words, given appropriate construction in terms of hafting, balance and mass, effective projectiles could have been constructed for use with bows or spear throwers using any of the stone tips included in Shea's database, or indeed, using tips that far exceed those in size, as in the case of the Australian leilira-tipped spears. If this is true, and no reason has so far been advanced why it should not be, then alternative indices or traces must be explored to better determine the types of delivery systems used in the past and thereby reconstruct their origins and importance over the course of human evolution.

The next section examines the value of fracture impact size measured for three different diagnostic impact fracture (DIF) types as a means of inferring the weapon delivery systems used in the past.

Impact Fracture Size Experiment

Diagnostic impact fractures have been the focus of intensive archaeological and experimental research to identify the diagnostic traces left by impacts as well as the presence of projectile tips in archaeological assemblages (Barton and Bergman 1982; Flenniken and Fisher et al. 1984; Flenniken and Raymond 1986; Towner and Warburton 1990; Dockall 1997; Hutchings and Brüchert 1997; Knecht 1997; Shea 2006; Hunzicker 2008; Villa and Lenoir 2009; Sisk and Shea 2009; Lombard and Philipson 2010; Schoville 2010; Yaroshevish et al. 2010; Lombard et al. 2010; Pétillon et al. 2011). Much of this research is aimed at identifying the types of fractures left by different projectile delivery systems and different contact materials, and applying these findings to archaeological assemblages to identify artifacts that likely served as projectiles.

As a result of this history of projectile research, it is well known that mechanically projected missiles typically hit with more force than hand thrown or thrusting weapons (Hutchings and Brüchert 1997; Hughes 1998). If impact fractures that are proportional in size to impact force, then mechanically projected weapons should generate larger impact forces than hand delivered weapons. In fact Fisher et al. (1984) remarked that impact scars from different experimental weapon systems were of different sizes, but

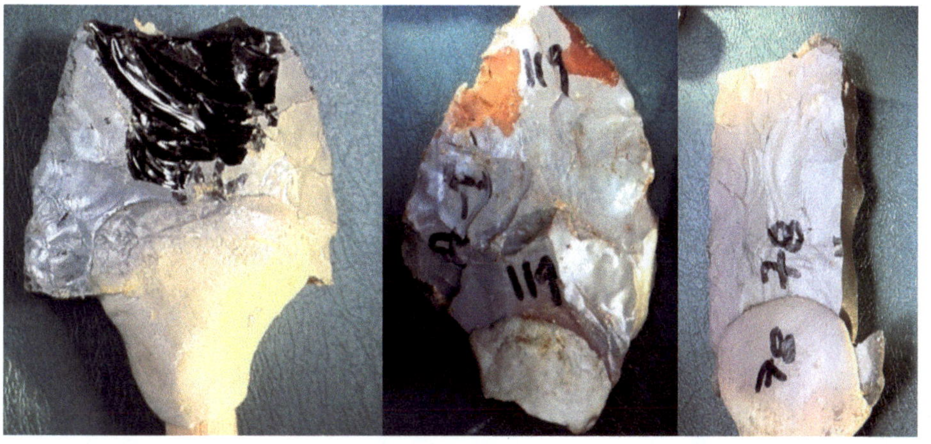

Fig. 13.7 DIFs revealed on spray painted points, Left to right: spinoff, laterals, burin

did not explore this in any detail. Impact fracture size could therefore potentially provide a convenient archaeological means of differentiating mechanical from hand delivery if DIF size differs significantly between projection systems.

To test this proposition, 154 obsidian or flint points were launched into racks of beef ribs with the meat remaining in four different ways: thrown by hand, thrown with a spear thrower, shot from a bow and stabbed with a thrusting spear. Using the same main-shaft and detachable fore-shaft system as that employed in the TCSA/TCSP experiments above, each tip was repeatedly launched until a DIF was generated. All shots were made by the author at a constant distance of 5 m from the target except for thrusting spears which were used at point blank. All points were painted with grey spray paint before use to easily identify DIFs generated upon impact on any margin and in areas of existing retouch (Fig. 13.3). DIFs were classified as one of three types, following the work of Pétillon et al. (2011) and Yaroshevish et al. (2010). These were spinoffs/flutes, lateral fractures, and burins (or pseudo-burins), as shown in Fig. 13.7. Spinoffs/flutes are hereafter referred to simply as spinoffs. The DIF type and length was recorded for each point as well as the combinations of DIFs. DIF length was recorded as the maximum length along the axis of fracture propagation.

A first observation is that the frequency of each DIF type differed markedly, as shown in Fig. 13.8. Spinoffs were found to be by far the most common DIF type resulting from impacts with bovid ribs, with 84% of points showing spinoffs either on their own or in combination with other DIF types. Laterals were the next most common DIF type with 36% of points showing laterals on their own or in combination. Burins were much rarer, with only 18% of points showing burin impact fractures alone or in combination.

A second observation is that raw material type made a significant difference to fracture size, with obsidian fractures being larger for all fracture types (*t*-tests: Spinoffs: $p = 0.019$; Laterals: $p = 0.01$; Burins: $p = 0.004$). Differences between the raw materials were quite significant and DIFs on obsidian were in the order of double the length of those on flint (obsidian mean = 10.1 ± 7.7 mm, flint mean = 5.8 ± 4.8 mm).

When the size of impact fractures is compared, variable results were obtained for each fracture type (Fig. 13.9). Spinoffs showed no significant difference in length between the different weapon delivery systems (Table 13.2). Laterals on the other hand showed significant overall differences between the four systems ($p = 0.007$), but only marginally significant differences between mechanical and hand projected weapons ($p = 0.08$) (Table 13.2). Burins returned significant results for all four weapons systems ($p = 0.017$) and for mechanical versus hand projected projectiles ($p = 0.02$) (Table 13.2). This means that archaeologists may be able to use burin fracture length measured on points of the same raw material to infer the presence of mechanically projected weapons when burin scars are particularly large. In this experiment, burin spalls larger than c.15 mm were only created on mechanically projected darts and arrows, and hence this cutoff provides a valuable threshold to focus attention on impact scar size on archaeological points and in future experiments. Experimental testing on equivalent raw materials to those found in archaeological assemblages will help calibrate for the effects of brittleness and quality/graininess on fracture size. Further testing of the effects of range, velocity and overall mass on fracture size would also help refine and calibrate these comparisons.

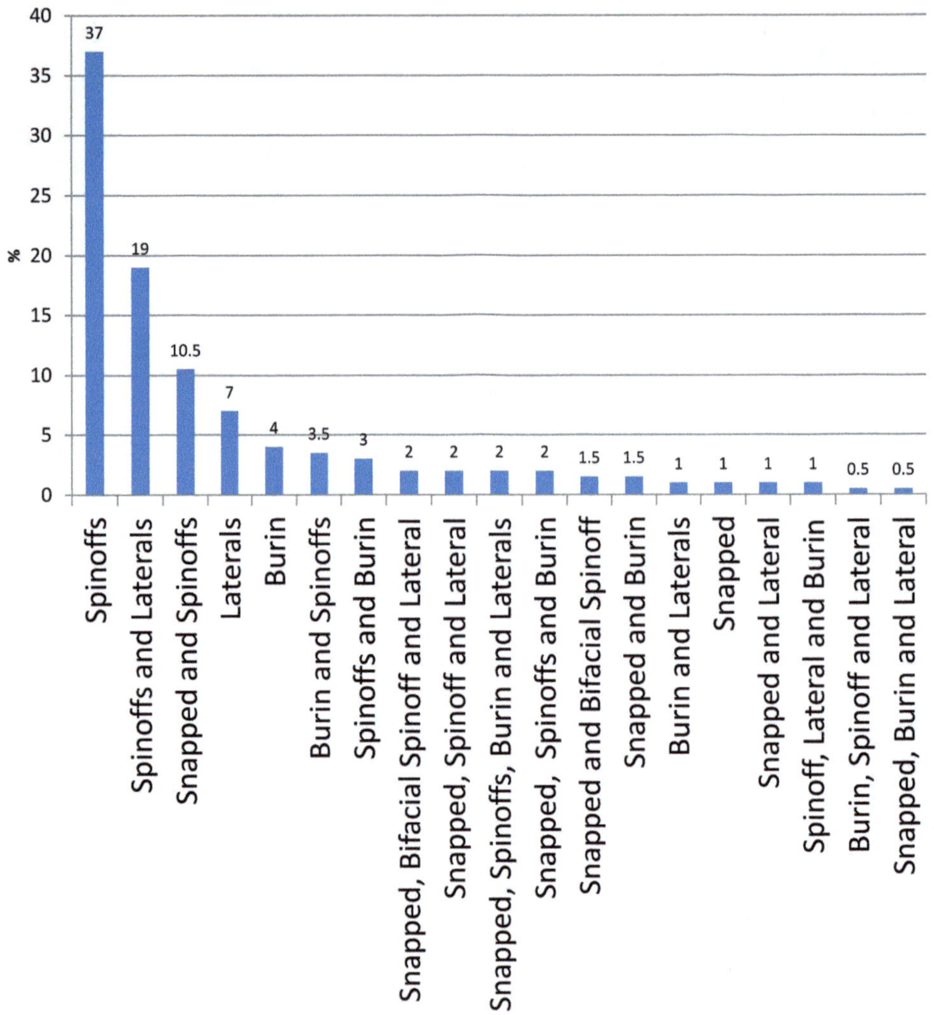

Fig. 13.8 Frequency of different combinations of DIFs on the 154 experimental points

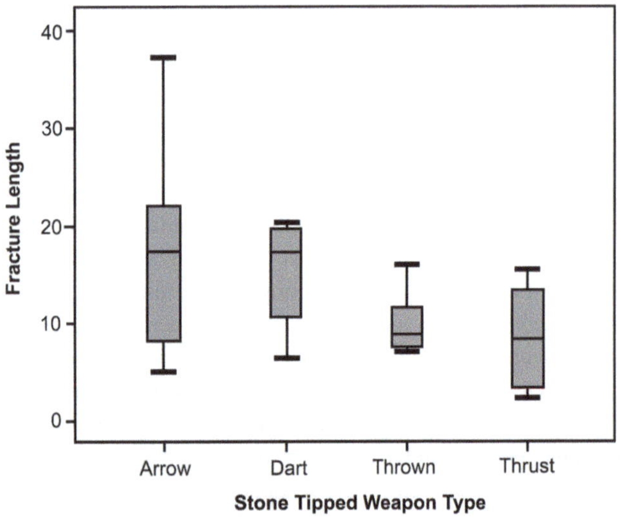

Fig. 13.9 Differences in burin impact fracture size for flint-tipped weapons. Differences between arrow/dart and thrown/thrusting are significant (p = 0.02)

Table 13.2 ANOVA tests for differences in fracture length for three different DIF types when compared between arrows, darts, hand thrown spears and thrusting spears

DIF Type	Arrows, darts, thrown and thrusting	Hand versus mechanically projected
Spinoffs	ANOVA, df = 3/172, F = , p = 0.998	df = 174, p = 0.878
Laterals	df = 3/70, F = 4.32, p = **0.007**	P = 0.08
Burins	df = 3, 30, F = 3.969, p = **0.017**	P = **0.02**

P values in bold are significant

Discussion

The analyses presented in this paper suggest that simple proxies like TCSA and TCSP, while attractive in offering a simple handle on past projectile design and use, in reality are unlikely to provide much valuable information about either of these issues. Both measures fail to provide a strong correlation with penetration depth and experimental results indicate that velocity and overall mass of the projectile are much better determinants of penetration depth, even for points of very different size. TCSA/TCSP also do not provide a valuable measure of the mechanical limits on projectile design. As the Australian example showed above, a huge range of point types and sizes can be employed within a single delivery system. The precise ways in which TCSA and TCSP of points affects the construction of these alternative spear and arrow designs is something that warrants future experimental work to develop a model of projectile design constraints and affordances.

Accepting that different components of projectile technology may have changed at different rates and in response to different technological and foraging stimuli means we must exercise much more caution when applying simple size measurements to infer the evolution of projectile technologies. The evolution of projectile technologies is now known to be multidirectional in at least some regions of the world, and this makes simple teleological schemes unsatisfactory descriptions of what could be a very complex evolution. The rarity of excellent organic preservation in the majority of archaeological sites in critical regions and time periods suggests unravelling such complexity will be very difficult in the majority of cases.

An alternative approach to identifying weapons systems using DIF size on archaeological points offers some positive preliminary results. Burin impact spalls appear more sensitive to differences in projectile delivery systems than other DIF types. The advantages of burin impact spalls over other DIF types also lies in the fact that they are easy to recognize, whereas laterals and spinoffs can sometimes be difficult to differentiate from existing retouch. The disadvantage of burins, however, is that they are the rarest impact fracture type and hence large point assemblages may be required to perform the analyses suggested here.

One potential complication to the use of DIF size lies in the fact that projectiles fastened with resin versus notching and tying can result in drastic differences in fracture rates on points. Points fastened with brittle resins are more likely to break out of the haft, saving the point tip, whereas those points that are notched and tied are more likely to be damaged catastrophically (Akerman 1978). Original experiments by the author comparing damage rates on brittle adhesives such as spinifex resin and pine pitch versus notched and tied points revealed that brittle resins result in less frequent DIFs and far fewer catastrophic breaks on points, consistent with Akerman's findings.

In addition to problems of obtaining enough burin impact fractures in archaeological assemblages and the effects of brittle resins on DIF frequency, other factors may cause major complications to determining projectile type. The strength/poundage, type of bone impacted, angle of impact and raw material type may all effect fracture size, and such variables may be very difficult to take into account. Controlled experiments will help determine exactly how each of these variables interact, but may not help determine how DIFs were created on individual archaeological specimens.

One important new technique has emerged that enables estimation of a crucial variable, that of impact velocity. Hutchings' (2009, 2011) new approach estimates fracture velocity from the angle of divergence between Wallner lines and fracture ripples on an impact fracture surface, and has been experimentally verified and applied to archaeological specimens (Hutchings 1999). Hutchings' research on the speeds at which wide range of fractures are propagated shows that mechanically projected points such as those from bows and spear throwers enter the "dynamic fracture" range and generate fracture speeds much in excess of those created on simple projectiles, knapping, accidental breakage and trampling. While best suited to cryptocrystalline siliceous rocks such as flint and obsidian, further work to determine the potential of this approach on other stone types including those with some degree of graininess (such as silcrete) is worth undertaking. Future research to determine whether DIF size and fracture velocity are related may prove valuable in the search for robust measures of impact speeds and hence projectile delivery systems.

Conclusion

When and where complex projectile technology first appeared is unresolved and will be the focus of much future research. While TCSA/TCSP is found not to offer much of value in this search, analysis of comparatively little-studied DIFs offers some promise in helping determine the type and evolution of projectile systems. When used in combination with Hutching's analysis of Wallner lines and fracture wings, these two approaches may yet offer valuable insights into the velocity and mass of past projectile delivery systems. Ultimately, however, multiple lines of evidence are needed to make further progress in discovering the evolution of weapon delivery systems, involving not only measurement of impact scars and the angles of divergence of Wallner lines and fracture ripples, but also microwear and residue studies, identification of hafting traces and the analysis of faunal assemblages for clues as to past prey selection and impact damages on bones.

Furthermore, such studies should also keep in mind the conflicting aspects of projectile design such as range, accuracy, penetration power and aerodynamics as these factors can all have significant effects on projectile construction and point size (Christenson 1986). Like any controlled and sustained study of the mechanics of fracture in different circumstances, continued research on impact fractures is likely to lead to the great improvement in understanding and new analytical techniques for exploring the history of projectile use.

Acknowledgements The invitation to participate in the International Symposium on Prehistoric Weapons in Mainz in September 2011 turned my long held interest in points into the chance to conduct my own projectile experiments, for which I am most grateful to Radu Iovita and Katsuhiro Sano. The workshop was a great success, and certainly much more than "a marriage of dogs".

Appendix

Details of the 105 points used in the experiment

ID	Type of projectile	Retouch type	Typology	Penetration depth (cm)	TCSA	TCSP	Total projectile weight
1	Dart	Bifacial	Bifacial point	29	292.5	103.94	216.4
2	Dart	Bifacial	Bifacial point	24	292.5	103.94	211.4
3	Dart	Bifacial	Bifacial point	29	292.5	103.94	201.4
4	Arrow	Unifacial	Unifacial point	30	288	84.16	86.1
5	Arrow	Unifacial	Mousterian point	26.5	287	90.64	91.1
6	Arrow	Bifacial	Bifacial point	33.5	287	99.29	81.1
7	Dart	Bifacial	Bifacial point	24	287	99.29	211.4
8	Arrow	Unifacial	Mousterian point	35	287	90.64	81.1
9	Dart	Bifacial	Bifacial point	21	286	102.21	211.4
10	Dart	Bifacial	Bifacial point	24.5	273	98.79	211.4
11	Dart	Unifacial	Unifacial point	24	273	87.01	201.4
12	Dart	Unifacial	Mousterian point	19	266.5	89.54	216.4
13	Arrow	Unifacial	Mousterian point	28	264	94.11	91.1
14	Arrow	Bifacial	Bifacial point	32	260	95.41	86.1
15	Dart	Unifacial	Mousterian point	20.5	258	92.24	221.4
16	Dart	Bifacial	Bifacial point	20.5	255	109.56	201.4
17	Arrow	Bifacial	Bifacial point	29	255	109.56	91.1
18	Arrow	Bifacial	Bifacial point	47.5	252	96.74	81.1
19	Dart	Bifacial	Bifacial point	30.5	246	95.01	221.4
20	Dart	Unifacial	Mousterian point	34.5	240.5	82.22	201.4
21	Arrow	Unifacial	Bifacial point	33.5	240	86.64	81.1
22	Dart	Bifacial	Bifacial point	31	231	94.82	196.4
23	Dart	Bifacial	Bifacial point	24.5	222	88.2	211.4
24	Arrow	Unifacial	Mousterian point	36.5	220.5	101.2	81.1
25	Dart	Bifacial	Bifacial point	32.5	220	85.65	221.4
26	Arrow	Unifacial	Unifacial point	34	217	72.77	81.1
27	Dart	Unifacial	Mousterian point	25.5	216	79.26	211.4
28	Dart	Unifacial	Mousterian point	22	216	79.26	201.4
29	Dart	Bifacial	Bifacial point	22	210	84.87	201.4
30	Arrow	Bifacial	Bifacial point	34	210	84.87	81.1
31	Arrow	Unifacial	Bifacial point	32.5	209	81.9	86.1
32	Arrow	Unifacial	Mousterian point	32.5	209	81.9	81.1
33	Arrow	Unifacial	Leilira	21.5	208	73.23	111.1
34	Arrow	Unifacial	Levallois point	25	203.5	80.04	76.1
35	Dart	Bifacial	Bifacial point	23.5	198	95.07	206.4
36	Dart	Unifacial	Leilira	31.5	198	73.8	211.4
37	Dart	Unifacial	Mousterian point	29	196	67.59	196.4
38	Arrow	Bifacial	Bifacial point	33	195	87.65	91.1
39	Dart	Bifacial	Bifacial point	24.5	192.5	82.68	211.4
40	Dart	Unifacial	Mousterian point	31.5	190	80.94	191.4
41	Arrow	Unifacial	Unifacial point	31.5	190	80.94	81.1
42	Dart	Bifacial	Leilira	19.5	187	80.99	211.4

(continued)

ID	Type of projectile	Retouch type	Typology	Penetration depth (cm)	TCSA	TCSP	Total projectile weight
43	Arrow	Unifacial	Unifacial point	28.5	187	74.49	91.1
44	Dart	Unifacial	Levallois point	29.8	180	77.18	201.4
45	Arrow	Bifacial	Stemmed Bifacial point	33.5	175.5	85.9	71.1
46	Arrow	Unifacial	Mousterian point	35.7	175	75.31	76.1
47	Dart	Bifacial	Bifacial point	22	175	80.62	201.4
48	Arrow	Unretouched	Levallois point	25.3	170	73.44	76.1
49	Arrow	Unifacial	Mousterian point	41.5	166.5	78.14	91.1
50	Dart	Bifacial	Bifacial point	30.5	162	80.49	191.4
51	Arrow	Bifacial	Bifacial point	32.5	154	92.34	76.1
52	Arrow	Unifacial	Mousterian point	44.5	152	79.23	71.1
53	Arrow	Bifacial	Bifacial point	32	148.5	75.17	81.1
54	Arrow	Bifacial	Bifacial point	39	147	88.54	71.1
55	Dart	Unifacial	Unifacial point	25.5	140	73.48	186.4
56	Dart	Bifacial	Bifacial point	30.5	136	75.15	191.4
57	Arrow	Unifacial	Mousterian point	35	130.5	63.13	76.1
58	Arrow	Bifacial	Bifacial point	39.5	130.5	68.26	76.1
59	Dart	Bifacial	Folsom point	21	126	77.25	201.4
60	Arrow	Bifacial	Bifacial point	32	121.5	64.89	61.1
61	Dart	Bifacial	Bifacial point	30.6	120	68	201.4
62	Dart	Unifacial	Unifacial point	28	120	55.24	196.4
63	Arrow	Bifacial	Bifacial point	32.3	120	62.48	71.1
64	Dart	Bifacial	Bifacial point	62	117.1665		186
65	Dart	Bifacial	Folsom point	31	116	66.24	201.4
66	Dart	Unifacial	Unifacial point	32.5	112.5	91.09	196.4
67	Arrow	Bifacial	Bifacial point	31.6	108.5	68.02	81.1
68	Dart	Bifacial	Kimberley point	30.6	104	61.05	211.4
69	Arrow	Bifacial	Bifacial point	37.5	103.5	58.41	61.1
70	Arrow	Unifacial	Levallois point	25.5	101.5	61.2	71.1
71	Arrow	Bifacial	Kimberley point	45.5	100	59.36	66.1
72	Arrow	Bifacial	Bifacial point	28.6	100	59.36	76.1
73	Dart	Bifacial	Bifacial point	32	99	70.22	196.4
74	Dart	Unifacial	Mousterian point	27.5	98	59.3	191.4
75	Dart	Bifacial	Kimberley point	22	94.5	60.82	191.4
76	Dart	Unifacial	Indian MP Tanged point	34.5	93	64.24	191.4
77	Dart	Bifacial	Bifacial point	26	87.5	57.3	181.4
78	Dart	Bifacial	Bifacial point	31	87	62.76	191.4
79	Dart	Unifacial	Leilira	24.5	84.5	42.06	216.4
80	Arrow	Bifacial	Bifacial point	38.8	84	55.56	71.1
81	Dart	Bifacial	Bifacial point	37.5	84	55.56	71.1
82	Arrow	Bifacial	Notched Bifacial point	37	80.5	53.85	76.1
83	Arrow	Bifacial	Bifacial point	39.5	78	57.27	71.1
84	Dart	Bifacial	Bifacial point	28.5	77	52.15	186.4
85	Arrow	Unifacial	Indian MP Tanged point	34	72	50.83	76.1
86	Arrow	Unretouched	Pointed blade	45	70	44.41	61.1
87	Dart	Bifacial	Bifacial point	17.5	69	51.88	181.4
88	Arrow	Bifacial	Notched bifacial point	27.5	52.5	46.51	86.1
89	Arrow	Bifacial	Bifacial point	43	51	41.61	56.1
90	Dart	Bifacial	Bifacial point	36	48	40	186.4

(continued)

ID	Type of projectile	Retouch type	Typology	Penetration depth (cm)	TCSA	TCSP	Total projectile weight
91	Arrow	Bifacial	Kimberley point	34	45	41.18	66.1
92	Arrow	Bifacial	Stemmed Bifacial point	42	40	37.73	56.1
93	Dart	Bifacial	Stemmed Bifacial point	21	36	33.94	176.4
94	Arrow	Bifacial	Notched Bifacial point	35	35	34.4	61.1
95	Dart	Bifacial	Kimberley point	22	32.5	32.8	181.4
96	Dart	Bifacial	Kimberley point	38	32.5	32.8	181.4
97	Dart	Bifacial	Bifacial point	22	30	34	176.4
98	Dart	Bifacial	Stemmed bifacial point	28	30	34	176.4
99	Arrow	Bifacial	Bifacial point	52	28	32.24	61.1
100	Dart	Bifacial	Tanged bifacial point	29.5	28	32.24	176.4
101	Arrow	Unifacial	Unifacial point	43	26	28.26	56.1
102	Arrow	Bifacial	Stemmed bifacial point	50	26	30.52	56.1
103	Arrow	Bifacial	Bifacial point	38.6	19.5	28.63	56.1
104	Dart	Unretouched	Pointed blade	31.5	19.5	27.31	186.4
105	Arrow	Unretouched	Pointed blade	33.2	19.5	27.31	71.1

References

Akerman, K. (1978). Notes on the Kimberley stone-tipped spear focusing on the point hafting mechanism. *Mankind, 11*(4), 486–490.

Barton, R. N. E., & Bergman, C. A. (1982). Hunters at Hengistbury: Some evidence from experimental archaeology. *World Archaeology, 14*(2), 237–248.

Blitz, J. H. (1988). Adoption of the bow in North America. *North American Archaeologist, 9*(2), 123–145.

Boëda, E., Bonilauri, S., Connan, J., Jarvet D., Mercier, N., Toby, M., et al. (2008). Middle Palaeolithic bitumen use at Umm el Tlel around 70000BP. *Antiquity, 82*(318), 853–861.

Buchanan, B., Collard, M., Hamilton, M. J., & O'Brien, M. J. (2011). Points and prey: A quantitative test of the hypothesis that prey size influences early Paleoindian projectile point form. *Journal of Archaeological Science*, *38*(4), 852–864.

Catellan, P. (1997). Hunting during the Upper Paleolithic: Bow, spearthrower, or both? In H. Knecht (Ed.), *Projectile technology* (pp. 213–240). New York: Plenum Press.

Christenson, A. L. (1986). Projectile point size and projectile aerodynamics: An exploratory study. *Plains Anthropologist, 31*(112), 109–128.

Costa, A. G. (2012). Were there stone-tipped armatures in the South Asia Middle Paleolithic? *Quaternary International, 269*, 22–30. doi:10.1016/j.quaint.2011.01.044.

Cundy, B. (1989). *Formal variation in Australian spear and spear thrower technology* (Vol. 546). Oxford: British Archaeological Reports.

Dockall, J. (1997). Wear traces and projectile impact: A review of the experimental and archaeological evidence. *Journal of Field Archaeology, 24*, 321–331.

Fischer, A., Vemming Hansen, P., & Rasmussen, P. (1984). Macro and micro wear traces on lithic projectile points: Experimental results and prehistoric examples. *Journal of Danish Archaeology, 3*, 19–46.

Flenniken, J. J., & Raymond, A. W. (1986). Projectile point typology: Replication experimentation and technological analysis. *American Antiquity, 51*, 603–614.

Friis-Hansen, J. (1990). Mesolithic cutting arrows: Functional analysis of arrows used in the hunting of large game. *Antiquity, 64*(244), 494–504.

Grünberg, J. M. (2002). Middle Palaeolithic birch-bark pitch. *Antiquity, 76*, 15–16.

Hughes, S. S. (1998). Getting to the point: Evolutionary change in prehistoric weaponry. *Journal of Archaeological Method and Theory, 5*, 345–408.

Hunzicker, D. A. (2008). Folsom projectile technology: An experiment in design, effectiveness and efficiency. *Plains Anthropologist, 53*, 291–311.

Hutchings, W. K. (1999). Quantification of fracture propagation velocity employing a sample of Clovis channel flakes. *Journal of Archaeological Science, 26*, 1437–1447.

Hutchings, W. K., & Brüchert, L. W. (1997). Spearthrower performance: Ethnographic and experimental research. *Antiquity, 71*, 890–897.

Knecht, H. (1997). Projectile points of bone, antler and stone: Experimental explorations of manufacture and use. In H. Knecht (Ed.), *Projectile technology* (pp. 191–212). New York: Plenum Press.

Lombard, M., & Clark, J. L. (2008). Variability and change in Middle Stone Age hunting behaviour: Aspects from the lithic and faunal records. In S. Badenhorst, P. Mitchell, & J. C. Driver (Eds.), *Animals and people: Archaeozoological papers in honour of Ina Plug* (pp. 46–56). Oxford: Archaeopress.

Lombard, M., & Phillipson, L. (2010). Indications of bow and stone-tipped arrow use 64,000 years ago in KwaZulu-Natal, South Africa. *Antiquity, 84*, 635–648.

Lombard, M., Wadley, L., Jacobs, Z., Mohapi, M., & Roberts, R. G. (2010). Still Bay and serrated points from Umhlatuzana Rock Shelter. *Kwazulu-Natal, South Africa Journal of Archaeological Science, 37*, 1773–1784.

Mackay, A. (2010). *History and selection in the late Pleistocene archaeology of the Western Cape, South Africa*. PhD dissertation, Australia: Australian National University.

McCall, G. (2007). Behavioral ecological models of lithic technological change during the later Middle Stone Age of South Africa. *Journal of Archaeological Science, 34*, 1738–1751.

Moncel, M.-H., Chacón, M. G., Coudenneau, A., & Fernandes, P. (2009). Points and convergent tools in the European Early Middle Palaeolithic site of Payre (SE, France). *Journal of Archaeological Science, 36*, 1892–1909.

Newman, K., & Moore, M. (2013). Ballistically anomalous stone projectile points in Australia. *Journal of Archaeological Science, 40*, 2614–2620.

Pargeter, J. (2007). Howiesons poort segments as hunting weapons: Experiments with replicated projectiles. *South African Archaeological Bulletin, 62*, 147–153.

Pétillon, J.-M., Bignon, O., Bodu, P., Cattelain, P., Debout, G., Langlais, M., et al. (2011). Hard core and cutting edge: Experimental manufacture and use of Magdalenian composite projectile tips. *Journal of Archaeological Science, 38*, 1266–1283.

Schoville, B. J. (2010). Frequency and distribution of edge damage on Middle Stone Age lithic points, Pinnacle Point 13B, South Africa. *Journal of Human Evolution, 59*, 378–391.

Shea, J. (1997). Middle Palaeolithic spear technology. In H. Knecht (Ed.), *Projectile technology* (pp. 79–106). New York: Plenum Press.

Shea, J. J. (2006). The origins of lithic projectile point technology: Evidence from Africa, the Levant, and Europe. *Journal of Archaeological Science, 33*, 823–846.

Shea, J. J., & Sisk, S. M. (2010). Complex projectile technology and *Homo sapiens* dispersal into Western Eurasia. *PaleoAnthropology, 2010*, 100–122.

Sisk, M., & Shea, J. J. (2009). Experimental use and quantitative performance analysis of triangular flakes (Levallois points) used as arrowheads. *Journal of Archaeological Science, 36*, 2039–2047.

Sisk, M., & Shea, J. J. (2011). The African origin of complex projectile technology: An analysis using tip cross-sectional area and perimeter. *International Journal of Evolutionary Biology, 2011*, 1–8.

Thomson, D. F. (1949). Arnhem Land: Explorations among an unknown people part III: On foot across Arnhem Land. *The Geographical Journal, 114*, 53–67.

Thieme, H. (1997). Lower Palaeolithic hunting spears from Germany. *Nature, 385*, 807–810.

Towner, R. H., & Warburton, M. (1990). Projectile point rejuvenation: A technological analysis. *Journal of Field Archaeology, 17*, 311–321.

Villa, P., & Lenoir, M. (2009). Hunting and hunting weapons of the Lower and Middle Paleolithic of Europe. In J.-J. Hublin & M. P. Richards (Eds.), *The evolution of hominin diets: Integrating approaches to the study of Palaeolithic subsistence* (pp. 59–85). Dordrecht: Springer.

Villa, P., Soressi, M., Henshilwood, C. S., & Mourre, V. (2009). The Still Bay points of Blombos Cave (South Africa). *Journal of Archaeological Science, 36*, 441–460.

Wadley, L., & Mohapi, M. (2008). A Segment is not a Monolith: evidence from the Howiesons Poort of Sibudu, South Africa. *Journal of Archaeological Science, 35*, 2594–2605.

Yaroshevich, A., Kaufman, D., Nuzhnyy, D., Bar-Yosef, O., & Weinstein-Evron, M. (2010). Design and performance of implemented projectiles during the Middle and the Late Epipaleolithic of the Levant: Experimental and archaeological evidence. *Journal of Archaeological Science, 37*, 368–388.

Chapter 14
Penetration, Tissue Damage, and Lethality of Wood- Versus Lithic-Tipped Projectiles

Paul E. Salem and Steven E. Churchill

Abstract Lithic projectile points are a universal component of the hunting tool kits of archeologically- and historically-known foragers. Recent experimental work with ballistic gelatin targets has shown that lithic-tipped projectiles do not have a marked penetration advantage over those with simple sharpened wooden points, leading to the suggestion that investment in the production of lithic points may serve social rather than economic motives. Here we report on experimental work with wood- and stone-tipped arrows fired into calibrated ballistic gel. While the stone-tipped arrows underperformed with respect to penetration, they far exceeded the wood-tipped arrows in the volume of gelatin destroyed. These results suggest that the total volume of tissue destroyed by a projectile is as or more important than its penetration depth, that adding a lithic point increases the lethality of a projectile, and that decisions about projectile armatures were motivated by economic rather than social concerns.

Keywords Bow and arrow • Costly signaling • Lithic points • Tissue disruption

Introduction

Lithic projectile points are ubiquitous in the archeologically record of prehistoric foragers (at least from the late Middle Paleolithic onwards), as are lithic and metal points in the tool kits of historically-known and extant hunter-gatherers. These points are generally small, rigid, and sharp – properties that presumably serve to increase penetration depth and tissue damage of the projectiles that they tip (Ellis 1997), whether they be spear-thrower darts or arrows. Lithic tips are brittle and thus prone to use-damage, which increases the maintenance costs of lithic-tipped projectiles relative to those constructed of tougher, more durable wood or bone. It is generally assumed that lithic points enhance the effectiveness of projectile weapons (see for example Thomas 1978; Friis-Hansen 1990; Straus 1990; Churchill 1993; Ellis 1997; Shea 2006) such that the improvement in hunting returns they confer outweighs the energetic and time costs of raw material procurement, point manufacture, and upkeep. This assumption has recently been challenged (Waguespack et al. 2009).

It has been argued that penetration depth is the critical variable determining the lethality of projectile weapons, and that incapacitation of the target requires penetration depths on the order of 15 cm (if the target is human) to 20 cm (for large ungulate prey) (Hughes 1998). Recent experimental work by Waguespack et al. (2009), however, found no functionally-significant difference in penetration depth of wood- versus stone-tipped projectiles. While arrows tipped with lithic points did penetrate ballistic gel 10% deeper on average than those tipped with wood, both types of arrows attained depths that were likely lethal (235.0 ± 10.6 mm vs. 212.7 ± 5.5 mm for stone- and wood-tipped arrows, respectively: Waguespack et al. 2009). Similar results were obtained in recent experimental work by Anderson (2010), who found that a sharpened, fire-hardened wooden tip obtained penetration depths in ballistic gelatin that were slightly above the median depth produced by replicas of eleven different lithic points of variable length, width and thickness. If the penetration performance of stone-tipped and untipped wooden projectiles is effectively equivalent, what benefits accrue to users of lithic or metal points that compensate for the greater costs of producing and maintaining them?

P.E. Salem · S.E. Churchill (✉)
Department of Evolutionary Anthropology, Duke University, Box 90383Durham, NC 27708, USA
e-mail: churchy@duke.edu

P.E. Salem
e-mail: psalem@gmail.com

S.E. Churchill
Evolutionary Studies Institute, University of the Witwatersrand, Private Bag 3, Johannesburg, 2050 WITS, South Africa

Radu Iovita and Katsuhiro Sano (eds.), *Multidisciplinary Approaches to the Study of Stone Age Weaponry*, Vertebrate Paleobiology and Paleoanthropology, DOI 10.1007/978-94-017-7602-8_14

Waguespack et al. (2009) suggest that social, symbolic factors may explain the ubiquitous use of brittle stone points in prehistory – that these costly to produce and difficult to maintain armatures served as an example of costly signaling (following Bliege Bird and Smith 2005), within the context of a behavior (hunting) which may itself be a form of costly signaling (Hawkes 1991; Hawkes et al. 2001; Bliege Bird et al. 2001). This is an intriguing idea, and we find it likely that this may account for some of the human predilection to invest in this artifact type. However, as a universal explanation for investment in stone projectile points across great expanses of time and space, the hypothesis is problematic. First, it assumes that hunter-gatherer groups across the inhabited world and over the last 40 ka have all largely opted to engage in the same practice of costly signaling – that is, that the use of lithic points to satisfy reproductive rather than economic goals is a human universal. The hypothesis would predict an inverse relationship between lithic raw material abundance and use of stone points (the less costly the production of lithic points, the lower their value as costly signals), yet regional studies of lithic assemblages relative to raw material availability suggest considerably more complex relationships (see for example Kelly 1988; Andrefsky 1994; Hiscock 2009). Second, since the tipping of weapon armatures with hafted lithics predates the origins of projectile weapons (Shea 1988, 1997; Boëda et al. 1999; see also Rios-Garaizar 2016 and Wilkins and Schoville 2016), the hypothesis implies that some groups of archaic humans (such as Neandertals) had converged on the same costly signaling behavior. Given uncertainty as to whether pre-modern humans engaged in abstract social signaling systems prior to contact by modern humans (see Mellars 2004, 2005 vs. d'Errico et al. 1998; Zilhão et al. 2010), the convergence of different human species on exactly the same costly signaling behavior seems unlikely. Finally, the hypothesis fails to account for the preference of modern sport bow hunters (who presumably are not engaged in the same costly signaling dynamic) for triangular broadhead points over simple pointed tips (field points). This preference in hunting (but not target practice) holds despite accuracy problems with broadheads relative to field points (see, for example, Bowhunting.com 2008).

It is also unclear as to why prehistoric flint knappers invested effort in producing symmetrical points (that is, symmetrical about the point long axis in ventral or dorsal view). It has been suggested that point symmetry improves flight dynamics and projectile accuracy (references in Churchill 1993). However, Odell and Cowan (1986) showed that arrows tipped with unmodified, irregular lithic points performed nearly as well as those with retouched, symmetrical points in penetrating animal carcasses. Although the arrows with symmetrical points penetrated slightly deeper on average (107.5 ± 65.5 mm versus 90.0 ± 58.0 mm for asymmetrical points), the difference in performance was not statistically significant. It could be argued that the investment of lithic reduction time in producing finely crafted, symmetrical points is also a form of costly signaling. However, symmetrical points appear to accumulate tip damage more slowly with repeated usage (Odell and Cowan 1986), such that an overall reduction in weapon maintenance time may be the payoff for a larger initial investment in point production (as might be predicted by the tech investment model of optimal foraging theory: see Bettinger et al. 2006; Bettinger 2009).

We hypothesize that stone-tipped projectiles, by virtue of their greater cross-sectional area and sharpness, produce greater tissue damage (and thus greater hemorrhaging) in prey than do untipped projectiles. This is supported by claims from ethnohistorically-known bow hunters that stone points produce more lethal wounds (Ellis 1997). Among modern sport bow hunters (here in North Carolina, at least), there is an oft expressed sentiment that "the larger the wound, the faster the kill," both because the greater bleeding induced by larger wounds tends to drop prey in a shorter amount of time, and because it produces a more obvious blood trail for tracking wounded prey. This is supported by experimental work on broadheads and field tips fired into pig carcasses (Karger et al. 1998: 498), in which:

> the broadhead caused starlike and sometimes gaping wound tracks, and in some instances, postmortem bleeding was considerable. The field tip and most other arrowheads caused incision-like wounds that were partially filled by the shaft of the arrow. The wounds were sharply cut, and tissue destruction from bruising or tearing was insignificant. Frequently, bleeding did not occur until the arrow was removed.

Additional experimental work with replicas of lithic points demonstrate that they tend to create wider lesions (in ballistic gelatin at least) than do simple wooden tips (Anderson 2010).

Lethality and hemorrhaging potential of armatures may weigh heavily in hunting success with bow and arrow, since even among modern hunters using compound bows fully 50–68% of arrow-wounded deer are never harvested (Causey et al. 1978; Stormer et al. 1979; Langenau 1986; see also Gaudzinski-Windheuser 2016, and Noe-Nygaard 1974 on nonfatal wounding of prey by Upper Paleolithic/Mesolithic and Neolithic hunters, respectively). To test the hypothesis that stone-tipped projectiles produce larger (and thus more lethal) wounds, we fired tipped and untipped mass-standardized arrows at constant velocity into calibrated ballistic gel. To evaluate possible performance differences between symmetrical and asymmetrical points, we also tested the wounding potential of symmetrical bifacial points versus asymmetrical pieces of pointed debitage.

Materials and Methods

To evaluate the performance of wood- and stone-tipped arrows we adopted a protocol used for modern firearms (Fackler 1988a), using a 27.2 kg (60 lbs) draw-weight crossbow (Barnett® Phantom Jr.) fired from a bench rest. Arrows were fired into a standard sized block (15 by 15 by 41 cm) of 10% calibrated VYSE® ballistic gelatin, prepared following standard methods (Fackler 1988a) to approximate the average density of mammalian tissue (1 gm cm^{-3}: Katch et al. 1967). Ballistic gelatin closely simulates the density and viscosity of mammalian muscle tissue, and is the industry standard for evaluating penetration and wound properties in the field of wound ballistics (Salisbury and Cronin 2009; Shepherd et al. 2009).

Experimental Set-up

All of the arrows used in this study were standardized to a total weight (arrow plus armature) of 19.5 ± 0.1 gm. We used commercially available arrows (40.8 cm Barnett® Phantom Jr. crossbow bolts) with a mass of 18 gm and an external shaft diameter of 8.8 mm. Wood-tipped arrows were constructed of simple sharpened dowel rods with a diameter of 7.9 mm, cut to a tip angle of approximately 40° (similar to that of a hunting field point), and inserted into the distal end of the arrow shaft. For the symmetrical stone points we used the lithic "arrowheads" that can be inexpensively purchased in museum gift shops. While these cheap souvenirs are not especially good replicas of projectile points, they do possess the properties of small size and *relatively* good bilateral symmetry about the point long axis (at least in the frontal plane – several of these pieces were somewhat curved in lateral view). Point selection was constrained by the basal thickness of the candidate "arrowheads," as many of the most symmetrical points had proximal ends that were too thick to permit hafting. The three points used in this study were made on obsidian and had an average length of 34.9 ± 6.1 mm, an average maximum width of 19.6 ± 1.8 mm, and an average maximum thickness of 4.5 ± 2.2 mm. Dimensions of the individual armatures are provided in Table 14.1.

The asymmetrical points were pieces of dacite debitage produced by a modern flintknapper. Pieces were selected that were roughly comparable in size to the symmetrical points (Table 14.1), and which although asymmetrical still had a discernable long axis that terminated in a point. The average dimensions of these four points were 40.1 ± 5.3 mm (length), 24.8 ± 3.7 mm (maximum width) and 4.2 ± 1.3 mm (maximum thickness). Stone tips were hafted to wooden dowel rod link-shafts with Liquid Nails® adhesive, which were then inserted into the arrow shafts (Fig. 14.1). Lighter stone tips received longer sections of wooden dowel rod inside the hollow aluminum shaft of the arrow to standardize overall mass without affecting arrow length. Total arrow length was standardized to 44.6 ± 0.1 cm.

Table 14.1 Dimensions (mm) of the lithic points used in the experiment

Tip[a]	Type	Length	Width	Thickness	TCSA[b]
b	Asymmetrical	43.5	29.6	5.7	84.4
c	Asymmetrical	45.7	24.9	4.9	61.0
d	Asymmetrical	36.0	24.0	3.2	38.4
e	Asymmetrical	35.3	20.6	3.0	30.9
f	Symmetrical	38.8	17.6	2.3	20.2
g	Symmetrical	27.8	20.6	4.5	46.4
h	Symmetrical	38.0	20.7	6.6	68.3

[a]Letters correspond to the labels provided in Fig. 14.1
[b]Tip cross-sectional area (mm^2)

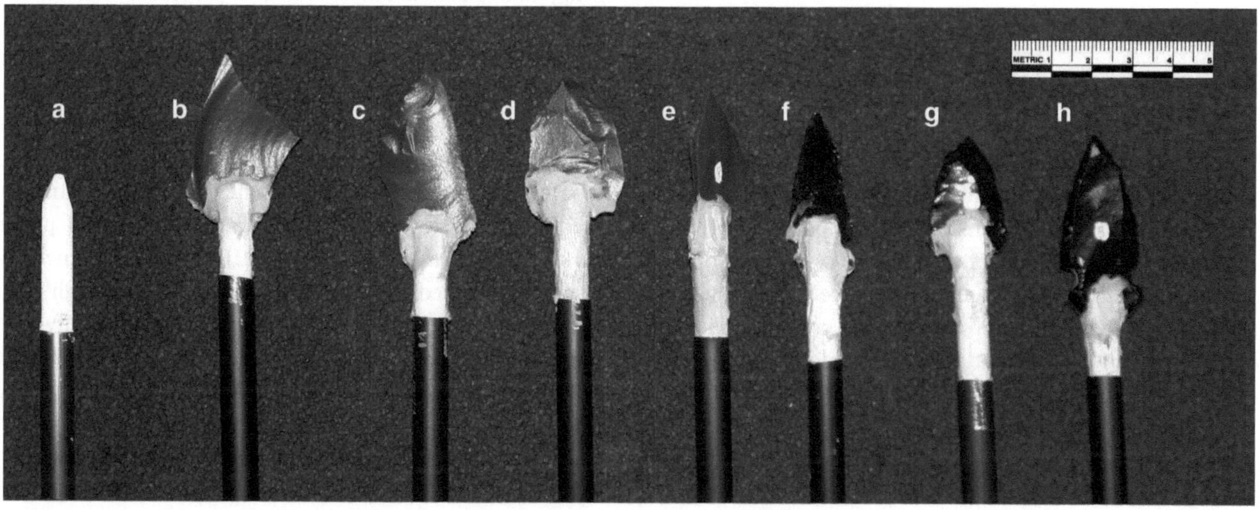

Fig. 14.1 Arrows used in the experiment: **a** one of the wood-tipped arrows; **b–e** asymmetrical points; **f–h** symmetrical points. Points **b**, **f** and **h** broke on the first shot, point **c** fractured on its fourth firing (see text)

To determine projectile velocity, arrows were fired through a Shooting Chrony® F-1 chronograph positioned 2.15 m from the distal end of the crossbow. The chronograph failed to register 27% of the shots, thus mean measures of velocity, kinetic energy, and momentum are based on a subset of the 44 gel-penetrating shots performed during the experiment. The mean velocities of wood- and stone-tipped arrows did not differ significantly (wood-tipped arrows: 45.8 ± 0.4 m s^{-1}, n = 13; stone-tipped arrows: 45.5 ± 0.6 m s^{-1}, n = 19; $p = 0.227$). Given equivalent mass and velocity of the stone- and wood-tipped arrows, momentum and kinetic energy was effectively constant between the two conditions. The distance from the distal end of the crossbow (the "muzzle") to the target surface of the ballistic gelatin was 5 m.

Ballistic gelatin was mixed in a 1:9 ratio (gelatin/water) by mass and prepared according to the simplified method outlined in Jussila (2004): 1 kg of gelatin was added to 4.5 l of 23 °C water, after which another 4.5 l of 75 °C water was added (temperature calculated to bring the overall temperature of the mixture to 45 °C). After thorough mixing the gelatin was poured into a mold and allowed to harden at room temperature for 24 h, and then refrigerated at 4 °C for at least 24 h. To verify the density of the ballistic gelatin, we fired a 0.177 caliber (4.5 mm) BB into the block from a Crosman® model 2100 air rifle at 180 m s^{-1}. BB penetration of 85 ± 5 mm is considered acceptable. However, since the velocity of the BBs could not be precisely controlled (average velocity 184.4 ± 3.5 m s^{-1} for four calibrating shots), we adjusted the penetration depth standard by BB velocity (measured by the chronograph) using the regression equation provided by Jussila (2004). All of the gelatin blocks used met the testing standard. All arrow shots at the ballistic gelatin were conducted within 20 min of removing the gel from refrigeration (to prevent undue changes in gel density caused by re-warming of the block).

Experimental Trials

We began the experiment with three arrows tipped with wooden points, three tipped with symmetrical stone points, and four tipped with asymmetrical stone points (Fig. 14.1). All of the arrows were pooled and were selected at random for firing into the gelatin blocks. To reduce possible bias, we selected and loaded the arrows into the crossbow after the shooter was in position and had his eye to the scope. This ensured that the shooter was blind as to the type of arrow being fired on any given trial.

We sought to conduct six trials for each armature (following Waguespack et al. 2009, who conducted six trials for each point *type*), which would have produced 18 wood-tip, 18 symmetrical lithic-tip, and 24 asymmetrical lithic tip trials. While a greater number of trials would be desirable, each gelatin block could only endure about four trials before needing to be retired, and thus the expense and time of block production was a limiting factor. Damage to the points incurred during testing (see below) further reduced the number of usable trials, such that we ultimately collected data on 18 wood-tip, 7 symmetrical lithic-tip, and 18 asymmetrical lithic-tip trails.

Quantification of Penetration and Volume Tissue Damage

Once fired into the gelatin block, arrows were marked on the shaft at the surface of the gel before extraction, and measurements of penetration depth were taken directly on the extracted arrow shaft. Trials in which an arrow entered or intersected the wound channel from a previous trial were discarded. Penetration depth was multiplied by the tip cross-sectional area (TCSA) of the arrow to calculate the total volume of tissue disrupted (TVTD). For the stone points, TCSA was calculated as 0.5 (maximum width × maximum thickness) following Shea (2006). For the wooden points, TCSA was calculated as πr^2. TVTD was taken as an indicator of the relative effectiveness of the stone and wooden tips, since the volume of tissue disrupted by a projectile should be proportional to its lethality.

Results

Only five of the ten arrows survived the experiment. One of the wood-tipped arrows missed the gelatin block on its first firing, and examination revealed a previously undetected defect (in the form of a slight bend) in the arrow shaft: this arrow was thus retired from service. Of the stone-tipped projectiles, three were damaged on the first or second shot, either as the result of a miss (one arrow) or upon striking the gel (but with downward trajectories that caused the point to impact the hard substrate below the gelatin: two arrows). These three arrows included two with symmetrical points and one with an asymmetrical point. One final asymmetrical stone tip fractured inside the gel block on its fourth trial. Two design features appeared to have factor heavily into arrow and tip performance (and breakage). First, geometric deviation of the long axis of the point from the center axis of the arrow shaft appears to have an effect on the accuracy of the overall projectile and the durability of the tip even when fired into a relatively soft medium. Asymmetrical tips that were shaped in such a way that the long axis of the point could not be aligned with the center axis of the arrow either missed the gelatin block completely (shattering the tip on the backstop) or broke apart on impact with the gelatin. Second, symmetrical points with even slight curvature in lateral perspective either snapped on striking the gelatin (no doubt due to bending moments created by axially loading a curved flake) or deviated in flight (perhaps due to aerodynamic effects or mass imbalance at the front of the projectile). Problematic amounts of point curvature appears to be an artifact of using replica "arrowheads," and no doubt would be less of a problem with actual projectile points made by more skilled flint knappers. The stone tips that survived for the duration of the experiment were all relatively small in size and hafted with their long axes in line with the arrow shaft axis. Penetration data was collected for two wood-tipped (18 trials), three asymmetrical (18 trials) and one symmetrical (7 trials) stone-tipped arrows.

Mass standardized wooden projectiles were found to penetrate deeper than stone tipped projectiles in the calibrated ballistic gelatin (Table 14.2). The wood-tipped arrows penetrated, on average, almost 1 cm deeper than the symmetrical stone-tipped arrow, and almost 1.5 cm deeper than the asymmetrical stone-tipped arrows – an improvement of roughly 9% over the average stone-tipped arrow. This difference is statistically significant ($t = 4.895$, $df = 41$, $p < 0.001$), although practically quite small. Penetration performance of symmetrical versus asymmetrical stone points was not statistically significant ($t = 1.176$, $df = 23$, $p = 0.256$).

As would be expected in an experiment in which arrow mass and velocity were controlled, variation in penetration depth was not attributable to variation in kinetic energy (KE) or momentum (p). KE and p were calculated for each shot based on the recorded velocity (v) of that shot and the mass (m) of the arrow ($KE = mv^2/2$; $p = mv$). Regressions of both variables on penetration depth were insignificant (KE: $r^2 = 0.003$, $p = 0.774$; p: $r^2 = 0.003$, $p = 0.781$).

With respect to tissue disruption (TVTD), however, there was a significant difference between the wooden and stone-tipped arrows. The stone tips provided a 56% improvement in tissue disruption, destroying a mean volume of 13,734 mm^3 compared to a mean of 7794 mm^3 for the wooden tips ($t = 25.518$, $df = 41$, $p < 0.001$). Although stone tipped projectiles did not penetrate as deeply as wooden tipped projectiles of equal mass (at least in our experiment), they did disrupt a larger volume of tissue (and penetrated almost as deeply). The asymmetrical points also outperformed, slightly, the symmetrical tips (Table 14.2), obtaining 7.2% greater volume damage on average.

Table 14.2 Mean tip cross-sectional area, penetration depth and tissue damage by arrow tip type

Tip type	TCSA[a] (mm^2)	Penetration depth (mm)	TVTD[b] (mm^3)
Wood			
Mean	49.0	159.1	7794.3
SD	0.0	6.9	337.6
n	2	18	18
Symmetrical stone			
Mean	87.9	149.5	13,135.7
SD	–	7.2	628.8
n	1	7	7
Asymmetrical stone			
Mean	97.1	145.1	14,093.2
SD	2.6	11.2	1,156.1
n	3	18	18

[a]Tip cross-sectional area (only for those points which produced penetration data: see text)
[b]Total volume tissue disruption (=TCSA * Penetration depth)

Discussion

In previous experimental work, wooden points were found to penetrate less deeply than (Waguespack et al. 2009), or comparably to (Anderson 2010), stone points. Anderson (2010) used cast replicas of lithic points of various shapes and sizes, and found that a wooden point out-performed some of the lithic points, but under-performed relative to others, and that overall its penetration depth was close to the median of the lithic points. In contrast to both of these studies, the wooden-tipped arrows used in this study generally out-performed those with lithic points (although there was overlap in the observed penetration depths between the two arrow types). What might account for this difference in mean penetration depths?

There is no doubt that energy transfer is an important determinant of the wounding capabilities of projectile weapons. For firearms, there is a long-standing idea that transfer of momentum is the principle agent of tissue damage in wound ballistics (Hatcher 1935). However, experimental studies and theoretical considerations show that the transfer of kinetic energy is the primary determinant of wound severity (Mendelson 1991) and that the role of momentum is likely to be negligible (Sellier and Kneubuehl 1994; Karger and Kneubuehl 1996). Bullet penetration into tissues or tissue simulants (such as ballistic gel) is proportional to the kinetic energy of the round, and inversely proportional to the cross-sectional area of the round (Sperrazza and Kokinakis 1968). Similar physical relationships must determine the wounding performance of lower velocity projectiles fired from a bow or spear-thrower (see Hughes 1998). Projectile shape is also a key variable, since it determines the amount of drag the projectile experiences as it passes through tissues: elliptical cylinders produce the lowest drag coefficients while triangular cylinders produce the highest (Hughes

1998). Furthermore, irregular protrusions which disrupt the smooth profile of the projectile also serve to increase the drag coefficient (Hughes 1998) and impede penetration. These and other considerations dictate the nature of the relationship between point morphology and performance, and provide an avenue for examining variation in point form in prehistory (see for example Milks et al. 2016; Clarkson 2016; Shott 2016).

Given that the mass and velocity (and thus kinetic energy and momentum) did not differ significantly between arrow types in this experiment, differences in mean penetration depth must be a function of differences in tip shape and cross-sectional area. The mean cross-sectional area of the lithic points (94.8 ± 5.0 mm^2, n = 4) was almost twice that of the wooden tips (49.0 ± 0.0 mm^2, n = 2), which no doubt caused greater energy loses as the points cut through and displaced the gelatin. The mean TCSA of the asymmetrical points was 10% larger than that of the symmetrical points, and not surprisingly they penetrated less deeply than the symmetrical points (cf. Odell and Cowan 1986). However, variation in TCSA only accounts for about 35% of the variation in penetration depth ($r^2 = 0.3518$: Fig. 14.2), so other factors must also be at play. Some of the variance observed in penetration depth may also be attributable to variation in arrow velocity, and thus variation in KE: however, the correlation between KE and penetration in our data was not significantly different than zero ($r^2 = 0.0028$). Point shape was also likely a factor, as the lithic points had cross-sectional shapes that were closer to that of a triangular cylinder than the rounder wooden tips. Furthermore, the asymmetrical points were also more likely to possess protrusions and other irregularities that would be expected to increase drag, which may have contributed to the lower average penetrated depth of these points relative to the symmetrical stone points (Table 14.2).

While the observed differences in penetration are likely partly attributable to these differences in point morphology, we suspect that variation in angle of arrow penetration also played a role. We noted that the stone-tipped arrows tended to penetrate the gel with greater obliquity (that is, less perpendicular to the target face), and thus they likely lost some of their energy in lateral displacements of the elastic gelatin – imparting motion to the gelatin without contributing to penetration. Differences between the wood- and stone-tipped arrows in striking angle is likely in turn a function of differences in tip symmetry (recall that even our "symmetrical" tips were imperfectly symmetrical, and symmetrical only in one plane, such that all of them had some degree of curvature when seen in lateral perspective). The more asymmetrical stone points may have experienced more erratic flight

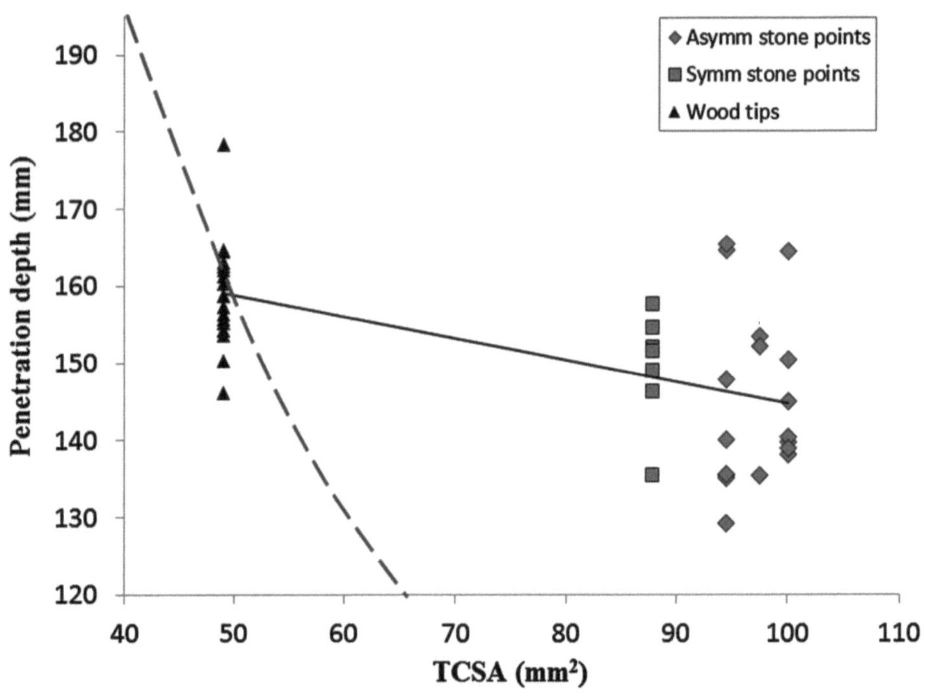

Fig. 14.2 Arrow penetration depth versus tip cross-sectional area (TCSA) for wood-tipped (triangles), symmetrical stone-tipped (squares) and asymmetrical stone-tipped (diamonds) arrows. Solid line least-squares regression line ($y = -0.2823x + 172.97$; $r^2 = 0.3518$). Dashed line second order polynomial ($y = 0.0503x^2 - 8.2353x + 442.57$) defining a line of equal tissue volume damage across a range of penetration depths and TCSAs that is equal to the mean TVTD of the wood-tipped arrows. Data points falling to the right of the dashed line denote arrows that disrupted a volume of gelatin greater than the average for wood-tipped arrows

trajectories either because of mass imbalances at the distal end of the projectile, or because of shape asymmetries at the tip (if the arrow is acting as a front-ruddered projectile).

The mean penetration depths obtained by both the wood- and stone-tipped arrows fell short of the 20 cm that has been argued to be critical for bringing down large prey (this value is based on observations of lethal arrow wounds in large prey and on experimental work with large animal carcasses: Guthrie 1983; Friis-Hansen 1990). The mean KE carried by our arrows (20.3 ± 0.4 J, n = 32) falls between experimentally reproduced values for Sioux (13.5 J) and Apache (25.9 J) bows and arrows (Bergman et al. 1988), which suggests that the penetrating power of primitive self bows was fairly limited (which no doubt drove the development of more powerful backed and composite bows: Bergman 1993). However, the arrows fired by Waguespack et al. (2009) into uncovered ballistic gel surpassed the 20 cm critical depth (wooden-tipped arrows: 21.3 ± 0.6 cm, n = 7; stone-tipped arrows: 23.5 ± 1.1 cm, n = 7). Both the Waguespack et al. study and this study employed (cross) bows with a 60 lb draw weight, and although the average mass of the arrows used in the former study was 9–20% greater than ours (22.7 ± 2.1 g, n = 6 for the wood-tipped arrows; 24.7 ± 1.6 g, n = 6 for the stone-tipped arrows: Waguespack et al. 2009), the KE of the arrows is unlikely to have differed markedly between the two studies (given equivalent bow draw weights, the heavier arrows would have been launched at slightly lower velocities). The greater penetration depths observed by Waguespack et al. (2009) may in part be a function of reduced bow-to-target distance (1.1 m vs. 5 m in our study), and hence less energy loses to aerodynamic drag before impact. However, if drag effects of this magnitude are playing out over a 3.9 m difference in distance, then primitive bows fired at a distance of 25.8 m (the average distance in hunting by modern foragers: Churchill 1993) would stand little chance of penetrating the target.

The penetrating performance of our arrows can be put into perspective by examining penetration values of modern firearms discharged into calibrated ballistic gel. Mass and velocity data for two handgun rounds reported by Mehremic and Karabegovic (2008) allows for calculation of KEs of 268 and 300 J for these rounds. Despite carrying KEs an order of magnitude greater than our arrows, and having cross-sectional areas (73.1 and 63.6 mm^2, respectively) that are similar to our arrows, these bullets only penetrated to 32 and 35 cm (respectively). Military rifle rounds that travel at much higher velocities, have very small cross-sectional areas (on the order of 24 mm^2), and carry KEs on the order of 1500–1850 J, still only penetrate calibrated gel to a depth of 45–49 cm (based on velocity and penetration depth provided by Dougherty and Matthews 2007, and assuming a 5.56 mm NATO round with a mass of 4.02 g). Given the modern ballistic data, penetrations of 21–23 cm for low KE arrows seem high, and we suspect that differences in gelatin density between the studies may be to blame. Proper calibration of the gelatin assures a density comparable to muscle tissue (Fackler 1988a), and allows comparison of results across studies. While projectiles fired at live prey or carcasses will often be stopped or deflected by striking bone, the ballistic gelatin provides a reasonable model of the penetrating behavior of projectiles through animal soft tissues. Based on the methodological description of the experimental work by Waguespack et al. (2009), it does not appear that they used calibrated gel. Accordingly, the lower penetration values obtained in our study are more likely to reflect the actual penetration performance of primitive projectile weapons in actual hunting contexts. However, it is important to note that, while the ballistic gel serves as a reasonable model for muscle tissue, it does not provide the same soft tissue (skin, tendon) or hard tissue (bone) impediments to projectile penetration as does an actual animal. Waguespack et al. (2009) obtained penetration depths that were between 3.1% (wooden arrows) and 4.5% (stone-tipped arrows) shallower when they fired their arrows through ballistic gel draped in caribou hide than they observed with uncovered gel. Thus while our results provide a basis for comparing projectile systems, it is likely that performance in the field would be somewhat less than we observed experimentally.

While wooden-tipped arrows penetrated further in our experiment, the data suggest that the stone-tipped arrows were far more lethal (Table 14.2, Fig. 14.2). Admittedly, the penetration differences between the two arrow types were fairly negligible in functional terms (despite being statistically significantly different), with the wooden arrows only penetrating about 1.2 cm deeper on average than the stone-tipped arrows. Both types of arrows failed to achieve the 20 cm penetration depths argued to be necessary to drop large prey (Guthrie 1983; Friis-Hansen 1990), and it is unclear how much a penetration increase of ca. 1 cm would improve hunting success rates. Considering the combined results of this experiment and those of Waguespack et al. (2009) and Anderson (2010), it can reasonably be concluded that wood- and stone-tipped arrows perform comparably with respect to tissue penetration. Indeed, if penetration depth were the only variable determining hunting success rates with projectile weapons, there would appear to be little return benefit to offset the higher production and maintenance costs of using lithic points (as argued by Waguespack et al. 2009). However, lethality likely depends on more than just projectile penetration.

Projectile lethality is a function of tissue damage. With modern projectiles (firearms), tissue damage occurs as the result of two processes – crushing and tissue displacement (Fackler 1988b) (Fragmentation of the round might be considered by some to be a third process: see Patrick 1989). Crushing is produced directly in front of the bullet by the forces generated by the deceleration of the round as it contacts and passes through tissues. Crushing produces a permanent cavity in the impacted tissues, and the amount of wounding is proportional to the penetration depth of the bullet (Fackler 1988b). Tissue displacement occurs as the result of centrifugal forces which accelerate tissues adjacent to the bullet path away from the round, producing a temporary cavity (cavitation) that stretches and tears tissues before collapsing as the energy is dissipated in the body (Fackler 1988b; Bartlett et al. 2000). The relative severity (and lethality) of these two injury processes depends on the tissues involved: fairly elastic tissues such as muscle, gut wall and lung handle cavitation well, and are primarily damaged by crushing; wounding of inelastic tissues like liver occur primarily by tissue displacement (Fackler et al. 1984). The forces that produce crushing and cavitation are generated by the deceleration of the bullet, and are thus proportional to the mass and velocity of the round (and thus its KE and p). Tissue displacement in particular requires a substantial transfer of energy from bullet to target (Amato et al. 1974; Bartlett et al. 2000). Primitive long range projectile weapons operate at low velocities (relative to modern firearm rounds) and relatively low KE (Hughes 1998). It is unlikely that an arrow strike deposits sufficient KE to produce any real lateral displacement of adjacent tissues, and for arrows tipped with wooden points the formation of a permanent cavity through crushing is likely to be the only significant source of wounding. Therefore lethality of wood-tipped arrows is primarily determined by their depth of penetration.

Lithic projectile points, by virtue of their generally greater TCSA, crush, slice and generally disrupt a greater volume of tissue than do wood-tipped missiles. Given the greater volume damage of lithic-tipped projectiles, it is clear that, while penetration depth continues to be an important component of performance (since vital organs and major arteries tend to lie deep within the body cavity), tissue disruption is an equally (and possibly more) important component. In the absence of striking a major blood vessel, the extent of bleeding created by an arrow wound is a function of the volume of tissue traumatized, which in turn is a function of the diameter of the permanent cavity (proportional to TCSA or tip perimeter) and the depth of penetration (Friis-Hansen 1990).

Lithic points also add another important dimension that improves their lethality: cutting. The sharp edges of brittle lithic armatures improve tissue disruption in two ways. First, these edges are likely to reduce drag effects as the missile passes through skin, muscle and organs. The perimeters of lithic armatures tend to be larger than those of the foreshaft to which they are hafted (Hughes 1998), which cuts a hole through skin and other tissues sufficient for the foreshaft and shaft to follow with less friction. Stone tips likely also produce a hole that reduces the "hilt effect" or haft-drag problem (see Guthrie 1983; Hughes 1998), whereby the binding agents used in hafting may experience drag as the proximal end of the point penetrates elastic tissues like skin (of course, it is the use of lithic armatures in and of itself that produces the haft-drag problem, since wooden self arrows lack hafting agents. However, some wooden tips may be slotted into an arrow shaft of greater diameter than the tip, which would create at least a small hilt effect). The greater penetration depths of lithic versus wood-tipped arrows reported by Waguespack et al. (2009) may be a function of drag reduction in the stone-tipped arrows, especially if the TCSA area of the points was similar to that of the wood-tipped points in their study (point dimensions were not provided). The addition of even a few microliths to antler points (adding sharp edges without significantly altering point cross-sectional area) has been shown to almost double their penetration ability in animal carcasses (Pétillon et al. 2011). Second, sharp lithic edges are more likely than wooden points to nick and cut arteries while passing through tissues. Arteries are elastic (Shadwick 1998), and unless punctured by the point of a wooden tip, are likely to be pushed aside by wooden points as the arrow creates its wound track. Accordingly, greater hemorrhaging, leading to faster collapse of a struck animal or a more conspicuous blood trail, is to be expected with lithic-tipped arrows.

The results of this experiment, in conjunction with reports of the practical experiences of recent bow hunters (Friis-Hansen 1990; Ellis 1997), suggest that the lethality of projectiles is greatly enhanced by arming them with broad, sharp points. The preference by modern sport bow hunters for broadheads over field points in hunting, despite their reduced accuracy, suggests that hunters are willing to pay a cost for the performance gains that broadheads provide (to modern bow hunters, that cost is paid through a reduction of accuracy or an additional time investment in tuning the bow and the arrows for broadhead use). Ethnographic accounts also indicate that hunters are willing to suffer the increased raw material procurement, production and maintenance costs associated with using stone projectile points, but that they do so with the understanding that these points perform better (at least in some circumstances, such as with larger prey) than do sharpened wooden or osseous armatures (Ellis 1997). Improvements to the lethality of hunting weapons directly reduce handling costs across a spectrum of prey body sizes, and thus there are strong economic incentives for developing both armatures and projectile systems that kill or immobilize prey faster. Given the enhanced ability of stone-tipped projectiles to disrupt impacted tissues and induce bleeding in prey, there is little reason to invoke social, reproductive motives for their adoption and use among prehistoric and recent hunters. This is not, however, to argue that there is no social element inherent in the design and manufacture of lithic projectile armatures, as it certainly seems reasonable that regional and temporal variation in projectile point design elements, or what might be considered "style," reflects a concern with conveying information about social identity among the foragers who made and used these artifacts (Wiessner 1983; McElreath et al. 2003). Humans invest much of their material culture – no matter how utilitarian – with symbolic meaning, and projectile points are no exception. However, the results obtained here suggest that practical (performance) rather than social (reproductive) concerns may be the primary factor behind the choice of stone over wood or bone as a raw material for projectile tips.

Acknowledgements Funding for this project was provided by an Undergraduate Research Support grant to PES from Trinity College of Arts & Sciences, Duke University. We thank Kelly Ostrofsky for technical assistance with the experiments.

References

Amato, J. J., Billy, L. J., Lawson, N. S., & Rich, N. M. (1974). High velocity missile injury: An experimental study of the retentive forces of tissue. *American Journal of Surgery, 127*, 454–459.

Anderson, D. (2010). How to make a bigger hole: An experimental analysis of projectile point morphology in wound creation. *Furthering Perspectives: Anthropological Views of the World, 4*, 37–56.

Andrefsky, W, Jr. (1994). Raw-material availability and the organization of technology. *American Antiquity, 59*, 21–34.

Bartlett, C. S., Helfet, D. L., Hausman, M. R., & Strauss, E. (2000). Ballistics and gunshot wounds: Effects on musculoskeletal tissues. *Journal of the American Academy of Orthopaedic Surgeons, 8*, 21–36.

Bergman, C. A. (1993). The development of the bow in Western Europe: A technological and functional perspective. In G. L. Peterkin, H. M. Bricker, & P. Mellars (Eds.), *Hunting and animal exploitation in the later Palaeolithic and Mesolithic of Europe* (Vol. 4, pp. 95–105).

Bergman, C. A., McEwen, E., & Miller, R. (1988). Experimental archery: Projectiles velocities and comparison of bow performances. *Antiquity, 62*, 658–670.

Bettinger, R. L. (2009). *Hunter-gatherer foraging: Five simple models*. Clinton Corners, NY: Eliot Werner Publications Inc.

Bettinger, R. L., Winterhalder, B., & McElreath, R. (2006). A simple model of technological intensification. *Journal of Archaeological Science, 33*, 538–545.

Bliege Bird, R., & Smith, E. A. (2005). Signaling theory, strategic interaction, and symbolic capital. *Current Anthropology, 46*, 221–248.

Bliege Bird, R., Smith, E. A., & Bird, D. W. (2001). The hunting handicap: Costly signaling in human foraging strategies. *Behavioral Ecology and Sociobiology, 50*, 9–19.

Boëda, E., Geneste, J. M., Griggo, C., Mercier, N., Muhesen, S., Reyss, J. L., et al. (1999). A Levallois point embedded in the vertebra of a wild ass (*Equus africanus*): Hafting, projectiles and Mousterian hunting weapons. *Antiquity, 73*, 394–402.

Bowhunting.com (2008). Field point accuracy with broadheads. http://www.bowhunting.com/publisher/Bowhunting-Knowledge/2008/9/2/field-point-accuracy-with-broadheads. Accessed October 11, 2011.

Causey, K., Kennamer, J. E., Logan, J., & Chapman, J. I, Jr. (1978). Bowhunting white-tailed deer with succinylcholine chloride-treated arrows. *Wildlife Society Bulletin, 6*, 142–145.

Churchill, S. E. (1993). Weapon technology, prey size selection, and hunting methods in modern hunter-gatherers: Implications for hunting in the Palaeolithic and Mesolithic. In G. L. Peterkin, H. M. Bricker, & P. Mellars (Eds.), *Hunting and animal exploitation in the later Palaeolithic and Mesolithic of Europe* (Vol. 4, pp. 11–24).

Clarkson, C. (2016). Testing archaeological approaches to determining past projectile delivery systems using ethnographic and experimental data. In R. Iovita & K. Sano (Eds.), *Multidisciplinary approaches to the study of Stone Age weaponry* (pp. 189–201). Dordrecht: Springer.

D'Errico, F., Zilhão, J., Julien, M., Baffier, D., & Pelegrin, J. (1998). Neanderthal acculturation in Western Europe? A critical review of the evidence and its interpretation. *Current Anthropology, 39 Suppl*, S1-S44.

Dougherty, P. J., & Matthews, B. K. (2007). Comparison of M-16A2 and M-4 wounding potential. *Military Medicine, 172*, 871–874.

Ellis, C. J. (1997). Factors influencing the use of stone projectile tips: An ethnographic perspective. In H. Knecht (Ed.), *Projectile technology* (pp. 37–78). New York: Plenum Press.

Fackler, M. L. (1988a). Ordnance gelatin for ballistic studies. Detrimental effect of excess heat used in gelatin preparation. *American Journal of Forensic Medicine and Pathology, 9*, 218–219.

Fackler, M. L. (1988b). Wound ballistics: A review of common misconceptions. *Journal of the American Medical Association, 259*, 2730–2736.

Fackler, M. L., Surinchak, J. S., Malinowski, J. A., & Bowen, R. E. (1984). Wounding potential of the Russian AK-74 assault rifle. *Journal of Trauma, 24*, 263–266.

Friis-Hansen, J. (1990). Mesolithic cutting arrows: Functional analysis of arrows used in the hunting of large game. *Antiquity, 64*, 494–504.

Gaudzinski-Windheuser, S. (2016). Hunting lesions in Pleistocene and Early Holocene European bone assemblages and their implications for our knowledge on the use and timing of lithic projectile technology. In R. Iovita & K. Sano (Eds.), *Multidisciplinary approaches to the study of Stone Age weaponry* (pp. 77–100). Dordrecht: Springer.

Guthrie, R. D. (1983). Osseous projectile points: Biological considerations affecting raw material selection and design among Palaeolithic and Palaeoindian people. In J. Clutton-Brock, & C. Grigson (Eds.), *Animals and archaeology. Volume 1. Hunters and their prey.* (Vol. BAR International Series 163, pp. 273–294). Oxford: British Archaeological Reports.

Hatcher, J. S. (1935). *Textbook of pistols and revolvers: Their ammunition, ballistics and use.* Plantersville: SC Small Arms Technical Publications Company.

Hawkes, K. (1991). Showing off: Tests of an hypothesis about men's foraging goals. *Ethology and Sociobiology, 12*, 29–54.

Hawkes, K., O'Connell, J. F., & Blurton Jones, N. G. (2001). Hadza meat sharing. *Evolution and Human Behavior, 22*, 113–142.

Hiscock, P. (2009). Reduction, recycling, and raw material procurement in Western Arnhem Land, Australia. In B. Adams & B. S. Blades (Eds.), *Lithic materials and Paleolithic societies* (pp. 78–94). Oxford: Wiley Blackwell.

Hughes, S. S. (1998). Getting to the point: Evolutionary change in prehistoric weaponry. *Journal of Archaeological Method and Theory, 5*, 345–408.

Jussila, J. (2004). Preparing ballistic gelatine—Review and proposal for a standard method. *Forensic Science International, 141*, 91–98.

Karger, B., & Kneubuehl, B. P. (1996). On the physics of momentum in ballistics: Can the human body be displaced or knocked down by a small arms projectile? *International Journal of Legal Medicine, 109*, 147–149.

Karger, B., Sudhues, H., Kneubuehl, B. P., & Brinkmann, B. (1998). Experimental arrow wounds: Ballistics and traumatology. *Journal of Trauma, 45*, 495–501.

Katch, F., Michael, E. D., & Horvath, S. M. (1967). Estimation of body volume by underwater weighing: Description of a simple method. *Journal of Applied Physiology, 23*, 811–813.

Kelly, R. L. (1988). The three sides of a biface. *American Antiquity, 53*, 717–734.

Langenau, E. E, Jr. (1986). Factors associated with hunter retrieval of deer hit by arrows and shotgun slugs. *Leisure Science, 8*, 417–438.

McElreath, R., Boyd, R., & Richerson, P. J. (2003). Shared norms and the evolution of ethnic markers. *Current Anthropology, 44*, 122–130.

Mehremic, S., & Karabegovic, I. (2008). Gunshot wounds – Contribution to the biomechanical analysis. *University Review, Mechanic of Materials, 2*, 38–42.

Mellars, P. (2004). Neanderthals and the modern human colonization of Europe. *Nature, 432*, 461–465.

Mellars, P. (2005). The impossible coincidence. A single-species model for the origins of modern human behavior in Europe. *Evolutionary Anthropology, 14*, 12–27.

Mendelson, J. A. (1991). The relationship between mechanisms of wounding and principles of treatment of missile wounds. *Journal of Trauma, 31*, 1181–1202.

Milks, A., Dinnis, R., & Pope, M. (2016). Morpho-metric variability of Early Gravettian tanged "Font-Robert" points, and functional implications. In R. Iovita & K. Sano (Eds.), *Multidisciplinary approaches to the study of Stone Age weaponry* (pp. 135–146). Dordrecht: Springer.

Noe-Nygaard, N. (1974). Mesolithic hunting in Denmark illustrated by bone injuries caused by human weapons. *Journal of Archaeological Science, 1*, 217–248.

Odell, G. H., & Cowan, F. (1986). Experiments with spears and arrows on animal targets. *Journal of Field Archaeology, 13*, 195–212.

Patrick, U. W. (1989). *Handgun wounding factors and effectiveness.* Quantico, VA: U.S. Department of Justice, Federal Bureau of Investigation.

Pétillon, J.-M., Bignon, O., Bodu, P., Cattelain, P., Debout, G., Langlais, M., et al. (2011). Hard core and cutting edge: Experimental manufacture and use of Magdalenian composite projectile tips. *Journal of Archaeological Science, 38*, 1266–1283.

Rios-Garaizar, J. (2016). Experimental and archeological observations of Northern Iberian Peninsula Middle Paleolithic Mousterian point assemblages. Testing the potential use of throwing spears among Neanderthals. In R. Iovita & K. Sano (Eds.), *Multidisciplinary approaches to the study of Stone Age weaponry* (pp. 213–225). Dordrecht: Springer.

Salisbury, C. P., & Cronin, D. S. (2009). Mechanical properties of ballistic gelatin at high deformation rates. *Experimental Mechanics, 49*, 829–840.

Sellier, K. G., & Kneubuehl, B. P. (1994). *Wound ballistics and the scientific background.* Amsterdam: Elsevier Health Sciences.

Shadwick, R. E. (1998). Elasticity in Arteries: A similar combination of rubbery and stiff materials creates common mechanical properties in blood vessels of vertebrates and some invertebrates. *American Scientist, 86*, 535–541.

Shea, J. J. (1988). Spear points from the Middle Paleolithic of the Levant. *Journal of Field Archaeology, 15*, 441–450.

Shea, J. J. (1997). Middle Paleolithic spear point technology. In H. Knecht (Ed.), *Projectile technology* (pp. 79–106). New York: Plenum Press.

Shea, J. J. (2006). The origins of lithic projectile point technology: Evidence from Africa, the Levant, and Europe. *Journal of Archaeological Science, 33*, 823–846.

Shepherd, C. J., Appleby-Thomas, G. J., Hazell, P. J., & Allsop, D. F. (2009). The dynamic behaviour of ballistic gelatin. *Proceedings of the American Physical Society Topical Group on Shock Compression of Condensed Matter, 1195*, 1399–1402.

Shott, M. J. (2016). Survivorship distribution in experimental spear points: Implications for tool design and assemblage formation. In R. Iovita & K. Sano (Eds.), *Multidisciplinary approaches to the study of Stone Age weaponry* (pp. 245–258). Dordrecht: Springer.

Sperrazza, J., & Kokinakis, W. (1968). Ballistic limits of tissue and clothing. *Annals NY Academy of Science, 152*, 163–167.

Stormer, F. A., Kirkpatrick, C. M., & Hoekstra, T. W. (1979). Hunter-inflicted wounding of white-tailed deer. *Wildlife Society Bulletin, 7*, 10–16.

Straus, L. G. (1990). The original arms race: Iberian perspectives on the Solutrean phenomenon. In J. Kozlowski (Ed.), *Feuilles de Pierre: Les industries à pointes foliacées du paléolithique supérieur européen. EURAL 42* (pp. 425–447). Liège: Univ. Liège.

Thomas, D. H. (1978). Arrowheads and atlatl darts: How the stones got the shaft. *American Antiquity, 43*, 461–472.

Waguespack, N. M., Surovell, T. A., Denoyer, A., Dallow, A., Savage, A., Hyneman, J., et al. (2009). Making a point: Wood- versus stone-tipped projectiles. *Antiquity, 83*, 786–800.

Wiessner, P. (1983). Style and social information in Kalahari San projectile points. *American Antiquity, 48*, 253–276.

Wilkins, J., & Schoville, B. J. (2016). Edge damage on 500-thousand-year-old spear tips from Kathu Pan 1, South Africa: The combined effects of spear use and taphonomic processes. In R. Iovita & K. Sano (Eds.), *Multidisciplinary approaches to the study of Stone Age weaponry* (pp. 101–117). Dordrecht: Springer.

Zilhão, J., Angelucci, D. E., Badal-García, E., d'Errico, F., Daniel, F., Dayet, L., et al. (2010). Symbolic use of marine shells and mineral pigments by Iberian Neandertals. *Proceedings of the National Academy of Sciences, 107*, 7621–7622.

Chapter 15
Experimental and Archeological Observations of Northern Iberian Peninsula Middle Paleolithic Mousterian Point Assemblages. Testing the Potential Use of Throwing Spears Among Neanderthals

Joseba Rios-Garaizar

Abstract The use of ranged weapons among Neanderthals is an important issue in paleoanthropology, due to its implications for understanding the adaptive advantages of modern humans as opposed to Neanderthals. This debate has been hindered by the existence of some preconceived ideas, such as Mousterian points being too bulky to be used as projectile points. In the last years we have analyzed several Middle Paleolithic assemblages in Northern Iberian Peninsula that included Mousterian points with impact traces. One of the main features of these points was that they were substantially lighter than expected, which made them appropriate as archeological reference to test if Neanderthal groups used these kinds of points as throwing spear tips. We developed an exploratory experiment to test if they were suitable for throwing, and to identify which variables were more important to demonstrate it. Finally we discuss the results from an evolutionary and historical perspective.

Keywords Projectile • Northern Iberian Peninsula • Middle Paleolithic • Experimental archeology • Use-Wear analysis

Introduction

The idea of Neanderthals using projectile weapons is commonly rejected by archeologists and paleoanthropologists. This rejection is founded on some biomechanical assumptions, on general ideas of Neanderthal capabilities (Churchill and Rhodes 2009) and also on the idea of how a Mousterian point should be (see for example the description made by Shea and Sisk 2010: "… these points are so large that they must have been attached to thick, heavy thrusting spears or hand-cast spears").

This question is not a minor one. Projectile use and distance hunting are major landmarks in human evolution because they enable better safety, success and prey selection and the possibility of hunting in new environments (Rhodes and Churchill 2009). These improvements would have caused major transformations in the economies and social organization of past populations, and they would also have given some advantages to these populations in a competitive scenario. Usually this improvement is associated with modern humans and observed as a major advantage that leaded to the ultimate extinction of Neanderthals.

The claim of projectile weapon use among Neanderthals is not new. Starting with the Schöningen spear (Thieme 1997, 1999), in recent years several authors have presented data from different sites and assemblages that apparently strengthen this idea (Callow 1986; Plisson and Beyries 1998; Shea 2006; Galván Santos et al. 2007–2008; Moncel et al. 2009; Villa et al. 2009; Villa and Lenoir 2006, 2009; Lazuén 2012), but the lack of effective demonstration reduces the discussion to the field of prejudices and skepticism.

To advance in this field a multi–proxy approach to the problem of projectile use among Neanderthals must integrate the analysis of fatigue traces (impact and hafting traces) with morphologic and metric analysis. The results must be experimentally tested and extensively contrasted with archeological assemblages. Indirect lines of evidence, such as those generated from the faunal record or Neanderthal diet, can inform us also about hunting strategies, prey selection or overall dependence from animal origin nourishment (Bocherens 2009). Also Neanderthal biomechanics can offer some insights about the performance capabilities of Neanderthals in activities as throwing or stabbing (Trinkaus 2008; Churchill and Rhodes 2009; Rhodes and Churchill 2009; Shaw et al. 2012).

J. Rios-Garaizar (✉)
Traceology Group, Consorcio CENIEH (Centro Nacional de Investigación Sobre La Evolución Humana), Paseo Sierra de Atapuerca, 3, 09002, Burgos, Spain
e-mail: joseba.rios@cenieh.es

Radu Iovita and Katsuhiro Sano (eds.), *Multidisciplinary Approaches to the Study of Stone Age Weaponry*, Vertebrate Paleobiology and Paleoanthropology, DOI 10.1007/978-94-017-7602-8_15

Also a cautionary note must be made about the vision of Neanderthal culture as a monolithic set of behaviors that did not vary very much over time. This perspective drives interpretation to an overgeneralization that converts concrete (historical) features into universal ones. This idea of variability of Neanderthal behavior and its relevance for the understanding of Neanderthal evolution has been successfully explored in recent years in the fields of lithic technology, subsistence practices, settlement dynamics, etc. (Gardeisen 1999; Costamagno et al. 2006; Delagnes et al. 2007; Jaubert and Delagnes 2007), concluding that Neanderthal behavior varied quite a lot through time and space, and that this variation cannot be interpreted as simple adaptations to different environments but as complex cultural choices (Rios-Garaizar 2009). From this perspective the use of different kinds of pointed weapons must be studied and interpreted from a discrete perspective, in order to achieve general interpretations and to demonstrate whether Neanderthals used point tipped weapons, whether they used them as projectile points, and why sometimes they chose this alternative, and how this information will help understand the changes in Neanderthal behavior.

Mousterian Point Morphology

In recent years point morphological variation has been analyzed from different perspectives, including the use of different indexes such as TCSA: Tip cross section area: ½ Width * Thickness; or TCSP: Tip Cross Sectional Perimeter: $4\sqrt{s}$ where $s = (½\ \text{Width})^2 + (½\ \text{Thickness})$ for bifacial points and $\text{Width} + 2 * \sqrt{((½\ \text{Width})^2 + \text{Thickness}^2)}$ for unifacial points (Hughes 1998; Villa and Lenoir 2006; Shea 2006; Sisk and Shea 2011). These indexes are useful for describing the lightness/heaviness of the points, elements that are directly related with ballistic features. Trying to describe points only from this perspective will neglect other important features also related with ballistics and penetration capability, such as weight, curvature or tip section/plan angles that can also undergo numerical analysis and comparisons. Nor can we forget another important question, of hafting and the relationship between hafting area (usually proximal part of the point) thickness and the minimum shaft diameter. In fact the shaft dimensions may be more important for throwing than the point itself. All these elements must be included in the description, and why not in the results presentation, in order to better assess Mousterian point variability.

Archeological Record

Northern Iberian Peninsula, concretely the region around the Bay of Biscay has been subject of an intense archeological investigation during 20th century and the beginning of the 21st century. Grace to this investigations several Middle Paleolithic sequences have been identified and

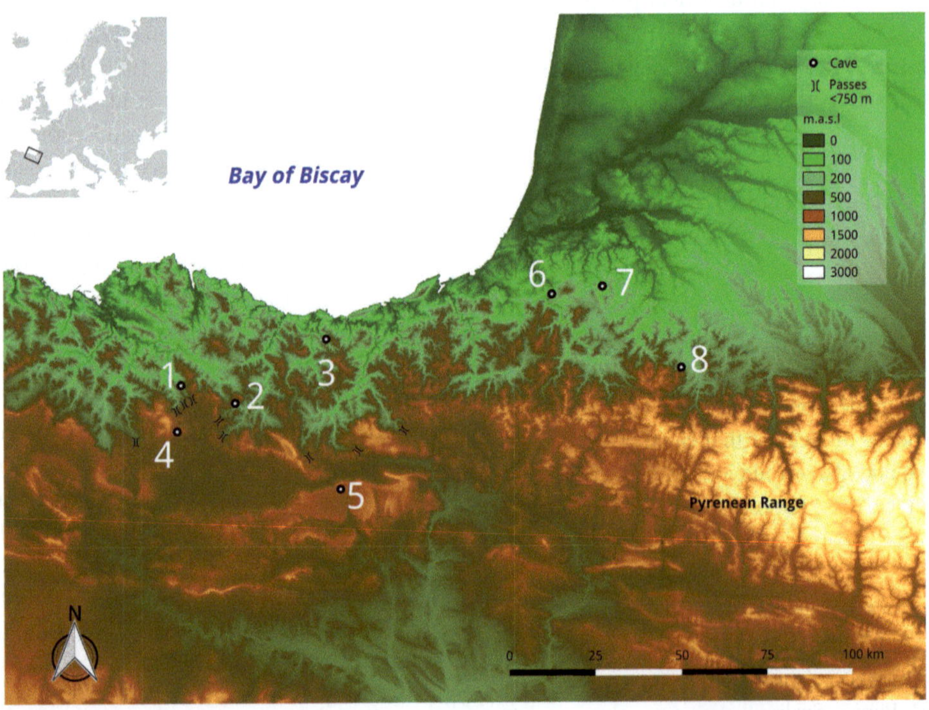

Fig. 15.1 Mousterian sites around the Bay of Biscay (N. Iberian Peninsula) cited in the text. 1: Axlor; 2: Lezetxiki; 3: Amalda; 4: Arrillor; 5: Mugarduia Norte; 6: Abri Olha; 7: Isturitz; 8: Gatzarria

excavated. Among them the most important sites are Isturitz, Gatzarria, Albri Olha, Lezetxiki, Axlor and Arrillor (Rios-Garaizar 2012a) (Fig. 15.1). In the last years we have studied directly the lithic collections of Amalda and Axlor (Rios-Garaizar 2005, 2010, 2012a; González et al. 2005, 2006). The main results of these analyses are the identification of an important technological variability among different sites and levels (Rios-Garaizar 2009) with a special incidence in the presence/absence of weapon points (Rios-Garaizar 2012b).

Amalda presents an isolated and undated Mousterian Level (Level VII) with a Levallois-Discoid industry dominated by sidescrapers and denticulates with few (4%), non-standardized, Mousterian points. Technological features include a complex management system of flint resources. This material is imported from close (15–25 km) and distant (35–90 km) localities as already made tools, including points. After this there is an *in situ* ramification of production oriented versus the production of small flakes which are used in different kind of activities including butchery or woodworking (Rios-Garaizar 2010). Local raw materials are also used to produce big tools as handaxes, cleavers or big pseudo-levallois points. These tools were probably used in a wide range of activities including first phases of carcass processing or woodworking. Faunal record is mainly composed by chamois, with few remains of larger fauna as red deer, bison, auroch or horse. The abundance of chamois has been interpreted by some authors as a result of carnivore generated taphocenosis (Yravedra 2007) but probably a big part of chamois assemblage was introduced by humans (Altuna and Mariezkurrena 2010). The site has been interpreted as a seasonal occupation where a full range of different activities, including hunting, carcass processing and weapon fabrications were developed (Rios-Garaizar 2010, 2012a). Beside Amalda there is another cave (Amalda III) where some points were found in a test excavation (Altuna et al. 1995).

Axlor site was discovered in the 30s and excavated by J. M. Barandiarán between 1967 and 1974 and by González, Ibañez and Rios-Garaizar between 2000 and 2008 (González et al. 2005). At least 5 Mousterian levels have been identified. Although there are some discrepancies between Barandiaran and recent excavations' stratigraphy (González et al. 2005) a succession of Levallois technology levels (VIII, VII) and Quina technology levels (VI–III) have been observed. Dating is only available for the upper part of the sequence, with a date of 42,010 ± 1280 BP (González and Ibáñez 2002) for level D (IV).

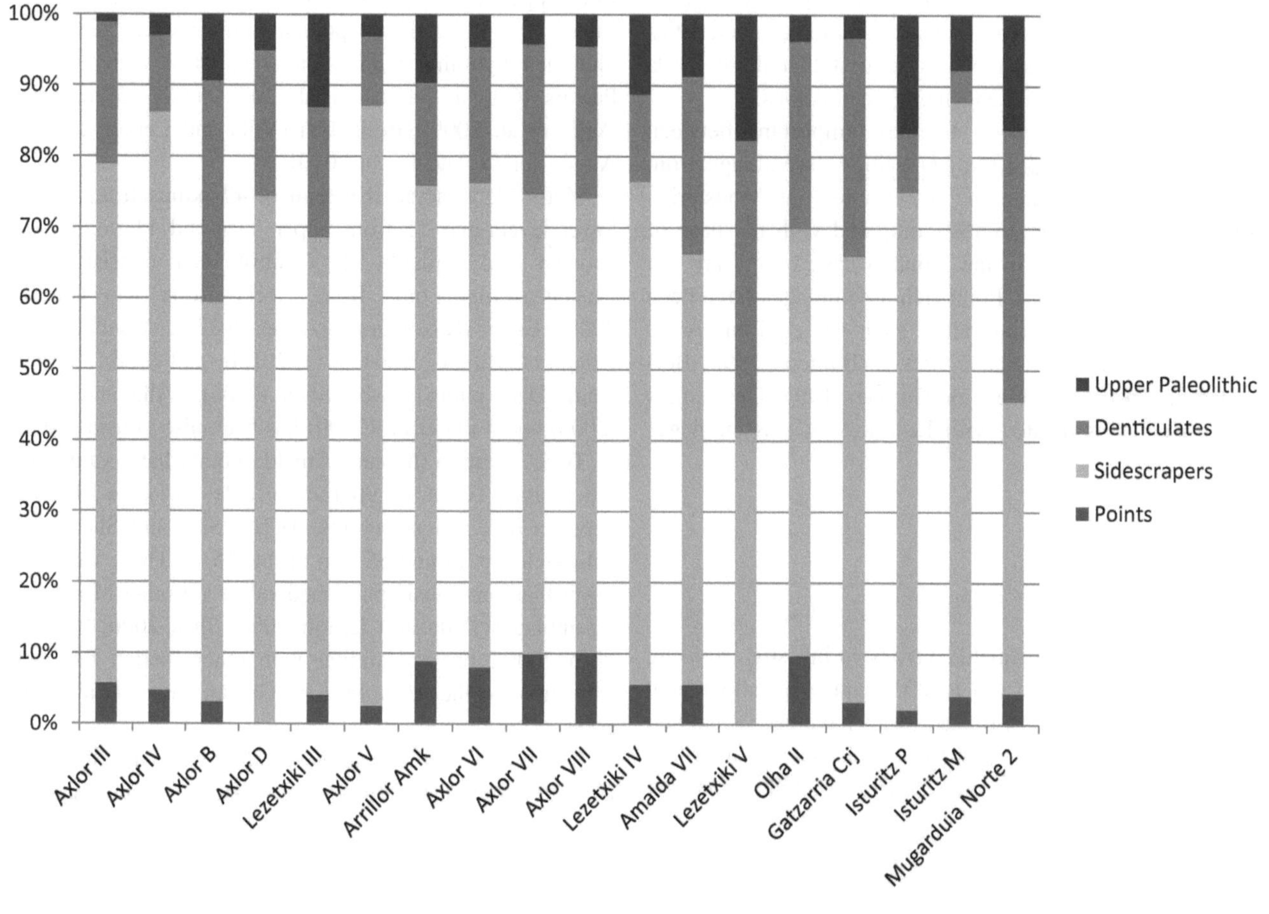

Fig. 15.2 Proportion between four main categories of formal tools in the Mousterian levels around Bay of Biscay

In lower levels flint is imported from distant localities (>30 km, <60%) as already made objects, some of them points. As in Amalda an *in situ* ramification of production was developed to obtain small size (<2 cm) Levallois flakes. Levallois technology was used also to exploit local materials but the objective of this production was to obtain big flakes (Rios-Garaizar 2012a). Faunal assemblage is characterized in these levels by the presence of red deer which was partially processed in the site (Altuna 1989). Another important feature of these levels is the presence of fireplaces, some of them with evidences of re-use. All these evidences suggest that the human occupations were quite stable, and that the site was probably used as a residential camp.

Upper levels show significant differences. Raw material provisioning is now concentrated on distant flint (>30 km, >80%) which is introduced in the site as already made thick Quina sidescrapers. These tools were involved in a complex ramified management consisting in intense resharpening and small flake production (Rios-Garaizar 2005, 2012a) for which bone retouchers were intensively used (Mozota 2012). Faunal assemblage is now composed by goats, red deer, bos/bison and horses (Castaños 2005). There are evidences of intense *in situ* processing of faunal remains but probably other activities as hide processing were also present. The absence of fireplaces and the characteristics of lithic and faunal assemblages suggest that these levels were formed by repeated ephemeral occupations.

Typological differences are also important between these levels. The Upper sequence lithic assemblage from Axlor is far richer than the lower one, but Mousterian points are quite scarce (3.8%) compared with the base of the sequence (7.8%). In the Amalda case points are not very important (4.04%). Point rich assemblages have been identified in neighboring sites such as Arrillor Amk; Lezetxiki IV or Abri Olha 2 Asfk-1 (Baldeón 1993; Bermúdez de Castro and Sáenz de Buruaga 1997; Deschamps 2010) usually associated with Levallois rich assemblages (Fig. 15.2).

Point Analysis

Points from Axlor and Amalda have been investigated using a binocular microscope (up to 80×) and a metallographic microscope (up to 200×). Impact traces have been identified following well known systematics (Fischer et al. 1984; Dockall 1997). Macroscopic traces include complex fractures (step terminating bending fractures, spin-off fractures and impact burination) other traces, less diagnostic, were considered only in association with other traces (oblique fractures in lateral edges, distal and proximal crushing). The most diagnostic microscopic traces are longitudinal striations. The absence of other kind of use, as scraping or cutting, was also considered.

The analyzed sample consisted in seven points or point fragments from Axlor level N (8% of retouched tools), two from level M, two from level D and one from level B. Some points recovered in old excavations from levels VIII (N = 2), VII (N = 1), VI (N = 2) and III–IV (N = 2) were also analyzed. This makes a total of 14 points or point fragments from Axlor lower (Levallois) levels and five from upper (Quina) levels. In Amalda only four points were recovered and analyzed. Finally two more points from Amalda III site were included in the use wear analysis.

Preservation of samples was good but chemical alteration affected the pieces form Amalda and Axlor upper levels. In Axlor lower levels thermic alterations caused by fireplaces were high-moderate. Micro use-wear was thus difficult to ascertain. On the contrary Diagnostic Impact Fractures were easy to identify and interpret (Fig. 15.3). That was the case for seven points recovered from Axlor lower levels, three from upper levels and three from Amalda, which makes a rough 50% of analyzed points. Similar traces have been identified by other authors in other sites, such as Abric Pastors (Galván Santos et al. 2007–2008), Oscurusciuto (Villa et al. 2009), Bouheben (Villa and Lenoir 2006) or Angé (Soressi and Locht 2010).

Metric data included a total of 42 points from Axlor's lower levels and 21 from upper ones and 11 points from Amalda and Amalda II. Control data consist in 39 Chatelperronian points from Labeko Koba (Rios-Garaizar 2008), Ekain (Rios-Garaizar et al. 2012a), Aranbaltza and Ollagorta (Rios-Garaizar et al. 2012b) and Morin 10 (own data). Data published by Shea (2006), Villa and Lenoir (2006) and Villa et al. (2009) has been also considered.

TCSA from Axlor and Amalda points have values that fell around TCSA of 90 mm^2 (Table 15.1), i.e., halfway between dart and spear tips. TCSP (Sisk and Shea 2011) values cluster around 55 mm (Table 15.2). These values are lower than previously published values for other Mousterian assemblages (Vila and Lenoir 2006; Shea 2006; Sisk and Shea 2009), but are similar with other Iberian Peninsula collections as the Abric Pastors (Galván Santos et al. 2007–2008).

However if we compare these values with Chatelperron or Gravette points, we observe that TCSA mean values of

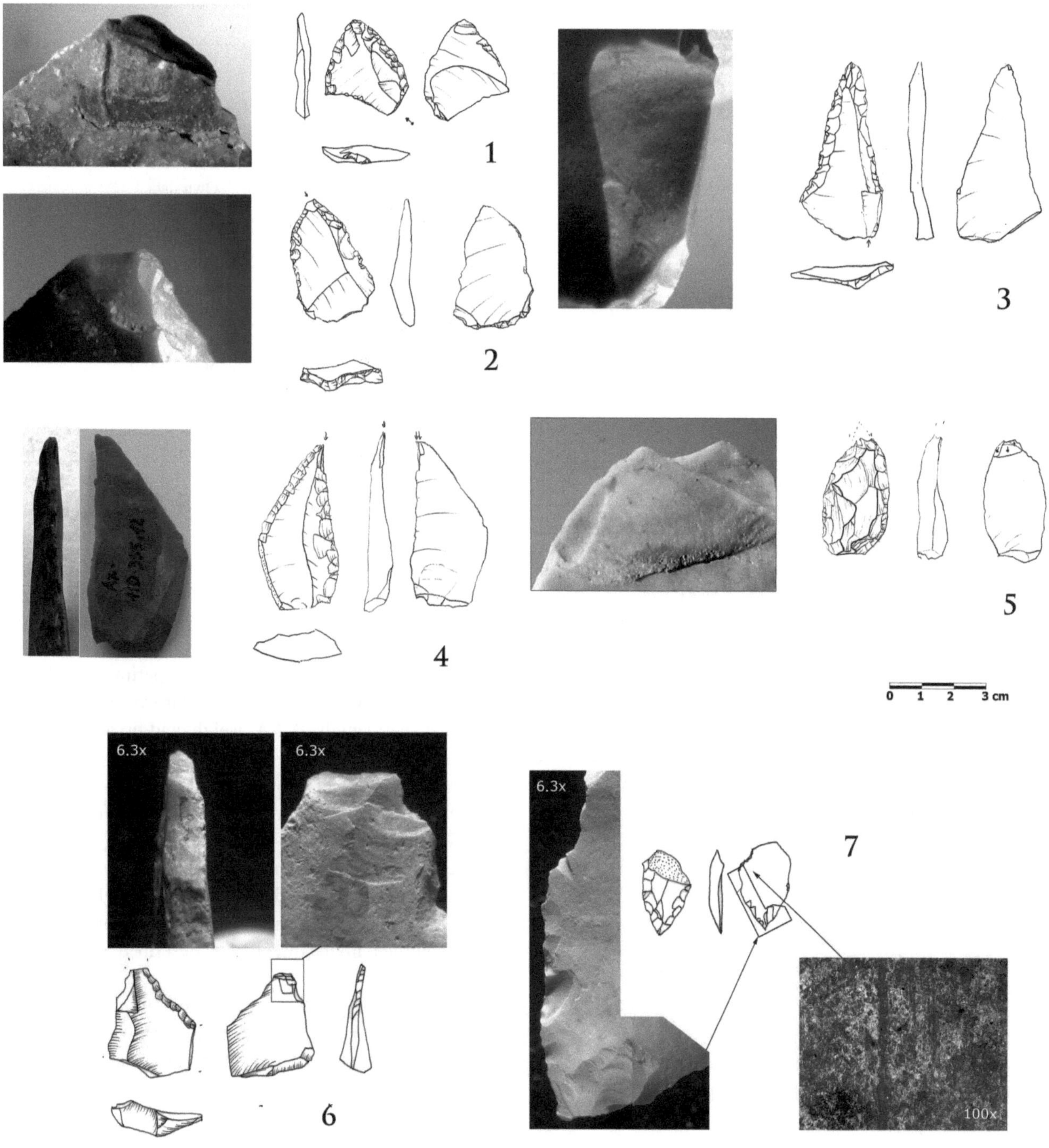

Fig. 15.3 Mousterian points with diagnostic impact traces: 1, 2, 5, 6, and 7: Flute-like fractures; 4: Burin-like fracture; 3: Proximal flute-like fracture; 7: Micro-striation. 1–5: Axlor; 6–7: Amalda

Table 15.1 TCSA data (mm^2) from different Mousterian assemblages and control assemblages (in italics, ethnographic data recorded by Shea 2006)

	Mean	SD	Min	Max	n	References
Bouheben	165	67.2	50	322	70	Villa and Lenoir (2006)
Oscurusciuto	85	24.7	52.5	112.5	4	Villa et al. (2009)
Le Moustier Mousterian points	266	64	195	378	12	Shea (2006)
Amalda	70.18	31.21	24	120	11	Own data
Axlor lower	89.83	56.23	19.5	283.5	42	Own data
Axlor (upper)	99.45	53.16	25.5	240	21	Own data
All mousterian points	88.43	52.17	19.5	283.5	74	Own data
Chatelperron points (Shea 2006)	63	25	28	130	33	Shea (2006)
Chatelperron points (own)	39.5	19.52	13	110	39	Own data
Gravette points	41	19	8	88	40	Shea (2006)
Arrowheads	33	20	8	146	118	Shea (2006)
Dart tips	58	18	20	94	40	Shea (2006)
Spear tips	168	89	50	392	28	Shea (2006)

Table 15.2 TCSP data (mm) from analyzed Mousterian assemblages and measured Chatelperronian points

	Mean	SD	Min	Max	n	References
Amalda	51.49	13.14	33	73.36	11	Own data
Axlor lower	49.86	20.61	29.23	128.52	21	Own data
Axlor (upper)	57.67	14.37	27.31	88.51	42	Own data
All mousterian points	54.18	16.25	27.31	128.52	74	Own data
Chatelperron points	32.65	7.03	21.04	51.73	39	Own data

analyzed Mousterian points are double UP mean values, which fall clearly halfway between arrow and dart points (Table 15.1).

The existence of indisputable impact traces on these light points leads us to consider that they were effectively used as throwing spear points.

Can Mousterian Points Be Effectively Thrown?

One of the main questions about Mousterian points is not whether they were used as weapon tips, but whether they were used only for close hunting or also for distance hunting. Differences in impact damage created by thrusting or throwing have been investigated through controlled experiments (Shea et al. 2001; Iovita et al. 2016) but little effort has been made to demonstrate that spear-throwing was a hunting technique practiced by Neanderthals (Rieder 2009).

We have conducted a prospective experiment to demonstrate that Mousterian points can be thrown and to test the effect that shaft length, TCSA, weight and tip angles have in flight ability.

A collection of 24 Mousterian points was made using the Levallois technique with Morocco flint (chalcedony). The overall morphologies fitted the expected variability of Middle Paleolithic Mousterian points; TCSA values ranged between 55 and 240, with different angles, weights and basal thickness (Table 15.3). These morphologies are similar to those identified in the previously described archeological record.

Eight of these points were selected, mounted on pinewood *pre-hampes and* glued with a birch/wax/ocher combination (Figs. 15.4 and 15.5). Then they were attached to 3 different shaft classes (pinewood, Ø15 mm, 220/200/180

Table 15.3 Different dimensions of experimental points

No.	Length (mm)	Width (mm)	Thick. (mm)	TCSA (mm^2)	TCSP (mm)	Tip plan angle	Tip section angle	Basal thick	Weight (g)
EXP4	31	30	9	135	64.98	70	35	6	6.4
EXP7	32	23	5	57.5	48.07	75	20	4	4.3
EXP11	53	26	9	117	57.62	50	30	8	14.2
EXP13	73	24	6	72	50.83	30	20	5	14.4
EXP15	43	25	7	87.5	53.65	45	80	5	7.6
EXP17	50	23	8	92	51.01	60	35	6	9.5
EXP18	34	36	12	216	79.26	75	40	12	12.5
EXP25	40	27	7	94.5	57.41	60	45	7	10.1

Fig. 15.4 Experimental points

lengths; 206, 188 and 170 g respectively). These dimensions fall inside the range of modern throwing spears (Oakley et al. 1977). The length is within the range of Schöningen spears, but not the diameter which is at least 50% lower (Thieme 1999).

The throws were made by an inexperienced thrower (male, 1.70 m and 70 kg) with constant atmospheric conditions (Wind: NW/13.6 km/h, 12.7 °C, 75% Humidity).

A total of 69 throws were made using different combinations of points and hafts (23 with each shaft length). All

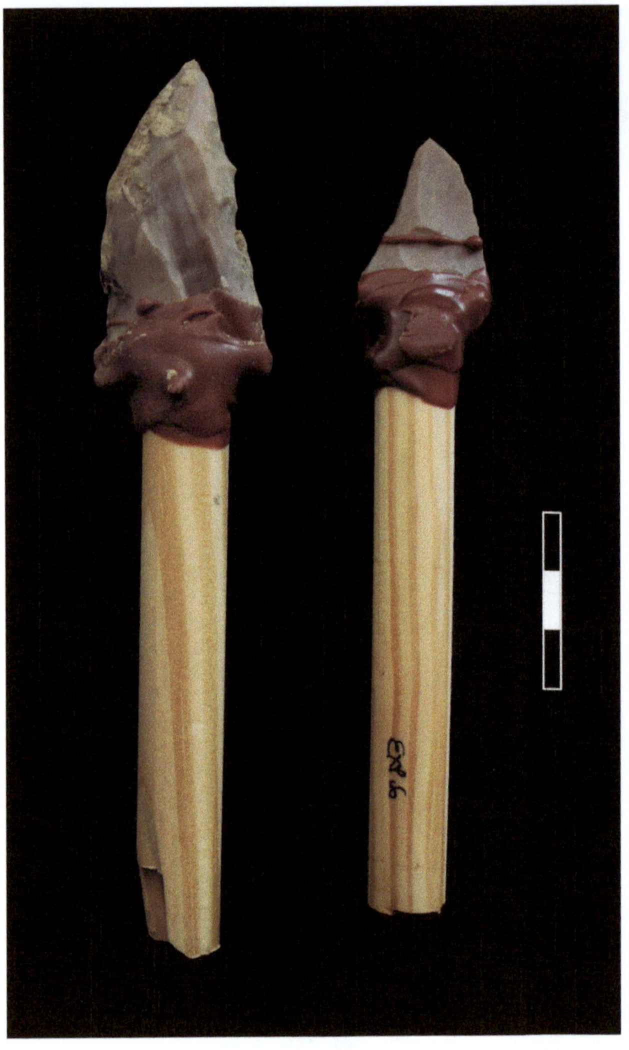

Fig. 15.5 Pre-hampe hafted experimental points

the throws were aimed at the same point. Several data from each throw was recorded, including landing point coordinates, distance and trajectory.

The throws dispersed in an angle of 45° with a maximum of 17 m of distance between them (Fig. 15.6). The majority of the distances are situated between 12 and 17 m. Differences can be observed between different shaft classes; long shafts allowed further shots than shorter ones, with appreciable differences between 2.2 and 1.80 m shafts (Table 15.4).

Table 15.4 Throw distances (m) corresponding with different shaft lengths (SL)

SL (cm)	All	2.2	2.0	1.8
N	69	23	23	23
Min	5.83	9.48	7.81	5.83
Max	21.93	21.93	18.60	18.86
Sum	978.76	389.54	300.81	288.40
Mean	14.18	16.93	13.07	12.53
Std. error	0.406	0.633	0.556	0.549
Variance	11.42	9.24	7.11	6.93
Stand. dev	3.38	3.04	2.66	2.633
Median	13.45	17.49	13	12.80

We cannot observe significant differences between the distance flown and the TCSA of points (Fig. 15.7, Table 15.5). Numbers indicate that there is no clear trend (i.e., the lighter the point further it flights). Standard errors are greater than in the cases of the shafts, and variance can reach 12 m. Weight has also no significant impact on the distance covered by the javelin, but the best values are obtained by points between 6.4 and 9.5 g (Fig. 15.7). In both cases (TCSA and weight), low and high values did not achieve good results.

With respect to trajectories, the observations are more qualitative. We recorded whether the javelin hit the ground with the point, assuming if so that the javelin followed an arched trajectory with good orientation of the point and no twisting or balancing. Shorter shafts were less prone to twisting. Heavier points also helped to keep the point in the right direction, and TCSA values between 87.5 and 117 mm^2 were the most suitable. Both tip plan and section angles have little impact on trajectory (Fig. 15.8).

The results of this experiment are exploratory, but some conclusions can be drawn. The main idea is that there is no problem for hand-thrown spears tipped with Mousterian points. Moreover, they easily reached distances of more than 10 m, which is sufficient to increase effectiveness in hunting. We also identified the main variables that results in good aiming. Shaft length is more decisive than any variable corresponding to tip morphology. Weight of point is important for maintaining balance, and medium TCSA values seem best suited for this kind of throwing.

We also observed that the point very rarely fractures when hitting the ground; no points sustained any fractures,

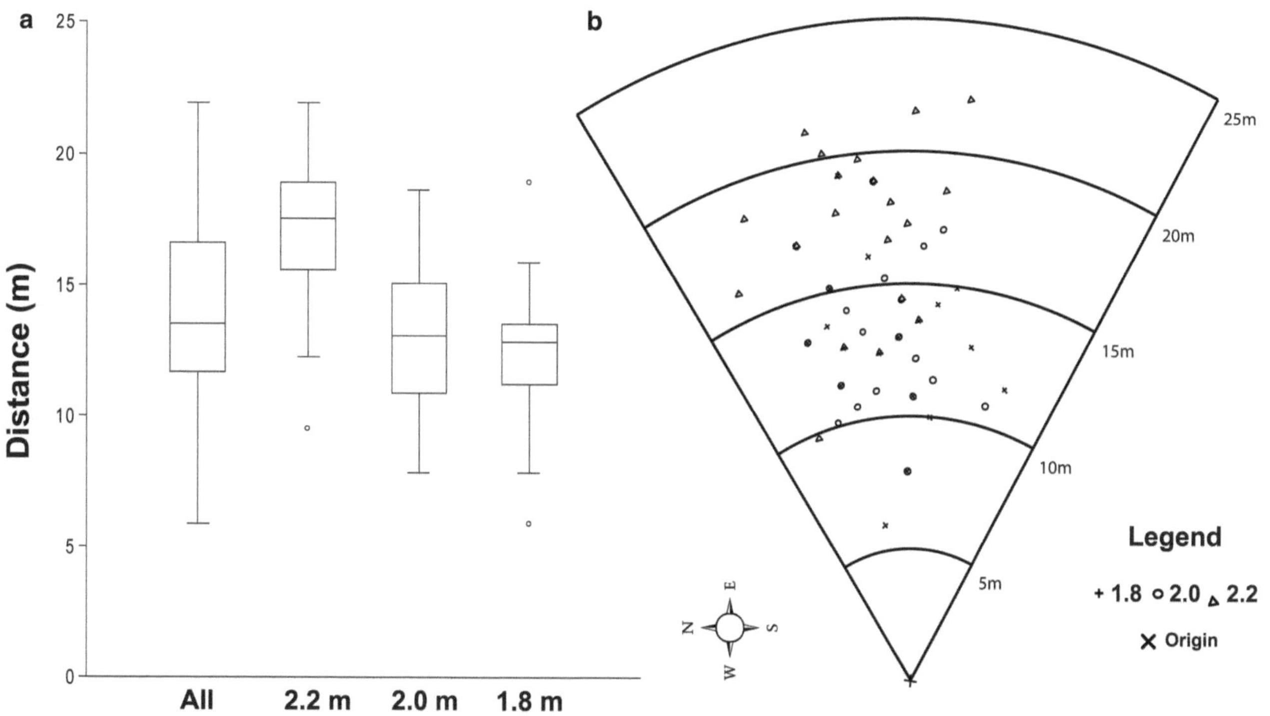

Fig. 15.6 Distance and orientation of throws corresponding with different shaft lengths

Table 15.5 Throw distances (m) corresponding with different TCSA values

TCSA (mm^2)	57.5	87.5	92	94.5	117	135	216
N	7	16	4	8	8	23	3
Min	5.83	9.84	13	7.81	10.63	9.48	12.53
Max	13.45	20.5	18.86	16.27	14.86	21.93	13.45
Sum	72.04	247.17	61.82	96.73	102.29	359.90	38.78
Mean	10.29	15.44	15.45	12.09	12.78	15.64	12.92
Std. error	1.006	0.881	1.275	0.889	0.580	0.662	0.273
Variance	7.09	12.43	6.50	6.32	2.69	10.09	0.22
Stand. dev	2.66	3.52	2.55	2.51	1.64	3.17	0.47
Median	11.18	15.96	14.97	12.06	13.11	15.52	12.80

even one that was thrown 23 times. Hafting was easy to make with materials such as birch or ocher, which have been documented among Neanderthals (Grünberg 2002; Roebroeks et al. 2012). This glue absorbs the impact and spinning forces and breaks before the shaft or point sustain any damage.

Further experiments must investigate the combination of flight and impact; to that end we will need to take into account more variability in shaft morphology, the kinetic energy generated, the penetrability of the points, etc. This will be absolutely necessary to see whether point morphology variation has a major effect on spear effectiveness. We

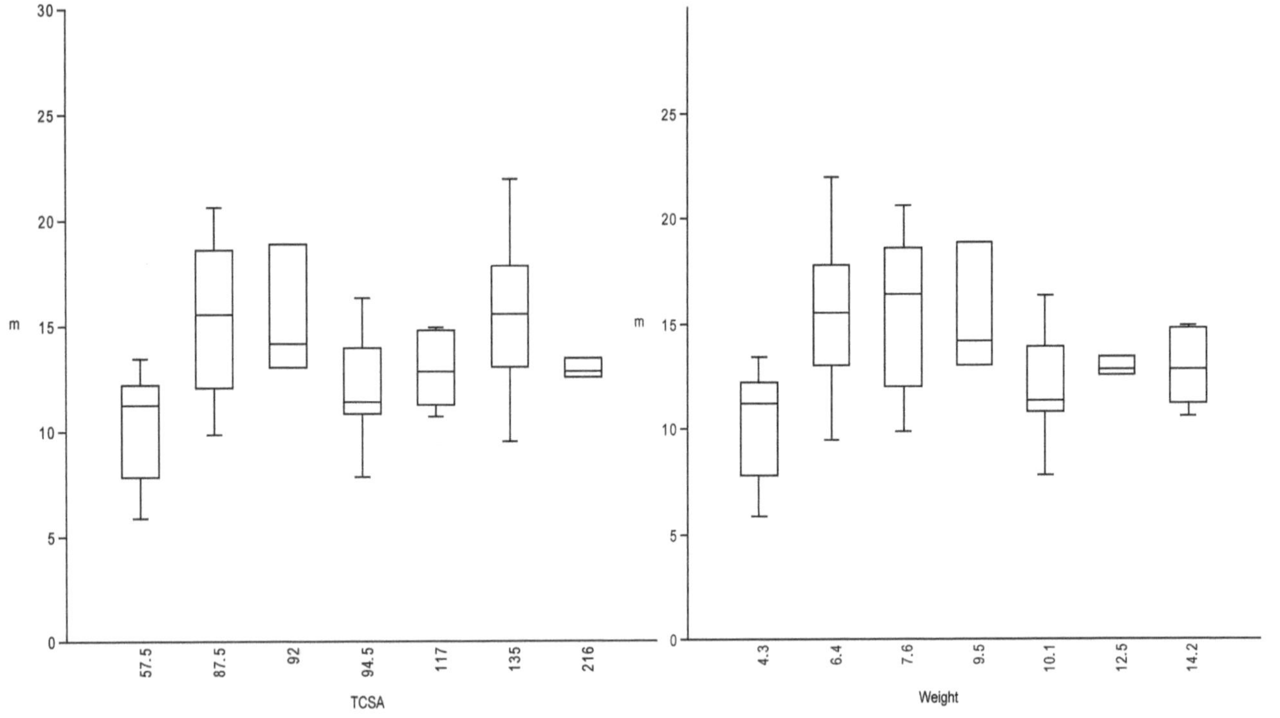

Fig. 15.7 Distance of throws corresponding with different TCSA and weight values

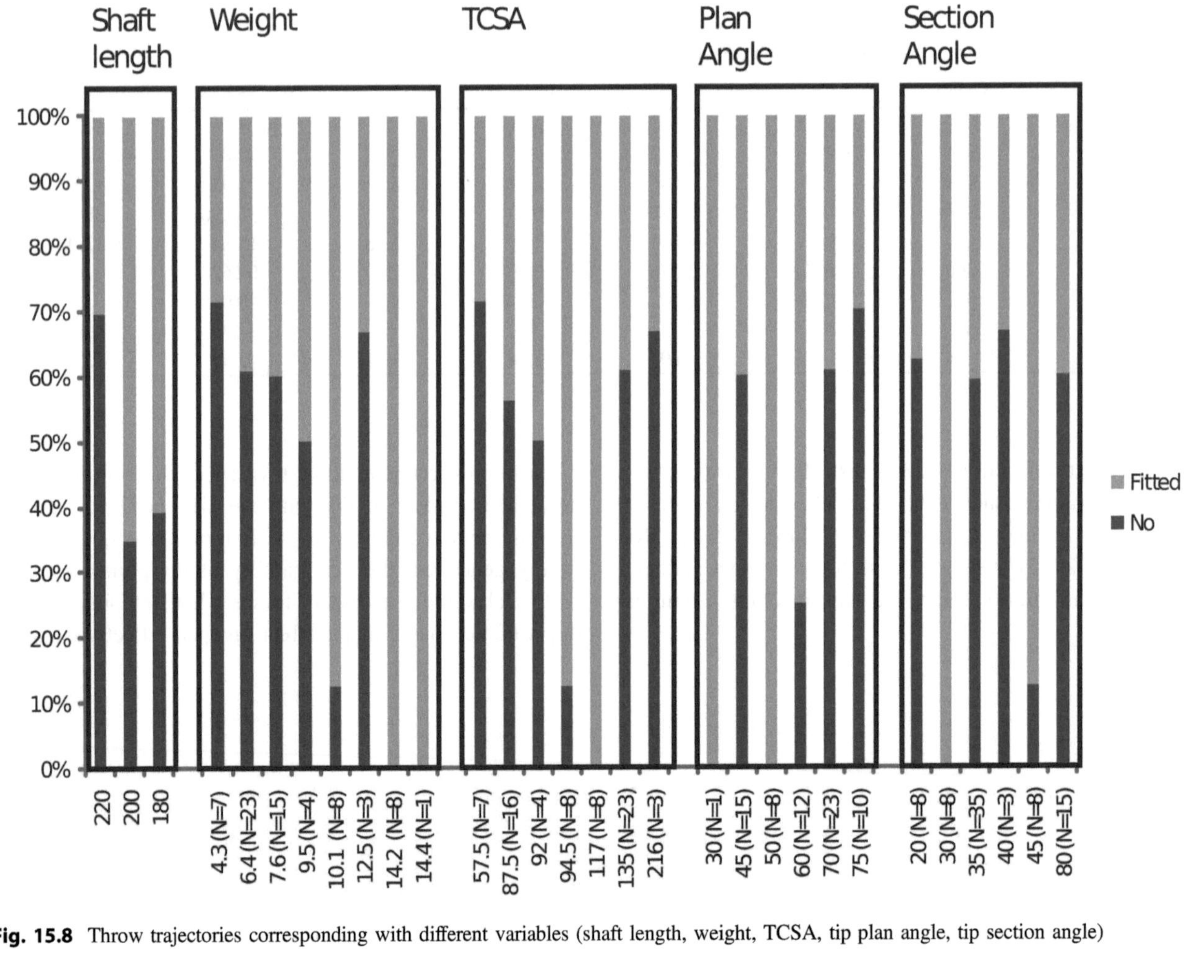

Fig. 15.8 Throw trajectories corresponding with different variables (shaft length, weight, TCSA, tip plan angle, tip section angle)

also need to seek differences in the traces generated by throwing or thrusting impacts if we want to compare these results with the archeological record.

Discussion

Weapon use among Neanderthals does not seem to be subject to homogeneous behavior during Middle Paleolithic. Mousterian point variability is the probable consequence of a variation in weapon types, hunting strategies, etc. The ability of Neanderthals to use range weapons has been repeatedly denied in order to increase the technological gap with modern humans, thereby assisting the explanation of the Neanderthals' demise. Little or no proof supported this idea, and in recent years the evidence of weapon use in the Middle Paleolithic has significantly increased. The discussion is focused now on demonstrating or refuting the possibility that Neanderthals used ranged weapons. In this discussion the mental template of a bulky Mousterian point and some metrical data used to describe Mousterian points as population (see Shea 2006) formed the idea that they were mainly thrusting spear tips. Conversely, there is also the contrary view that postulates that Neanderthals used ranged weapons, without any substantial proof.

The analysis of the archeological record showed that there is a great variability not only in point morphology, but also in the role played by these points. These differences are probably related with changes in mobility, tool management and hunting strategies (Rios-Garaizar 2009). In any case the point assemblage considered (74 points), all of them coming from different, non-synchronic, Eastern Cantabrian region sites, shows that point variability is quite important and, at least in this region, points are lighter than previously stated with TCSA values around 90 mm^2 and TCSP values around 55 mm. Very likely, this opens the possibility of range weapon use in this concrete space-temporal frame.

Previous experiments suggested that, although light Levallois points were capable to penetrate into targets, they were not properly designed for projectile use (Sisk and Shea 2009). Nevertheless the points used in this experiment rarely reached the lower values found in Mousterian sites described here.

Through an experimental approach we have explored the possibility of throwing spears armed with Mousterian points. First results seem quite positive and a hunting range of at least 10 m can be easily accepted. We have also explored the variables that are important for good performance in flight. Shaft length seems to be decisive, but other variables such as tip weight or TCSA are also important. These results suggest that Mousterian points were well designed for close throwing hunting. However it needs to be demonstrated that thrown spears achieved enough energy to be able to penetrate animal skins. More experiments testing the already proposed variables must be performed, and clear differences (in point morphology or in the fatigue traces) must be determined before we can state that Neanderthals were regular range weapon users.

From an evolutionary perspective this question is quite important. From our perspective the argument of ranged weapon use cannot be simplistically brought into play to discuss adaptive differences between Neanderthals and modern humans. Neanderthals were efficient hunters and they probably used close-ranged weapons. However we cannot ignore that in EUP in Europe much more lighter points and new kinds of weapons (bone points, multi-composite points) were developed, indicating improvements in hunting techniques and strategies.

References

Altuna, J. (1989). La subsistance d'origine animal pendant le Moustérien dans la région Cantabrique (Espagne). In M. Pathou & L. G. Freeman (Eds.), *L'Homme de Neandertal. La Subsistance. Actes du colloque international de Liège* (Vol. 6, pp. 41–43). Liège: ERAUL.

Altuna, J., Armendariz, Á., Etxeberria, F., Mariezkurrena, K., Peñalver, X., & Zumalabe, F. J. (1995). *Carta arqueológica de Gipuzkoa II. Cuevas* (Munibe Suplemento, Vol. 10). San Sebastián: Sociedad de Ciencias Aranzadi.

Altuna, J., & Mariezkurrena, K. (2010). Tafocenosis en yacimientos del País Vasco con predominio de grandes carnívoros. Consideraciones sobre el yacimiento de Amalda. *Zona Arqueológica, 13*, 214–228.

Baldeón, A. (1993). El yacimiento de Lezetxiki (Gipuzkoa, Pais Vasco). *Los niveles musterienses. Munibe, 45*, 3–97.

Bermúdez de Castro, J. M., & Sáenz de Buruaga, A. (1997). Étude préliminare du site Pléistocène supérieur à hominidé d'Arrillor (Pays Basque, Espagne). *L'Anthropologie, 103*, 633–639.

Bocherens, H. (2009). Neanderthal dietary habits. Review of isotopic evidence. In J.-J. Hublin & M. Richards (Eds.), *The evolution of hominin diets* (pp. 241–250). The Netherlands: Springer.

Callow, P. (1986). The flint tools. In P. Callow & J. Cornford (Eds.), *La Cotte de St. Brelade* (pp. 251–314). Norwich: Geo Books.

Castaños, P. (2005). Revisión actualizada de las faunas de macromamíferos del Würm antiguo en la Región Cantábrica. In R. Montes Barquín & J. A. Lasheras Corruchaga (Eds.), *Actas de la Reunión científica: Neandertales Cantábricos. Estado de la cuestión* (pp. 201–207). Madrid: Ministerio de Cultura.

Churchill, S., & Rhodes, J. (2009). The evolution of human capacity for "Killing at a distance". The human fossil evidence for the evolution of projectile weaponry. In J.-J. Hublin & M. Richards (Eds.), *The evolution of hominin diets* (pp. 201–210). The Netherlands: Springer.

Costamagno, S., Liliane, M., Cédric, B., Bernard, V., & Bruno, M. (2006). Les Pradelles (Marillac-le-Franc, France): A Mousterian reindeer hunting camp? *Journal of Anthropological Archaeology, 25*, 466–484.

Delagnes, A., Jaubert, J., & Meignen, L. (2007). Les technocomplexes du Paléolithique moyen en Europe occidentale dans leur cadre diachronique et géographique. In B. Vandermeersch & B. Maureille

(Eds.), *Les Néandertaliens. Biologie et culture* (pp. 213–230, Collection des documents préhistoriques Vol. 23). Paris: Comité des travaux historiques et scientifiques.

Deschamps, M. (2010). Le Vasconien: Révision de sa signification à partir des industries lithiques d'Olha I et II, d'Isturitz et de Gatzarria. *Paléo, 21*, 103–126.

Dockall, J. E. (1997). Wear traces and projectile impact: A review of the experimental and archaeological evidence. *Journal of Field Archaeology, 24*, 321–331.

Fischer, A., Hansen, P. V., & Rasmussen, P. (1984). Macro and micro wear traces on lithic projectile points. *Journal of Danish Archaeology, 3*, 19–46.

Galván Santos, B., Hernández Gómez, C. M., & Francisco Ortega, I. (2007–2008). Elementos líticos apuntados en el Musteriense Alcoyano el Abric del Pastor (Alicante). *Veleia, 24–25*, 367–383.

Gardeisen, A. (1999). Middle Palaeolithic subsistance in the West Cave of 'Le Portel' (Pyrénées, France). *Journal of Archaeological Science, 26*, 1145–1158.

González Urquijo, J. E., & Ibáñez Estévez, J. J. (2002). Abrigo de Axlor (Dima). *Arkeoikuska, 2001*, 90–93.

González Urquijo, J. E., Ibáñez Estévez, J. J., Rios-Garaizar, J., & Bourguignon, L. (2006). Aportes de las nuevas excavaciones en Axlor sobre el final del Paleolítico Medio. In V. Cabrera Valdés, F. Bernaldo de Quirós & J. M. Maíllo (Eds.), *Ante el Centenario de la Cueva de El Castillo: el Ocaso de los Neandertales* (pp. 269–291). Santander: UNED.

González Urquijo, J. E., Ibáñez Estévez, J. J., Rios-Garaizar, J., Bourguignon, L., Castaños Ugarte, P., & Tarriño Vinagre, A. (2005). Excavaciones recientes en Axlor. Movilidad y planificación de actividades en grupos de neandertales. In R. Montes Barquín & J. A. Lasheras Corruchaga (Eds.), *Actas de la Reunión científica: Neandertales Cantábricos. Estado de la cuestión* (pp. 527–539, Monografías del Museo Nacional y Centro de Investigación de Altamira, Vol. 20). Madrid: Ministerio de Cultura.

Grünberg, J. M. (2002). Middle Palaeolithic birch-bark pitch. *Antiquity, 76*, 15–16.

Hughes, S. (1998). Getting to the point: Evolutionary change in prehistoric weaponry. *Journal of Archaeological Method and Theory, 5*, 345–408.

Iovita, R., Schönekeß, H., Gaudzinski-Windheuser, S., & Jäger, F. (2016). Identifying weapon delivery systems using macrofracture analysis and fracture propagation velocity: A controlled experiment. In R. Iovita & K. Sano (Eds.), *Multidisciplinary approaches to the study of Stone Age weaponry* (pp. 13–27). Dordrecht: Springer.

Jaubert, J., & Delagnes, A. (2007). De l''space parcouru à l''space habité au Paléolithique moyen. In B. Vandermeersch & B. Maureille (Eds.), *Les Néandertaliens. biologie et culture* (pp. 263–282), Collection des documents préhistoriques Vol. 23. Paris: Comité des travaux historiques et scientifiques.

Lazuén, T. (2012). European neanderthal stone hunting weapons reveal complex behavior long before the appearance of modern humans. *Journal of Archaeological Science, 39*, 2304–2311.

Moncel, M.-H., Chacón, M. G., Coudenneau, A., & Fernandes, P. (2009). Points and convergent tools in the European Early Middle Palaeolithic site of Payre (SE, France). *Journal of Archaeological Science, 36*, 1892–1909.

Mozota, M. (2012). *El hueso como materia prima: el utillaje óseo del final del Musteriense en el sector central del Norte de la Península Ibérica. Ph. D. dissertation*. Spain: Universidad de Cantabria.

Oakley, K. P., Andrews, P., Keeley, L. H., & Clark, J. D. (1977). A reappraisal of the Clacton spearpoint. *Proceedings of the Prehistoric Society, 43*, 13–30.

Plisson, H., & Beyries, S. (1998). Pointes ou outils triangulaires? Données fonctionnelles dans le Moustérien levantin. *Paléorient, 24*, 5–16.

Rhodes, J. A., & Churchill, S. E. (2009). Throwing in the Middle and Upper Paleolithic: Inferences from an analysis of humeral retroversion. *Journal of Human Evolution, 56*, 1–10.

Rieder, H. (2009). Zur Qualität der Schöninger Speere als Jagdwaffen - Aus der Sicht der Sportwissenschaft. In H. Thieme (Ed.), *Die Schöninger Speere. Mensch und Jagd vor 400000 Jahren* (pp. 159–162). Hannover: Theiss.

Rios-Garaizar, J. (2005). Características de la producción lítica al final del Paleolítico Medio en el País Vasco. El caso del nivel B de Axlor (Dima, Bizkaia). In R. Montes Barquín & J. A. Lasheras Corruchaga (Eds.), *Actas de la Reunión Científica: Neandertales Cantábricos. Estado de la Cuestión. Monografías del Museo Nacional y Centro de Investigación de Altamira n° 20* (pp. 333–348). Madrid: Ministerio de Cultura.

Rios-Garaizar, J. (2008). Nivel IX (Chatelperroniense) de Labeko Koba (Arrasate-Gipuzkoa): Gestión de la industria lítica y función del sitio. *Munibe (Antropologia-Arkeologia), 59*, 25–46.

Rios Garaizar, J. (2009). Variabilidad tecnológica en el Paleolítico Medio de los Pirineos Occidentales: una expresión de las dinámicas históricas de las sociedades neandertales. *Treballs d''rqueologia, 14*, 172–195.

Rios Garaizar, J. (2010). *Organización económica de las sociedades Neandertales: el caso del nivel VII de Amalda (Zestoa, Gipuzkoa)* (pp. 15–37). LXV: Zephyrus.

Rios-Garaizar, J. (2012a). *Industria Lítica y Sociedad en la Transición del Paleolítico Medio al Superior en Torno al Golfo de Bizkaia*. Santander: PUbliCan – Ediciones de la Universidad de Cantabria.

Rios-Garaizar, J. (2012b). Técnicas de caza en el Paleolítico Medio del País Vasco. *Isturitz. Cuadernos de Sección. Prehistoria-Arqueologia, 12*, 7–37.

Rios-Garaizar, J., Arrizabalaga, Á., & Villaluenga, A. (2012a). Haltes de chasse du Châtelperronien de la Péninsule Ibérique. Labeko Koba et Ekain (Pays Basque Péninsulaire). *L'Anthropologie*. doi: http://dx.doi.org/10.1016/j.anthro.2012.10.001.

Rios-Garaizar, J., Libano Silvente, I., & Garate Maidagan, D. (2012b). El yacimiento chatelperroniense de Aranbaltza (Barrika, Euskadi). *Munibe (Antropologia-Arkeologia), 63*, 81–92.

Roebroeks, W., Sier, M. J., Nielsen, T. K., De Loecker, D., Parés, J. M., Arps, C. E. S., et al. (2012). Use of red ochre by early Neandertals. *Proceedings of the National Academy of Sciences*. doi:10.1073/pnas.1112261109.

Shaw, C. N., Hofmann, C. L., Petraglia, M. D., Stock, J. T., & Gottschall, J. S. (2012). Neanderthal humeri may reflect adaptation to scraping tasks, but not spear thrusting. *PLoS ONE, 7*, e40349.

Shea, J. J. (2006). The origins of lithic projectile point technology: evidence from Africa, the Levant, and Europe. *Journal of Archaeological Science, 33*, 823–846.

Shea, J. J., Davis, Z., & Brown, K. (2001). Experimental tests of Middle Palaeolithic spear points using a calibrated crossbow. *Journal of Archaeological Science, 28*, 807–816.

Shea, J. J., & Sisk, M. L. (2010). Complex projectile technology and *Homo sapiens* dispersal into Western Eurasia. *PaleoAnthropology, 2010*, 100–122.

Sisk, M. L., & Shea, J. J. (2009). Experimental use and quantitative performance analysis of triangular flakes (Levallois points) used as arrowheads. *Journal of Archaeological Science, 36*, 2039–2047.

Sisk, M. L., & Shea, J. J. (2011). The African origin of complex projectile technology: An analysis using tip cross-sectional area and perimeter. *International Journal of Evolutionary Biology, 2011*. doi:10.4061/2011/968012.

Soressi, M., & Locht, J. L. (2010). Les armes de chasse de Neandertal Première analyse des pointes moustériennes d'Angé. *ARCHÉO-PAGES, 28*, 6–11.

Thieme, H. (1997). Lower Palaeolithic hunting spears from Germany. *Nature, 385*, 807–810.

Thieme, H. (1999). Altpaläolithische Holzgeräte aus Schöningen, Lkr. Helmst edt: bedeutsame Funde zur Kulturentwicklung des frühen Menschen. *Germania, 77*, 451–487.

Trinkaus, E. (2008). Behavioral implications of the Muierii 1 early modern human scapula. *Annuaire Roumain d'Anthropologie, 48*, 27–41.

Villa, P., Boscato, P., Ranaldo, F., & Ronchitelli, A. (2009). Stone tools for the hunt: Points with impact scars from a Middle Paleolithic site in southern Italy. *Journal of Archaeological Science, 36*, 850–859.

Villa, P., & Lenoir, M. (2006). Hunting weapons of the Middle Stone Age and the Middle Palaeolithic: Spear points from Sibudu, Rose Cottage and Bouheben. *Southern African Humanities, 18*, 89–122.

Villa, P., & Lenoir, M. (2009). Hunting and hunting weapons of the Lower and Middle Paleolithic of Europe. In J.-J. Hublin & M. Richards (Eds.), *The evolution of hominin diets* (pp. 59–85). The Netherlands: Springer.

Yravedra Sainz de los Terreros, J. (2007). Nuevas contribuciones en el comportamiento cinegético de la Cueva de Amalda. *Munibe: Antropología y arqueología, 58*, 43–88.

Part IV
Weapons as Curated Technologies

Chapter 16
More to the Point: Developing a Multi-faceted Approach to Investigating the Curation of Magdalenian Osseous Projectile Points

Michelle C. Langley

Abstract The majority of osseous projectile points recovered from archaeological sites were intentionally discarded by their owners in prehistory because they were considered no longer usable. This usability being determined by both functional (physical ability to effectively penetrate game) and cultural (ideals about form and efficiency) constraints. While a significant amount of research into Magdalenian osseous projectile points has been undertaken, very few studies have considered the processes which lead to their discard. This paper highlights this underdeveloped avenue of research and outlines potential methods of investigating osseous projectile point reduction and curation.

Keywords Chaîne opératoire • Curation • Discard • Magdalenian • Osseous • Projectile • Reduction

Introduction

As Dibble (1995: 303) succinctly put it, "artifacts are analyzed to understand not only why and how they were manufactured, but also why they were thrown away". The majority of osseous projectile points recovered from archaeological sites were intentionally discarded by their owners in prehistory because they were considered no longer usable. This useability was determined by both functional (physical effectiveness to penetrate) and cultural (ideals about form and efficiency) constraints. Identifying how intensively implements were repaired before discard, that is, how intensely they were curated, will allow us to draw inferences about potential cultural and environmental factors that may have impacted on the subsistence and technological choices of the populations under study.

Unfortunately, osseous point repair and curation (unlike their lithic counterparts) is an issue that has only been briefly mentioned in analyses of implements from various prehistoric cultural contexts (for example: Tyzzer 1936; Guthrie 1983; Knecht 1991, 1993a, b, 1997; Redmond and Tankersley 2005; Liolios 2006; Moore and Schmidt 2009; though see Liolios 1999; Pétillon 2002; Christensen and Chollet 2005). Consequently, development of a robust methodology for the identification of reduced implements and the contextual investigation of osseous point curation in prehistory remains in its infancy.

The aim of this paper is to highlight this underdeveloped avenue of research and to outline potential methods for investigating osseous projectile point curation. It will focus on examples from the Magdalenian as its rich assortment of osseous projectile points provides the perfect starting point for establishing the potential of investigating the rejuvenation and curation of these implements. Furthermore, the most progress in establishing osseous point curation research has been in the study of Upper Paleolithic, primarily Aurignacian and Magdalenian, assemblages. The purpose of this paper is not to provide an exhaustive outline of how the investigation of curation in osseous projectile points can be undertaken, but simply to provide a starting point for discussions.

Curation and Use Life

Curation, as originally outlined by Binford (1979: 263), is "the practice of maximizing the utility of tools by carrying them between successive settlements". Over the following years, the use of the concept of curation in technological

M.C. Langley (✉)
Institute of Archaeology, University of Oxford, Oxford, OX2 PG1, UK
e-mail: michelle.langley@anu.edu.au

and

Archaeology and Natural History, School of Culture History and Language, College of Asia and the Pacific, Australian National University, Acton ACT 0200, Australia

Radu Iovita and Katsuhiro Sano (eds.), *Multidisciplinary Approaches to the Study of Stone Age Weaponry*, Vertebrate Paleobiology and Paleoanthropology, DOI 10.1007/978-94-017-7602-8_16

analyses came to involve a number of different aspects including: the production of implements in advance of use; the design of implements for multiple uses; the transport of implements from location to location; tool recycling; efficiency; and tool maintenance (see Bamforth 1986 and Odell 1996 for discussions). The aspect of curation that is of interest in this paper is how projectile points were maintained (repaired, resharpened, reworked etc.) and ultimately, to determine at what point Magdalenian hunters eventually discarded these points and why. From here, the term curation will be used to refer to this process of tool maintenance.

A concept which is often related to curation is "use life" . Use life describes the length of service of tool classes in systemic context (Schiffer 1976; Shott 1989). It can be measured through a variety of methods, including the number of strokes or shots fired (e.g., Gallagher 1977; Hayden 1979; Weedman 2000); however, Shott (1989) argues that use life is probably best expressed as a function of time owing to the fact that this treatment ensures agreement between researchers and also because formation rates of archaeological assemblages are likewise examined in this way (see Shott, 2016 for discussion of this point).

Shott and Sillitoe (2005) have noted that archaeologists often equate use life with curation, "on the logic that things that last long times are highly curated and things that are used and wear out or break quickly are little curated" (2005: 654). Theoretically, however, curation and use life are distinct concepts: use life simply being how long an implement lasts and curation being how extensively an implement is reduced to exhaustion. Therefore, an implement can be both short lived and highly curated or long lived and little curated.

The curation (resharpening, reworking, retooling) of equipment including projectile points, is an important part of technological systems as the investment of labor in this aspect may increase the use life of individual implements, and, in turn, increase the total efficiency of the whole technological system (Binford 1977). Rates of curation of individual implements or whole tool categories can be influenced by shifts in mobility strategies caused themselves by changes in environmental and/or social conditions (Rondeau 1996), as well as differences in cultural notions of form and efficiency (Nelson 1997).

Differing rates of curation for both individual implements and tool types also influence the size and composition of archaeological assemblages; with well-curated types being discarded less often than poorly curated ones for equal amounts of use (Ammerman and Feldman 1974; Shott 1989, 2016; Shott and Sillitoe 2005). In particular, Ammerman and Feldman (1974) have shown that the frequency of a tool class (such as antler bilaterally barbed points or stone scrapers) is the joint product of its use frequency and use life, and that use life is, in itself, an independent factor conditioning the content of archaeological assemblages. In Magdalenian contexts, while the scarcity of spearthrowers (Averbouh and Cattelain 2002) and knapping hammers (Averbouh 1999) have already been considered in this light, osseous projectile points remain to be considered.

It is therefore apparent that determining the range and average use life of a projectile point type, how they are curated, if they were curated differentially through time and across space and at what point they are commonly discarded may provide essential data for building a comprehensive understanding of both the social and economic factors which result in the archaeological assemblages which are available for study today.

Magdalenian Osseous Projectile Points

The Magdalenian is the last of the Upper Paleolithic archaeologically defined cultures to appear in the European Pleistocene archaeological record and dates to between 20,500 and 14,000 cal BP being divided into three phases – Lower, Middle and Upper (Langlais et al. 2012). It is found throughout Western Europe, but concentrated in France, Spain, Belgium and Germany with sites identified in a wide variety of topographical settings, from the limestone rock-shelters and caves of the Périgord to the relatively open river valleys and plains of northern Europe. Its assemblages are rich in osseous (bone, antler, ivory – but overwhelmingly antler in the majority) weapons technologies including items such as *sagaies* (termed "unbarbed" points here and include single bevelled, double bevelled and fork-based points), unilaterally and bilaterally barbed points (*harpons* in the French literature), self-barbed points, Half Round Rods (HRRs or *baguettes demi rondes* in the French literature), *Foënes* and composite antler/lithic projectile points. These implements are indicative of this archaeological culture and, as an implement that is central to obtaining sustenance, provide a source of information concerning various aspects of Magdalenian social and economic life.

Research into Magdalenian osseous projectile points has resulted in a thorough understanding of how these implements were manufactured (Julien 1977, 1982, 1999; Allain and Rigaud 1986, 1992; Weniger 1992, 1995, 2000; Averbouh 2005; Pétillon 2005, 2006, 2008, 2016; Rigaud 2006; Pétillon et al. 2011) and has demonstrated their efficiency in exploiting both terrestrial and aquatic species (Stodiek 1990, 1991, 1993, 2000; Rozoy 1992; Pokines 1993, 1998; Pokines and Krupa 1997; Pétillon 2005, 2006; Letourneux and Pétillon 2008; Pétillon et al. 2011). While ethnographic survey and experimentation has demonstrated the superiority of osseous projectile points over their lithic counterparts in terms of ease of maintenance, durability in colder climates and reliability against large and/or dangerous animals (see Ellis 1997; Elston and Brantingham 2002 for discussion), how long the absolute use life of these implements extends and the physical parameters which lead to their discard remains undetermined.

Previous Work in Investigating Magdalenian Osseous Point Curation

Most studies of Magdalenian points have been limited to simply observing that many implements, particularly barbed points, have been subject to rejuvenation in their life history (Julien 1977, 1982; Arndt and Newcomer 1986; Barandiaran 1987; Dobres 1995; Weniger 2000; Pétillon 2005, 2006). Elsewhere it has been recognized that these implements were subject to repair (sometimes extensively), as Michèle Julien (1982: 134) noted: "the calculations carried out on the overall lengths of the harpoons and the proportions of components compared to others, only reveal their state of abandonment rather than the initial design of the instrument" (authors translation). Weniger (2000: 82) similarly remarks that "original forms are very rare. Even points that seem to be quite well preserved have been reworked because of former damages and do not display a primary form. The Magdalenian barbed points were well prized and once damaged one tried to repair the object or at least to transform it into another secondary form that could be useful".

Michèle Julien took the first steps towards developing a quantitative approach to determining the extent of osseous point repair in her extensive 1982 study of Magdalenian barbed points. In this study she used four metric variables to distinguish between two sub-sets of points: (1) "*à extrémité primaire*", those points preserving their initial dimensions and (2) "*à extrémité secondaire*", those points whose dimensions are reduced through utilization and repair. Julien and Orliac (2003) further examined the rejuvenation of Magdalenian barbed points through the examination of the extensive La Vache collection. The authors found that 22.3% (55/246) of these implements exhibited traces of reworking, often to more than one segment (distal, proximal, mesial [including barbs]) on the same implement rejuvenated. Detailed description of how this repair was undertaken in prehistory (i.e., ground, shaved with a lithic blade, with care to create a new, smooth surface or less carefully only creating a new sharp point) is not included however.

Similarly, Christensen and Chollet (2005) in their study of the osseous technological assemblage from Grotte du Bois-Ragot took particular note of those barbed points which exhibited evidence for maintenance (resharpening of the distal extremity and the removal of barbs or base protuberances). This last study is the most detailed thus far undertaken regarding the particularities of barbed point maintenance.

Pétillon (2002) further developed this approach through focusing on Magdalenian fork-based points and using four quantitative and qualitative variables (mesio-distal length, mesial section, distal section and the present/absence of traces of repair). Through the use of these variables, Pétillon determined that the morphometric variability in fork-based points was owing to the reduction of these implements through utilization and repair. He further worked up this idea in his 2006 monograph, in which he proposed a sequence through which the fork-based points recovered from Isturitz may have passed. This schema begins with an intact point and outlines whether a point would be repaired, recycled or abandoned according to the damage sustained during use. This outline is the closest thing to a study of the later stages of the *chaînes opératoire* of osseous projectile points thus far assembled; though, it is restricted to fork-based points (Pétillon 2006: 170, Fig. 152).

Inter-site and interregional comparison of how intensely osseous point assemblages were repaired before discard, similarly remains at a preliminary stage with only two studies having considered this aspect. Dobres (1995) reportedly studied between 15 and 20% of the barbed point assemblage from La Vache and found that about 40% of those in her sample showed evidence for repair which she notes, in both absolute numbers and proportionally, is significantly more artifact repair than seen at any other site in the region. It should be noted; however, that Julien and Orliac (2003) found that only 22.3% of the barbed points from this same site exhibited evidence for maintenance. Pétillon's (2002) study of fork-based points, meanwhile, found that the points from Isturitz, Gourdan and Lortet were unable to be distinguished between the sites.

Research into Magdalenian osseous point curation has consequently been largely restricted to identifying that these implements were subject to repair, often extensively, and that this process needs to be recognized when undertaking typological analyses. It is yet to tackle whether there are significant differences in the methods used to repair these implements and in the rate of discard between sites, regions and even time slices (e.g., Middle versus Late Magdalenian). Determining differences or similarities between sites, regions and periods will enable us to construct a deeper understanding of both the social and technological choices Magdalenian populations made while interacting with their environment.

Developing Osseous Point Curation Research

Investigating how Magdalenian osseous projectile points were curated by their owners, will require the integration of several distinct but related datasets. Owing to difficulties often encountered in differentiating between initial and extensively reworked forms, emphasis must be placed on collecting large comparative datasets through which population trends can be identified. Experimental and use wear data will also play a central role in this analysis. Other lines of evidence, including assemblage composition, the location of decorations and striations on an implement, and ethnographic data will full fill a supportive role. It is hoped that through the integration of several lines of evidence, at least a basic understanding of how these implements were curated as a collective as well as differences and similarities between sites and regions can be achieved.

Metric Analysis and Population Distributions

While the identification of points as having been repaired and rejuvenated is common, the development of a *quantitative* approach which determines how intensively these implements were repaired and reduced is yet to be constructed. For the analysis of osseous projectile points, as with their lithic counterparts, it is difficult to ascertain where along the use-continuum a particular artifact is without knowledge of the original form of the particular point, and, as Weniger (2000) has remarked, the extensive reworking of points throughout their use life obscures the boundaries between point types and makes morphometric analysis difficult. It is these two issues which have primarily led researchers to believe that identifying the extent of curation of these implements is troublesome.

It is true that by looking at a single artifact it is often impossible to ascertain more than the minimum about its restoration: has it been resharpened/reworked or not? Which sections have been reworked and how? Unfortunately, owing to the nature of osseous point production study of the manufacturing debris is not able to provide the detailed data required for this issue (see Averbouh 2001). However, these problems might be circumvented by approaching the problem from an alternative view point, that is, by focusing on assemblage population distributions.

Analysis of extensive lithic projectile point datasets has allowed researchers to determine the approximate average original length of projectile points and make inferences about the extent to which these implements were curated (e.g., Grimes and Grimes 1985; Shott and Sillitoe 2005; Buchanan 2006). A similar approach to osseous points might prove insightful, the potential for which can be highlighted by data previously presented in the work of Michèle Julien (1982) and Jean-Marc Pétillon (2002).

Both of these studies produced histograms presenting the distribution of total length for the assemblages under study. Julien (1977: 179, 1982: 26) presents histograms of the maximum length for both unilateral and bilaterally barbed points, and interprets these data as the combined result of raw material constraints on initial manufacture, intentional design and the impact of point utilization and repair stating that the short points were the result of repeated use and repair. While keeping in mind that some of this variation is no doubt the result of intentional design (smaller implements for smaller game etc.), these data can be further developed in order to demonstrate the potential of dataset collected and organized particularly for this kind of analysis (Fig. 16.1).

The distribution for unilateral and bilaterally barbed points can be split into three groups: 31–81 mm, 81–161 mm, and 161–330 mm for unilateral (with peaks at 60 mm, 100 mm and 140 mm) and 31–81 mm, 81–141 mm, and 141–230 mm for bilaterally barbed points (with peaks at 70 mm, 100 mm, and 120 mm). These apparent groupings could be interpreted in the following manner: the longest points (161–330 mm for unilateral and 141–230 mm for bilateral) may reflect those implements that are closest to their original manufactured state, i.e., points that were discarded after little use (accidental loss of semi-new points or deliberate abandonment of little used points) and/or cached "complete" points. It is true that if a point is expected to undergo several resharpening episodes throughout its use life, then it would have to be long enough and thick enough to allow loss of edge due to resharpening (Christenson 1986), and that newly manufactured points should be quite rare in the archaeological record.

The middle range points (81–161 mm for unilateral and 81–141 mm for bilateral) includes the large majority of pieces examined in the analysis, and therefore may reflect the most common range in which points were discarded. The last grouping (31–81 mm for unilateral and 31–81 mm for bilateral) include the fewest points in the sample, and therefore may represent those cases which were more intensively reduced owing to what Binford (1979: 267) has termed "situational circumstances". That is, if while out in the field hunters did not have access to sufficient replacement points, and may have therefore reduced their points more intensively than would otherwise be normal habit. Another explanation is provided by recent research by Averbouh (2005) who has argued that osseous points were manufactured during the winter months, a suggestion backed up by ethnographic analogy. It might be therefore suggested that, as in human groups known through ethnography, people came together during the periods of the year when resources were plentiful, either at good fishing locations or along reindeer migration routes (Weniger 1989). When this period was drawing to a close at the end of the autumn, supplies of new antler points may have been running low, and existing points may have had to be reduced more extensively than would be otherwise desired. These more intensively reduced points would then be discarded and replaced once the

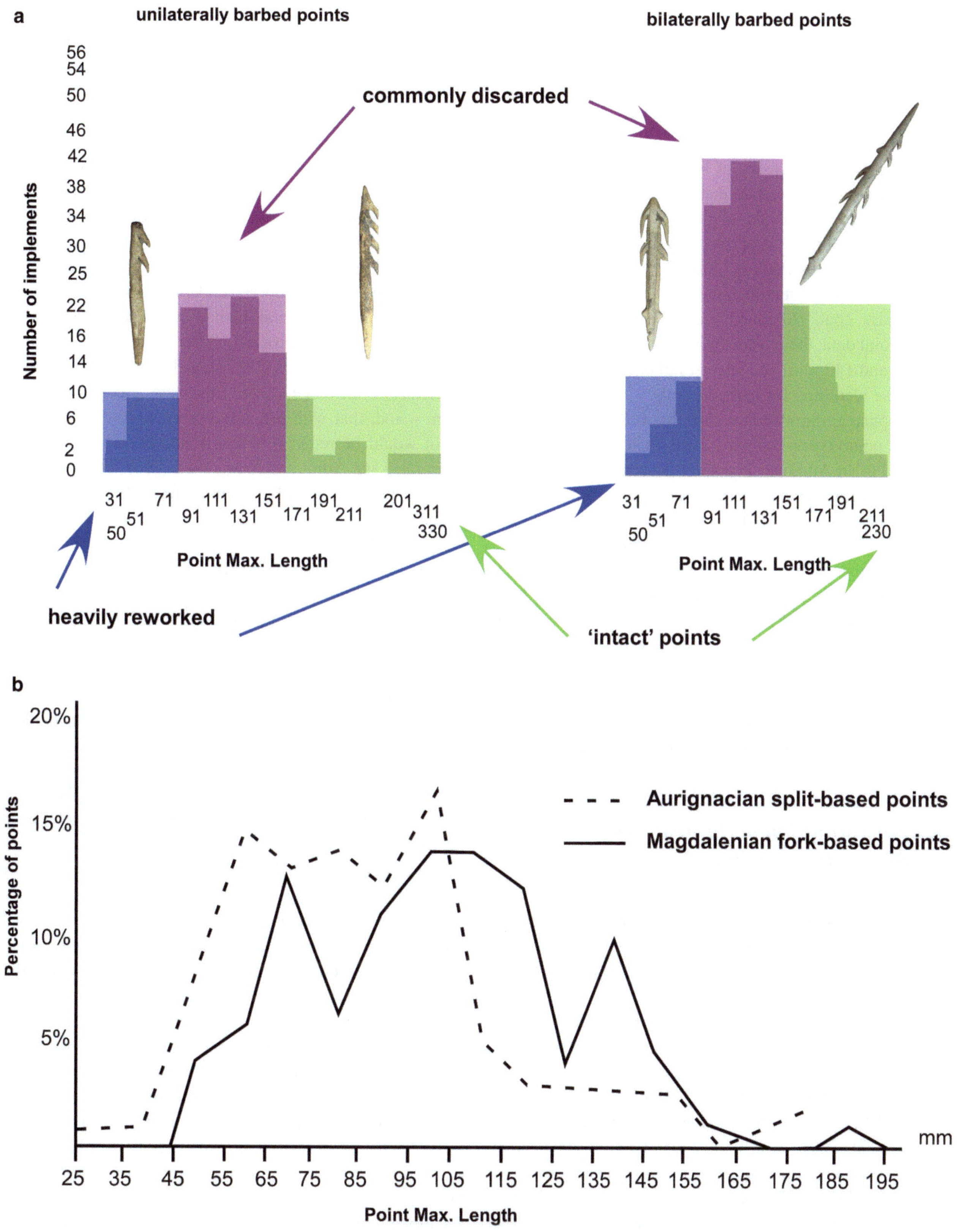

Fig. 16.1 Examples of population distribution data: **a** re-interpretation of Julien's (1982) barbed point maximum length data; **b** comparison of maximum length data for Magdalenian fork-based points (data from Pétillon 2002) and Aurignacian split-based points (data from Knecht 1991)

hunters had access to newly made points during the winter, and would result in a number of heavily reworked points entering the archaeological record.

While this is only an initial reinterpretation of previously collected data not intended for this kind of analysis, and therefore requires extensive testing along with the separation of regional data and classes of implements from the overall dataset, it does demonstrate that data collected specifically for this kind of analysis might prove worthwhile. This suggestion is further supported by a recent analysis of fork-based points by Pétillon (2002) in which identifying the consequences of point reduction through use and repair were the primary goal. He found that, as with Julien's (1982) barbed point data, there were three peaks in point total length (70, 100 and 140 mm) and interestingly these peaks coincide with those identified for barbed points.

The apparent correlation within the Magdalenian data for different point types becomes even more intriguing when compared to an Aurignacian dataset. Knecht (1991) studied a number of Aurignacian osseous projectile point types (also see Liolios 1999), and as part of this extensive analysis, presented the total length data for the split-based points studied. When the total length data for the "whole" points are plotted (n = 107), again three peaks can be identified. However, unlike the Magdalenian point types, these peaks fall at 55, 84 and 105—significantly shorter than those from the Magdalenian. Do these *very* preliminary data indicate that there was a significant difference in how points were curated between the Aurignacian and Magdalenian? Further in depth analysis which disentangles regional or site specific trends is the only way to find out.

Manufacturing Stigmata, Use Wear and Rejuvenation Stigmata

Through the examination of use traces, it will be possible to identify the range and most common methods of point rejuvenation within sites, regions and time periods as well as differences between point types.

Extensive research has been undertaken focused on the manufacturing methods and techniques used in prehistory to produce an osseous projectile point (e.g., Clark and Thompson 1953; Newcomer 1977; Allain and Rigaud 1986; Arndt and Newcomer 1986; Stodiek 1990, 1991, 1993, 2000; Rigaud 2004, 2006; Averbouh 2005; Pétillon 2005, 2006; Liolios 2006; Pétillon et al. 2011). Manufacturing stigmata (Newcomer 1974, 1977; Knecht 1991, 1993a, b, 1997), as well as the range and cause of impact fractures and use wear have been thoroughly examined through experiments and studies of archaeological collections (e.g., Tyzzer 1936; Arndt and Newcomer 1986; Stodiek 1990, 1991, 1993, 2000; Pétillon 2005, 2006; Buc 2011).

A number of researchers have also noted stigmata indicative of the resharpening and reworking on barbed, double-bevelled and forked-based projectile points in Magdalenian contexts (Julien 1982; Arndt and Newcomer 1986; Stodiek 1991, 1993, 2000; Weniger 2000; Pétillon 2005, 2006) and similar wear on points from Aurignacian and Paleoindian contexts as well as replicated experimental points have also been identified (Knecht 1991, 1993a, b, 1997; Redmond and Tankersley 2005; Moore and Schmidt 2009; Buc 2011).

While describing these manufacturing and use traces is too extensive to be properly undertaken here, a couple of examples for Magdalenian barbed points can illustrate how this analysis would proceed. After extensive analysis of these implements, Julien (1982) concluded that at the time of manufacture the distal section (tip) were of conical, piercing form – and, that with use and rejuvenation – the tip of the point would be reduced in length and become spatulate in form (also see Arndt and Newcomer 1986). Thus, while the original point is usually carefully formed, with smoothly ground surfaces up to a sharp tip, points which have been resharpened show one or more resharpening facets, often with striations caused by either the uneven edge of a lithic tool, a grindstone with coarse inclusions or the point was ground on a grindstone in conjunction with an abrasive (see Newcomer 1974 for example with bone points) (Figs. 16.2 and 16.3).

Additionally, it is often possible to observe where barbs or the scars of barbs broken on impact have been intentionally removed through grinding (Fig. 16.2). As Julien (1982: 134–135) originally observed, "…one can see scars carefully abraded from where the barbs were attached to the side… Where the attachment traces have completely disappeared, it is often possible to guess that these points were bilateral in the past thanks to the double protuberances of the base (bulb) and the type of decoration" and that "…in the rare cases, it is possible to suppose that the base was re-made, by shortening the length of the barbs and perhaps the barbed portion by removing one of the proximal barbs" (author's translation).

Julien's observations of Magdalenian barbed points further determined that these implements were sometimes extensively reworked, occasionally so much so that they crossed typological boundaries. Attempting to identify the frequency of this occurrence and whether it is more common in particular sites, regions or time periods would be interesting for determining differing approaches to typological boundaries in prehistory in addition to reduction methods. As Julien herself states, her data suggest that (at least some) Magdalenians were not averse to changing the functional properties of a point as it was reduced through breakage and rejuvenation. But how widespread was this approach? Determining the frequency and range of point rejuvenation

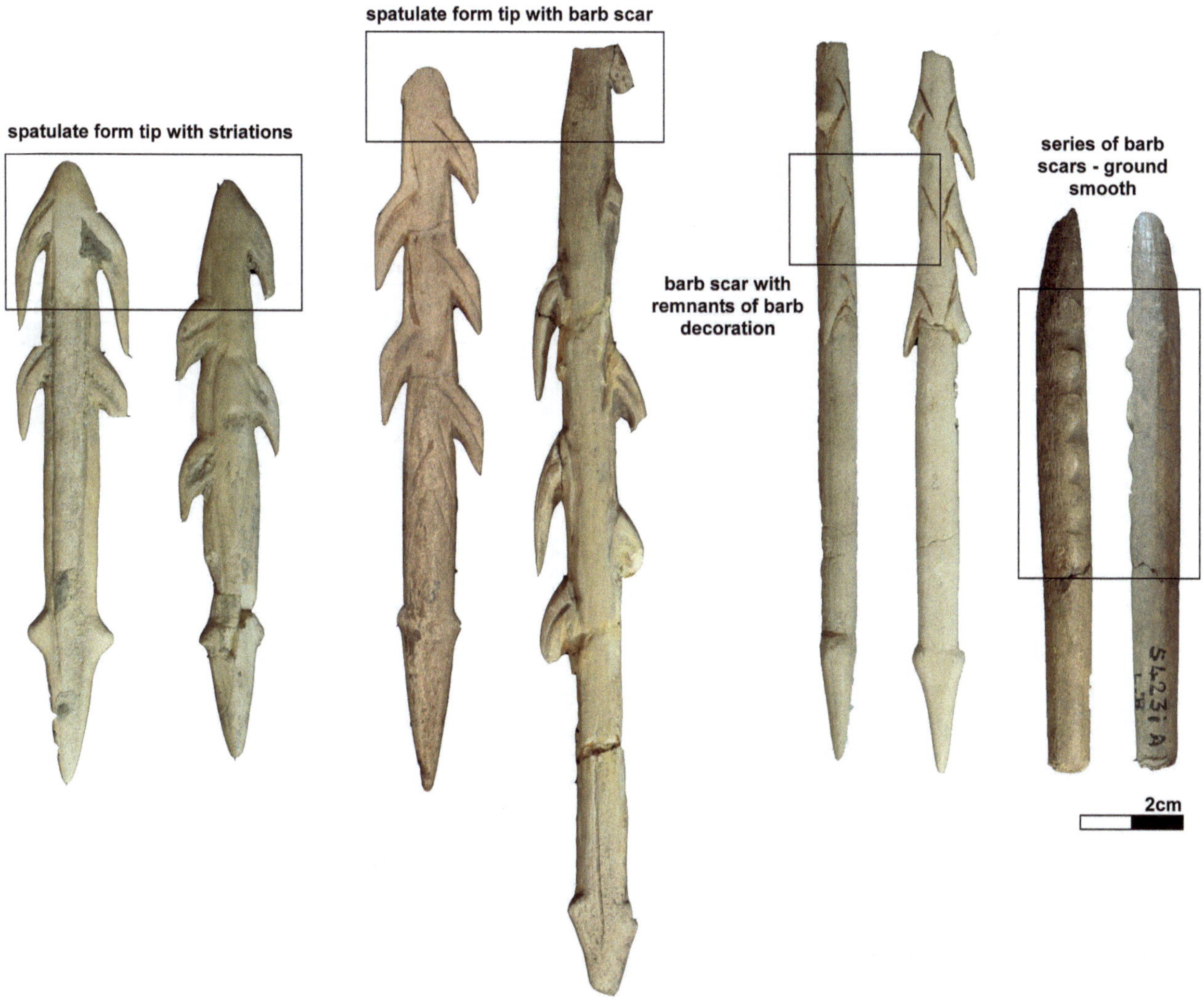

Fig. 16.2 Examples of evidence for maintenance. From left to right, two examples from La Vache, one from Courbet, two from La Madeleine, and the last from Laugerie-Basse (Photos: M. C. Langley by permission of the Trustees of the British Museum and the *Musée d'Archaéologie National*, Saint-Germaine-En-Laye)

methods in different assemblages may provide the insights in point curation which enables a greater understanding of this technological system.

Decoration and Striations

Features commonly associated with Magdalenian osseous projectile points are incised linear striations and engraved decoration. Striations are frequently present on the basal section (proximal, distal or proximal and distal sections of the base) of a point to aid hafting (Allain and Rigaud 1986; Julien 1999; Weniger 2000; Pétillon 2008) and have also been identified on the barbed section of points (along with grooves) which Julien (1999) argues was to assist the adherence of lithic bladelets to the antler shaft. Many of these points, particularly barbed points, were also decorated with engraved geometric patterns and figurative depictions (Conkey 1980; Allain and Rigaud 1986; Straus 1992). Ethnographic studies have shown that projectile points can have significance in boundary maintenance (Wiessner 1983) and archaeologists have suggested that the proliferation of art and the extensive decoration of weapons reflect an increase in social tensions and a considerable amount of social competition during the Magdalenian (Geist 1978; Bahn 1982). Additionally, and of the most interest for the current topic, the creation of distinctive barb types and decorations suggests that these points were intended for an extended use life as well as being the product of different communities of practice (Laurent 1974).

Fortunately, in some cases these features can be used to help identify rejuvenated points. For example, Fig. 16.3 shows three examples where decoration can be used to help

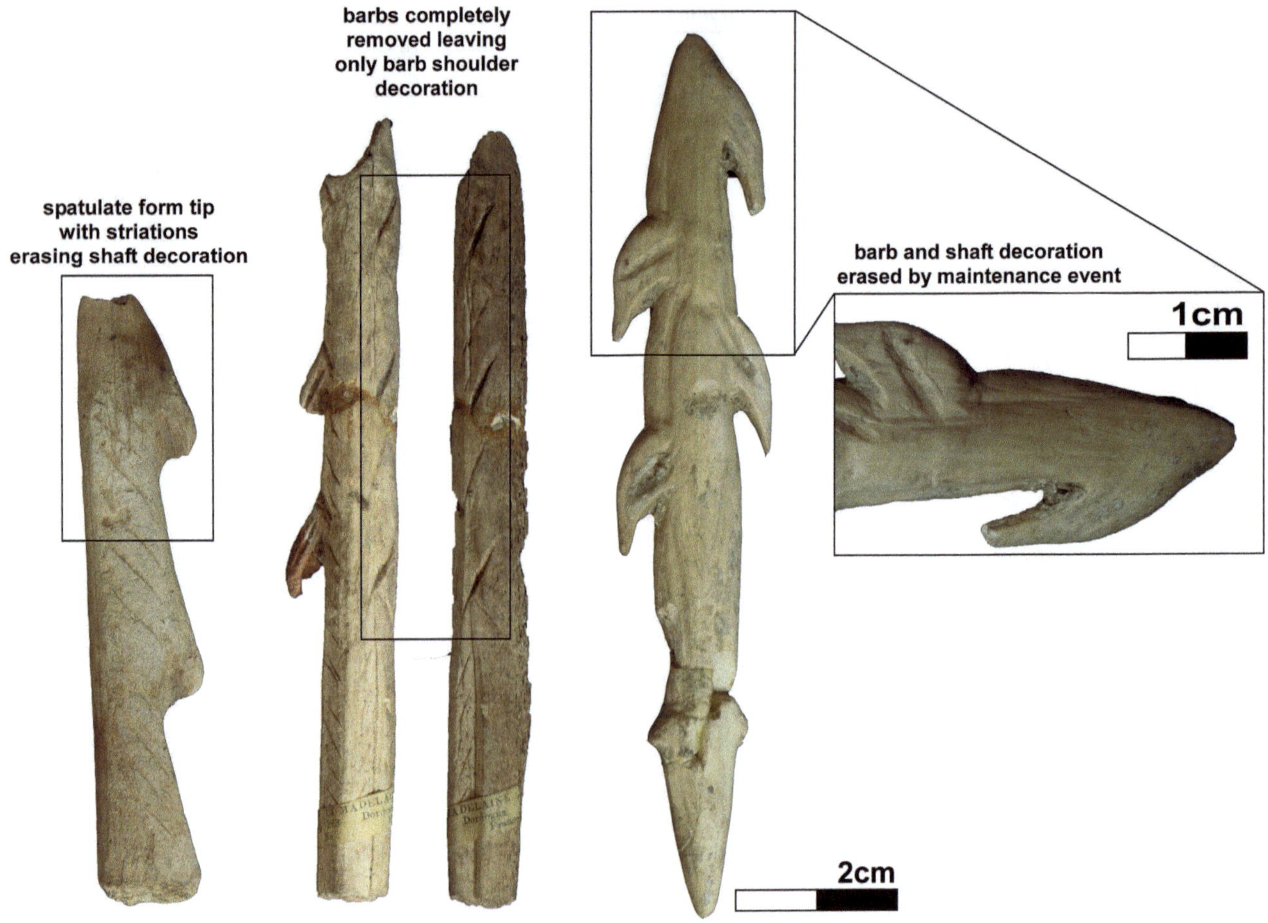

Fig. 16.3 Examples of where shaft and barb decoration can be used to indicate maintenance. First and third examples from La Vache. Middle example from La Madeleine (Photos: M. C. Langley by permission of the *Musée d'Archaéologie National*, Saint-Germaine-En-Laye)

identify implements which have undergone maintenance. The first is an example of where striations from a resharpening event have encroached on shaft decoration (in this case, parallel oblique lines running down the superior face of the shaft). The second, a case where multiple barbs have been removed (ground smooth) leaving only the decoration associated with these features to indicate that they had once existed. The final example is a bilaterally barbed point which has also undergone maintenance to its distal extremity. Striations from this event are clearly visible and run down to the first barb on the left, consequently erasing the decoration on the top right barb (single line down middle of barb) and partially erasing the curved line engraved into the left and right sides of the shaft between barbs. This is particularly visible for what remains of the curved line on the left hand side of the shaft, just above the first barb.

Similarly, the location of striations (for the attachment of lithic barbs and aiding of hafting) on both the shaft and basal sections and their relative placement to other sections of the point may indicate if an implement has been reworked from its original state, though the erasure of original striations and the addition of new striations as part of this rejuvenation process will complicate this identification. Careful observation of point sections may; however, produce results.

Assemblage Composition

The composition of an archaeological assemblage and its comparison with neighboring spatial and temporal assemblages can be highly informative in terms of identifying different approaches to the curation of osseous and other technologies across space and through time.

Data that could be expected to be informative for investigating curation rates of osseous projectile points may include the proportions of different point types recovered in an assemblage (unbarbed versus barbed for example). As Binford (1973, 1977: 34) has argued, "important items are maintained and curated, thus their entry into the archaeological record, in terms of frequency, is inversely proportional to the level of maintenance and hence their technological importance, other things being equal." That is, that the negative evidence, those point types that are rare in an assemblage, may reflect point types that were highly curated against other point types which were less so. However, popularity of a point type and its frequency of use during prehistory must also be taken into consideration.

Quantity of point sections with impact fractures present in an assemblage in comparison to one another (e.g., proximal sections versus medial-proximal sections) along with analysis of the frequency of evidence for rejuvenation within and between section populations will also be informative. For example, when considering simple points, a proportionally higher quantity of proximal (base) fragments with an impact fracture on the distal end may indicate that this section frequently broke during use. Consequently, entirely "new" bases may be worked onto used points to continue their use life. As these "new" bases will have erased all indications of any previous base/s, the only evidence of this reworking event will be the broken sections brought back to the site in the haft (see Chadelle et al. 1991; Pétillon 2006), as well as the overall morphology of the point becoming squatter. The examination of projectile point section frequency in assemblages may therefore provide evidence for implement rejuvenation, even when rejuvenation stigmata are not overtly evident on the implement itself.

Previous work by Marcia-Anne Dobres, published fifteen years ago presents an example of where broad assemblage comparison may be insightful in examining curation. Dobres (1995) studied the osseous material from several Magdalenian sites, including needles, awls, polishers, *baguettes*, unbarbed and barbed points. She reportedly studied between 15 and 20% of the harpoon assemblage from La Vache and found that about 40% of those in her sample showed evidence of repair which she notes, in both absolute numbers and proportionally, is significantly more artifact repair than seen at any other site in the region. She found that, in her study sample of sites, different artifacts were repaired at different sites: harpoons at La Vache and Montfort and unbarbed points at Bédeilhac, Les Eglises, Montfort, Massat and La Vache. This study demonstrates the potential that in depth comparative analyses of unbarbed and barbed point curation within and between sites provides for producing a contextualized understanding of osseous weapons repair during the Magdalenian.

Finally, identifying and considering the presence/absence of artifacts which could be argued to be associated with the reworking and retooling of osseous projectile points, such as burins, ochre, bladelets, and grindstones may provide insightful information concerning the tool kit utilized to undertake this process and the sites where this work was undertaken.

These analyses will be complicated by post-depositional factors. As Allain and Rigaud (1986) have pointed out, implements with short proportions such as the Lussac-Angles points are often recovered intact while those with longer shaft proportions are rarely so. The post-depositional breakage of longer points will necessitate the investigation of possible refitting of point sections; however, this may not be possible as many museum collections from early excavations will not be include all point fragments recovered from the deposit as well as collections being spread between several museums. These factors will muddy the waters for investigating the curation of such implements from old excavations; however, those collections from (relatively) newly excavated sites will allow this kind of analysis to be carried out with greater confidence.

Furthermore, Arndt and Newcomer (1986) conclude that severely damaged points may be under-represented in archaeological assemblages owing to several factors, one of which being that they may have been resharpened or recycled into other tools. However, in the case of the latter occurrence, despite the barbed or simple point being reworked (e.g., into a wedge) or simply reused in another context (e.g., as an *outils intermediaries*), they are still recognizable as having once been projectile points and can therefore be included in analyses (see for examples: Deffarge et al. 1977; Allain and Rigaud 1986; Pétillon 2006). Each of these factors needs to be taken into account in future analyses of curation.

Experimental Studies

Experimental projectile point studies will be a key aspect of investigating the use life and curation of osseous projectile points. Through experimentation it will possible to determine the absolute use life of a point, that is, determine the point at which they become functionally unusable (termed "maximum utility" by Shott and Sillitoe 2005 or"potential utility" by Shott 1989). While several past researchers reportedly resharpened a number of points during their experiments (Arndt and Newcomer 1986; Knecht 1991, 1993a, b, 1997; Stodiek 1991, 2000; Nuzhnyi 1993), the precise number of points is often not specified and examples often appear to have been resharpened only once. Therefore, while these studies did demonstrate the effectiveness and durability of osseous projectile points, as well as ease of maintenance, they did not investigate the absolute use life (maximum utility) of the replicated points.

Experiments designed to determine the maximum utility of different osseous projectile point types by use, rejuvenation and reuse of a point to exhaustion (when the point is no longer able to function) are needed. Once the absolute use life (maximum utility) of a projectile point is determined quantitatively, we will be able to compare these data with archaeological collections and determine whether a given archaeological population was reducing their projectile points to absolute exhaustion or whether there was another stage at which points were commonly discarded – that is, demonstrate whether there was a cultural ideal about when a point was exhausted and no longer usable (termed "realized utility" by Shott and Sillitoe 2005). Of course, determining the absolute use life of a given projectile point type (double bevel based, fork-based, bilaterally barbed point, etc.) will require the careful consideration of the peculiarities of maintenance for that type. For example, barbed points may have barbs removed (ground down, leaving little or no trace) – either because they broke during use, or because the distal extremity required resharpening and the owner, rather than producing a shorter, spatulate form tip, preferred to remove one or more barbs to create a longer, more conical form tip. Familiarity with the types of maintenance evidence for the subject projectile point type within archaeological assemblages will allow for a more informed experiment procedure, which in turn, will result in the better identification of their absolute use life.

Parietal and Mobile Art

Images of material culture in both parietal and mobile art provide an additional source of information for both the appearance and use of implements in Pleistocene contexts (see Welch 1996 for an Australian example). Projectile points are depicted in both Magdalenian parietal and mobile art and include multiple examples from Lascaux, Niaux, Cosquer, La Madeleine, Isturitz and La Garenne, with additional examples found at Laugerie-Basse, La Colombière, and Cougnac (Sieveking 1987; Baffier 1990; Allain and Rigaud 1992). Self-barbed points along with both unilateral and bilaterally barbed points and possibly HRRs may

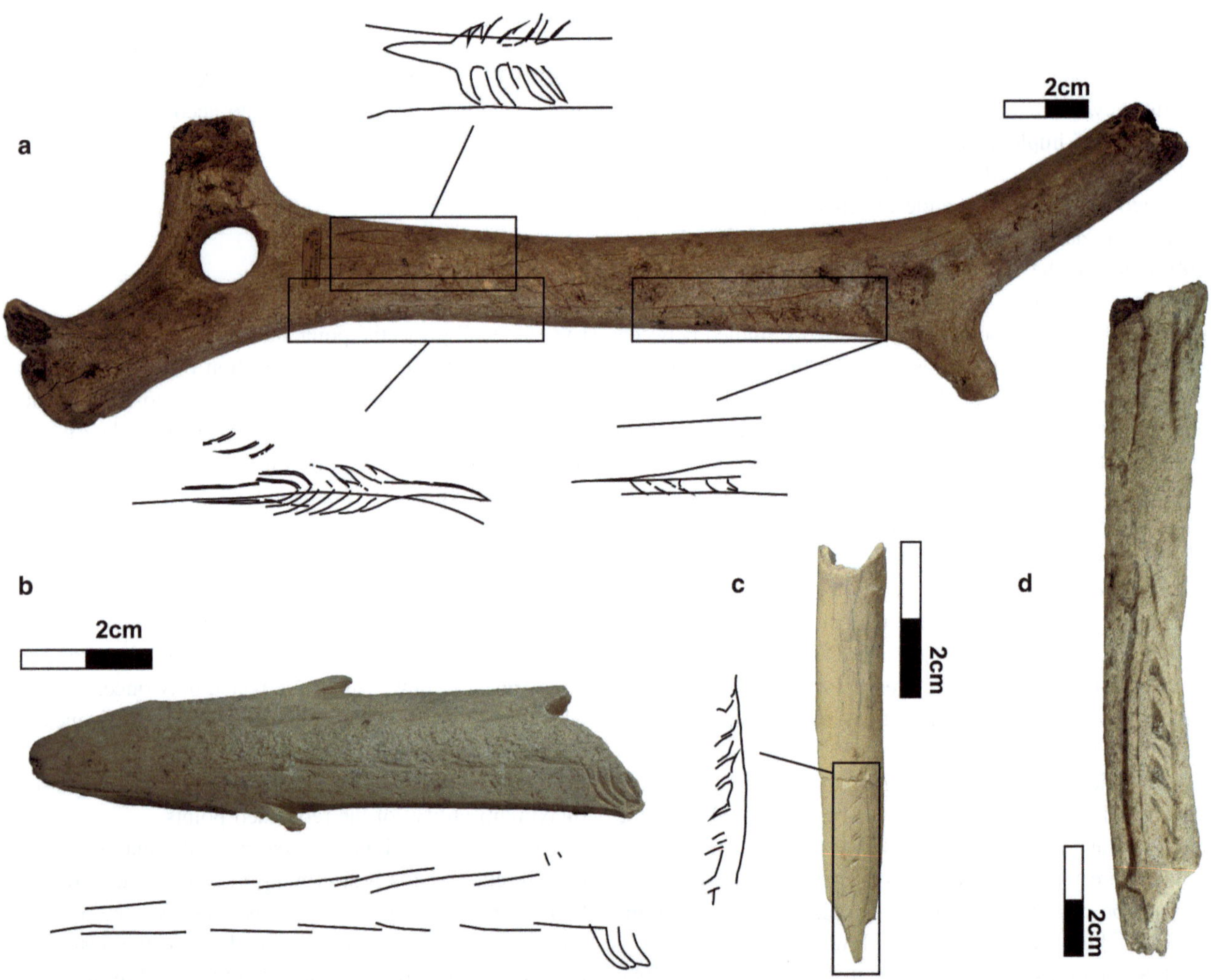

Fig. 16.4 Engravings of uni- and bilaterally barbed points **a** La Madeleine **b** Laugerie-Basse **c** La Vache **d** Isturitz (Photos: M. C. Langley by permission of the Trustees of the British Museum and the *Musée d'Archaéologie National*, Saint-Germaine-En-Laye)

be depicted, sometimes with feathering on the proximal end of the spear shaft indicating fletching. Unfortunately, the identification of projectiles in Magdalenian parietal and mobile art must be undertaken with the greatest caution, and in the majority, those examples that are commonly agreed to be projectiles are too schematized to provide any useful detail about the cultural ideal for point dimensions or form. There are; however, a few examples which are insightful.

These examples are in the form of engravings, the first found on an incomplete *bâton* recovered from La Madeleine (Sieveking 1987: 22), one found on fragment of bird bone from La Vache (Graziosi 1960: 191), another on a fragment of antler recovered from Isturitz (Saint-Périer 1936), and the last on a barbed point itself from Laugerie-Basse (Fig. 16.4). Each of these present depictions of barbed points – a unilaterally barbed point in the case of the La Vache, La Madeleine and Isturitz pieces, and bilaterally barbed points for the *bâton* from La Madeleine and the barbed point fragment from Laugerie-Basse. In each case the points are depicted in a more realistic fashion than those identified in parietal art, and interestingly, on all the pieces the unilaterally barbed points are shown to have relatively short distal parts while the bilaterally barbed points have very long distal parts. Julien (1982) estimated that the original dimensions of both uni- and bilaterally barbed points would have been around 15% of the total length of the point at the time of manufacture. These depictions; however, suggest that the initial dimensions of the distal part of uni- and bilaterally barbed points may have been significantly longer as well as differing between uni- and bilaterally barbed point types. This idea is currently being investigated through analysis of archaeological collections (Langley 2014).

Ethnography

While the comparison of modern hunter-gatherer cultures to Pleistocene peoples is problematic (Hiscock 2008), it has been argued that ethnographic studies provide the only glimpse of lithic and other technologies in an ongoing cultural context (Dibble 1995).

Probably the most extensive ethnography of osseous technologies is provided by Osgood's (1940) ethnographic account of Ingalik material culture in which he reports the material, construction, place and time of manufacture, manufacturer, method of use, place and time of use, use and length of life of a number of implements. A number of additional ethnographies touch on the use of antler and other osseous projectile points among ethnographically known cultures (e.g., Cantwell 1889; Murdoch 1892; Morice 1894; Nelson 1899; Curtis 1911; Emmons 1911; Skinner 1911; Stefánsson 1914; Davidson 1934; Osgood 1936, 1937, 1971; Birket-Smith and De Laguna 1938; Rausch 1951; Giddings 1952; Leechman 1954; McKennan 1965, 1981; Binford 1979; De Laguna and McClellan 1981; McClellan 1981; MacGregor 1985; Rogers and Smith 1981; Townsend 1981; Betts 2007); however, all of these studies lack the detail necessary to evaluate the use life of these technologies and not one reports maintenance activities, though we know that these implements were resharpened on a number of occasions, at least in some cultures (Nunamiut for example: L. Binford via A. Johnson pers. comm. 2011).

This situation is the result of most of the usable ethnographic data being recorded during the 19th and early 20th centuries, when ethnographic hunter-gatherer communities had already been significantly altered by contact with Europeans and other agricultural groups. Their osseous weaponry was quickly replaced with metal types and their hunting methods changed to suit the use of steel traps and shotguns. Additionally, ethnographers were more concerned with recording the details of manufacture and use of osseous projectile points and not with their maintenance and curation.

Ethnographic studies which address the rejuvenation and use life of lithic tools; however, do exist (e.g., Indigenous Australia: Cooper 1954; Tindale 1965; Gould et al. 1971; Hayden 1977, 1979; Papua New Guinea Highlands: Shott and Sillitoe 2004, 2005; and south-central Ethiopia: Gallagher 1977; Weedman 2000, 2002a, b, c; Shott and Weedman 2007). These studies show that lithic tools, at least, are extensively resharpened throughout their use life and that, because of this continuous process of rejuvenation, the morphology of a tool at the time of discard may be significantly different than when it was first used (see Gould 1980 for an example of wooden spears). It was found in these studies that "the diminution of the size through resharpening is an important consideration for eventual discard of a stone tool" (Dibble 1995: 308). Logic allows us to expect this to also be the case for osseous tools.

Gallagher (1977: 411) reports that the Ethiopian scrapers used to work hide were "…used and resharpened until so little of the piece protrudes from the handle that the proper angle for scraping is not possible". Additionally, it was observed that people preferred to continually resharpen their tools in the haft to the point of exhaustion rather than manufacture a new tool – this prolonged use and curation of time expensive hafted technologies was undertaken to recoup large manufacture and maintenance costs (Hayden 1979; Shott and Sillitoe 2004, 2005). Osseous projectile points, as another hafted technology and as an elastic raw material would be particularly predisposed to rejuvenation while still hafted, particularly if the damage to the point was restricted to the distal portion.

These studies all demonstrate that significant morphological change can occur in tools throughout their use life and furthermore, that the diminution of size through resharpening is an important consideration for eventual discard of a tool. This last observation is perhaps the most interesting for the current study in that, osseous projectile points may be resharpened to a stage when their length and/or width becomes too small to continue being effective as projectiles before being discarded.

While ethnographies cannot provide data pertaining to how osseous points may be maintained and reduced, they do provide supporting information concerning the seasonal organization of manufacturing activities (Kniffen 1940), which has been identified archaeologically for Magdalenian osseous projectile points (Averbouh 2005). Additionally, it has been reported that ethnographic osseous points used in warfare (Osgood 1940) or that have successfully killed game (Marshall 1996) may be thrown away or destroyed, while others were the object of exchange and gifts resulting in a single village having an highly varied assemblage of arrows (Heath and Chiara 1977; Wiessner 1983). A number of ethnographies further report that "wherever possible, both the shafts and the arrowheads were recovered and used again" (McKennan 1965: 36; also see Keeley 1982; Wiessner 1983). These studies provide an indication of the range of processes which lead to the formation of archaeological assemblages of osseous projectile points.

Future Directions: What Questions Can We Ask?

The preceding sections outlined a number of avenues of research which may produce data for identifying how Magdalenian's curated their osseous projectile points. Now that the kinds of data which might be informative have been identified, what kinds of questions can be addressed once these data have been collected?

First, we might be able to identify differences in the approach to osseous projectile points during different time slices (e.g., Middle versus Late Magdalenian) or between regions during the same period, and in combination with other lines of data (lithic, faunal, etc.) try to determine whether these differences were the result of social and/or environmental conditions. Individual small sites could be compared, particularly, against major aggregation or multiple occupation sites such as Mas d'Azil and Isturitz, to see if curation rates and styles were more varied in the latter as already shown in the presence/frequency of decorated points (Conkey 1980) – ultimately indicating differing approaches to the curation of these implements between sites and/or territories. Identifying if significant differences in the curation rates exist between various point types, such as simple and barbed forms, may prove informative about how these points were used in prehistory. Furthermore, determining gaps between the maximum utility and realized utility of point types could shed some light on cultural ideals regarding point form and efficiency.

Additional questions that we might ask include: were osseous projectile points curated significantly differently to their lithic counterparts within the same assemblages?; how often did point types cross typological boundaries as they were reworked and did their function change accordingly?; and can identifying curation rates inform us further about manufacture and maintenance scheduling?

All of these avenues of research provide opportunities to extend our knowledge of Magdalenian technological and social systems. It is hoped that research currently underway using both the examination of archaeological assemblages and projectile experiments will provide the first insights into how the Magdalenians curated their osseous projectile points.

Conclusion

Despite observations that Magdalenian points, particularly barbed points, frequently exhibit evidence for extensive resharpening and reworking, analysts of Magdalenian osseous projectile points have framed their interpretations in terms of initial design rather than maintenance and discard. Therefore, while it seems to be widely recognized that osseous points were rejuvenated or recycled throughout prehistory, including the Magdalenian (Julien 1982; Barandiaran 1987; Weniger 2000), very little quantitative research has been undertaken to thoroughly investigate this aspect of osseous technology. Instead, studies have been restricted to either identifying artifacts which exhibit evidence of resharpening, reworking, retooling or recycling (Arndt and Newcomer 1986; Knecht 1991, 1993a, b, 1997; Weniger 2000; Christensen and Chollet 2005; Pétillon 2005, 2006; Liolios 2006), or making qualitative observations about the necessity or ease in which these points can be rejuvenated for reuse during replicative experiments and their general change in overall morphology (Tyzzer 1936; Knecht 1991, 1993a, b, 1997; Pokines 1998; Pétillon et al. 2011).

A careful reassessment of archaeological collections of Magdalenian osseous projectile points using a combination of qualitative and quantitative methods is called for in the hope that regional differences and diachronic change indicative of changing approaches to curation (maintenance) might become evident when approached through the integration of several lines of evidence. Being able to identify the processes involved in the rejuvenation and curation of these implements is vital to building a cohesive understanding of not only the Paleolithic technological system of which they are a part, but also the assemblage formation processes which result in the collections from which we build our interpretations.

Acknowledgements I would like to thank the valuable advice given to my from my PhD supervisor, Nick Barton, as well as that given by Sébastien Lacombe, Meg Conkey, Jean-Marc Pétillon, and Chris Clarkson. I am also appreciative for permission to study the collections curated in the British Museum and the Musée d'Archéologie Nationale, Saint-Germain-en-Laye. This research would be unable to be undertaken without this advice and also without the help of a Clarendon Scholarship (University of Oxford). Finally, I would like to thank Radu Iovita and Katsuhiro Sano for their invitation to participate in the Stone Age Weaponry workshop and the three anonymous reviewers for their most constructive and useful comments.

References

Allain, J., & Rigaud, A. (1986). Décor et Fonction. Quelques exemples Tirés du Magdalénien. *L'Anthropologie, 90*, 713–738.

Allain, J., & Rigaud, A. (1992). Les petites pointes dans l'industrie osseuse de La Garenne: Fonction et figuration. *L'Anthropologie, 96*, 135–162.

Ammerman, A., & Feldman, M. (1974). On the 'making' of an assemblage of stone tools. *American Antiquity, 39*, 610–616.

Arndt, S., & Newcomer, M. H. (1986). Breakage patterns on prehistoric bone points: An experimental study. In D. A. Roe (Ed.), *Studies in the Upper Palaeolithic of Britain and Northwestern Europe* (pp. 165–173). International Series 269. Oxford: British Archaeological Reports.

Averbouh, A. (1999). Un fragment de percuteur sur partie basilaire de la grotte Magdalénienne d'Enlène (Ariège). *Bulletin de la Société Préhistorique Française, 96*, 497–504.

Averbouh, A. (2001). Methodological specifics of the techno-economic analysis of worked bone and antler: Mental refitting and methods of application. In A. M. Choyke & L. Bartosiewicz (Eds.), *Crafting bone: Skeletal technologies through time and space* (pp. 111–121). Oxford: Archaeopress.

Averbouh, A. (2005). Collecte du bois de renne et territoire d'exploitation chez les groups Madgaleniens des Pyrenees Ariegeoises. In D. Vialou, J. Renault-Miskovsky & M. Patou-Mathis (Eds.), *Comportements des hommes du paleolithqiue moyen et superieur en europe: Territoires et milieux* (pp. 59–70). Liege: ERAUL.

Averbouh, A., & Cattelain, P. (2002). De l'analyse fonctionnelle des propulseurs par l'économie de leur production. *Bulletin du Cercle Archéologique Hesbaye-Condroz, 26*, 63–73.

Baffier, D. (1990). Lecture technologique des représentations paléolithiques liées a la chasse et au gibier. *Paléo, 2*, 177–190.

Bahn, P. G. (1982). Inter-site and inter-regional links during the Upper Palaeolithic: The Pyrenean Evidence. *Oxford Journal of Archaeology, 1*, 247–268.

Bamforth, D. B. (1986). Technological efficiency and tool curation. *American Antiquity, 51*, 38–50.

Barandiaran, I. (1987). Industria ósea Paleolítica de la Cueva Del Juyo, excavaciones de 1978 y 1979. In I. Barandiaran, L. G. Freeman, J. G. Echegaray & R. G. Klein (Eds.), *Excavaciones en la Cueva Del Juyo* (pp. 161–194). Santander: Centro de Investigación y Museo de Altamira, Monografías 14.

Betts, M. W. (2007). The Mackenzie Inuit whale bone industry: Raw material, tool manufacture, scheduling, and trade. *Arctic, 60*, 129–144.

Binford, L. R. (1973). Interassemblage variability: The Mousterian and the "functional" argument. In C. Renfrew (Ed.), *The explanation of culture change: Models in prehistory* (pp. 227–254). London: Duckworth.

Binford, L. R. (1977). Forty-seven trips: A case study in the character of archaeological formation processes. In R. V. Wright (Ed.), *Stone tools as cultural markers: Change evolution and complexity* (pp. 24–38). New Jersey: Humanities Press.

Binford, L. R. (1979). Organization and formation processes: Looking at curated technologies. *Journal of Anthropological Research, 35*, 255–272.

Birket-Smith, K., & De Laguna, F. (1938). *The Eyak Indians of the Copper River Delta, Alaska*. Kolenhaven: Lenin and Munksgaard.

Buc, N. (2011). Experimental series and use-wear in bone tools. *Journal of Archaeological Science, 38*, 546–557.

Buchanan, B. (2006). An analysis of Folsom projectile point resharpening using quantitative comparisons of form and allometry. *Journal of Archaeological Science, 33*, 185–199.

Cantwell, J. C. (1889). Exploration of the Kowak River Alaska. Ethnological notes. In M. A. Healey (Ed.), *Report of the cruise of the revenue Marine Steamer Corwin in the Arctic Ocean 1884* (pp. 75–98). Washington D.C.: Government Printing Office.

Chadelle, J. P., Geneste, J. M., & Plisson, H. (1991). *Processus fonctionnels de formation des assemblages technologiques dans les sites du Paleolithique superieur. Les pointes de projectiles lithiques du Solutréen de la grotte de Combe-Saunière (Dordogne, France)* (pp. 275–287). Juan-Les-Pins: APDCA.

Christensen, M., & Chollet, A. (2005). L'industrie sur bois de cervidé et os des niveaux magdaléniens et aziliens du Bois-Ragot: étude préliminaire. *Mémoire de la Société Préhistorique Française* XXXVIII (pp. 223–257).

Christenson, A. L. (1986). Projectile point size and projectile aerodynamics: an exploratory study. *Plains Anthropologist, 31*, 109–128.

Clark, J. G. D., & Thompson, M. W. (1953). The groove and splinter technique of working antler in Upper Palaeolithic and Mesolithic Europe. *Proceedings of the Prehistoric Society, 19*, 148–160.

Conkey, M. W. (1980). The identification of prehistoric hunter-gatherer aggregation: The case of Altamira. *Current Anthropology, 21*, 609–630.

Cooper, H. (1954). Material culture of the Australian Aborigines. *Records of the South Australian Museum, 11*, 91–103.

Curtis, E. S. (1911). *The North American Indian 8*. Norwood: Plimpton Press.

Davidson, D. S. (1934). Australian spear traits and their derivations. *Journal of the Polynesian Society, 43*, 143.

Deffarge, R., Laurent, P., & de Sonneville-Bordes, D. (1977). Sagaies et ciseaux du Magdalenien superieur du Morin, Girondi: Un essai de definition. In H. Camps-Fabrer (Ed.), *Methodologie Appliquee a L'Industrie de L'Os Prehistorique* (pp. 99–110). Paris: Editions du Centre National de la Recherche Scientifique.

De Laguna, F., & McClellan, C. (1981). Ahtna. In J. Helm (Ed.), *Handbook of North American Indians* (Vol. 6, pp. 641–663). Subarctic Washington, DC: Smithsonian Institution Press.

Dibble, H. L. (1995). Middle Paleolithic scraper reduction: Background, clarification, and review of the evidence to date. *Journal of Archaeological Method and Theory, 2*(4), 299–368.

Dobres, M.-A. (1995). Gender and prehistoric technology: On the social agency and technical strategies. *World Archaeology, 27*, 25–49.

Ellis, C. J. (1997). Factors influencing the use of stone projectile tips: An ethnographic perspective. In H. Knecht (Ed.), *Projectile technology* (pp. 37–74). New York: Plenum Press.

Elston, R. G., & Brantingham, P. J. (2002). Microlithic technology in northern Asia: A risk-minimizing strategy of the Late Paleolithic and Early Holocene. In R. G. Elston & S. L. Kuhn (Eds.), *Thinking small: Global perspectives on microlithization* (pp. 103–116). Arlington: American Anthropological Association.

Emmons, G. T. (1911). The Tahltan Indians. *University of Pennsylvania, the Museum, Anthropological Publications, 4*, 1–120.

Gallagher, J. (1977). Contemporary stone tools in Ethiopia: Implications for archaeology. *Journal of Field Archaeology, 4*, 407–414.

Geist, V. (1978). *Life strategies, human evolution, environmental desight. Toward a biological theory of health*. New York: Springer-Verlag.

Giddings, J. L. (1952). *The artic woodland culture of the Kobuk River*. Philadelphia: The University Museum, University of Pennsylvania.

Gould, R. (1980). *Living archaeology*. Cambridge: Cambridge University Press.

Gould, R., Koster, D., & Sontz, A. (1971). The lithic assemblage of the Western Desert Aborigines of Australia. *American Antiquity, 36*, 149–169.

Graziosi, P. (1960). *Palaeolithic art*. London: Faber and Faber.

Grimes, J. R., & Grimes, B. G. (1985). Flakeshavers: Morphometric, functional and life-cycle analyses of Paleoindian unifacial tool class. *Archaeology of Eatern North America, 13*, 35–57.

Guthrie, R. D. (1983). Osseous projectile points: Biological considerations affecting raw material selection and design among Paleolithic and Paleoindian peoples. In J. Clutton-Brock & C. Grigson (Eds.), *Animals and archaeology* (pp. 273–294). International Series 163. Oxford: British Archaeological Reports.

Hayden, B. (1977). Stone tool function in the Western Desert. In R. Wright (Ed.), *Stone tools as cultural markers: Change, evolution and complexity* (pp. 178–188). Canberra: Humanities Press.

Hayden, B. (1979). *Paleolithic reflections: Lithic technology and ethnographic excavation among the Australia Aborigines*. New Jersey: Humanities Press.

Heath, E. G., & Chiara, V. (1977). *Brazilian Indian archery*. Manchester: Simon Archery Foundation.

Hiscock, P. (2008). *The Archaeology of ancient Australia*. New York: Routledge.

Julien, M. (1977). Harpons unilatéraux et bilatéraux: Evolution morphologique ou adaptation differénciée? In H. Camps-Fabrer (Ed.), *Méthodologie appliquée a l'industrie de l'os préhistorique* (pp. 177–189). Paris: Editions du Centre National de la Recherche Scientifique.

Julien, M. (1982). *Les harpons Magdaléniens. Supplement a Gallia Préhistoire 17*. Paris: Editions du Centre National de la Recherche Scientifique.

Julien, M. (1999). Une tendance créatrice au Magdalénien: à propos de stries d'adhérence sur quelques harpons. In M. Julien, A. Averbouh, D. Ramseyer, C. Bellier, D. Buisson, P. Cattelain, M. Patou-Mathis & N. Provenzano (Eds.), *Préhistoire d'Os: Recueil d'Études sur l'industrie osseuse préhistorique* (pp. 133–142). Aix-en-Provence: Publications de l'Université de Provence.

Julien, M., & Orliac, M. (2003). Les harpons et les elements barbeles. In J. Clottes, H. Delporte, & D. Buisson (Eds.), *La Grotte de La Vache (Ariège)* (pp. 221–274). Paris: Reunion des Musees Nationaux/CTHS.

Keeley, L. (1982). Hafting and retooling: Effects on the archaeological record. *American Antiquity, 47*, 798–809.

Knecht, H. (1991). *Technological innovation and design during the Early Upper Paleolithic: A study of organic projectile technologies*. Ph.D. Dissertation, New York University.

Knecht, H. (1993a). Splits and wedges: The techniques and technology of Early Aurignacian antler working. In H. Knecht, A. Pike-Tay & R. White (Eds.), *Before Lascaux: The complex record of the Early Upper Paleolithic* (pp. 137–162). London: CRC Press.

Knecht, H. (1993b). Early Upper Paleolithic approaches to bone and antler projectile technology. In G. Larsen Peterkin, H. M. Bricker, & P. Mellars (Eds.), *Hunting and animal exploitation in the later Palaeolithic and Mesolithic of Eurasia* (pp. 33–47). USA: Archeological Papers of the American Anthropological Association No. 4.

Knecht, H. (1997). Projectile points of bone, antler, and stone: Experimental explorations of manufacture and use. In H. Knecht (Ed.), *Projectile technology* (pp. 191–212). New York: Plenum Press.

Kniffen, F. B. (1940). *Pomo geography* (Vol. 36). Berkeley: University of California Publications in American Archaeology and Ethnology.

Langlais, M., Costamango, S., Laroulandie, V., Pétillon, J.-M., Discamps, E., Mallye, et al., (2012). The evolution of Magdalenian societies in South-West France between 18,000 and 14,000 calBP: Changing environments, changing tool kits. *Quaternary International 272–273*, 138–149.

Langley, M. C. (2014). Magdalenian antler projectile point design: Determining original form for uni- and bilaterally barbed points. *Journal of Archaeological Science, 44*, 104–116.

Laurent, P. (1974). On the morphology of barbs: Observation préliminaires sur la morphologie des harpons du Magdalénien supérieur. In H. Camps-Fabrer (Ed.), *Premier colloque international sur l'industrie de l'os dans la Préhistoire* (pp. 187–191). Aix-en-Provence: Universite de Provence.

Leechman, D. (1954). *The Vanta Kutchin*. National Museum of Canada.

Letourneux, C., & Pétillon, J.-M. (2008). Hunting lesions caused by osseous projectile points: Experimental results and archaeological implications. *Journal of Archaeological Science, 35*, 2849–2862.

Liolios, D. (1999). *Variabilité et caractéristiques du travail des matiéres osseuses au début de l'Aurignacien: Approche technologique et économique*. Paris: Université Paris X-Nanterre.

Liolios, D. (2006). Reflections on the role of bone tools in the definition of the Early Aurignacian. In O. Bar-Yosef & J. Zilhao (Eds.), *Towards a definition of the Aurignacian: Proceedings of the symposium Held in Lisbon, Portugal, June 25–30, 2002* (pp. 37–51). Lisboa: Instituto Portugues de Arqueologia.

Marshall, I. (1996). *A history and ethnography of the Beothuk*. London: McGill-Queen's University Press.

MacGregor, A. G. (1985). *Bone, antler, ivory and horn: The technology of skeletal materials since the Roman Period*. London: Croom Helm.

McClellan, C. (1981). History of research in the Subarctic Cordillera. In J. Helm (Ed.), *Handbook of North American Indians* (Vol. 6, pp. 35–42). Washington, DC: Smithsonian Institution.

McKennan, R. A. (1965). *The Chandalar Kutchin*. Arctic Institute of North America Technical Paper 17.

McKennan, R. A. (1981). Tanana. In J. Helm (Ed.), *Handbook of North American Indian* (Vol. 6, pp. 562–576). Subarctic Washington, DC: Smithsonian Institution.

Moore, C. R., & Schmidt, C. W. (2009). Paleoindian and early Archaic organic technologies: A review and analysis. *North American Archaeologist, 30*, 57–86.

Morice, A. G. (1894). Notes archaeological, industrial and sociological on the Western Denes. *Transactions of the Canadian Institute, Session, 1892–93*, 4.

Murdoch, J. (1892). *Ethnological results of the point barrow expedition*. Washington: Bureau of Ethnology, Smithsonian Institution, Ninth Annual Report.

Nelson, E. W. (1899). The Eskimo about Bering Strait. In *18th Annual Report of the Bureau of American Ethnology*, 1896–1897, Washington.

Nelson, M. C. (1997). Projectile points: Form, function, and design. In H. Knecht (Ed.), *Projectile technology* (pp. 371–384). New York: Plenum Press.

Newcomer, M. H. (1974). Study and replication of bone tools from Ksar Akil (Lebanon). *World Archaeology, 6*(2), 138–153.

Newcomer, M. H. (1977). Experiments in Upper Paleolithic bone work. In H. Camps-Fabrer (Ed.), *Méthodologie appliquée a l'industrie de l'os préhistorique* (pp. 293–301). Paris: Editions du Centre National du Recherche Scientifique.

Nuzhnyi, D. (1993). Projectile weapons and technical progress in the Stone Age. In P. C. Anderseon, S. Beyries, M. Otte & H. Plisson (Eds.), *Traces et fonction: Les gestes retrouvés* (pp. 41–54). Liege: Université de Liège, Service de Préhistoire (ERAUL, 50).

Odell, G. H. (1996). Economizing behavior and the concept of "curation". In G. H. Odell (Ed.), *Stone tools: Theoretical insights into human prehistory* (pp. 51–80). New York: Plenum Press.

Osgood, C. (1936). *Contributions to the ethnography of the Kutchin*. New Haven: Yale University Publications in Anthropology 14.

Osgood, C. (1937). *The ethnography of the Tanaina*. New Haven: Yale University Publications in Anthropology 16.

Osgood, C. (1940). *Ingalik material culture*. New Haven: Yale University Publications. in Anthropology.

Osgood, C. (1971). *The Han Indians*. New Haven: Yale University.

Pétillon, J.-M. (2002). Typologie et utilisation: l'exemple des pointes à base forchue magdaléniennes. In M. Patou-Mathis, P. Cattelain & D. Ramseyer (Eds.), *L'industrie osseuse pré- et protohistorique en europe. Approches techniques et fonctionnelles* (pp. 53–62). Liege: Actes du Colloque 1.6 XIV Congres de l'UISPP 2–8 Septembre 2001.

Pétillon, J.-M. (2005). Tir expérimental de pointes à base fourchue en bois de renne. In V. Dujardin (Ed.), *Industrie osseuse et parures du solutréen au Magdalénien en europe* (pp. 243–256). Paris: Société Préhistorique Française.

Pétillon, J.-M. (2006). *Des Magdaléniens en arms. Technologie des armatures de projectile en Bois de Cervidé du Magdalénien superiéur de la Grotte d'Isturitz (Pyrénées-Atlantiques)*. Treignes: Centre d'Études et de Documentation Archéologiques.

Pétillon, J.-M. (2008). What are these barbs for? Preliminary study on the function of the Upper Magdalenian barbed weapon tips. *Palethnologie, 1*, 69–102.

Pétillon, J.-M. (2016). Thirty years of experimental research on the breakage patterns of Stone Age osseous points. Overview, methodological problems and current perspectives. In R. Iovita & K. Sano (Eds.), *Multidisciplinary approaches to the study of Stone Age weaponry* (pp. 47–63). Dordrecht: Springer.

Pétillon, J.-M., Bignon, O., Bodu, P., Cattelain, P., Debout, G., Langlais, et al. (2011). Hard core and cutting edge: Experimental manufacture and use of Magdalenian composite projectile tips. *Journal of Archaeological Science, 38*, 1266–1283.

Pokines, J. (1993). *Antler points from El Juyo (Santander, Spain): Form, manufacture and parameters of use*. Department of Anthropology, University of Chicago, Chicago.

Pokines, J. (1998). Experimental replication and use of Cantabrian Lower Magdalenian antler projectile points. *Journal of Archaeological Science, 25*, 875–886.

Pokines, J., & Krupa, M. (1997). Self-Barbed antler spearpoints and evidence of fishing in the Late Upper Paleolithic of Cantabrian Spain. In H. Knecht (Ed.), *Projectile technology* (pp. 241–262). London: Plenum Press.

Rausch, E. S. (1951). The Nunamuit Eskimo and mammals of the Anaktuvak Pass Region. *Arctic, 4*, 147–195.

Redmond, B. G., & Tankersley, K. B. (2005). Evidence of Early Paleoindian bone modification and use at the Sheriden Cave Site (33WY252), Wyandot Country, Ohio. *American Antiquity, 70*, 503–526.

Rigaud, A. (2004). Fiche débitage du bois de renne au Magdalénien. L'exemple de La Garenne (Indre, France). In D. Ramseyer (Ed.), *Matières et Techniques, Industrie de l'Os Préhistorique, Cahier XI* (pp. 79–87). Paris: Editions S.P.F.

Rigaud, A. (2006). Étude technologique des baguettes demi-rondes de Labastide (Hautes-Pyrénées). *Archaéologie des Pyrénées Occidentales et des Landes, 25*, 229–246.

Rogers, E. S., & Smith, J. G. E. (1981). Environment and culture in the Shield and Mackenzie Borderlands. In J. Helm (Ed.), *Handbook of North American Indians* (Vol. 6, pp. 130–145). Subarctic Washington, DC: Smithsonian Institution.

Rondeau, M. F. (1996). When is an Elko? In G. H. Odell (Ed.), *Stone tools: Theoretical insights into human prehistory* (pp. 229–243). New York: Plenum Press.

Rozoy, J. G. (1992). Expérimentation de lancer de sagaies avec le propulseur. *Bulletin de la Société Royale Belge d'Études Géologiques et Archéologiques – Les Chercheurs de la Wallonie, 32*, 169–184.

Saint-Périer, R. (1936). *La Grotte d'Isturitz II: La Magdalénien de la Grande Salle*. Paris: Masson.

Schiffer, M. B. (1976). *Behavioral archaeology*. New York: Academic Press.

Sieveking, A. (1987). *A catalogue of Palaeolithic art in the British Museum*. London: British Museum Publications.

Shott, M. (1989). Diversity, organization and behavior in the material record: An ethnographic examination. *Current Anthropology, 30*, 283–301.

Shott, M. J. (2016). Survivorship distribution in experimental spear points: Implications for tool design and assemblage formation. In R. Iovita & K. Sano (Eds.), *Multidisciplinary approaches to the study of Stone Age weaponry* (pp. 245–258). Dordrecht: Springer.

Shott, M., & Sillitoe, P. (2004). Modeling use-life distributions in archaeology using New Guinea Wola ethnographic data. *American Antiquity, 69*, 339–355.

Shott, M., & Sillitoe, P. (2005). Use life and curation in New Guinea experimental used flakes. *Journal of Archaeological Science, 32*, 653–663.

Shott, M. J., & Weedman, K. J. (2007). Measuring reduction in stone tools: An ethnoarchaeological study of Gamo hidescrapers from Ethiopia. *Journal of Archaeological Science, 34*, 1016–1035.

Skinner, A. (1911). Notes on the Eastern Cree and Northern Salteaux. *Anthropological Papers of the American Museum of Natural History, 9*, 1–177.

Stefansson, V. (1914). The Stefansson-Anderson Expedition of the American Museum: Preliminary ethnological report. *Anthropological Papers of the American Museum of Natural History, 14*, 1–376.

Stodiek, U. (1990). Jungpaläolithische speerschleudern und speereein rekonstrucktionswersuch. Experimentelle Archäologie in Deutschland. *Archäeologische Mitteilungen aus Nordwestdeutschland, 4*, 287–297.

Stodiek, U. (1991). Erste Ergebnisse experimenteller Untersuchengen von Geweihgeschoßspitzen des Magdalénien. *Archäologische Mitteilungen aus Nordwestdeutschland Beiheft, 6*, 245–256.

Stodiek, U. (1993). *Zur Technologie der jungpaläolithischen Speerschleuder. Eine Studie auf der Basis archäologischer, ethnologischer und experimenteller Erkenntnisse*. Tübingen: Archaeologia Venatoria.

Stodiek, U. (2000). Preliminary results of an experimental investigation of Magdalenian antler points. In C. Bellier, P. Cattelain & M. Otte (Eds.), *La chasse dans la préhistoire* (pp. 70–78). Bruxelles: Société Royale Belge d'Anthropologie et de Préhistoire/Service de Préhistoire de l'Université de Liege/Centre d'Études et de Documentation Archéologiques.

Straus, L. G. (1992). *Iberia before the Iberians. The Stone Age prehistory of Cantabrian Spain*. Albuquerque: University of New Mexico Press.

Tindale, N. (1965). Stone implement making among the Nakako, Ngadadjara and Pitjandjara of the Great Western Desert. *Records of the South Australian Museum, 15*, 131–164.

Townsend, J. B. (1981). Tanaina. In J. Helm (Ed.), *Handbook of North American Indians 6* (pp. 623–626). Washington: Smithsonian Institution.

Tyzzer, E. E. (1936). The “simple bone point” of the shell-heaps of the northeastern Algonkian Area and its probable significance. *American Antiquity, 1*, 261–279.
Weedman, K. J. (2000). *An ethnoarchaeological study of stone scrapers among the Gamo People of Southern Ethiopia*. Gainesville: Department of Anthropology, University of Florida.
Weedman, K. J. (2002a). An ethnoarchaeological study of stone-tool variability among the Gamo hide workers of southern Ethiopia. In S. Beyries & F. Audoin-Rouzeau (Eds.), *Le Travail du Cuir de la Préhistoire* (pp. 131–142). Antibes: Editions APDCA.
Weedman, K. J. (2002b). On the spur of the moment: Effects of age and experience on hafted stone scraper morphology. *American Antiquity, 67*, 731–744.
Weedman, K. J. (2002c). Gender and stone tools: An ethnographic study of the Konso and Gamo hideworkers of southern Ethiopia. In L. Frink & K. J. Weedman (Eds.), *Gender and hide production* (pp. 175–196). Walnut Creek: Altamira.
Welch, D. M. (1996). Material culture in Kimberley rock art, Australia. *Rock Art Research, 13*, 104–123.
Weniger, G.-C. (1989). The Magdalenian in Western Central Europe: Settlement pattern and regionality. *Journal of World Prehistory, 3*, 323–372.
Weniger, G. C. (1992). Function and form: An ethnoarchaeological analysis of barbed points from northern hunter-gatherers. In F. Audouze (Ed.), *Ethnoarchéologie: Justification, problèmes, limites* (pp. 257–268). Juan-les-Pins: XIIemes reconctres internationales d’archaéologie et d’historie d’Antibes, Centre national de recherches archéologiques en Belgique.
Weniger, G. C. (1995). *Widerhakenspitzen des Magdalénien Westeuropas*. Mainz: Ein Vergleich mit ethnohistorischen Jägergruppen Nordamerikas.
Weniger, G. C. (2000). Magdalenian barbed points: Harpoons, spears and arrowheads. In C. Bellier, P. Cattelain & M. Otte (Eds.), *La chasse dans la préhistoire* (pp. 79–87). Bruxelles: Société Royale Belge d’Anthropologie et de Préhistoire/Service de Préhistoire de l’Université de Liege/Centre d’Études et de Documentation Archeologiques.
Wiessner, P. (1983). Style and social information in Kalahari San projectile points. *American Antiquity, 48*, 253–276.

Chapter 17
Survivorship Distributions in Experimental Spear Points: Implications for Tool Design and Assemblage Formation

Michael J. Shott

Abstract How long points last is a performance attribute just as important as how well they fly and how deeply they penetrate targets. I analyze longevity data in a set of experimental North American Paleoindian Folsom spear-point replicas described by Hunzicker (Plains Anthropologist, 53:291–311, 2008) and previously analyzed for other purposes by Shott et al. (Lithic Technol, 32:203–217, 2007). My goal is to demonstrate the value, descriptively and analytically, of the evidence of longevity encoded in spear points and to consider how they can be estimated in archaeological assemblages. This is possible even though, unlike in experimental data, it cannot be observed or measured directly. At least dimly, results point the way toward the ability to estimate how long tools were used before they failed, how to estimate the distribution of this quantity for populations of points, and how to analyze such distributions.

Keywords Curation • Longevity • Resharpening • Weibull • Gompertz • Reduction measures

Introduction

No one can doubt the skill that is reflected in complex modern objects like computers. Yet such things are the product of many hands assisted at every step by sophisticated instruments and processes. The modern mind is less inclined to recognize the virtuosity of our stone-age ancestors but ancient spear points, to cite one example of Paleolithic artifice, are marvels of invention. Made by skilled but solitary hands from refractory stone to standards tested and refined by hard experience, points were designed with a remarkable range of performance requirements in mind. Their haft elements had to accommodate secure connection to shaft or foreshaft. In their size and mass, points had to conform to the requirements for weight distribution, flight characteristics, and accuracy. Entering the target, they had to penetrate to sufficient depth either to deliver poison or to cause a fatal wound. In many cases, they had to survive two or more such uses. At first glance, spear points may seem crude. Upon reflection, however, they emerge as impressive evidence of prehistoric artisanry (Hughes 1998; Wilhelmsen 2001; Ratto 2003).

Some spear points may have been designed to shatter upon impact or, more likely, penetration of the target. Yet prehistoric hunters could not always be so free in their use of points which, in many cases, were expected to be used repeatedly before replacement. Therefore, longevity and resistance to damage could be design attributes of spear points (Yaroshevich et al. 2016) just as easily as could flight characteristics or penetration (Beck 1998: 25). In brittle stone points hurled at considerable speed into animal targets composed not just of soft hide or flesh but also of hard bone (or, missing those targets, instead hitting trees, rocks or ground), and which, once struck, might strain both to escape the hunter and expel the spear, the probability of breakage and other damage could not have escaped their users' attention. That probability admitted, ancient hunters would have sought to minimize it, in part by designing points to minimize extent of damage and maximize the possibility of repair. In this way, spears could be used, resharpened and reused in cycles before loss or irretrievable failure. Longevity, then, could be an important performance attribute, and one way to promote it would be initial design whose size and form accommodated a considerable range of repair, reduction and reuse.

M.J. Shott (✉)
Department of Anthropology and Classical Studies, University of Akron, Akron, OH 44303, USA
e-mail: shott@uakron.edu

Radu Iovita and Katsuhiro Sano (eds.), *Multidisciplinary Approaches to the Study of Stone Age Weaponry*, Vertebrate Paleobiology and Paleoanthropology, DOI 10.1007/978-94-017-7602-8_17

Longevity and Curation

A brief excursion to define and distinguish two concepts before exploring longevity in detail. Most Paleolithic archaeologists are familiar, broadly, with the concept of curation, which many identify with longevity (or use life). Yet there is an important distinction between the two entities: longevity is an absolute measure of service life, curation a relative one. Binford (1973) and many other defined curation in various, ultimately ambiguous, ways. My own assay (Shott 1996; see also Elston 1992; Langley 2016) defined curation as the ratio of realized to maximum utility. This definition, of course, begs the question of what "utility" means. In archaeological thought, utility has several meanings (Elston 1992: 40–42; Kuhn 1994: 430–432; Shott 1996: 269–271); generally, it signifies the amount of use that a tool can supply in time, tasks performed, or other measures of use. This concept is equivalent to "use-life utility" (Elston 1992: 41), "number of uses" (Schiffer 1976: 54), and "remnant uselife" (DeBoer 1983: 26). For stone tools that were resharpened, maximum utility is the amount of reduction possible between first use and discard, realized utility simply the amount that each specimen underwent. Therefore, measuring utility requires knowing the amount of usable material that a tool contained originally (Kuhn 1994; Shott 1996: 270; Iovita 2009).

This is not the place to debate the reduction thesis which, in the event, is well documented in sources too numerous for tedious citation (see Fullagar 2016 as well). Rather, I assume the validity of the reduction thesis and, motivated by the belief that degree and pattern of reduction is both preserved in tools and itself a source of considerable theoretical insight to past behavior, use it to gain knowledge otherwise difficult to obtain.

At maximum curation, by definition all specimens survive to the point of maximum utility. At progressively lower curation rates fewer specimens survive to maximum utility and more fail or are discarded at lower values of realized utility. Curation is a relative measure simply because it measures longevity attained relative to maximum longevity possible. For instance, consider three tool types of 10 specimens each, in all of which maximum utility is 10 units of time, performance or reduction (Table 17.1). In Type 1, all specimens are used to maximum utility. This type is maximally curated, and its cumulative survivorship appears as Curve 1 in Fig. 17.1. In Type 2, one specimen fails at t_1, one at t_2, one at t_3 and so on to t_{10}. Its failure rate is constant (i.e., exponential), and its cumulative survivorship curve appears as Curve 2 in Fig. 17.1. In Type 3, most tools fail in t_1 or t_2 and only one survives to t_{10}. Type 3 is poorly curated and its cumulative survivorship appears as Curve 3 in Fig. 17.1.

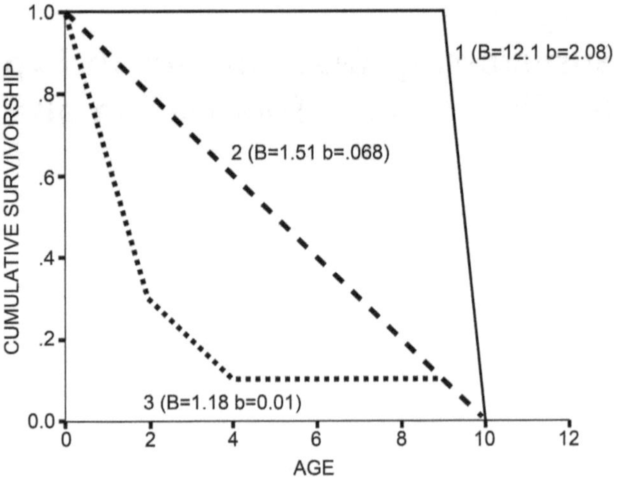

Fig. 17.1 Cumulative survivorship curves and Weibull (B) and Gompertz-Makeham (b) parameter estimates for Table 17.1's three hypothetical datasets. Curve 1 describes high curation rate, Curve 3 low curation rate

Table 17.1 Failure by time interval in three hypothetical tool types

Time	1	2	3
1	0	1	4
2	0	1	3
3	0	1	1
4	0	1	1
5	0	1	0
6	0	1	0
7	0	1	0
8	0	1	0
9	0	1	0
10	10	1	1

Two hypothetical tool types can be equally highly curated even if one has a maximum utility or longevity of, say, 10 years that most specimens approach and the second a maximum utility or longevity of merely 10 min that, again, most specimens approach. Similarly, a hypothetical type can be poorly curated if most specimens only last for five years while some last for 10, and another type well curated even if specimens only can endure for 10 min, provided only that most of them do so. Empirically, for instance, high curation can be found in briefly used informal tools (e.g., Shott and Sillitoe 2005).

Characterizing Longevity Distributions

Conceding the importance of longevity, it remains to characterize and measure the quantity. Longevity is service life measured by age, whether in units of time, use, or activity. It is a quality of individuals, whether people or things. Inquiring of someone's age is an ordinary question. The answer might be "12" or "62". In stone tools, longevity can be described as a simple mean value or as a fixed property of a type, e.g., "Spear points last for five days or five uses".

Yet in many cases there is much dispersion of use-life values around the mean. So perhaps "Spear points last from 1 to 12 months or 2 to 7 uses". In analysis, then, longevity is best treated as a distribution of values across cases. Asking about someone's age distribution is apt to draw a blank stare because populations, not individuals have such distributions. Unless all elements of a population, for instance spear points of a given type, experienced equal longevity, then the distribution of their longevity is an important characteristic of the population separate from the experience of its constituents. Then the mean value is less informative, and certainly no more valuable than knowing and explaining the distribution of individual observations around it.

Knowing the longevity of this or that particular tool rarely is important, but knowing the longevity distribution of populations of tools can be. Central tendency (e.g., mean, median) and dispersion (e.g., standard deviation) crudely summarize sets of numbers and may sufficiently characterize some longevity distributions. But two types might be equal in mean longevity yet differ significantly in their distributions. Distributions efficiently summarize the life-history experience of organisms or objects across their full age range, have explanatory value – discussed below – not always appreciated by archaeologists and facilitate comparison between populations (Pletcher et al. 2000; Shott and Sillitoe 2005). Longevity distributions can be characterized both graphically and mathematically.

Graphic Depiction

One way to represent longevity distributions is as cumulative survivorship curves. These curves summarize the longevity experience of closed cohorts from birth or manufacture through death or failure of the last member. They accomplish this by plotting age intervals on the abscissa against cumulative number surviving to that age on the ordinate. Curves are constrained to 100% survivorship (typically expressed as 1.00) at Age 0 and unchanging or declining survivorship at successive ages. Therefore, cumulative survivorship curves never can rise, but only remain unchanged (i.e., horizontal) or decline.

Cumulative survivorship curves can be compared between populations. For instance, Fig. 17.2 shows survivorship in several empirical modern human populations. Curves vary considerably in form but share a common range, from age 0 to approximately 80. Survivorship distributions also can be compared between populations – of people, animals, or things – that vary not only in mean longevity but in range or scale. As differences in mean longevity rise, however, comparison is hindered by the difference in scale. It would be difficult to compare the longevity experience of, say, fruit flies and elephants using a common absolute scale because the days or at most weeks of a typical fruit fly's life comprise a microscopic slice of a typical elephant's life span.

One solution calibrates different longevities to common scales, plotting cumulative survivorship against multiples of mean longevity. Pearl and Miner (1935) plotted such curves for various biological taxa and for automobiles; Kurtz (1930) did the same for equipment like telephone poles and water pumps. Another expresses realized longevity as a ratio to maximum longevity, i.e., curation distributions. Earlier work applied the method to ethnoarchaeological used flakes from New Guinea (Shott and Sillitoe 2005).

Why fuss about distributions, absolute or relative? Consider again Fig. 17.1 – whose distributions are hypothetical but plausible – and its archaeological implications by means of longevity and curation jointly. One implication concerns simply the number of tools discarded into the archaeological

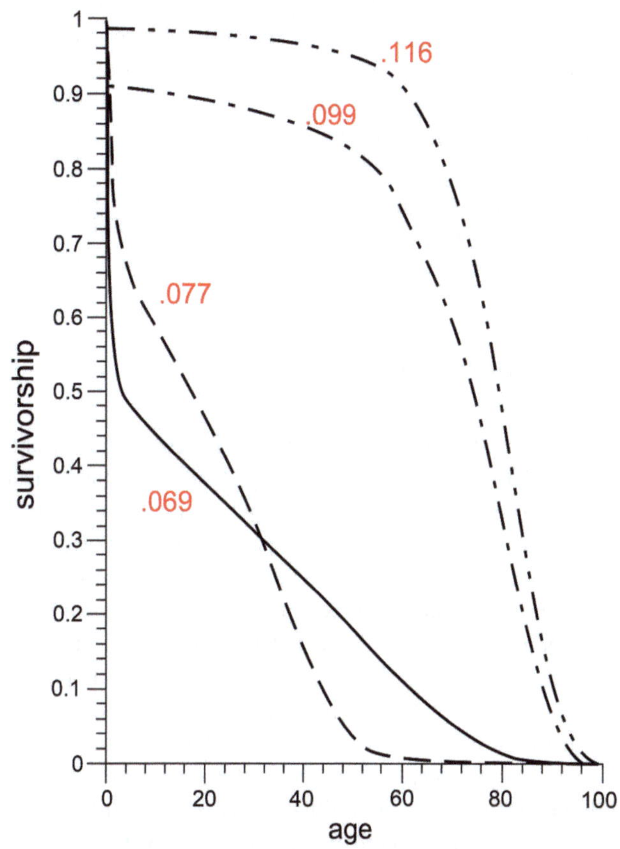

Fig. 17.2 Joint variation in survivorship and Gompertz-Makeham b parameters in human populations that vary from sub-Saharan Africa (concave-upward curves and lower b, lower left) to northern Europe (convex-upward curves and higher b, upper right)

record. All else equal, of course, the more extensively that tools are curated the fewer of them discarded. Another concerns amount of use that resides in tools. Specimens from Distribution 1 nearly are identical in use life; each one found represents about the same number of uses, at least as this quantity is measured by use life. Specimens from Distributions 2 and 3 represent increasingly variable use lives, and cannot be assumed equivalent in amount of use experienced. Two specimens in Distribution 3 represent only 1 unit of use, another specimen 9 units (i.e., the range reported in Table 17.1).

Consider as well the possible differences in spatial distribution of specimens in the archaeological record. If all artifacts-identical in form, material and other salient properties except longevity distribution–are in simultaneous and continuous use and they are used in Interval 1 at one site, Interval 2 at a second site and so on, seven sites will contain one or two specimens each from Distribution 3, four will contain one to four specimens each from Distribution 2 and only three sites will contain any specimens from Distribution 1, and one of those nearly all the specimens. This hypothetical permutation is simplistic but it shows how use-life distributions are important in assemblage formation. The same amounts of use of nine specimens each in three distributions yields highly variable assemblage size and composition. It would be tempting to interpret such differences as evidence of differences in activities performed between sites or some other behavioral or occupational factor, yet there would be no such difference in tool-using activity, merely in assemblage formation mediated by longevity distribution. The better we understand them, the more important distributions become.

Mathematical Description

Longevity distributions also can be both characterized and compared to one another by fitting them to mathematical models. These may seem esoteric when applied to prehistoric stone tools, but are informative in ways that archaeologists do not always appreciate. First, models describe data, as do simple statistics, but they also explain patterns in them by reference to causes like accident and attrition. That is, different ranges of values of model parameters implicate different causes of failure, or measure differences between the distributions. Second, models facilitate the comparison of sets of tool types or subsets of the same type made in different ways, places or of different materials. Third, models are parsimonious in both description and explanation, by reducing distributions of large data sets to few parameter estimates. They are particularly useful ways to study the formation of archaeological assemblages.

Early formation research (e.g., Ammerman and Feldman 1974; Aldenderfer 1981) did not inspire more extensive archaeological study (cf. Shott and Sillitoe 2004, 2005). But Surovell (2009) modeled the formation of North American Folsom Paleoindian assemblages in remarkably sophisticated terms considering the dearth of prior research. His pioneering study might refocus archaeological thought and effort upon formation processes and the models that govern longevity distributions, particularly in stone-tool assemblages. In this perspective, Surovell's explicit (2009: 75) assumption that tool discard followed an exponential distribution (cf. Shott and Sillitoe 2005) is less important than the framework for analysis provided.

Several mathematical models have been applied in engineering failure analysis and demographic analysis. I use the Weibull model for its generality, its theoretical content, its popularity in other fields, and because there is a growing body of Weibull application to experimental and archaeological specimens (e.g., Shott and Sillitoe 2004, 2005; Shott 2009). The Weibull cumulative distribution function is:

$$F(x) = 1 - \exp\left[-(x/\alpha)^{\beta}\right]$$

where age is measured in units of x and α is a scale parameter that is an approximate measure of mean longevity, such that it equals the age by which about 63% of individuals in a population or cohort fail (Dorner 1999: 37). β is commonly known as the shape parameter (i.e., it describes the shape of Weibull and also cumulative survivorship distributions) because it measures the shape of longevity distributions. α correlates with mean longevity, so gives little information not already known. For analytical purposes, β is the more informative parameter both because it provides information not otherwise given and because a considerable body of theory and analysis suggests that different ranges of its value implicates different causes of failure. In particular, the Weibull β parameter identifies different causes of failure, depending upon its value. If β is near a value of 1, failure is by chance. If β significantly exceeds 1, failure is by attrition. β significantly less than 1 indicates high burn-in failure or infant mortality. Thus, α and β both describe longevity distributions, but β also suggests their causes. Weibull distributions can be depicted graphically as plots of the Weibull transformation of failure – the natural logarithm of the antilog of cumulative proportion of failures by units of time or age – against cumulative age (e.g., Shott and Sillitoe 2004, 2005).

Causes of failure are not esoteric or abstruse, but instead fundamental to the thorough analysis of longevity distributions, and recent experimental work suggests the relevance to archaeological data of these concepts. For instance, Iovita (2011) reported damage that was minor to the point of minute accumulated on the edges of experimental points over several firings, followed by catastrophic failure in a subsequent firing. This result may seem to implicate chance, but the small, cumulative damage that contributed to eventual failure seems more consistent with attrition. Similarly, J.-M. Pétillon (personal communication, 2011; see also Pétillon 2016) noted that failure may be intrinsically a chance event in the sense that points may suffer little damage when penetrating nothing harder than skin and flesh. However, when they chance to strike bone, especially cortical bone, points may fail catastrophically. In this scenario, chance is expressed as the probability of striking bone. Yet attrition may be at play even when points strike softer material, provided only that the points' edges suffer slight damage that may accumulate as a function of time and uses until the last strike causes catastrophic failure.

Weibull's virtues include its "virtually limitless versatility" (Dorner 1999: 38) applied to many empirical data, and the explanatory value of its β parameter. Weibull is widely used in engineering and biological studies (e.g., Parker and Arnold 1997; Dorner 1999). It has been used sparingly by archaeologists (references in Shott 2002: 24) but some in my earlier research (Shott and Sillitoe 2004, 2005; Shott 2002, 2009). Yet Weibull's chief virtue is its explanatory power. When $\beta = 1$, failure rate is constant across the range of a longevity distribution, indicating equal contributions to failure of accident and attrition (McCool 1998; Dorner 1999: 36). Then, the Weibull reduces to an exponential distribution (e.g., Distribution 2 in Fig. 17.1). Accordingly, the Weibull distribution encompasses the exponential one that Surovell (2009) assumed. β values that significantly exceed 1 indicate proportionally greater contributions of attrition to failure (e.g., Fig. 17.1's Distribution 1) (Dorner 1999: 37). In this way, β describes longevity distributions with respect to causes and estimates the relative contributions of accident and attrition.

I also use the Gompertz or Gompertz-Makeham models because of their popularity in demography, their suitability as models of aging, and because they are robust; most other demographic models are "extensions of the Gompertz-Makeham model" (Wood et al. 2002: 147). The Gompertz-Makeham cumulative distribution function is:

$$1 - \exp\left[-\lambda x - a/b\left(e^{bx} - 1\right)\right]$$

where λ is age-independent risk of failure, and *a* and *b* measure the scale and shape or slope, respectively, of age- or time-related failure; Gompertz-Makeham λ measures the constant probability of chance failure (Wood et al. 2002: 146). Gompertz *a* sometimes is called "baseline mortality"; *b* measures rate of aging. Essentially, the Gompertz-Makeham model considers the joint effects of age-dependent mortality, which rises exponentially with time, and age-independent mortality, which is random with respect to time. For stone tools or other physical objects and for people, elephants or fruit flies, age-dependent mortality involves senescence, age-independent mortality the probability of death by accident or other causes unrelated to age, wear, or attrition.

Figure 17.2 shows Gompertz-Makeham *b* estimates for each modeled human population. There is a clear pattern of rising *b* with increasing curve convexity. Concave distributions – where mortality is high at young ages and death by aging or attrition is comparatively rare – yield a low *b* values. Convex distributions – as for modern Sweden, shown in the upper right of Fig. 17.2, with high survivorship rates, attrition more than accident, and the demographic equivalent of high curation – yield high *b* values. Concave distributions are analogous to longevity curves of tools that are poorly curated, because they reflect high failure rates at young ages and failure of most specimens to reach maximum utility. Convex curves indicate low initial failure rates followed by rising later attrition and many specimens approaching maximum utility. Hypothetical and archaeological data corroborated this pattern, so elsewhere (Shott and Sillitoe 2005: 658). I argued that Gompertz *b* is a correlate of curation in distributions such that higher *b* values indicate greater convexity and higher survivorship rates per age interval. The higher a population's survivorship, the more convex the shape of its curve, the more individuals live to greater ages and the higher the mean age at death. The more curated a tool type, the more convex its survivorship curve, the more specimens that survive to greater ages before discard and the higher the mean use life.

Longevity Scales

Therefore, Gompertz *b* correlates with curation rate. Unfortunately, it does not scale uniformly with it. Customarily, data are expressed on a relative scale for survivorship: proportions between 1 at birth for all members of a cohort and 0 at death of the last individual. But age is expressed in absolute values. Thus, survivorship curves are plotted with a relative scale for the ordinate and an absolute one for the abscissa. The shape of curves and values of estimated model parameters like Gompertz *b* are influenced by this format. In comparing curation rates between tool types, therefore, the shape of survivorship curves and the value of model parameters are scale-dependent. All else equal, the wider the range of age variation, the lower the *b* value. For example, a tool type whose specimens lasted for, say, anywhere between one and ten days might have an identical shape to its survivorship distribution and thus be as well or poorly curated as one whose specimens lasted from 1 to 100 days.

Yet the first type would have a higher *b* value purely because of the narrower age range of its longevity distribution.

Longevity, or use-life in archaeological terms (Ammerman and Feldman 1974; Schiffer 1976), can be measured in time of use, in whatever units of time. Raw time is valid in some cases but the age, for instance, of a pen might be less accurately measured by the time elapsed between first and last use, which includes very long intervals when the pen was not used at all, than by number of words or total length of script written. Similarly, months or longer could elapse between production of spear points and their final use, yet actual use of the weapons might only comprise mere seconds or minutes of that interval. Instead, in sporadically used things like pens and spear points, longevity might be measured in amount of use or number of uses. By the nature of their use, spear points lend themselves to longevity measured by number of firings.

It is all well and good to speak of longevity and curation as theoretical concepts. It is entirely different to measure them in archaeological specimens, where of course we cannot directly observe or measure either quantity (see Langley 2016 for methods applicable to bone tools). Conceding the concepts' importance, our challenge is to devise measures directly observable in tools (e.g., dimensions or morphometric landmarks) that can be validated by the control provided by experiments.

Things like number of shots are known in experimental data but not obviously in empirical archaeological ones. But number of uses might pattern in some way with number of resharpenings or amount of resulting reduction experienced by tools. In this way, the reduction thesis bears upon not only tool typology but, because reduction is at least an indirect measure of longevity, upon analysis of distributions as well. If archaeologists measure tools on scales that correlate with reduction, distributions of those measures are longevity distributions; if they can correlate reduction with utility obtained (e.g., degree of reduction from original size), then reduction distributions are curation curves (e.g., Shott 1996: Fig. 1).

Data

North America abounds in chipped stone bifaces, which are suitable subjects for analysis of longevity distribution. Accordingly, my subject is bifacial spear (or dart) and arrow points (simply "points" henceforth), the "pretty-facts" of American collectors and popular culture. Although replications of points are a popular pastime, few studies report their experimental use as hunting or other weapons, still fewer the number of uses experienced by each point before failure. Briefly, Odell and Cowan (1986) replicated a range of prehistoric North American point types and also flake points made on central North American Burlington chert, and fired them into animal targets at close range. Points were reused without rejuvenation until catastrophic or cumulative damage disabled them. Couch et al. (1999) fired Great Basin (e.g., Elko) replica spear points, mostly of obsidian, for distance over open ground until they failed. Truncer (1990) made and used eastern North American Perkiomens as spear points. Data from these sources were analyzed previously (Shott 2002), although I return to them in some respects here.

More recently, Cheshier and Kelly (2006) made typologically generic side-notched obsidian points that they fired into a deer carcass. These were arrow, not spear, points fired from a self bow; none was resharpened or otherwise rejuvenated during use; all were used until failure. Many points "shattered beyond the point of repair" (Cheshier and Kelly 2006: 357). Hunzicker (2008) fired replicas of Paleoindian Folsom points into cow carcasses at short range using a calibrated cross-bow (see Clarkson 2016; Iovita et al. 2016; Sano et al. 2016). Each specimen was used until it suffered catastrophic failure. Before it was reached, however, and unlike in other experiments, most points experienced one or more cycles of damage that could be made good by resharpening. Both at experiment's start and at each resharpening cycle, Hunzicker made a cast of the specimen, producing an accurate model of each point as it passed through from one to five or more resharpening cycles. Such resharpening of slightly damaged points sets apart Hunzicker's experiment, and may be faithful to some ancient practice. Hunzicker's data were used in an earlier morphometric study (Shott et al. 2007) and are undergoing further morphometric analysis. Here, I use the number of firings of Cheshier and Kelly and Hunzicker replicas and, for some analyses, the number of resharpenings experienced by each Hunzicker point (Table 17.2).

Table 17.2 Longevity distribution of firings (and resharpenings for Hunzicker). Censored cases in parentheses

Firings	Hunzicker		Odell and Cowan	Chesier and Kelly
	Uses	Resharp.		
1	3	3	2	21
2	4	7	2	12
3	0	4	2	8
4	6(+1)	4	3	6
5	3(+2)	0(+7)	3	1
6	2(+1)		1	1
7	0(+1)		–	1
8	0(+1)		–	–
9	0		–	–
10	0		–	–
11	0(+1)		–	–
Total	18 (+7)	18(+7)	13	25

Because all of Cheshier and Kelly's specimens were used to failure and none remained useful at experiment's end, their data are uncensored, a quality they share with portions of Odell and Cowan's and Couch et al.'s data. However, seven of Hunzicker's 25 Folsom spear points remained serviceable after (varying) numbers of uses and resharpenings, as shown in Table 17.2. These data are censored, because the experiment ended before those seven points were used, hence their full service life was truncated or censored. Both Gompertz and Weibull parameter estimates are sensitive to censoring.

Shott (2002: 97) summarized sources that provided similar data that nevertheless differed sufficiently in ways that preclude their analysis. For instance, Huckell (1982) used Clovis replicas experimentally as thrusting spears, Frison (1989) as thrown spears. Both reported some information on number of uses and failure of points but neither source reported systematic data. Woods (1987) made replicas of western North American spear points and, in order to study fracture type and distribution, deliberately induced breakage by hurling them against trees and rocks, not animal targets. Other sources provide similar data for technologically distinct point types (e.g., Shea et al. 2001: Table 1; Shea et al. 2002: Table 11; Sisk and Shea 2009: 2044 for Levallois flakes hafted as spear points). Cattelain (1997: 233) reported mean number of uses of European Upper Paleolithic Gravette point replicas from an experiment similar in design to Hunzicker's, but did not report number of uses by specimen. Nor did Flegenheimer et al. (2010) in experiments using fluted fishtail points on thrusting spears and atlatl-launched darts, or Burnett and Otárola-Castillo (2008), who conducted morphometric analysis to explore allometric shape change in 24 resharpened Elko Corner-Notched (a North American Great Basin type) replicas. Flenniken and Raymond (1986) used and reworked similar replicas, but again did not report number of firings per specimen which, in any event, were deliberately fired into "trees, soft loamy soil, and thick underbrush" (1986: 607). The focus of their experiment was the resharpening experienced by replicas, not their number of uses. Lombard and Pargeter (2008: 2525) reported some, but not complete, survivorship data from experiments with replicas of South African backed flakes hafted as composite points; Yaroshevich et al. (2010) reported similar data for experimental microlith points. Smallwood (2006) used five Clovis replicas, not necessarily to exhaustion or the point of irreparable damage. Waguespack et al. (2009) summarized some relevant data from other experiments.

Methods

In previous studies I used a spreadsheet method to estimate Weibull parameters, and McCool's (1998) method and unpublished tables to estimate confidence limits of β (Shott 2002, 2009; Shott et al. 2007; Shott and Sillitoe 2004, 2005). Maximum-likelihood estimation (MLE) of Weibull parameters now is easily accomplished on-line (Wessa 2008; ReliaSoft Corp. 2011), which distinguishes censored and uncensored cases (Weibull parameter estimates for both Odell and Cowan (1986) and Couch et al. (1999) data differ somewhat from those reported in Shott (2002) owing to use of different estimation methods). I estimated 90% confidence intervals of calculated Weibull β values using Pivotal (Phan and McCool 2009; McCool 2012), a generalization of the estimation procedure of McCool (1998) used previously (Shott and Sillitoe 2004; Shott 2009). Generally, larger samples yield narrow intervals.

I used WinModest (Pletcher et al. 2000), a computer-intensive method that produces maximum likelihood estimates (MLE) of Gompertz or Gompertz-Makeham parameters. WinModest estimates model parameters from starting values supplied either by default or estimated by the user. I estimated a and b in two ways: (1) from WinModest default starting values; and (2) from the intercept and slope, respectively, of regression of ln-mortality rate upon age (S. Pletcher, personal communication, 2004). In all cases, the approaches yielded identical results or nearly so, indicating that MLE parameter estimates are robust. Although WinModest accommodates censored data, it does so only by age interval, not by individual specimen, so cannot distinguish censored and uncensored data when, as in Hunzicker's data, some intervals contained entries of each type. I assume but cannot prove that this inability to control for censoring is insignificant.

Analysis

Previous research suggests that "Projectile points do not last very long" (Cheshier and Kelly 2006: 357; see also Waguespack et al. 2009: 787; cf. Flegenheimer et al. 2010). Yet this valid generalization masks considerable variation that may pattern with factors that include launching methods, materials, target characteristics, range, and the size, form and mechanical properties of the points themselves. Flegenheimer et al. (2010: Table 2), for instance, reported a mean of 18 uses before failure of hand-thrusted spears but only 7 of atlatl-launched darts. These and other factors must be explored in exhaustive experimental programs. My limited purpose here is to assess different ways to measure longevity, the distinction in practice between longevity and curation explored above in concept, and how best to estimate measures of longevity like number of uses or resharpenings that cannot be observed directly in archaeological data.

Ways of Measuring Longevity

Trivially, points were made to be used, by firing and in other ways. Although they were not made strictly to be resharpened, they may have been designed to accommodate some range or pattern of reduction in sustained use. Therefore, points may have experienced resharpening as a means to the end of prolonging longevity (see Fullagar 2016). Resharpening contributes to pattern and degree of reduction, which can be measured directly in experimental tools and estimated in archaeological specimens.

Hunzicker spear points were used as projectiles, but they also were resharpened. Therefore, their age or amount of use can be measured either in firings or in resharpenings. Table 17.2 shows frequency distributions for both quantities. Figure 17.3 compares their cumulative survivorship curves. Despite the correlation between the variables (Fig. 17.4; $r = 0.84$ $p < 0.01$; $r_s = 0.90$ $p < 0.01$), curves differ clearly in scale and arguably in form, because the curve for shots has a much longer lower tail. If anything, the firings curve is more concave-upward, which suggests lower curation despite the higher mean value. As Table 17.3 shows, the shots or firings distribution has both lower Gompertz *b* and Weibull β estimates (although both pairs of estimates' confidence limits overlap considerably). Shots and resharpenings have different longevity distributions. Paradoxically, shots are more important than resharpenings for many theoretical purposes, yet resharpenings or at least cumulative amount and degree of resharpening, can be more easily inferred from reduced stone tools.

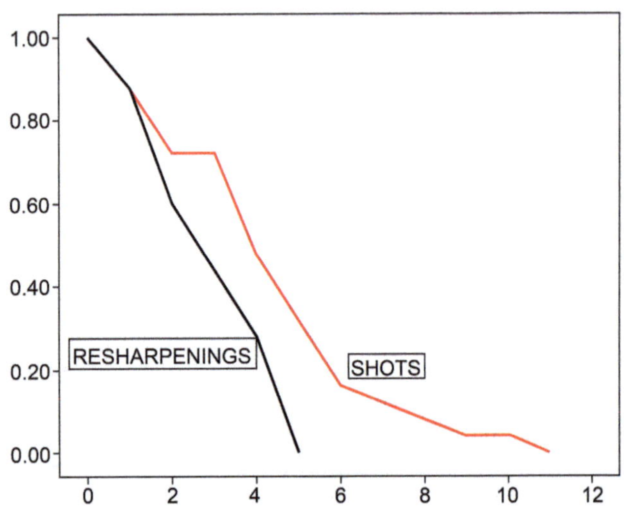

Fig. 17.3 Hunzicker longevity distribution measured in two different ways: resharpenings and shots. Difference between the curves illustrates the influence of aging measures upon survivorship distributions

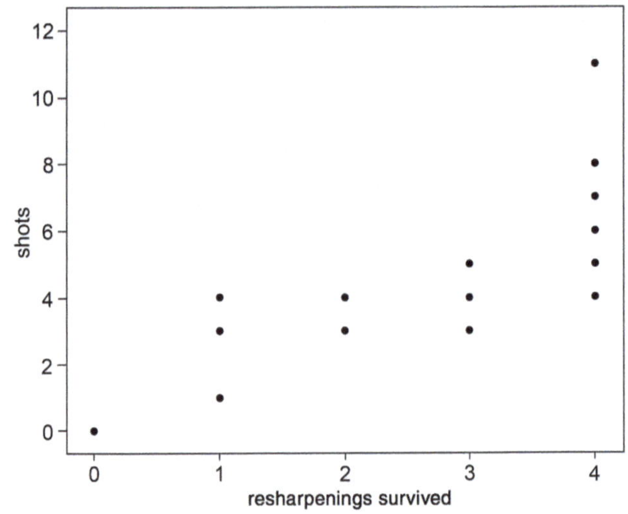

Fig. 17.4 Covariation between shots versus resharpenings in Hunzicker data, showing a positive relationship

Table 17.3 Weibull β and Gompertz *b* parameter estimates by sample, with 90% confidence intervals

Sample	– < β < –	LCI	b	UCI
Hunzicker, actual shots	1.01 < 1.56 < 2.05	0.22	0.36	0.59
Hunzicker, resharpenings	1.15 < 1.79 < 2.36	0.44	0.68	1.06
Cheshier and Kelly, shots	1.38 < 1.70 < 1.97	0.16	0.29	0.52
Odell and Cowan, shots	1.49 < 2.43 < 3.20	0.34	0.63	1.15
Hunzicker, estimated shots	1.33 < 2.06 < 2.71	0.43	0.64	0.95
Hunzicker, estimated resh.	1.52 < 2.35 < 3.09	0.72	1.07	1.60
Hunzicker, LTredn	0.88 < 1.36 < 1.79	0.10	0.21	0.48
Hunzicker, VOLredn	0.86 < 1.34 < 1.76	0.13	0.30	0.69

Longevity and Curation by Number of Uses

Figure 17.5 compares cumulative survivorship for Hunzicker and Cheshier and Kelly's points. The latter display a concave-upward distribution that suggests low curation,

Hunzicker's spears a more equivocal concave-upward distribution that suggests at least somewhat higher curation. The Cheshier and Kelly curve somewhat resembles Fig. 17.1's Distribution 3, suggesting comparatively low curation; the Hunzicker curve is somewhat ambiguous in form, but resembles Fig. 17.1's Curve 2 or 3. Arrow points have a considerably shorter mean longevity and, consistent with their survivorship pattern, a lower Gompertz *b* estimate (Table 17.3). This observation matches archaeological expectations for higher failure rate and corresponding higher archaeological abundance of arrows compared to darts, all else equal. Both distributions yield Weibull β estimates that exceed the exponential value of 1, although the spear estimate's wider 90% confidence limit barely exceeds this threshold; its encompassing of the arrow confidence limits suggests little meaningful difference between the two in Weibull β. [With respect to these data at least, Surovell's (2009) assumption of exponential discard distributions seems not unreasonable, although neither these nor other data are governed by exponential failure (e.g., Shott and Sillitoe 2005).] In both, therefore, attrition along with accident contributes to failure. Thus, Hunzicker's spears last longer on average and are somewhat more extensively curated than are Cheshier and Kelly's arrows. In particular, and contra my (labored?) distinction above, mean longevity and curation seem to vary in tandem.

Yet additional data reveal a more complex picture. Compared to Hunzicker's spear points, the Odell and Cowan data analyzed previously yield a concave-upward survivorship curve similar in form, if displaced leftward, to Hunzicker spears (Fig. 17.6); they show slightly lower mean longevity but considerably higher Gompertz *b* and Weibull β

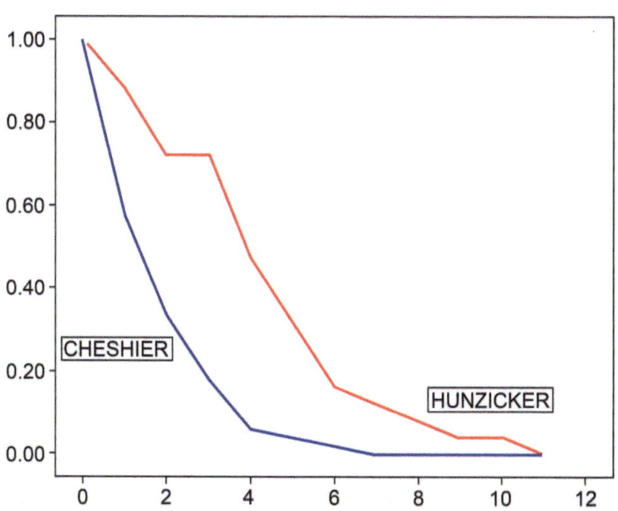

Fig. 17.5 Cumulative survivorship in experimental arrow ("Cheshier" for Cheshier and Kelly) and dart (Hunzicker) points. The arrow curve more nearly approximates Fig. 17.1's low-curation Curve 3, the dart curve more nearly Fig. 17.1's constant failure-rate Curve 2

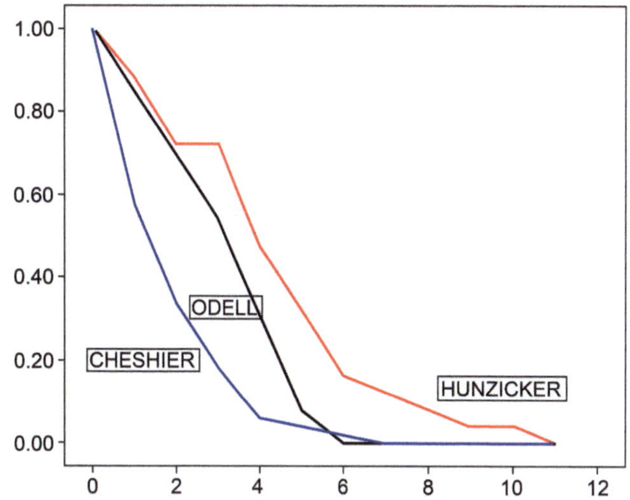

Fig. 17.6 Cumulative survivorship of Odell and Cowan compared to Cheshier and Kelly and Hunzicker data. The Odell and Cowan curve reflects shorter use-life but higher curation rates than the Hunzicker curve, illustrating the independence of use-life and curation rate

estimates (although the respective confidence limits for β overlap considerably) (Table 17.3). Mean longevity and curation do not co-vary in this comparison, suggesting that, as argued above, they can be independent quantities.

Estimating Unknown Longevity

Measured by shots or resharpenings, longevity in experimental data like Hunzicker's is known simply by observation. But archaeological data do not permit direct observation of number of uses, in which case use number or amount must be estimated. For stone tools this can be done in various geometric (e.g., Cardillo 2006; Iovita 2009; Hiscock and Tabrett 2010) or allometric (e.g., Buchanan 2006; Shott et al. 2007; Austin and Mitchell 2010) ways (see Langley 2016 for methods relevant to bone tools). Rather than review the relevant literature in detail, here I consider the unique value of Hunzicker's data derived from the information they preserved on specimen size and form at each resharpening cycle.

In Hunzicker's data, the simple allometric ratio of length to thickness (LT), directly observable in archaeological specimens, correlated significantly with amount of reduction experienced, as measured by volume reduced or number of resharpenings (Shott et al. 2007). LT, therefore, is a valid general reduction correlate. LT at the final stage of use of Hunzicker points (LTFinal) is directly observable in archaeological specimens; simply measure both maximum length and thickness, and divide the first by the second. LTFinal correlated significantly with both number of shots ($r = -0.67$ $p < 0.01$;

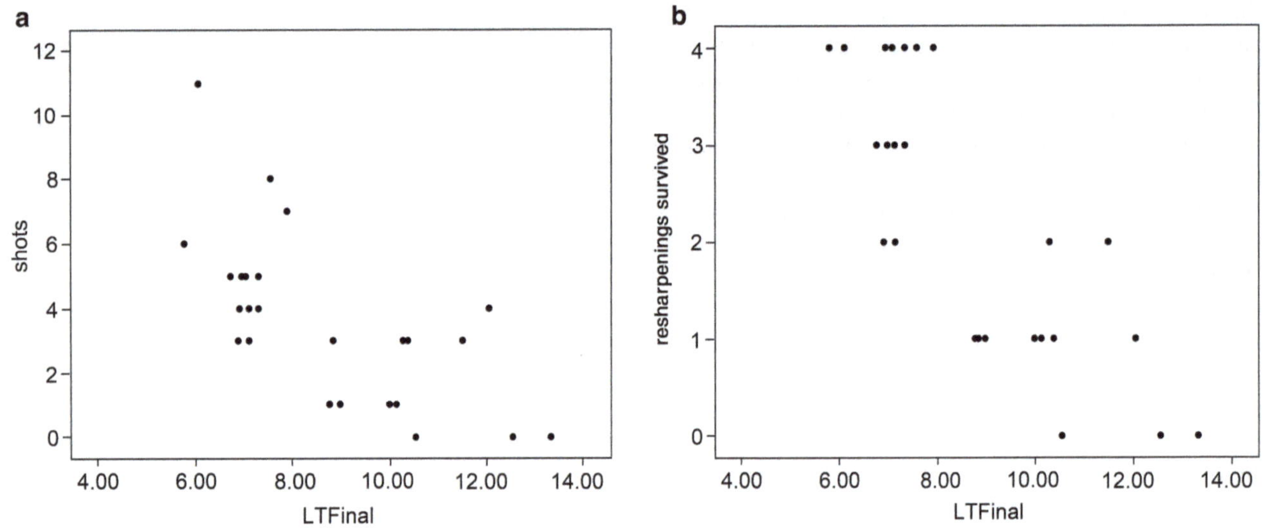

Fig. 17.7 Number of shots and number of resharpenings against LTFinal (LT ratio at final stage of use), showing inverse relationship between measures of age and LTFinal

$r_s = -0.70 < 0.01$) and particularly number of resharpenings ($r = -0.81$ $p < 0.01$; $r_s = -0.78$ $p < 0.01$) (Fig. 17.7). In this way, number of firings or shots was estimated as:

$$\text{Estimated uses} = 10.80 - 0.83 * \text{LTFinal}$$

Similarly, number of resharpenings was estimated as:

$$\text{Estimated resharpenings} = 6.97 - 0.55 * \text{LTFinal}$$

Table 17.4 shows estimated resharpenings and uses, which comprise separate longevity distributions to compare with actual values; Fig. 17.8 plots the distributions. The Gompertz *b* estimates for both estimated resharpenings and estimated uses significantly exceed their respective actual values although, again, confidence intervals overlap considerably. Weibull β estimates also exceed those for actual values; in this case, 90% confidence intervals for uses barely overlap, but the intervals for resharpenings do considerably. Unfortunately, then, estimating number of uses or resharpenings from allometric relationships yields parameter estimates that do not closely match actual values.

The scanning process used in our 3D morphometric program (Shott and Trail 2010) records volume (in a very American way: in^3). Therefore, we measured volume of each specimen in each resharpening cycle, and used the ratio of final to original volume as a measure of overall reduction. Another way to estimate amount of use and resharpening, then, expresses size at failure as a proportion of original size (i.e., LTFinal/LTOriginal and VOLFinal/VOLOriginal).

Table 17.4 Hunzicker longevity distributions by estimated uses and estimated resharpenings. Censored cases in parentheses

Number	Estimated uses/shots	Estimated resharpenings
1	3	4
2	1	5
3	5	5(+2)
4	3	4(+5)
5	6(+1)	–
6	0(+5)	–
7	0(+1)	–

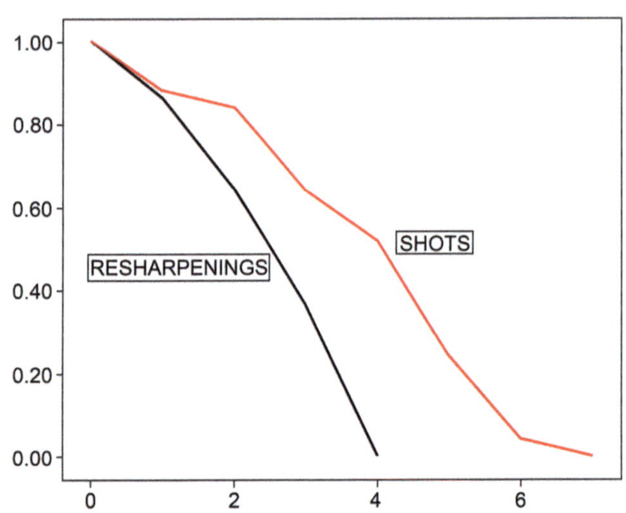

Fig. 17.8 Cumulative survivorship measured by estimated shots and by resharpenings, Hunzicker spears

I subtracted these proportions from 1 so that resulting values would ascend with amount of use and reduction, then multiplied by 100, calling the resulting variables LTredn and VOLredn. Thus,

$$\mathrm{LTredn} = (1 - \mathrm{LTFinal}/\mathrm{LTOriginal}) * 100$$

and

$$\mathrm{VOLredn} = (1 - \mathrm{VOLFinal}/\mathrm{VOLOriginal}) * 100.$$

Distributions for both variables appear in Table 17.5 and Fig. 17.9. Gompertz *b* estimates are very similar between actual uses and longevity estimated from VOLredn, and only slightly lower for LTredn estimates; confidence intervals of estimates are similar. Similarly, Weibull β estimates are slightly lower than, but occupy similar 90% confidence intervals to, actual number of uses. Because model parameter estimates differ between number of uses and number of resharpenings, LTredn and VOLredn do not closely resemble the latter. Tentatively, results suggest that at least the distribution of number of uses can be approximated from LTredn and VOLredn distributions despite the difficulty, noted above, in estimating uses from reduced tools.

Unfortunately, neither LTredn nor VOLredn are themselves directly observable in archaeological context, because they require both LTFinal and VOLFinal (which are observable) and LTOriginal and VOLOriginal (which are not). Yet original length or volume might be estimated by allometric or geometric means (e.g., Shott et al. 2007), or from the dimensions of unused cache specimens, at least for types that are highly standardized by original size and form, but absent such data cannot be used to generate longevity distributions.

Finally, I estimated amount of both use and resharpening as fractions of mean LTFinal or mean VOLFinal, i.e., relative measures like Pearl and Miner (1935), but where the

Table 17.5 Hunzicker longevity distributions LTredn and VOLredn. Censored cases in parentheses

Value	LTredn	VOLredn
05	5	3
10	0	2
15	1	0
20	1	1
25	2	3
30	1	1
35	3	2
40	2	0
45	2(+1)	1
50	1(+4)	1
55	0(+2)	2(+2)
60		0(+2)
65		2(+3)

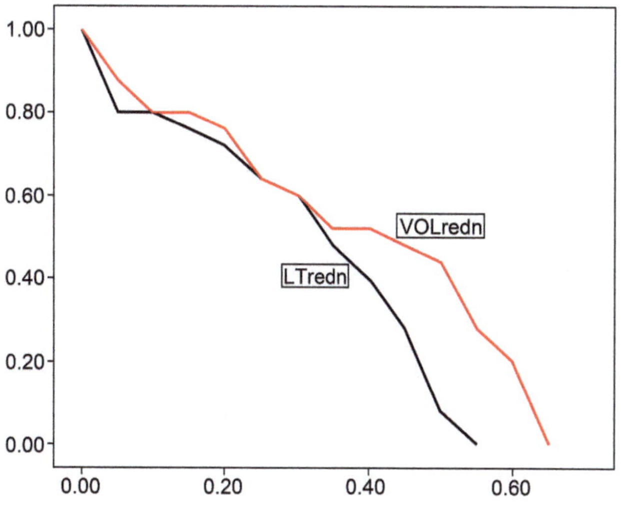

Fig. 17.9 Cumulative survivorship measured by LTredn ((1 − LTFinal/LTOriginal) * 100) and VOLredn ((1 − VOLFinal/VOLOriginal) * 100) measures

quantity is scaled to mean dimensions at discard, not to mean longevity. This measure can be taken from archaeological specimens, so requires no knowledge of actual values or method to estimate them. Unfortunately, resulting survivorship distributions and model parameter estimates resembled those for regression estimates of uses and resharpenings upon LTFinal and considerably exceeded corresponding estimates for actual values. On available evidence, estimates of longevity from intervals of mean final dimensions are unsatisfactory.

Summary

Experiments permit archaeologists directly to observe what we can only estimate in archaeological data. Amongst the things we might wish directly to observe is number of uses and number and degree of resharpening experienced by projectile points in the course of their sometimes repeated use. Equally important, we might wish to compile and study distributions of values for use or resharpening across sets of points. Graphic expression of distributions, at least as cumulative survivorship curves, describe the range and form of a point population's life-history. Fitting distributions to the Weibull model, especially estimation of its β parameter, identifies either burn-in, chance, or attrition as major causes of point failure. The Gompertz or Gompertz-Makeham *b* parameter is at least a relative or comparative measure of curation rate.

Hunzicker's (2008) replica Folsom points fail chiefly by attrition, but the 90% confidence limits of their estimated Weibull β parameter barely exceed the threshold for chance failure. Although obsidian arrow points do not last nearly as long on average, they too fail mostly by attrition, as do experimental spear points analyzed previously. Although the latter have slightly lower mean longevity, Gompertz *b* estimates suggest higher curation than for Folsom replicas; curation and mean longevity are different quantities.

Spear-point longevity can be measured by number of shots or number of resharpenings. Distributions of these measures yield different model parameter estimates in Hunzicker data, resharpening, by its considerably higher Gompertz *b* estimate, suggesting higher curation. How we measure longevity matters. Both longevity quantities correlate well with the allometric reduction measure LT (Shott et al. 2007), so, at least crudely, can be estimated from that archaeologically observable ratio. Resulting estimated longevity distributions for shots and resharpenings are similar to one another but their Weibull β estimates are considerably higher than corresponding estimates from known shots and resharpenings. For both estimated distributions, Gompertz *b* estimates are higher than corresponding estimates from known data. Finally, longevity estimated from LTredn or VOLredn (which themselves require estimation in archaeological data) yield Weibull β and Gompertz *b* estimates similar to, if somewhat lower than, known data.

Future Directions

Why all the huffing and puffing about distributions? Longevity distributions are worth analyzing even if only for experimental data, because they implicate causes of failure and estimate curation rates. At the very least, their use can standardize comparison between experimental data sets. The greatest challenge, however, lies, in archaeological context, where no longevity measure is known. This and earlier studies (e.g., Shott et al. 2007) identify significant patterning between longevity and what can be measured directly on archaeological points, but in this study longevity distributions estimated from correlates or proxies differ somewhat from known longevities.

Archaeologically observable or inferable measures approximate longevity (and curation) distributions. Therefore, we can estimate or reconstruct the longevity and perhaps the curation distributions of prehistoric tool types. Their graphic expression in cumulative survivorship curves and their fit to mathematical models suggest processes and causes of failure, with implications for assemblage models, for archaeological pattern and abundance of types, and for patterns of association between them, possibly even for performance characteristics of types.

Analytically, priorities include methods to determine statistically significant (dis)similarity between model parameter estimates for different samples, and further study of the scaling relationship between Gompertz *b* parameter and degree of curation. For instance, Gompertz-Makeham *b* values for Fig. 17.2's human populations vary over a narrower range than do comparable estimates for Fig. 17.1's hypothetical tool populations, despite the general similarity in distributions' forms. Perhaps the difference owes to the data sets' different age ranges, which differ by nearly an order of magnitude. But in our (or at least my) current state of ignorance, much remains unknown about scaling in model parameter estimates.

More broadly, accepting the thesis that durability or resistance to failure was an important performance attribute of points, some implications for future research on distributions emerge. First, we require controlled experiments in which both firing technology and targets are held constant so that the effects of raw material, point size, and performance attributes like tip acuity, tip section area, and varieties of hafting configurations (both on points and shafts/foreshafts) may be explored (Hughes 1998; Wilhelmsen 2001; Ratto 2003). Some such experiments may involve points made to generic, even arbitrary standards, but others should involve both replicas of specific prehistoric types but also of ranges of forms that might form transitions between recognized types. Given the sensitivity of especially Weibull parameter estimates to sample size, experiments should involve the largest samples practicable. In controlled experiments, we can begin to model the changing performance requirements that, in some cases at least, stimulated the guided evolution of one type from its antecedent. Second, we require actualistic experiments that replicate conditions of use more faithful to original conditions. In the latter case, the inability to control for independent factors like material, design and the like will require more and a wider range of experiments.

Similarly, a wider range of tool types should be replicated and used, including unifaces and various flake tools. In these cases, of course, longevity cannot be measured in firings, but perhaps in strokes, amount of material worked and the like. One small set of ethnoarchaeological flake-tool data suggests a surprising range and form of variation in longevity and survivorship (Shott and Sillitoe 2005), but a great deal more experimentation is required before clear patterns can be expected to emerge.

Obviously, work must be done to identify more and better geometric and allometric reduction measures (Hiscock and Tabrett 2010) and to document their ability to produce valid longevity estimates. Still, LTredn and VOLredn estimate longevities whose distributions approximate actual number of uses. We are far from accurate, precise and therefore validated estimates, but at least the path to that end is perceptible. Not bad for a start, but the most vital need is sustained research on these questions that can take us where this study only can point.

Acknowledgements Thanks are due to Radu Iovita and Katsuhiro Sano for their kind invitation to participate in the conference "Stone Age Weaponry," and to the Römisch-Germanisches Zentralmuseum, Forschungsinstitut für Archäologie, in Mainz, Germany. I thank David Hunzicker as well for generously sharing his data. Scott Pletcher provided WinModest and generously offered advice in its use.

References

Aldenderfer, M. S. (1981). Creating assemblages by computer simulation: The development and uses of ABSIM. In J. Sabloff (Ed.), *Simulation in archaeology* (pp. 67–117). Albuquerque, NM: University of New Mexico.

Ammerman, A., & Feldman, M. (1974). On the 'making' of an assemblage of stone tools. *American Antiquity, 39*, 610–616.

Austin, R., & Mitchell, S. (2010). Reconstructing the life histories of Bolen hafted bifaces from a north Florida archaeological site. *Early Georgia, 38*, 3–27.

Beck, C. (1998). Projectile point types as valid chronological units. In A. Ramenofsky & A. Steffen (Eds.), *Unit issues in archaeology: Measuring time, space and material* (pp. 21–40). Salt Lake City, UT: University of Utah Press.

Binford, L. R. (1973). Interassemblage variability: The Mousterian and the 'functional' argument. In C. Renfrew (Ed.), *The explanation of culture change: Models in prehistory* (pp. 227–254). London: Duckworth.

Buchanan, B. (2006). An analysis of Folsom projectile point resharpening using quantitative comparisons of form and allometry. *Journal of Archaeological Science, 33*, 185–199.

Burnett, P., & Otárola-Castillo, E. (2008). *Knapping on an idea: Gauging experimental and archaeological projectile point morphological variability using geometric morphometrics.* Paper presented at the 73rd Annual Meeting of the Society for American Archaeology, Vancouver.

Cardillo, M. (2006). Temporal trends in the morphometric variation of the lithic projectile points during the Middle Holocene of southern Andes (Puna Region): A coevolutionary approach. In H. Muscio & G. E. José Lopez (Eds.), *Theoretical and methodological issues in evolutionary archaeology: Toward an unified Darwinian paradigm* (pp. 13–20). British Archaeological Reports, International Series 1915.

Cattelain, P. (1997). Hunting during the Upper Paleolithic: Bow, spearthrower, or both? In H. Knecht (Ed.), *Projectile technology* (pp. 213–240). New York: Plenum.

Cheshier, J., & Kelly, R. L. (2006). Projectile point shape and durability: The effect of thickness-length. *American Antiquity, 71*, 353–363.

Clarkson, C. (2016). Testing archaeological approaches to determining past projectile delivery systems using ethnographic and experimental data. In R. Iovita & K. Sano (Eds.), *Multidisciplinary approaches to the study of Stone Age weaponry* (pp. 189–201). Dordrecht: Springer.

Couch, J., Stropes, T. A., & Schroth, A. B. (1999). The effect of projectile point size on atlatl dart efficiency. *Lithic Technology, 24*, 27–37.

DeBoer, W. R. (1983). The archaeological record as preserved death assemblage. In A. Keene & J. Moore (Eds.), *Archaeological hammers and theories* (pp. 19–35). New York: Academic.

Dorner, W. W. (1999). Using Microsoft Excel for Weibull analysis. *Quality Digest, 19*, 333–338.

Elston, R. G. (1992). Modeling the economics and organization of lithic procurement. In R. Elston (Ed.), *Archaeological investigations at Tosawihi, A Great Basin quarry* (Vol. 1, pp. 31–47). Silver City, NV: Intermountain Research.

Flegenheimer, N., Martínez, J. G., & Colombo, M. (2010). Un experimento de lanzamiento de puntas Cola de Pescado. In M. Berón, L. Luna, M. Bonomo, C. Montalvo, C. Aranda & M. Aizpitarte (Eds.), *Mamül Mapu: Pasado y Presente desde la Arqueología Pampeana* (pp. 215–232). Ayacucho, Argentina: Editorial Libros del Espinillo.

Flenniken, J. J., & Raymond, A. W. (1986). Morphological projectile point typology: Replication, experimentation and technological analysis. *American Antiquity, 51*, 603–614.

Frison, G. C. (1989). Experimental use of Clovis weaponry and tools on African elephants. *American Antiquity, 54*, 766–784.

Fullagar, R. (2016). Uncertain evidence for weapons and craft tools: Functional investigations of Australian microliths. In R. Iovita & K. Sano (Eds.), *Multidisciplinary approaches to the study of Stone Age weaponry* (pp. 159–166). Dordrecht: Springer.

Hiscock, P., & Tabrett, A. (2010). Generalization, inference and the quantification of lithic reduction. *World Archaeology, 42*, 545–561.

Huckell, B. B. (1982). The Denver Elephant Project: A report on experimentation with thrusting spears. *Plains Anthropologist, 27*, 217–224.

Hughes, S. S. (1998). Getting to the point: Evolutionary change in prehistoric weaponry. *Journal of Archaeological Method and Theory, 5*, 345–408.

Hunzicker, D. A. (2008). Folsom projectile technology: An experiment in design, effectiveness and efficiency. *Plains Anthropologist, 53*, 291–311.

Iovita, R. (2009). Ontogenetic scaling and lithic systematics: Method and application. *Journal of Archaeological Science, 36*, 1447–1457.

Iovita, R., Schöekeß, H., Gaudzinski-Windheuser, S., & Jäger, F. (2011). Controlled ballistic experiments with glass replicas of Levallois points. Manuscript on file, Römisch-Germanisches Zentralmuseum, Mainz, Germany.

Iovita, R., Schönekeß, H., Gaudzinski-Windheuser, S., & Jäger, F. (2016). Identifying weapon delivery systems using macrofracture analysis and fracture propagation velocity: A controlled experiment. In R. Iovita & K. Sano (Eds.), *Multidisciplinary approaches to the study of Stone Age weaponry* (pp. 13–27). Dordrecht: Springer.

Kuhn, S. L. (1994). A formal approach to the design and assembly of mobile toolkits. *American Antiquity, 59*, 426–442.

Kurtz, E. B. (1930). *Life expectancy of physical property based on mortality laws*. New York: Ronald.

Langley, M. C. (2016). More to the point: Developing a multi-faceted approach to investigating the curation of Magdalenian osseous projectile points. In R. Iovita & K. Sano (Eds.), *Multidisciplinary approaches to the study of Stone Age weaponry* (pp. 229–244). Dordrecht: Springer.

Lombard, M., & Pargeter, J. (2008). Hunting with Howiesons Poort segments: Pilot experimental study and the functional interpretation of archaeological tools. *Journal of Archaeological Science, 35*, 2523–2531.

McCool, J. I. (1998). Inference on the Weibull location parameter. *Journal of Quality Technology, 30*, 119–126.

McCool, J. I. (2012). The Program Pivotal.exe. In J. McCool (Ed.), *Using the Weibull distribution: Reliability, modeling, and inference* (pp. 213–234). New York: Wiley. doi:10.1002/9781118351994

Odell, G. H., & Cowan, F. (1986). Experiments with spears and arrows on animal targets. *Journal of Field Archaeology, 13*, 195–212.

Parker, W. C., & Arnold, A. J. (1997). Species survivorship in Cenozoic planktonic foraminifera: A test of exponential and Weibull models. *Palaios, 12*, 3–12.

Pearl, R., & Miner, J. R. (1935). Experimental studies on the duration of life, XIV: The comparative mortality of certain lower organisms. *Quarterly Review of Biology, 10*, 60–79.

Pétillon, J.-M. (2016). Thirty years of experimental research on the breakage patterns of Stone Age osseous points. Overview, methodological problems and current perspectives. In R. Iovita & K. Sano (Eds.), *Multidisciplinary approaches to the study of Stone Age weaponry* (pp. 47–63). Dordrecht: Springer.

Phan, L. D., & McCool, J. I. (2009). Exact confidence intervals for Weibull parameters and percentiles. *Proceedings of the Institution of Mechanical Engineers, Part O: Journal of Risk and Reliability, 223*, 387–394.

Pletcher, S. D., Khazaeli, A. A., & Curtsinger, J. W. (2000). Why do life spans differ? Partitioning mean longevity differences in terms of age-specific mortality parameters. *Journal of Gerontology, 55A*, B381–B389.

Ratto, N. (2003). *Estrategias de caza y propiedades de registro arqueológico en la Puna de Chascuil* (Dpto. de Tinogasta, Catamarca, Argentina). PhD dissertation, Depto. de Ciencias Antropológicas, Universidad de Buenos Aires, Argentina.

ReliaSoft Corporation. (2011). iTools: Simple 2-parameter Weibull analysis. http://www.weibull.com/itools/. Accessed September 9, 2011.

Schiffer, M. B. (1976). *Behavioral archaeology*. New York: Academic.

Sano, K., Denda, Y., & Oba, M. (2016). Experiments in fracture patterns and impact velocity with replica hunting weapons from Japan. In R. Iovita & K. Sano (Eds.), *Multidisciplinary approaches to the study of Stone Age weaponry* (pp. 29–46). Dordrecht: Springer.

Shea, J. J., Brown, K., & Davis, Z. (2002). Controlled experiments with Middle Paleolithic spear points: Levallois points. In J. Mathieu (Ed.), *Experimental archaeology: Replicating past objects, behaviors, and processes* (pp. 55–72). BAR International Series 1035. Oxford, UK: Archaeopress.

Shea, J., Davis, Z., & Brown, K. (2001). Experimental tests of Middle Palaeolithic spear points using a calibrated crossbow. *Journal of Archaeological Science, 28*, 807–816.

Shott, M. J. (1996). An exegesis of the curation concept. *Journal of Anthropological Research, 52*, 259–280.

Shott, M. J. (2002). Weibull estimation of uselife distribution in experimental spear-point data. *Lithic Technology, 27*, 93–109.

Shott, M. J. (2009). Stone-tool demography: Reduction distributions in North American Paleoindian tools. In S. Lycett & P. Chauhan (Eds.), *New perspectives on old stones: Analytical approaches to Palaeolithic technologies* (pp. 275–293). Dordrecht: Springer/Kluwer.

Shott, M. J., Hunzicker, D. A., & Patten, B. (2007). Pattern and allometric measurement of reduction in experimental Folsom bifaces. *Lithic Technology, 32*, 203–217.

Shott, M. J., & Sillitoe, P. (2004). Modeling use-life distributions in archaeology using New Guinea Wola ethnographic data. *American Antiquity, 69*, 339–355.

Shott, M. J., & Sillitoe, P. (2005). Use life and curation in New Guinea cxperimental used flakes. *Journal of Archaeological Science, 32*, 653–663.

Shott, M. J., & Trail, B. W. (2010). Exploring new approaches to lithic analysis: Laser scanning and geometric morphometrics. *Lithic Technology, 35*, 195–220.

Sisk, M. L., & Shea, J. J. (2009). Experimental use and quantitative performance analysis of triangular flakes (Levallois points) used as arrowheads. *Journal of Archaeological Science, 36*, 2039–2047.

Smallwood, A. M. (2006). *Use-wear analysis of the Clovis biface collection from the Gault site in Central Texas*. Unpublished MA thesis, Texas A&M University, College Station, TX, USA.

Surovell, T. A. (2009). *Toward a behavioral ecology of lithic technology: Cases from Paleoindian archaeology*. Tucson, AZ: University of Arizona.

Truncer, J. J. (1990). Perkiomen points: A study in variability. In R. Moeller (Ed.), *Experiments and observations on the terminal archaic of the Middle Atlantic Region* (pp. 1–62). Bethlehem, CT: Archaeological Services.

Waguespack, N. M., Surovell, T. A., Denoyer, A., Dallow, A., Savage, A., Hyneman, J., et al. (2009). Making a point: Wood- *versus* stone-tipped projectiles. *Antiquity, 83*, 786–800.

Wessa, P. (2008). Maximum-likelihood Weibull distribution fitting (v1.0.2) in Free Statistics Software (vt.1.23–45). http://www.wessa.net/rwasp_fitdistrweibull.wasp. Accessed September 8, 2011.

Wilhelmsen, K. H. (2001). Building the framework for an evolutionary explanation of projectile point variation: An example from the central Mississippi River Valley. In T. Hunt, C. Lipo & S. Sterling (Eds.), *Posing questions for a scientific archaeology* (pp. 97–144). Westport, CT: Bergin & Garvey.

Wood, J. W., Holman, D. J., O'Connor, K. A., & Ferrell, R. J. (2002). Mortality models of paleodemography. In R. Hoppa & J. Vaupel (Eds.), *Paleodemography: Age distributions from skeletal samples* (pp. 129–168). Cambridge, UK: Cambridge University.

Woods, J. C. (1987). *Manufacturing and use damage on pressure-flaked stone tools* (M.A. thesis). Idaho State University.

Yaroshevich, A., Kaufman, D., Nuzhnyy, D., Bar-Yosef, O., & Weinstein-Evron, M. (2010). Design and performance of microlith implemented projectiles during the Middle and the Late Epipaleolithic of the Levant: Experimental and archaeological evidence. *Journal of Archaeological Science, 37*, 368–388.

Yaroshevich, A., Zaidner, Y., & Weinstein-Evron, M. (2016). Projectile damage and point morphometry at the Early Middle Paleolithic Misliya Cave, Mount Carmel (Isral): Preliminary results and interpretations. In R. Iovita & K. Sano (Eds.), *Multidisciplinary approaches to the study of Stone Age weaponry* (pp. 119–134). Dordrecht: Springer.

Part V

Weapons as Cultural and Cognitive Markers

Chapter 18
Morphological Diversification of Stemmed Projectile Points of Patagonia (Southernmost South America). Assessing Spatial Patterns by Means of Phylogenies and Comparative Methods

Marcelo Cardillo and Judith Charlin

Abstract The aim of this work is to model patterns of morphological variation in Middle-Late Holocene stemmed projectile points from Patagonia through comparative methods. With this purpose, we explore the potential of different analytical strategies using projectile point shapes, obtained by means of geometric morphometrics. Phylogenetic and spatial variations were used to model morphological patterns on different scales. Morphological data comes from digitized images of projectile points from different areas of Patagonia. Morphometric characters were obtained using landmark and semilandmark descriptors. Mean shape by area was computed and used in cladistic analysis to model diversification trends. Then, phylogenetic and geographical coordinates were estimated for each data set and used as predictor variables in multiple regression procedures. Results suggest that historical patterns of shape change are channeled by spatial dimension. Pattern of mobility and interaction among human populations in Patagonia in the Middle-Late Holocene are discussed in light of these results.

Keywords Cladistic • Geometric morphometrics • Middle-Late Holocene • Shape changes

Introduction

The projectile points of Patagonia show a wide range of metric and morphological variation, particularly during the Middle-Late Holocene (Nami 1984; Ratto 1994; Franco et al. 2005, 2009, 2010; Gómez Otero et al. 2009; Cardillo and Charlin 2010; Charlin and González-José 2012) (Fig. 18.1). Initially, this variation was employed as a tool to define cultural identity as well as the interaction and replacement of hunter-gatherer populations in Patagonia, mainly within a culture-history typological perspective (Menghin 1952; Bórmida 1964). More recently, and from a regional approach, patterns of spatial variation in metric and morphological characters have been identified, and linked to functional aspects, design changes, and life-history of artifacts (Ratto 1992, 1994; Franco et al. 2005, 2009, 2010; Gómez Otero et al. 2009; Charlin and González-José 2012; see Rots 2016). In southern Patagonia, Franco and collaborators have observed spatial differentiation in stem size and hafting technique within a same projectile point type (Franco et al. 2005, 2009). In the northern part of the region, the existence of multiple metric and morphological variants of triangular projectile points has been indicated, contrasting with the greater homogeneity observed further South (Gómez Otero et al. 2009; Franco et al. 2010).

These sources of variation have also begun to be studied by means of different geometric morphometric methods, exploring morphological variation as a continuous quantitative phenomenon (Franco et al. 2009; Castiñeira et al. 2009, 2011, 2012; Charlin and González-José 2012). The diverse methods show the existence of regional differences in the design of projectile points in the Middle-Late Holocene, mainly along a north-south axis. We believe these variations may be explained by large-scale processes such as geographic distance, environmental variability, and temporal differentiation between human populations, among others. Thus our particular interest is to frame the observed patterns on a more inclusive spatial scale in order to explore trends in morphological change related to geographical and temporal factors. This broader scale accords with a biogeographical perspective, and the application of comparative methods (Harvey and Pagel 1991; Mace and Pagel 1994) allows hypotheses about technological evolution to be generated

M. Cardillo (✉) · J. Charlin
Consejo Nacional de Investigaciones Científicas y Técnicas, Instituto Multidisciplinario de Historia y Ciencias Humanas, Universidad de Buenos Aires, Saavedra 15 5th floor, 1083ACA Ciudad Autónoma de Buenos Aires, Argentina
e-mail: marcelo.cardillo@gmail.com

J. Charlin
e-mail: judith.charlin@gmail.com

Radu Iovita and Katsuhiro Sano (eds.), *Multidisciplinary Approaches to the Study of Stone Age Weaponry*, Vertebrate Paleobiology and Paleoanthropology, DOI 10.1007/978-94-017-7602-8_18

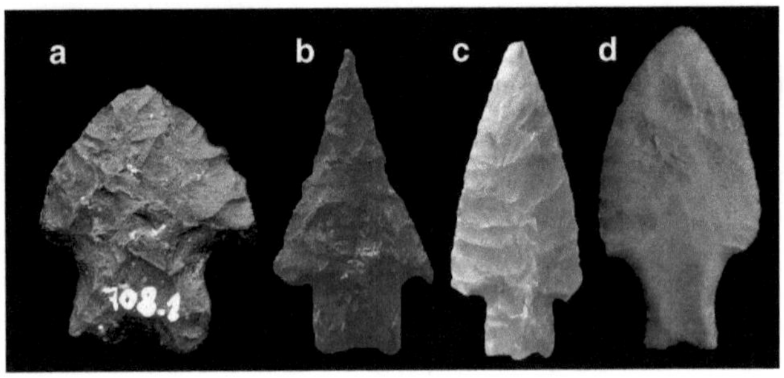

Fig. 18.1 Variation among stemmed projectile points in the Middle-Late Holocene (**a**, **b**, **c**). Fishtail projectile point from the Pleistocene-Holocene transition (**d**, taken from Flegenheimer 2009)

and contrasted. Prior analyses carried out on projectile points within this perspective suggest a clinal tendency in the morphological variation of stemmed Middle-Late Holocene points between 40° and 52° southern latitude (Cardillo and Charlin 2010). This first approximation showed a significant correlation between the mean shape of stemmed points grouped in six latitudinal strips and their corresponding spatial and phylogenetic (cladistic) distances.

The aim of the current study is to make use of morphological information obtained from a new set of samples to get better spatial coverage, and compare our data with previous results. Spatial and phylogenetic autocorrelation will be analyzed by means of statistical methods. This will allow us to generate a more accurate model on the *tempo* and *mode* of morphological evolution of stemmed projectile points in Patagonia.

Objectives and Hypotheses

In archaeology, phylogenetic reconstruction is a useful tool to generate hypotheses regarding *tempo* and *mode* of technological change (O'Brien and Lyman 2003; Lipo et al. 2005; among others), under the assumption that culture conforms and evolutionary system with a hierarchy of genealogical units analogous to the genealogical hierarchy of organic evolution (Boyd et al. 1997). Different approaches like maximum parsimony, distance-based, maximum likelihood and Bayesian statistics have been applied to explore hypothesis about the evolution of basketry (Jordan and Shennan 2009), tapestry motifs (Tehrani and Collard 2002), ceramics (Harmon et al. 2006), lithics (Darwent and O'Brien 2006; Buchanan and Collard 2007, 2008, 2010; Mesoudi and O'Brien 2008a, b; Cardillo 2009; Lycett 2009) and languages (Gray and Atkinson 2003; Atkinson et al. 2008). Since some of them are related to specific hypotheses about rates of change, the basic principle for application is based on the observation that culture constitutes an independent system of inheritance (but in many cases related to genetic one) (Cavalli-Sforza and Feldman 1981; Durham 1991). Ethnographic and experimental observations indicate that despite of the important role of horizontal transmission (between biologically unrelated persons and cultural groups) in blending cultural information, an accurate phylogenetic estimation is possible in many cases (Collard et al. 2005; Collard 2006; see also Greenhill et al. 2009 and Muscio 2010 for examples of phylogenetic methods under different levels of borrowing). This signal is expressed in the form of a branching pattern of trait modification of cultural features (Collard et al. 2005; Collard 2006). Within a comparative strategy, obtaining a phylogenetic tree is not an end itself but an additional stage in an analysis involving both the evaluation of results and the comparison with different lines of evidence (spatial, environmental, temporal, technological, among others). In this case we are interested in cladistic reconstruction as a way of generating a quantitative model of patterns of technological evolution throughout Patagonia. Within this perspective, space is not merely a scenario, but takes on an active role in the formation and evolution of variability (e.g., Cavalli-Sforza et al. 1994; Guglielmino et al. 1995; Pérez and Monteiro 2009). We expect that, during the Holocene, two evolutionary phenomena related to relative distance or closeness of cultural entities (human groups) will have taken place: parallelism or joint evolution, and divergence. On the one hand, parallelism can explain cases in which artifact populations from adjacent or nearby regions show similar trajectories of change (see Gould 2002 for further discussion). This is to be expected given the high probability of information exchange (horizontal transmission, *sensu* Cavalli-Sforza and Feldman 1981; Cavalli-Sforza et al. 1994) among human groups that inhabited neighboring spaces or areas within the same home range over time. On the other hand, artifact populations separated from one another by greater distances will tend to diverge because the vertical transmission processes will be more important than the horizontal ones, a product of increased isolation between human groups related to spatial distance (Cavalli-Sforza et al. 1994). In consequence, it is to be expected that the range of mobility of human populations in time and the spatial distance between them should generate "thresholds" in which one mechanism is more important than another. As suggested by recent research, it has been possible to observe this phenomenon from multiple lines of evidence, and it seems to be a relevant factor in the evolution of human populations in Patagonia (Cocilovo and Guichón 1986–1986; Pérez and Monteiro 2009; Pérez et al. 2010, 2011, among others). In this study we use a spatial scale of 1500 km, encompassing the whole of continental Patagonia (Fig. 18.2), to model shape variation in Middle-Late Holocene stemmed projectile points.

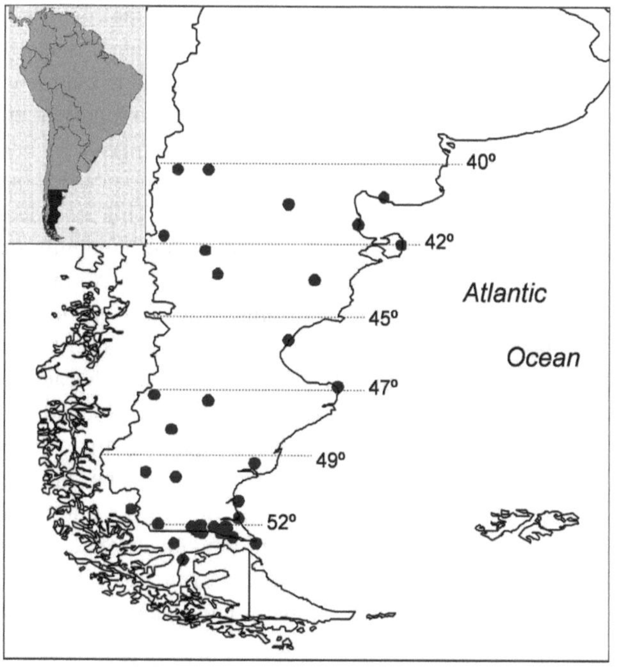

Fig. 18.2 Map of Patagonia with the location of the projectile point samples used in this work

We explore how spatial dimension mediates on the process of projectile points shape diversification. It is expected that the greater the independence between space and phylogeny, a historical model of evolution will be the most plausible to explain differences between divergent Hypothetical Taxonomic Units (HTU), whereas the model determined by geography is expected on different levels of correlation between both variables, resulting in parallel evolution between evolutionary units. The procedure of phylogenetic analysis employed is based on distances. The morphological distance between the different evolutionary units is later compared with the spatial (latitudinal) data to evaluate the role of geography in the patterns of the observed groupings.

Materials and Methods

Morphological data comes from digitized images of projectile points (n = 301) collected by us and published by other researchers from different Patagonian areas from 40° to 52° south (Fig. 18.2). The total sample was divided into latitudinal strips. We define eight latitudinal strips, represented by the average values of the analyzed shapes. Spatial coordinates were obtained from calculating the centroid of the spatial locations by latitude. A spatial subdivision according to longitude was not possible due to sample size.

For the analysis four steps common to different comparative methods were followed (Fig. 18.3): (a) acquisition of shape coordinates; (b) phylogenetic reconstruction; (c) spatial analysis; and (d) correlation between spatial and temporal vectors by multiple regression.

(a) The first step was obtaining shape data by geometric morphometric methods (Bookstein 1982; Rohlf 1993; Adams et al. 2004). Our own digital database was completed with scanned images of projectile points from other places of Patagonia published by other researchers. They were digitized by a flat scanner at 100 dpi resolution. Four landmarks (points with topological correspondence among objects) and 18 semilandmarks (sets of points related to each other) were established to describe homologous outlines (Bookstein 1996–1997). The raw coordinates were scaled and rotated to eliminate size-related information and then superimposed by the Procrustes method (Rohlf and Slice 1990). Next, the shape coordinates were grouped in eight groups according to latitude and the mean shape of each group was estimated. Seven latitudinal strips are the same size, with the eighth HTU comprising two degrees of latitude (43°–45°). This grouping was carried out to minimize the effect of the small sample size from these areas (see below). The Thin-Plate-Spline algorithm (Rohlf and Slice 1990) was used to depict the morphological change from the ancestral morphology (Fishtail projectile point) and the shape of each HTU.

(b) Phylogenetic reconstruction was used to generate a model of projectile point morphological diversification. To this end, a Neighbor Joining (NJ) method based on distances was utilized. Results were compared with those obtained by Maximum Parsimony (data not shown). The first method is commonly employed for continuous data (Saitou and Nei 1987). Although the method is closer to cluster analysis and phenetics, NJ is widely used in phylogenetic reconstruction, since the tree can be polarized to indicate the direction of change. Moreover, experimental studies show that NJ is either effective in recovering the true phylogeny or in many cases is significantly closer to the actual tree (Atteson 1997; Gascuel and Steel 2006; Mihaescu et al. 2009). Another advantage is that it assumes that evolutionary change has a constant rate, making it accurate to the study of continuous change. This is consistent with that observed in culture, where rates of evolution are much higher than the biological ones (Boyd and Richerson 1985). The algorithm used seeks to minimize the total length of the tree by interactively joining the closest HTUs starting out from a matrix of Euclidean distances. Later the uncertainty in tree reconstruction and HTU membership was evaluated by a bootstrap procedure (n = 5000). Only those branches with a support greater than 50% were deemed robust solutions and highlighted on the tree.

To polarize the tree the mean shape of Fishtail stemmed points assigned to the Pleistocene-Holocene boundary (*circa* 11,000 BP) from Bird (1988, 1946, 1988) collection was used.

Subsequently the NJ tree was utilized as a base to obtain new variables or vectors that represent the phylogenetic structure and to correlate HTU evolution at different topological levels. An autocorrelation signal along the branches of the tree is expected because the ancestor-descendent relationship models cladogenetic history of the evolutionary units. In this analysis, the first vectors describe the general pattern of diversification (early diversification events) and the other vectors the more local patterns that result in small clades (Peres-Neto 2006). Because of this decomposition, phylogenetic autocorrelation vectors are useful to describe tree topology at different levels.

To build these vectors, Moran's I autocorrelation method was utilized, based on the correlation between each pair of HTUs, measured by the quantity of nodes separating them (Diniz-Filho et al. 1998, 2007; Peres-Neto 2006). These vectors are standardized, which means that they are centered to a mean of zero and unit variance, and are uncorrelated. Those HTUs that possess the smallest distance between each other in all possible rearrangements of the tree in use (by permutation of branches) will show higher correlation values, whether negative or positive. These values indicate the same as in ordinary correlation, a positive or negative covariation between the HTUs throughout the phylogeny. Other procedures and measures of closeness, such as the number of nodes separating the HTUs can be estimated for similar purposes. The choice of this method instead of other kinds of distance analysis was based on the possibility of employing a similar metric to obtain the phylogenetic and spatial data.

(c) Spatial variables were obtained by Spatial Eigenvector methods (Griffith and Peres-Neto 2006; Rangel et al. 2010) which enable the decomposition of orthogonal (independent) variables from a matrix of Cartesian coordinates (average longitude and latitude), which in this case belong to the centroid of the different samples by latitudinal string. Starting out from a matrix of Euclidean distances, which is truncated to represent the minimum distance between sites, decomposition into independent vectors of different variance is carried out by the analysis of principal coordinates. This procedure allows the representation of the spatial variability on different scales, measured according to Moran's I autocorrelation index. In this way this method is superior to approximations based on linear distances among sites or the use of raw coordinates because it models all kind of spatial structures (see Borcard and Legendre 2002). As a general rule, the first vector explains broad spatial tendencies and the greatest spatial variance, whereas the following vectors are more local and have less informational value (especially the last ones). The Moran's I index is similar to Pearson's correlation, in that it has values close to one (positive or negative) when the correlation between two points is higher, and close to zero when this association is absent. As this information is obtained from the Cartesian coordinates alone, it is independent of the morphological data and can be incorporated as a factor or a covariate in regression models (Dormann 2007; Rangel et al. 2010). Therefore this procedure follows a similar logic to the one employed to represent both space and phylogeny as variables on different scales.

(d) The first phylogenetic and spatial autocorrelation vectors were utilized in a multiple regression analysis. For this purpose a multivariate multiple regression procedure based on distances was used, in which the p-value of the null hypothesis is obtained by permutations (Anderson 2001, 2003). This procedure is more robust and preferable to parametric regression when the samples are not distributed normally and have relatively small degrees of freedom. In this way, the null hypothesis of absence of spatial structure in the pattern of morphological diversification of stemmed points in the Middle-Late Holocene is brought into play.

The TPS statistical package was employed for the geometric morphometric analyses (Rohlf 2004, 2007). For the estimation of phylogeny by means of NJ and the visualization of deformations by Thin-Plate-Spline the program Past 2.14 (Hammer et al. 2001) was used. Spatial and temporal autocorrelation vectors were obtained by the program SAM (Rangel et al. 2010) and R 2.13 (R development core team 2007) respectively. The multiple regression was carried out with the program DISLIM *v5* (Anderson 2001, 2003).

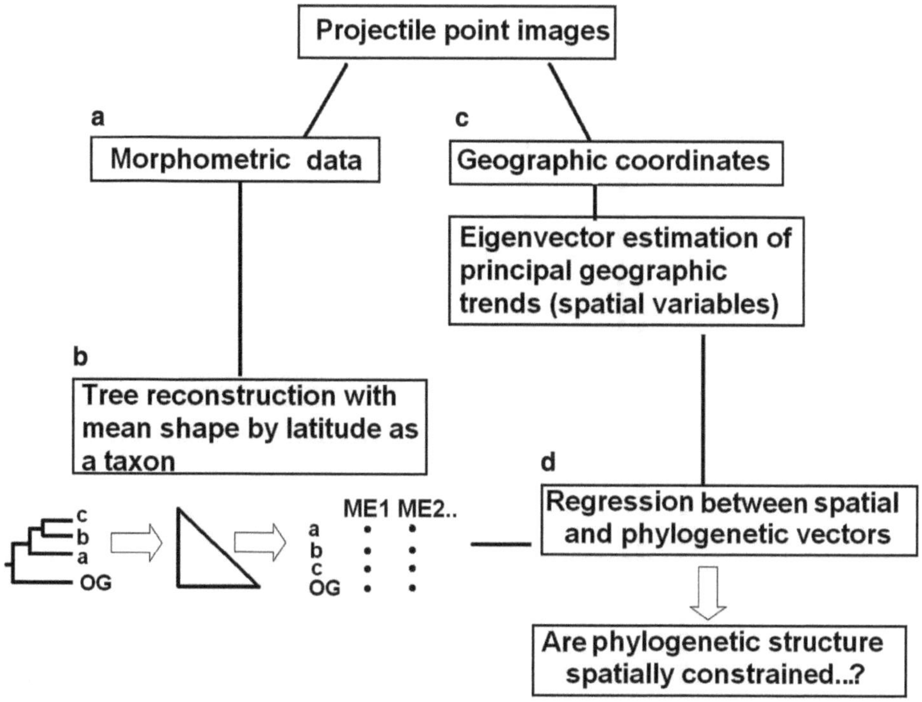

Fig. 18.3 Structure of the comparative analysis between phylogenetic and spatial evidence used here to contrast the null hypothesis of independence (see text for references)

Analysis and Results

The NJ analysis shows two large groups of morphologies separated by a 100% bootstrap support (n = 5000): the first group (A) joins the latitudes 40°, 41°, 42° and 47° S; and the second one (B) comprises 51°, 50°, 52° and 43°–45° S (Fig. 18.4). Group A joins the assemblages with more expanded stems -especially at the level of the neck- and shorter blades, while group B shows the opposite pattern, with more elongated blades and smaller stems. The bootstrap values indicate that only some branches are highly supported, which points to a certain level of ambiguity in the phylogenetic reconstruction. Yet a tendency is observed in Northern and Southern Patagonian groups to converge on morphologies with similar deformation patterns. The NJ tree indicates that in general terms the average shape of projectile points follows a spatially grouped pattern. The groups with best bootstrap support are those from high latitudes (50°, 51°, and 52°) while at middle and low latitudes a greater variability is observed in the results. Among these latter, the clades with best support are those of latitude 40° and 47°, and they have a slightly higher value than the average for the base of this clade (40°, 41°, 42°, 47°) (Fig. 18.4).

Another tree was obtained using Maximum Parsimony (not shown), with results agreeing with NJ. In this case, the reconstruction of ancestral shapes for each node shows that the greatest change takes place at the beginning of the diversification process, towards the base of the tree.

The NJ tree was then used to generate phylogenetic autocorrelation vectors (Fig. 18.5). The pattern obtained suggests the existence of a phylogenetic structure in the data. The first Moran's I eigenvector (ME) shows two large groups with high autocorrelation (the correlation between clades A and B is negative). The second and third ME point out autocorrelation values on a lower scale: these fluctuate between positive and negative Moran's I and with different intensity (Fig. 18.5).

Using a similar criterion, spatial analysis generated two vectors. The first one describes the general tendency (79%) of spatial variation in the data (Fig. 18.6a), while the second covers the remaining 21% (Fig. 18.6b). The first filter indicates that as distance increases, the spatial relation between the points changes gradually (the greater the distance, the greater the difference). The second one indicates local variation patterns, with fluctuations in positive-negative correlation.

Finally, phylogenetic and spatial information was used to explore the geographically related diversification pattern.

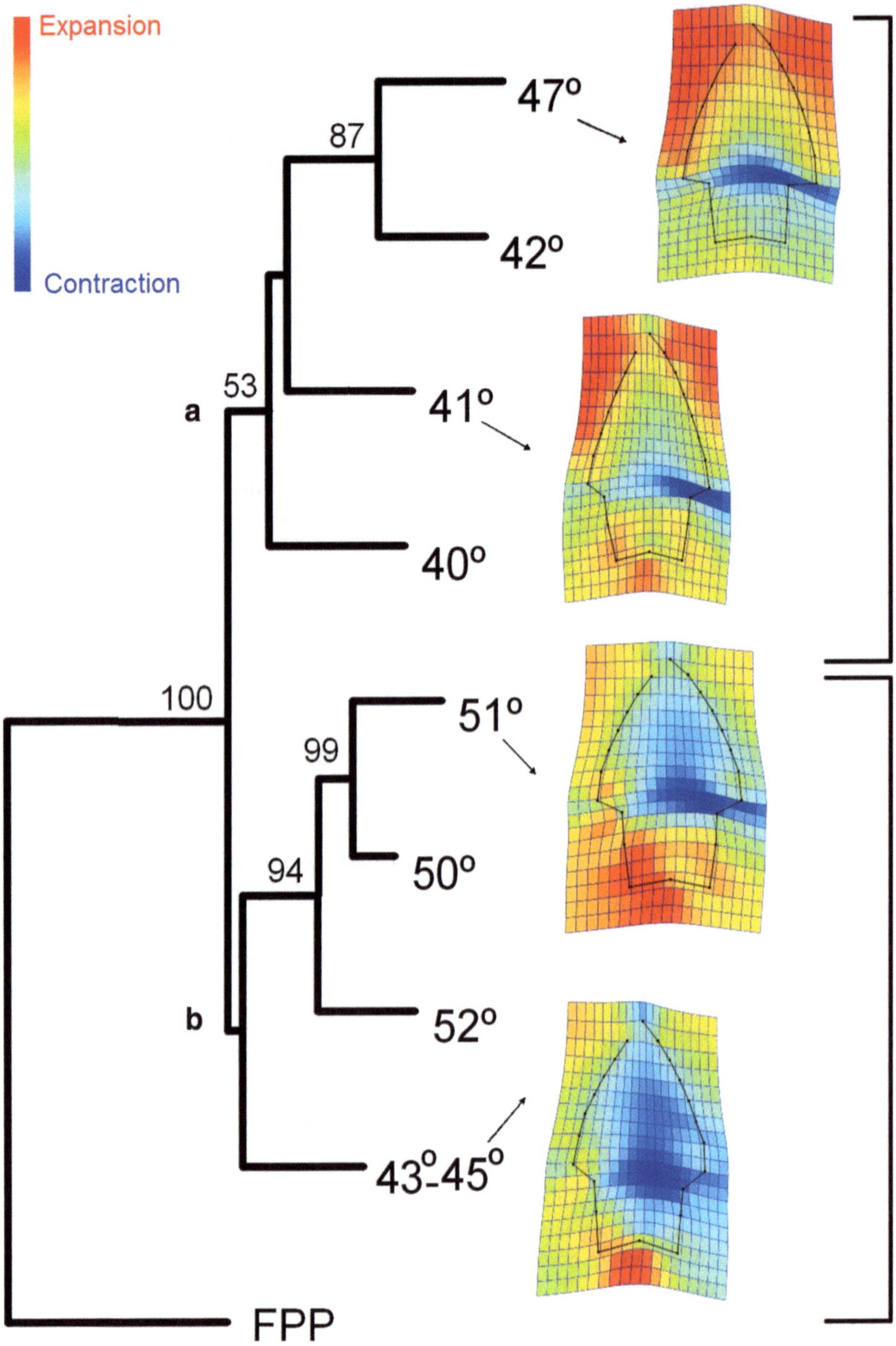

Fig. 18.4 Neighbor joining tree. The morphological variation of each HTU with respect to FPP is represented via Thin-Plate-Spline deformation grids. The numbers at the base of the branch indicate the support of those above 50%. Two large clades separated by 100% of the bootstrap support (**a** and **b**) are observed

The correlation matrix between the first spatial filter (SF) and phylogenetic Moran's I eigenvector (PME) shows significant correlations between some of them (Table 18.1).

The highest correlation is observed between the first PME and the second SF ($r = 0.82$, $p < 0.05$), and between the second PME and the first SF ($r = 0.82$, $p < 0.05$). This suggests a spatial structure pattern on both the macroscale and the regional level. It means that the phylogenetic pattern is spatially channeled at different levels (see discussion). Finally, multiple regression between the phylogenetic and the two spatial vectors, taken as independent variables, shows significant results (*pseudo-F* = 11.44, $p = 0.0006$).

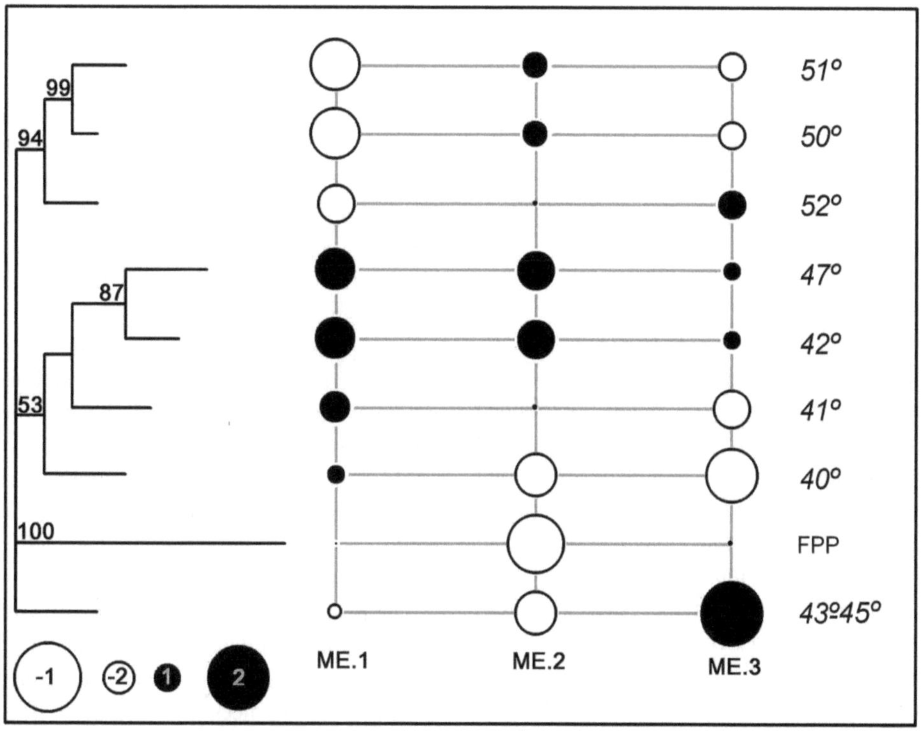

Fig. 18.5 Phylogenetic autocorrelation pattern. The size of the circles indicates greater positive (black) or negative (white) correlation between adjacent HTUs

Fig. 18.6 Spatial correlation vectors obtained from geographical coordinates. The first vector **a** describes 79% of the spatial variation, the second one **b** the remaining 21%

Table 18.1 Correlation matrix between the first spatial and phylogenetic vectors. Upper triangle: *p*-values, lower triangle: Pearson *r* values. SF: Spatial Filters; PME: Phylogenetic Moran's I Eigenvector

	PME1	PME2	PME3	SF1	SF2
PME1		1	1	0.29	0.01
PME2	0.0		1	0.01	0.44
PME3	0.0	0.0		0.72	0.35
SF1	−0.40	0.82	0.14		1
SF2	0.82	0.3	0.35	0.0	

The difference between the sum of the squares (Ss) explained by the regression model (Ss = 14.24) and the residual one (Ss = 3.73) indicates that both filters taken together explain 79% of the phylogenetic variation represented by the first two PME, while 21% remains unexplained by these factors.

Discussion

Results suggest that, at least for the Middle-Late Holocene, variability in the morphology of projectile points is explained by the occurrence of geographical (spatial) and historical macroscale-related mechanisms. Two large clades with high support are observed, which suggests the evolution of the shape of projectile points towards more elongated blades and contracted stems in one clade (Northern Patagonia), and more expanded blades with wider stems in the other (Southern Patagonia). This separation would have happened early in the evolutionary history of the projectile point populations, as the topology of the tree suggests. As it was pointed out by previous analyses (Cardillo and Charlin 2010), the grouping pattern indicates clusters that follow a geographical order, both on the broad 1500 km scale as well as in more regional order of magnitude (e.g., the 50°, 51°, 52° clade). The regression between the first two phylogenetic vectors and the wide-scale (SF1) and more local-scale (SF2) spatial factors indicates that the spatial model explains 79% of phylogenetic variability. This allows rejection of the null hypothesis of the independence between the diversification pattern and geography. Thus we can argue that the divergence into two large groups may be a phenomenon channeled by spatial distance. As it was mentioned before, this phenomenon can be related to processes linked to mobility and information flow between human populations, the balance between the vertical and horizontal transmission of information as well as environmental factors. We understand that distance increases the probability of both neutral (*sensu* Dunnell 1978) and adaptive variation becoming fixed, generating variation in designs. These differences could be random, without any adaptive or cultural value as Binford (1963) or Morrow and Morrow (1999) suggest. The latter authors observed a clinal pattern or metric variation gradient along South America in the attributes of Fishtail projectile points. A similar pattern was observed for this type of projectile point by Castiñeira et al. (2012) using geometric morphometric methods but on a smaller scale, restricted to Southern South America. Therefore it may be possible that the groups most distant from each other, in Northern (40°–49°) and Southern Patagonia (50°–52°), show a clear tendency towards technological divergence throughout the Holocene.

Given the considerable environmental variability in Patagonia, it is also to be expected that the pattern we have observed here may be linked to ecological mechanisms. The Patagonian environment is highly conditioned by latitude (Clapperton 1993), so it may be expected that environmental and spatial variables will be interrelated. In this way, the connection between adaptive factors (changes in design related to performance requirements) as well as purely random ones (chance variations, transmitted vertically) may explain the observed variability.

The spatial vectors can be included as covariables into a regression analysis to model the relationship between the pattern of diversification and the environment, when the purely spatial variation is controlled (Peres-Neto 2006). This would be useful to account for the "pure" environment-related variation, since we suppose that projectile point design was influenced by performance requirements in different environments.

The results obtained are concordant with the observations made of other Middle-Late Holocene coastal technologies (Cardillo 2011), in which there is an increase in differences related to the distances. In this case results are partly related to environmental factors that, together with neutral random variation, would have led to a trend towards technological divergence between Northern and Southern Patagonia. We believe this phenomenon is similar to that observed through other lines of evidence, which suggest an increase in the phenotypical regionalization of human populations in the Late Holocene (Pérez et al. 2011). Differences between hunter-gatherer and horticultural populations in Patagonia have been recorded in dental and cranial structures (Béguelin and Barrientos 2006; Pérez and Monteiro 2009; Pérez et al. 2011) as well as in molecular information (Lalueza et al. 1997). This research suggests that the ecological dimension (environment, diet) and geographical space could explain the differences between human groups in the studied area as well as in large tracts of South America (Bernal et al. 2010; Pérez et al. 2011).

It should be noted that the debate regarding the existence of variations between Northern and Southern Patagonia is a long-standing one (Orquera 1987; Borrero 2001). It has mainly been focused on the Santa Cruz River (50° S) being posited as a biogeographical barrier, since this river is the largest in Southern Patagonia (Borrero and Borrazzo 2011 and references therein). Despite of the debate over some of the cultural traits that mark the differences, the distinctiveness of the archaeology South of the Santa Cruz River can be demonstrated, including a preference for dark volcanic rocks, a decrease in the use of blades, the presence of non-standardized end-scrapers, a low ratio of end to side-scrapers (Orquera 1987; Borrero 2001; Franco 2002; Cardillo 2011) and, as has been recently demonstrated, a differential distribution of rock art motifs (Charlin and Borrero 2012).

Beyond these considerations concerning the regional archaeological record and within the general pattern, some discrepancies must be noted. There is no exact fit between geography and phylogeny, since some groups include points distant from each other, and other HTUs composed by adjacent strip show a low general support. This could be due to the morphological variability and the sample size, since the mean shape estimated by latitude could be affected by "outlier" shapes. The 43°–45° HTU shows a broader grouping that potentially encloses greater variation. At the same time it is possible that some morphologies assigned to the Middle-Late Holocene may be even earlier and they show different variation patterns, affecting the estimation of the mean shape. As the observed pattern also averages variation relative to the life-history of the projectile points, it is possible that different rates of maintenance, recycling, and discard among HTUs may give confuse results, at least in part. However, this can be controlled in subsequent analyses by choosing assemblages more temporally controlled by means of multiple radiocarbon datings, as well as incorporating a larger number of characters in the phylogenetic reconstruction, which will allow a better support of the resulting branches to be obtained.

We believe that, though the results point to the importance of the spatial dimension in channeling diversification process, this does not invalidate the importance of building a historical model. Rather, it suggests that, for this purpose, it is necessary to bear the spatial scale in mind. In the same way, more exact historical models (for instance, for North or South Patagonia) could be built by combining metric (length, width, thickness) and morphological (i.e., following the method developed by Catalano et al. 2010; Goloboff and Catalano 2011) information or discrete variables (presence or absence of characters, technical attributes) within the character matrices (Goloboff et al. 2006). Also, varying levels of reuse and recycling can bias the results to some extent, since successive use, damage and rejuvenation events change projectile point shape, as has been observed in many experimental and allometric studies (Flenniken and Raymond 1986; Andrefsky 2006; Buchanan 2006; Shott and Ballenger 2007; Shott et al. 2007; Hunzicker 2008). Because the stem is the portion less affected by reduction and life-history (Charlin and González 2012; Thulman 2012), it could be used as an independent shape module for phylogenetic purposes. In the same way, the variation produced by the life-history of projectile points (Shott 2016) can be incorporated into a phylogenetic model as a character, as it is plausible that the reduction procedures may contain information related to ecological and historical mechanisms.

Conclusions

This work has sought to explore, via cladistic and spatial analyses, the morphological variability of stemmed Middle-Late Holocene projectile points. The general result agrees with what was expected within a model of isolation by distance. Within this model, the increasing spatial separation between evolutionary units generates the conditions through which processes of divergence are produced in a space with great environmental and ecological variability. Consequently, the results allow us to generate a model of the evolutionary history of stemmed projectile points in the Middle-Late Holocene, a history that should not be understood as divorced from adaptation and diversification processes among the human populations of Patagonia.

Acknowledgements We are grateful to Radu Iovita and Katsuhiro Sano, for inviting us to participate of the workshop "Multidisciplinary Approaches to the Study of Stone Age Weaponry" carried out in the Institut für Vor- und Frühgeschichte, University of Mainz. Thanks to Peter White and Jimena Alberti for their fruitful suggestions about English translation and to the reviewers whose comments have very much improved this paper. We are entirely responsible for any errors or omissions. This research was funded by Consejo Nacional de Investigaciones Científicas y Técnicas.

References

Adams, D., Rohlf, J., & Slice, E. (2004). Geometric morphometrics: Ten years of progress following the 'revolution'. *Italian Journal of Zoology, 71*, 5–16.

Anderson, M. J. (2001). A new method for non-parametric multivariate analysis of variance. *Austral Ecology, 26*, 32–46.

Anderson, M. J. (2003). *DISTLM* v3. *a FORTRAN computer program to calculate a distance-based multivariate analysis for a linear model.* New Zealand: Department of Statistics, University of Auckland.

Andrefsky, W, Jr. (2006). Experimental and archaeological verification of an index of retouch for hafted bifaces. *American Antiquity, 71*, 743–757.

Atkinson, Q. D., Meade, A., Venditti, C., Greenhill, S. J., & Pagel, M. (2008). Languages evolve in punctuational bursts. *Science, 319*, 588.

Atteson, K. (1997). The performance of neighbor-joining algorithms of phylogeny reconstruction. In T. Jiang & D. Lee (Eds.), *Lecture notes in computer Science* (pp. 101–110). Berlin: Springer.

Béguelin, M., & Barrientos, G. (2006). Variación morfométrica postcraneal en muestras tardías de restos humanos de Patagonia: Una aproximación biogeográfica. *Intersecciones en Antropología, 7*, 49–62.

Bernal, V., Pérez, S. I., González, P. N., Sardi, M., & Pucciarelli, H. M. (2010). Spatial patterns and evolutionary processes in southern South America: A study of dental morphometric variation. *American Journal of Physical Anthropology, 142*, 95–104.

Binford, L. R. (1963). 'Red ocher' caches from the Michigan Area: A possible case of cultural drift. *Southwestern Journal of Anthropology, 19*, 89–109.

Bird, J. (1938). *Antiquity and migrations of the early inhabitants of Patagonia* (Vol. XXVIII, pp. 250–275). Geographical Review.

Bird, J. (1946). The archaeology of Patagonia. In J. H. Steward (Ed.), *Handbook of South American Indians* (Vol. I, pp. 17–24). The marginal tribes Washington: Smithsonian Institution, Bureau of American Ethnology.

Bird, J. (1988). *Travels and archaeology in South Chile*. Iowa: University of Iowa Press.

Bookstein, F. (1982). Foundation of morphometrics. *Annual Review of Ecology and Systematics, 13*, 451–470.

Bookstein, F. L. (1996–1997). Landmarks methods for form without landmarks: Morphometrics of group differences in outline shape. *Medical Image Analysis, 1*, 225–243.

Borcard, D., & Legendre, P. (2002). All-scale spatial analysis of ecological data by means of principal coordinates of neighbour matrices. *Ecological Modelling, 153*, 51–68.

Bórmida, M. (1964). *Arqueología de la costa norpatagónica*. Madrid: Trabajos de Prehistoria, XIV.

Borrero, L. A. (2001). Cambios, continuidades, discontinuidades: Discusiones sobre arqueología Fuego-patagónica. In E. Berberián & A. Nielsen (Eds.), *Historia Argentina Prehispánica* (pp. 815–838). Córdoba: Editorial Brujas.

Borrero, L. A., & Borrazzo, K. (2011). La geografía cultural del sudoeste de Patagonia continental. In L. A. Borrero & K. Borrazo (Eds.), *Bosques, Motañas y Cazadores* (pp. 7–36). Buenos Aires: CONICET-IMICIHU.

Boyd, R., & Richerson, P. J. (1985). *Culture and the evolutionary process*. Chicago: University of Chicago Press.

Boyd, R., Richerson, P. J., Borgerhoff-Mulder, M., & Durham, W. H. (1997). Are cultural phylogenies possible? In P. Weingart, P. J. Richerson, S. D. Mitchell & S. Maasen (Eds.), *Human by nature. Between biology and the social sciences* (pp. 355–386). New York: Lawrence Erlbaum Associates.

Buchanan, B. (2006). An analysis of Folsom projectile point resharpening using quantitative comparisons of form and allometry. *Journal of Archaeological Science, 33*, 185–199.

Buchanan, B., & Collard, M. (2007). Investigating the peopling of North America through cladistic analyses of Early Paleoindian projectile points. *Journal of Anthropological Archaeology, 26*, 366–393.

Buchanan, B., & Collard, M. (2008). Phenetics, cladistics, and the search for the Alaskan ancestors of the Paleoindians: A reassessment of relationships among the Clovis, Nenana, and Denali archaeological complexes. *Journal of Archaeological Science, 35*, 1683–1694.

Buchanan, B., & Collard, M. (2010). A geometric morphometric-based assessment of blade shape differences among Paleoindian projectile point types from western North America. *Journal of Archaeological Science, 37*, 350–359.

Cardillo, M. (2009). Temporal trends in the morphometric variation of the lithic projectile points during the Middle Holocene of Southern Andes (Puna region). A coevolutionary approach. In H. Muscio & G. López (Eds.), *Theoretical and methodological issues in evolutionary archaeology: Toward an unified Darwinian paradigm* (pp. 13–20). Oxford: BAR International series.

Cardillo, M. (2011). Exploring latitudinal patterns of lithic technology variation in continental coastal Patagonia, Argentina. *Journal of Archaeological Science, 38*, 2675–2682.

Cardillo, M., & Charlin, J. (2010). Diversificación morfológica en las puntas de proyectil de Patagonia. Work presented at Primer Encuentro de Morfometría "Morfometría geométrica y estudios filogenéticos", Universidad de La Plata, La Plata.

Castiñeira, C., Cardillo, M., Charlin, J., Fernicola, J. C., & Baeza, J. (2009). Análisis morfométrico de los cabezales líticos "cola de pescado" del Uruguay. In O. Palacios, C. Vázquez, T. Palacios & E. Cabanillas (Eds.), *Arqueometría Latinoamericana* (pp. 360–366). Buenos Aires: Comisión Nacional de Energía Atómica.

Castiñeira, C., Cardillo, M., Charlin, J., & Baeza, J. (2011). Análisis de morfometría geométrica en puntas Cola de Pescado del Uruguay. *Latin American Antiquity, 22*, 335–358.

Castiñeira, C., Charlin, J., Cardillo, M., & Baeza, J. (2012). Exploring morphometric variations in Fishtail projectile points from Uruguay, Pampa and Patagonia. In L. Miotti, M. Salemme, N. Flegenheimer & T. Goebel (Eds.), *Southbound: Late Pleistocene peopling of Latin America, current research in the Pleistocene special publication* (Part 2, pp. 57–62). Texas: Center for the Study of the First Americans.

Catalano, S., Goloboff, P., & Giannini, N. (2010). Phylogenetic morphometrics (I): The use of landmark data in a phylogenetic framework. *Cladistics, 26*, 539–549.

Cavalli-Sforza, L., & Feldman, M. (1981). *Cultural transmission and evolution: A quantitative approach*. Princeton: Princeton University Press.

Cavalli-Sforza, L., Menozzi, P., & Piazza, A. (1994). *The history and geography of human genes*. Princeton: Princeton University Press.

Charlin, J., & Borrero, L. A. (2012). Rock art, inherited landscapes and human populations in Southern Patagonia. In J. McDonald & P. Veth (Eds.), *A companion to rock art*, (Chapter 22, pp. 381–398). Hoboken: Wiley-Blackwell.

Charlin, J., & González-José, R. (2012). Size and shape variation in Late Holocene projectile points of southern Patagonia. *A geometric morphometric study. American Antiquity, 77*, 221–242.

Clapperton, Ch. (1993). *Quaternary geology and geomorphology of South America*. Amsterdam: Elsevier.

Cocilovo, J., & Guichón, R. (1985–1986). Propuesta para el estudio de las poblaciones aborígenes del extremo austral de Patagonia. *Anales del Instituto de la Patagonia, 16*, 111–123.

Collard, M. (2006). Branching, blending, and the evolution of cultural similarities and differences among human populations. *Evolution and Human Behavior, 27*, 169–184.

Collard, M., Shennan, S. J., & Tehrani, J. (2005). Branching versus blending in macroscale cultural evolution: A comparative study. In C. P. Lipo, M. J. O'Brien, M. Collard & S. J. Shennan (Eds.), *Mapping our ancestors: Phylogenetic methods in anthropology and prehistory* (pp. 53–65). New York: Aldine Transaction.

Darwent, J., & O'Brien, M. (2006). Using cladistics to construct lineages of projectile points from northeastern Missouri. In C. P. Lipo, M. O'Brien, M. Collard & S. Shennan (Eds.), *Mapping our ancestors: Phylogenetic approaches in anthropology and prehistory* (pp. 185–208). New York: Aldine.

Diniz-Filho, J. A. F., Bini, L. M., Rodríguez, M. Á., Rangel, T. F. L. V., & Hawkins, B. A. (2007). Seeing the forest for the trees: Partitioning ecological and phylogenetic components of Bergmann's rule in European Carnivora. *Ecography, 30*, 598–608.

Diniz-Filho, J. A. F., De Santana, C. E. R., & Bini, L. M. (1998). An eigenvector method for estimating phylogenetic inertia. *Evolution, 52*, 1247–1262.

Dormann, C. F. (2007). Effects of incorporating spatial autocorrelation into the analysis of species distribution data. *Global Ecology and Biogeography, 16*, 129–138.

Dunnell, R. C. (1978). Style and function: A fundamental dichotomy. *American Antiquity, 43*, 192–202.

Durham, W. H. (1991). *Coevolution: Genes culture and human diversity*. Stanford: Stanford University Press.

Flegenheimer, N. (2009). *Catálogo para Réplicas de Las Primeras Ocupaciones Pampeanas* Necochea: Área Arqueología y Antropología, Municipalidad de Necochea.

Flenniken, J. J., & Raymond, A. W. (1986). Morphological projectile point typology: Replication, experimentation and technological analysis. *American Antiquity, 51*, 603–614.

Franco, N. (2002). Estrategias de utilización de recursos líticos en la cuenca superior del Río Santa Cruz. PhD dissertation, Universidad de Buenos Aires.

Franco, N., Cardillo, M., & Borrero, L. (2005). Una primera aproximación a la variabilidad presente en las puntas denominadas "Bird IV". *Werken, 6*, 81–95.

Franco, N., Castro, A., Cardillo, M., & Charlin, J. (2009). La importancia de las variables morfológicas, métricas y de micro-desgaste para evaluar las diferencias en diseños de puntas de proyectil bifaciales pedunculadas: Un ejemplo del sur de Patagonia continental. *Magallania, 37*, 99–112.

Franco, N., Gómez Otero, J., Guráieb, G., Goye, S., Cirigliano, N., & Banegas, A. (2010). Variaciones espaciales en diseños de puntas pedunculadas medianas en Patagonia Argentina: Una nueva aproximación. In J. R. Bárcena & H. Chiavazza (Eds.), Arqueología Argentina en el Bicentenario de la Revolución de Mayo, tomo I (Chapter 5, pp. 281–286). Mendoza: Facultad de Filosofía y Letras de la Universidad Nacional de Cuyo, INCIHUSA-CONICET & ANPCyT.

Gascuel, O., & Steel, M. (2006). Neighbor-joining revealed. *Molecular Biology and Evolution, 23*, 1997–2000.

Goloboff, P. A., & Catalano, S. A. (2011). Phylogenetic morphometrics (II): Algorithms for landmark optimization. *Cladistics, 27*, 42–51.

Goloboff, P. A., Mattoni, C. I., & Quinteros, A. S. (2006). Continuous characters analyzed as such. *Cladistics, 22*, 589–601.

Gómez Otero, J., Banegas, A., Goye, M. S., & Franco, N. V. (2009). Variabilidad morfológica de puntas de proyectil en la costa centro-septentrional de Patagonia Argentina: Primeros estudios y primeras preguntas. In *Octavo Congreso de Historia Social y Política de la Patagonia Argentino-Chilena. Las fuentes en la construcción de una Historia Patagónica* (pp. 110–118). Rawson: Secretaría de Cultura de la Provincia del Chubut.

Gould, S. K. (2002). *The structure of evolutionary theory*. Cambridge: Harvard University Press.

Gray, R. D., & Atkinson, Q. D. (2003). Language-tree divergence times support the Anatolian theory of Indo-European origin. *Nature, 426*, 435–439.

Greenhill, S. J., Currie, T. E., & Gray, R. D. (2009). Does horizontal transmission invalidate cultural phylogenies? *Proceedings of Royal Society, 276*, 2299–2306.

Griffith, D. A., & Peres-Neto, P. (2006). Spatial modeling in ecology: The flexibility of eigenfunction spatial analyses. *Ecology, 87*, 2603–2613.

Guglielmino, C. R., Viganotti, C., Hewlett, B., & Cavalli-Sforza, L. L. (1995). Cultural variation in Africa: Role of mechanism of transmission and adaptation. *Proceeding of the National Academy of Sciences, 92*, 7585–7589.

Hammer, Ø., Harper, D. A. T., & Ryan, P. D. (2001). PAST: Paleontological statistics software package for education and data analysis. *Palaeontologia Electronica, 4*, 9 http://palaeo-electronica.org/2001l/past/issue101.htm

Harmon, M. J., VanPool, T. L., Leonard, R. D., VanPool, C. S., & Salter, L. A. (2006). Reconstructing the flow of information across time and space: A phylogenetic analysis of ceramic traditions from prehispanic western and northern Mexico and the American southwest. In C. P. Lipo, M. J. O'Brien, S. Shennan & M. Collard (Eds.), *Mapping our ancestors: Phylogenetic approaches in anthropology and prehistory* (pp. 209–229). New Brunswick: Aldine Transaction.

Harvey, P. H., & Pagel, M. D. (1991). *The comparative method in evolutionary biology*. Oxford: Oxford University Press.

Hunzicker, D. (2008). Folsom projectile technology: An experiment in design, effectiveness and efficiency. *Plains Anthropologist, 53*, 291–311.

Jordan, P., & Shennan, S. J. (2009). Diversity in hunter-gatherer technological traditions: Mapping trajectories of cultural 'descent with modification' in Northeast California. *Journal of Anthropological Archaeology, 28*, 342–365.

Lalueza, C., Pérez-Pérez, A., Prats, E., Cornudella, L., & Turbon, D. L. (1997). Lack of founding Amerindian mitochondrial DNA lineages in extinct aborigines from Tierra de Fuego-Patagonia. *Human Molecular Genetics, 6*, 41–46.

Lipo, C., O'Brien, M., Collard, M., & Shennan, S. J. (Eds.). (2005). *Mapping our ancestors: Phylogenetic methods in anthropology and prehistory*. New York: Aldine Transaction.

Lycett, S. J. (2009). Understanding ancient hominin dispersals using artefactual data: A phylogeographic analysis of Acheulean handaxes. *PLoS ONE, 4*, 1–6.

Mace, R., & Pagel, M. (1994). The comparative method in anthropology. *Current Anthropology, 35*, 549–564.

Meghin, O. F. A. (1952). Fundamentos cronológicos de la prehistoria de Patagonia. *Runa, 5*, 23–43.

Mesoudi, A., & O'Brien, M. J. (2008a). The cultural transmission of Great Basin projectile-point technology I: An experimental simulation. *American Antiquity, 73*, 3–28.

Mesoudi, A., & O'Brien, M. J. (2008b). The cultural transmission of Great Basin projectile-point technology II: An agent-based computer simulation. *American Antiquity, 73*, 627–644.

Mihaescu, R., Levy, D., & Pachter, L. (2009). Why neighbor-joining works. *Algorithmica, 54*, 1–24.

Morrow, J. E., & Morrow, T. A. (1999). Geographic variation in fluted projectile points: A hemispheric perspective. *American Antiquity, 64*, 215–231.

Muscio, H. J. (2010). Transferencia Horizontal, Cladismo y Filogenias Culturales. In J. L. Escasena Carrasco, D. García Rivero & F. J. García Fernández (Eds.), *Clasificación y Arqueología: Enfoques y Métodos Taxonómicos a la Luz de la Evolución Darwiniana* (pp. 223–251). Sevilla:Universidad de Sevilla.

Nami, H. (1984). Algunas observaciones sobre la manufactura de las puntas de proyectil de El Volcán. *PREP: Informes de Investigación, 1*, 85–107.

O'Brien, M., & Lyman, R. (2003). *Cladistics and archaeology*. Salt Lake City: University of Utah Press.

Orquera, L. A. (1987). Advances in the archaeology of the Pampas and Patagonia. *Journal of World Prehistory, 1*, 333–413.

Peres-Neto, P. R. (2006). A unified strategy for estimating and controlling spatial, temporal and phylogenetic autocorrelation in ecological models. *Oecologia Brasiliensis, 10*, 105–119.

Pérez, S. I., & Monteiro, L. M. (2009). Nonrandom factors in modern human morphological diversification: A study of craniofacial variation in southern South American populations. *Evolution, 63*, 978–993.

Pérez, S. I., Diniz-Filho, J. A. F., Bernal, V., & González, P. N. (2010). Spatial regression techniques for inter-population data: Studying the relationships between morphological and ecological variation. *Journal of Evolutionary Biology, 23*, 237–248.

Pérez, S. I., Lema, V., Diniz-Filho, J. A. F., Bernal, V., González, P. N., Gobbo, D., et al. (2011). The role of diet and temperature in shaping cranial diversification of South American human populations: An approach based on spatial regression and rate tests. *Journal of Biogeography, 38*, 148–163.

R Development Core Team 9 (2007). R: A language and environment for statistical computing, reference index version 2.2.1. *R foundation for statistical computing, Vienna, Austria.* URL http://www.R-project.org

Rangel, T. F., Diniz-Filho, J. A. F., & Bini, L. M. (2010). SAM: A comprehensive application for spatial analysis in macroecology. *Ecography, 33*, 46–50.

Ratto, N. (1992). Técnicas de caza prehistóricas en ambientes de Patagonia (Tierra del Fuego, Argentina). *Palimpsesto, 1*, 37–49.

Ratto, N. (1994). Funcionalidad vs. adscripción cultural: Cabezales líticos de la margen norte del estrecho de Magallanes. In J. L. Lanata & L. A. Borrero (Eds.), *Arqueología de cazadores-recolectores. Límites, casos y aperturas* (pp. 105–120). Buenos Aires: Arqueología contemporánea 5. Edición Especial.

Rohlf, J. (1993). Relative warp analysis and an example of its application to mosquito wings. In L. F. Marcus, E. Bello & A. Garcia-Valdecasa (Eds.), *Contributions to morphometrics* (pp. 131–159). Madrid: Museo Nacional de Ciencias.

Rohlf, F. (2004). *TPSUtil. version 1.26*. Department of Ecology and Evolution. New York: State University.

Rohlf, F. (2007). *TPSRelw version 1.45*. Department of ecology and evolution, New York: State University.

Rohlf, F., & Slice, D. (1990). Extensions of Procrustes method for the optimal superimposition of landmarks. *Systematic Zoology, 39*, 40–59.

Rots, V. (2016). Projectiles and hafting technology. In R. Iovita & K. Sano (Eds.), *Multidisciplinary approaches to the study of Stone Age weaponry* (pp. 167–185). Dordrecht: Springer.

Saitou, N., & Nei, M. (1987). The neighbor-joining method: A new method for reconstructing phylogenetic trees. *Molecular Biology and Evolution, 4*, 406–425.

Shott, M. J. (2016). Survivorship distribution in experimental spear points: Implications for tool design and assemblage formation. In R. Iovita & K. Sano (Eds.), *Multidisciplinary approaches to the study of Stone Age weaponry* (pp. 245–258). Dordrecht: Springer.

Shott, M., & Ballenger, J. A. M. (2007). Biface reduction and the measurement of Dalton curation: A southeastern United States case study. *American Antiquity, 72*, 153–175.

Shott, M., Hunzicker, D. A., & Patten, B. (2007). Pattern and allometric measurement of reduction in experimental Folsom bifaces. *Lithic Technology, 32*, 203–217.

Tehrani, J., & Collard, M. (2002). Investigating cultural evolution through biological phylogenetic analyses of Turkmen textiles. *Journal of Anthropological Archaeology, 21*, 443–463.

Thulman, D. K. (2012). Discriminating Paleoindian point types from Florida using landmark geometric morphometrics. *Journal of Archaeological Science, 39*, 1599–1607.

Chapter 19
Hunting Technologies During the Howiesons Poort at Sibudu Cave: What They Reveal About Human Cognition in KwaZulu-Natal, South Africa, Between ~65 and 62 ka

Marlize Lombard and Lyn Wadley

Abstract Encounter hunting, especially of big game, is an activity firmly associated with people who lived in the Middle Stone Age. Most hunting is assumed to have taken place in groups, using spears of varying complexity. Recent data suggest that various meat-acquisition techniques were used, at least within the last 65 ka. Bow-and-arrow sets as well as snares appear to have complemented spear hunting. Many archaeologists have devoted a great deal of time to the study of lithic technologies required for the creation of spearheads and arrow tips. Rarely, however, have the cognitive correlates of Middle Stone Age meat-acquisition technologies been considered. Here we show that the mental concepts behind the meat-acquisition strategies are equally, or perhaps, more important than the technological complexity involved in manufacturing the necessary equipment. Notwithstanding this claim, it is also true that the longer the chain of operations involved in making composite weapons, the more likely it is that complex cognition was a prerequisite.

Keywords Bow hunting • Game snares • Human cognition • Middle Stone Age

M. Lombard (✉)
Department of Anthropology and Development Studies, University of Johannesburg, Auckland Park, PO Box 524Johannesburg 2006, South Africa
e-mail: mlombard@uj.ac.za

L. Wadley
Evolutionary Studies Institute, Centre of Excellence, Palaeosciences, University of the Witwatersrand, Johannesburg, PO WITS 2050, South Africa
e-mail: Lyn.Wadley@wits.ac.za

Introduction

The study of Middle Stone Age hunting weaponry or techniques is often approached technologically or from the point of view of meat acquisition. Yet, by exploring well-contextualized archaeological evidence theoretically and using new methodology, the theme can also reveal a great deal about levels of complexity of human cognition in the past. We cannot have an in-depth discussion here, but we summarize our interpretations of human cognition, based on archaeological evidence for the use of certain meat-procurement technologies. We focus on a single culture-stratigraphic context at Sibudu Cave: the Howiesons Poort, although the first use of the meat-getting strategies is likely to have been earlier, both at Sibudu and other sites. Inferring levels of human cognition from the Stone Age archaeological record is difficult, yet we propose that our results generate data that can be used with bridging theory to create hypotheses about human cognitive evolution. These can subsequently be refined or falsified.

The Howiesons Poort Industry at Sibudu

Sibudu is a large rock shelter on a cliff above the uThongathi River, about 15 km inland of the Indian Ocean and just over 100 m above sea-level (Fig. 19.1a, b). The south-west-facing cliff, hillside and shelter are shaded for most of the day and the shelter is surrounded by evergreen forest because of the cool conditions with low evapotranspiration rates. In contrast, on the other side of the river in view of Sibudu, there are sunny and warm, north-facing hill slopes that promote deciduous wood- and grassland, under suitable rainfall and temperature conditions (Fig. 19.1c). A riverine habitat at the foot of the cliff is likely to have been relatively constant during the times Sibudu was occupied by humans

Radu Iovita and Katsuhiro Sano (eds.), *Multidisciplinary Approaches to the Study of Stone Age Weaponry*, Vertebrate Paleobiology and Paleoanthropology, DOI 10.1007/978-94-017-7602-8_19

(Fig. 19.1d), and the mosaic of surrounding habitats suggests that the area could have supported a wide range of fauna in the past, as it did in historic times before the land was farmed (Wadley 2010a).

Sibudu's cultural sequence comprises a pre-Still Bay lithic assemblage (Fig. 19.1e), assemblages associated with the Still Bay, Howiesons Poort and Sibudu technocomplexes [the latter was previously informally referred to as post-Howiesons Poort and late Middle Stone Age (Lombard et al. 2012)], and a final Middle Stone Age assemblage (Fig. 19.1e) (Villa et al. 2005; Wadley 2005, 2006, 2007; Cochrane 2006; Delagnes et al. 2006; Villa and Lenoir 2006; Wadley and Jacobs 2006; Wadley and Mohapi 2008; Soriano et al. 2009; Mohapi 2012). The Middle Stone Age layers are directly overlain by Iron Age occupations; no Later Stone Age material is present. The Sibudu Howiesons Poort has ages between 64.7 ± 1.9 and 61.7 ± 1.5 ka (Jacobs et al. 2008) (Fig. 19.1e) and is characterized by blade technology and backed tools (made on blades), mostly segments (sometimes called crescents or lunates), that were blunted to a 90° angle along one lateral, leaving the other lateral as a sharp cutting edge (Fig. 19.1e) (Wadley and Mohapi 2008).

Sedimentological and mineralogical analyses of the Howiesons Poort layers show fairly high percentages of calcite, suggesting relatively high humidity compared to the younger layers at Sibudu (Pickering 2006; Schiegl and Conard 2006). The carbonized seed assemblage for ~65–62 ka mostly comprises evergreen taxa, implying closed forested environments (Wadley 2004; Sievers 2006). This interpretation is supported by evergreen forest taxa in the charcoal assemblage, such as *Podocarpus* (yellowwood), *Buxus* (boxwood) and *Curtisia* (assegai tree or Cape lancewood). The presence of these taxa suggests that available moisture was high during this period, but not necessarily higher than present (Allott 2006). Carbon isotope analyses of yellowwood and *Celtis* (stinkwood) charcoal also indicate conditions of elevated levels of water availability and humidity (Hall and Woodborne 2010). Although the charcoal assemblage at this time is dominated by yellowwood species, and the area appears to have been mainly a forested one, there is botanical evidence of a woodland/savanna landscape in the vicinity (Allott 2006), thus it is likely that a mosaic of habitats existed near the site during the Howiesons Poort occupations. This is partly due to the location of the site and the continual presence of the uThongathi River (Wadley 2006). Carbonized *Cyperaceae* (sedge) nutlets from taxa that grow near water are present throughout the archaeological sequence (Sievers 2006, 2011; Wadley et al. 2011; Sievers and Muasya 2011).

The faunal assemblage provides further evidence of environmental change through time in the area. For layers with ages of ~65–62 ka, it is dominated by small species preferring semi-closed/closed habitats such as *Philantomba monticola* (blue duiker), *Tragelaphus scriptus* (bushbuck), *Potamochoerus larvatus* (bushpig) and *Cercopithecus pygerythrus* (vervet monkey). In addition, some species, including *Syncerus caffer* (buffalo), *Equus capensis* (zebra), *Tragelaphus oryx* (eland) and *Connochaetes taurinus* (blue wildebeest), imply an open savanna/woodland near the site (Clark and Plug 2008; Clark 2011). The faunal data support the botanical data that point to a mosaic of vegetation types within the foraging range of Sibudu. A large variety of aquatic species including mammals, reptiles, water birds, fish, amphibians and molluscs have been identified (Plug 2004, 2006), reinforcing the suggestion that the uThongathi River was perennial. The micro-mammal species composition supports evidence of a cooler, humid forested environment. Two key species found in Sibudu, *Cricetomys gambianus* (giant rat) and *Rhinolophus clivosus* (Geoffroy's horseshoe bat) (Glenny 2006), both require humid conditions, and the giant rat cannot tolerate high temperatures (Skinner and Chimimba 2005). Multiple lines of evidence thus converge, indicating that during the Howiesons Poort occupations of Sibudu, local conditions were cool and moist, and that the site was surrounded by evergreen forests with patches of open, savanna/woodland.

Weapons for Encounter Hunting

Evidence for the Use of Spears and Bows and Arrows

During 2008 a suite of four papers in the *Journal of Archaeological Science* indicated considerable variability in hunting technologies and the possible use of bow-and-arrow technology during the Howiesons Poort at Sibudu. Micro-residue analysis revealed that most of the stone segments had been in contact with animal material, that they were hafted in different configurations on different materials, and that they could have been used as tips and/or barbs for arrows and/or spears as part of a diverse hunting technology (Fig. 19.2Ai) (Lombard 2008). This was followed by the publication of a slender, broken bone point, found with other worked bone from the same context (Backwell et al. 2008). The point was cautiously compared to the un-poisoned bone arrow points from Later Stone Age, Iron Age and historical Bushman contexts, and found to be morphologically similar (Fig. 19.2Bi–ii). Since there is no Later Stone Age occupation at Sibudu and the point cannot be the result of displacement from such layers, it was suggested that, if substantiated by future discoveries, it will push back the origin of bow and bone-tipped arrow technology by at least 20 ka (Backwell et al. 2008).

a

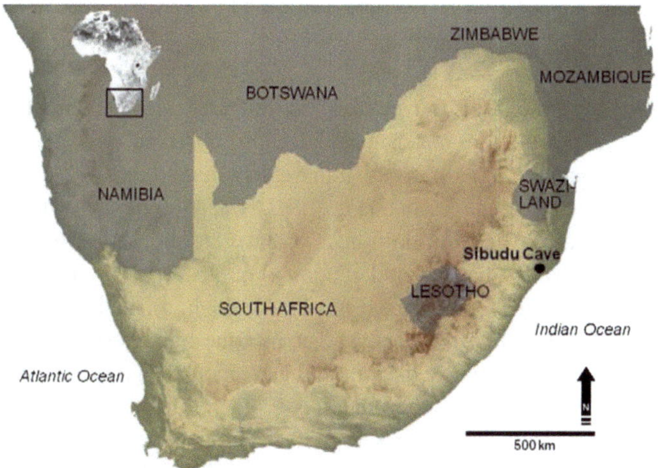

b

c

d

e

Layers	Age (ka)	Industry	Examples of characteristic artefacts
Co Bu	38.0 ± 2.6 39.1 ± 2.5	final MSA	
LBMOD MOD OMOD OMOD-BL RSp RD BSp SS P Ch2 Y1 B/Gmix	49.9 ± 2.5 49.1 ± 2.1 46.6 ± 2.3 47.6 ± 1.9 46.0 ± 1.9 49.4 ± 2.3 57.6 ± 2.1 59.6 ± 2.3 59.0 ± 2.2 58.3 ± 2.0 58.6 ± 2.1 58.2 ± 2.4	Sibudu	
GR 2 GS2 PGS	61.7 ± 1.5 63.8 ± 2.5 64.7 ± 1.9	Howie-sons Poort	
RGS	70.5 ± 2.0	Still Bay	
LBG LBG2 BS	72.5 ± 2.0 73.2 ± 2.3 77.2 ± 2.1	pre-Still Bay	Currently being analysed

◀ **Fig. 19.1** **a** Map showing Sibudu's location. **b** The shelter. **c** View into the shelter from the sunny and warm, north-facing hill slopes with deciduous wood- and grassland. **d** Riverine habitat at the foot of the cliff. **e** Sibudu cultural sequence with dated layers adapted from Wadley (2010c) [age estimations from single grain optically stimulated luminescence Dating, Jacobs et al. (2008)]

Experimental work with stone segments showed that they can be hafted in four different configurations to tip hunting weapons (Pargeter 2007). Macrofracture analyses imply that the frequencies of diagnostic impact fractures on Howiesons Poort samples from South Africa, including those on backed tools from Sibudu, compare well with those documented on experimental and European archaeological samples from the Holocene, known to have been used to tip weapons such as arrows (Fig. 19.2Aii–iii) (Lombard and Pargeter 2008). The morphometric results published in the final paper of that year (Wadley and Mohapi 2008), showed significantly different sizes and shapes for three groups of segments knapped from three rock types, quartz, hornfels and dolerite. As a result, there is a strong possibility that the small quartz segments, which have standardized shapes (short and deep) (Fig. 19.2Aiv/a), were hafted as transverse arrowheads. The elongated hornfels and dolerite segments seem better suited to diagonal and/or back-to-back hafting as weapon tips, but their elongated forms also make them suitable for barbs on a variety of weapons (Fig. 19.2Aiv/b) (Wadley and Mohapi 2008).

Although, at first, we were cautious about presuming that bow-and-arrow technology existed before 60 ka in KwaZulu-Natal, the 2008 results merited further exploration. This time ML (2011) focused only on the small (<20 mm long) quartz segments identified by Wadley and Mohapi (2008) as morphologically the most likely candidates for arrow tips (Fig. 19.2Aiv/a). Each tool was examined for a range of use-traces (e.g., macro-fractures, edge damage, edge rounding, polish, striations and micro-residues). The use-trace distribution and orientation on the tools was explicitly tested for: (a) transverse hafting and use-traces to support the hypothesis that bows and stone-tipped arrows were used at the site from ~65 ka (Fig. 19.2Ai/b) (e.g., Wadley and Mohapi 2008; Lombard and Phillipson 2010); (b) longitudinal hafting and use that might cast into doubt the employment of the artifacts as transversely hafted arrow tips (Fig. 19.2Ai/a) (as in Villa et al. 2010); and (c) diagonal hafting and use that will not contradict either scenario a or b, because this configuration can be equally successful as tips or barbs in either mechanically- or hand-delivered weaponry (Fig. 19.2Ai/c) (see Lombard 2008; Villa et al. 2010).

The results show that more than half of the backed quartz tools in the sample were hafted transversely. Most of these tools have scars or fractures along their sharp edges that are consistent with those observed on replicated backed tools used as transversely hafted arrows during hunting experiments (Fig. 19.2Av/a, j) (see Yaroshevich et al. 2010). In addition they have striations that indicate transverse hafting and motion (Fig. 19.2Av/a, c, j), traces of hafting adhesives (Fig. 19.2Av/b, h, i), and accompanying animal residues that support the hypothesis that they were used as inserts in hunting weapons (Fig. 19.2Av/d, e, f). The best-fit interpretation is, therefore, that several of the small quartz backed artifacts from the Sibudu Howiesons Poort were used as transversely hafted arrowheads (Lombard 2011). A number of the small quartz tools also display traces of having been hafted and used diagonally, supporting the interpretation that some pieces could have been used in an innovative way to supply wooden spears with barbs and/or cutting inserts (Fig. 19.2Ai/a, c) (e.g., Lombard 2008; Villa et al. 2010). Only a single tool showed signs of having been hafted and used longitudinally, perhaps as a knife insert for butchering (e.g., Semenov 1964), or as a longitudinal insert for an arrowhead or spear (e.g., Lombard and Parsons 2008; Pétillon et al. 2011). Such hafting would create cutting edges for lacerating prey after the weapon tip had made the initial puncture wound.

Experiments were also conducted with replicated bone points to record fracture patterns resulting from hunting with hand-delivered spears or a bow and arrows (Bradfield and Lombard 2011). The experimental fractures were compared to the fractures on the Sibudu bone point, which has fractures or damage in three locations. These potential use-traces include crushing on the tip (Fig. 19.2Biii/a), a medial hinge-terminating fracture (Fig. 19.2Biii/b), and a proximal snap fracture with a large (~5 mm long), unifacial step-terminating spin-off fracture (Fig. 19.2Biii/c). The crushing on the tip, and the fracture types and pattern on the proximal extremity are consistent with its use as a hafted arrow tip, based on the damage caused to experimental arrow tips. The medial hinge-terminating fracture may be consistent with post-depositional damage, but this inference remains to be tested. Although not conclusive, the experimental and comparative findings thus do not contradict the current interpretation that the bone point from Sibudu was used as an arrowhead.

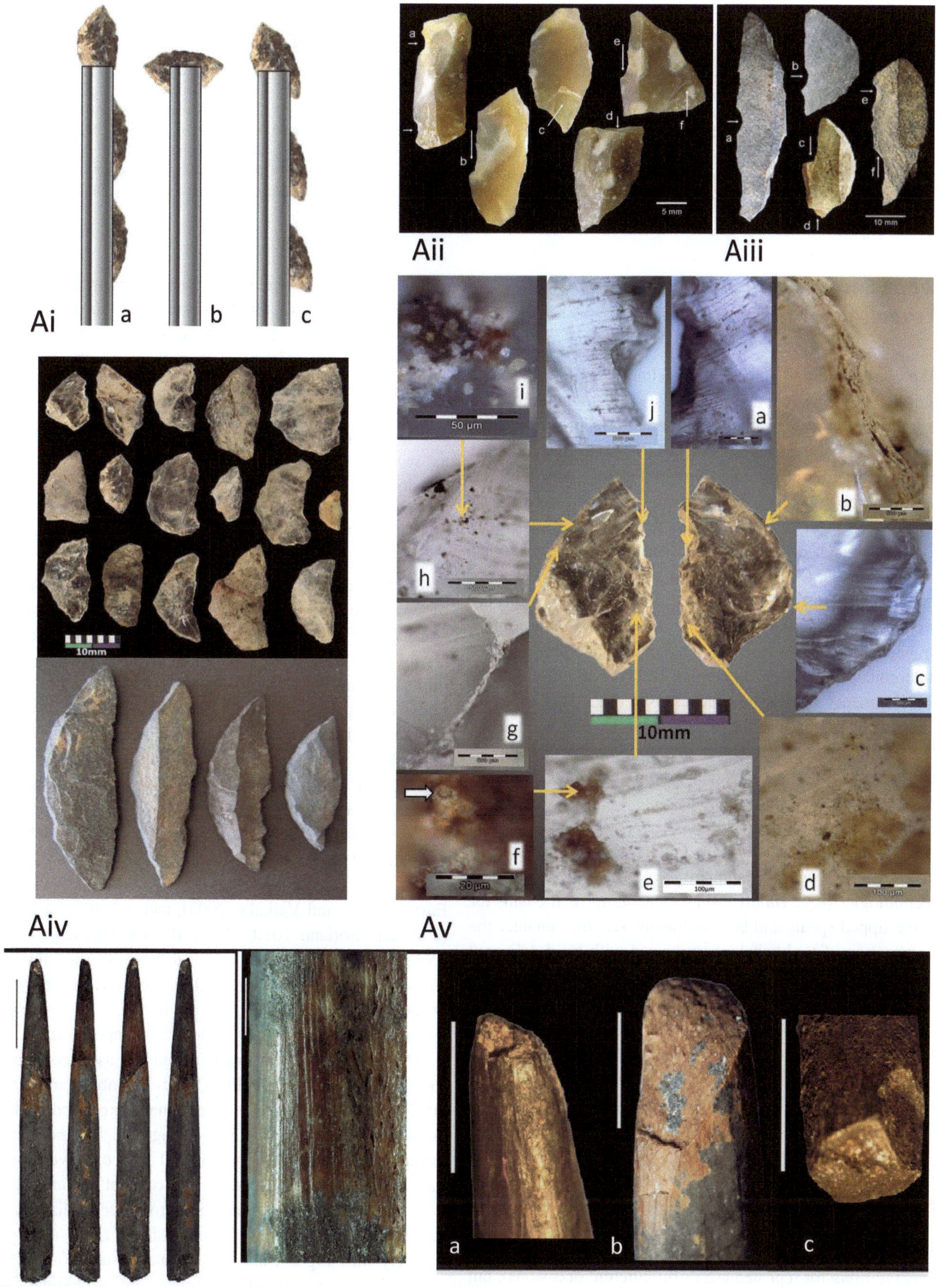
Ai
a
b
c
Aii
5 mm
Aiii
10 mm
Aiv
10mm
Av
50 µm
20 µm
100µm
10mm
Bi
Bii
Biii

◀ **Fig. 19.2** **Ai** Schematic representation of the different possible hafting angles for segments mentioned in the text; *a* longitudinal hafting, *b* transverse hafting, *c* diagonal hafting (Lombard 2011). **Aii** Experimentally documented impact fractures as a result of hunting activities; and **Aiii** similar fractures documented on Howiesons Poort segments (Lombard and Pargeter 2008). **Aiv** Howiesons Poort segments/backed artifacts with small quartz pieces at the top. **Av** A small quartz segment with use-traces that indicate its use as transversely hafted arrow tip; *a* slightly diagonal striations associated with impact scarring, *b* resinous residues, *c* transverse triations, *d* fatty residue, *e* animal tissue and fatty residue associated with transverse striations, *f* animal tissue with blood cell (white arrow), *g* scarring on the dorsal ridge associated with transverse striations, *h* resinous residues associated with transverse striations, *i* white starchy residue and ochre associated with the resin (Lombard 2011). **Bi** Refitted bone point from the Howiesons Poort layers at Sibudu Cave, and **Bii** close-up view showing fine longitudinal striations produced by scraping with a burin or unretouched stone edge (Backwell et al. 2008) (scale bars = 10 mm). **Biii** Close-up of fractures on the bone point; *a* tip crushing, *b* medial hinge-terminating fracture, *c* proximal unifacial step-terminating spin-off fracture initiating from a snap fracture (Bradfield and Lombard 2011) (scale bars = 10 mm)

What Encounter Hunting Weapons Reveal About Human Cognition

The above findings indicate that technologies for encounter hunting were varied and well-developed to exploit meat resources provided by the mosaic environment of the time. Hunters probably used an array of weapons that included hand-delivered spears and bow-and-arrow sets with an assortment of tip types. Evidence for hand-delivered spears with stone tips or barbs by ~65 ka in southern Africa comes as no surprise because they were used before 200 ka at other sites. The early use of bow-and-arrow sets, however, is previously unrecorded though mechanically-projected weaponry could be expected, given the extraordinary archaeological record of sub-Saharan Africa where modern anatomical features, and complex behavioral and cognitive developments converged precociously (e.g., Brooks et al. 2006; Shea 2006; Sisk and Shea 2009; Shea and Sisk 2010). Despite this emphasis, little attempt has been made to explore/explain the potential cognitive implications of mechanically-projected weaponry.

Analyzing complete chains of operation, i.e., all the operational units contained in cognigrams, is one way of investigating cognitive similarities or differences between technologies (e.g., Haidle 2011). Generating such chains for stone-tipped spears and bow-and-arrow sets thus enables the comparison of tool behavior associated with hand-delivered weaponry with that of mechanically-projected weaponry (Lombard and Haidle 2012, also see this publication for the complete series of annotated cognigrams, and guidelines for reading them). The effective chain of manufacture and use of a stone-tipped spear (or spear with stone inserts in other positions) demonstrates the cognitive component of composition (the encircled +) (Fig. 19.3a) (Lombard and Haidle 2012). The cognitive requirements for composition include; (a) the decoupling of a tool (e.g., a hammerstone) from the satisfaction of a basic need (e.g., hunger); the modularization of action units, where action units are completed separately from each other (e.g., making a sharp-edged stone artifact, or making fire/s for repeated use or on different occasions), but sequenced together they aim to satisfy a basic need; (b) the ability to combine several fully separate elements or materials (e.g., stone, glue, twine, wood, bone) to create a new functional concept (e.g., a stone-tipped spear). Composition thus represents an innovative concept in the problem-solution distance. It introduces new effects that combinations of materials and tools can have on each other, and these effects were absent from single-unit, un-hafted artifacts such as wooden spears (Fig. 19.3b). Composite tools signal development towards advanced technological, behavioral and cognitive modularization and flexibility (see Lombard and Haidle 2012; also Wadley et al. 2009; Ambrose 2010; Barham 2010; Wadley 2010b; Haidle 2010, 2011).

Such advanced modularization signifies the modification of cognitive tool behavior, opening the way towards a considerable expansion of problem solutions. The modular organization of thought-and-action processes constitutes an important simplification of complex, multifaceted operations; it facilitates solutions that can otherwise hardly be considered (Haidle 2009, 2011). We will most likely never know how, where, or how many times, the concept of a hafted tool was "invented", but it probably developed gradually over the past 300 ka and it radically changed the world of hominin technology. Stone-tipped, hand-delivered spears could have been used from ~285 ka in sub-Saharan Africa (McBrearty and Tryon 2005), ~270 ka in the Near East (Mercier and Valladas 2003), and ~200 ka in Europe (Villa and Soriano 2010). Thus, if some Howiesons Poort backed artifacts were used as spear barbs or tips, they have similar cognitive and behavioral implications as other composite tools; representing a long tradition in conceptual, technological and behavioral modular composition.

The effective chain of manufacture and use reconstructed for a bow-and-arrow set (Fig. 19.3c) (Lombard and Haidle 2012), shows a cognitive development expressed in technological symbiosis ("{" in the diagram). The concept of technological symbiosis includes all the cognitive requirements previously highlighted for the concept of composition, but, it incorporates the added ability to conceptualize a set of separate, yet inter-dependent composite or single-unit tools. Actively focusing on, and manipulating, such complementary tool sets represents the augmentation of modular flexibility (amplified conceptual, technological and behavioral modularization). It further increases the problem-solution distance, enabling the conceptualization of new technological categories representing yet another major increase in levels of behavioral and cognitive complexity and flexibility (Fig. 19.3b).

Such complementary tool sets unleash new tool properties that would have been inconceivable without the active, simultaneous manipulation of several tools. The individual components only reach their full potential when used in a symbiotic set. Not all complementary tool sets have to be as complex as a bow-and-arrow set, other examples of technological symbiosis can be found in the production and use of a spearthrower and dart, a hammer and chisel, or a fishing rod with line and hook. Once the concept of symbiotic technologies is understood, different elements and series of elements can be adapted and grouped in multiple ways, and in sequences of various length and complexity, to achieve diverse results. The increase in cognitive, and consequently behavioral, flexibility is the main evolutionary advantage of complementary tool sets or symbiotic technologies; this can hardly be overestimated (Lombard and Haidle 2012).

Mindful Procurement: Using Snares

Evidence for the Use of Snares

Another meat-acquisition technology, snaring, may be an even more reliable indicator of high levels of cognition. The concept of remote capture involved in the creation and use of snares implies enhanced working memory and complex cognition (Fig. 19.4). It is mostly impossible to recognize snares archaeologically because they were invariably made from materials that have not preserved. Only circumstantial evidence such as mortality profiles, taxonomic diversity and high frequencies of creatures that are susceptible to capture in snares can be used to infer their presence. Circumstantial evidence is not as desirable as direct evidence, but used cautiously it may augment our interpretations of meat acquisition (Wadley 2010a).

Indirect signs of past snare use might be found in archaeofaunal assemblages with high proportions of prey that would be difficult to capture otherwise. Nocturnal animals, or those that are non-migratory with small home ranges, solitary and shy, living in dense forest or woodland, seem ideal targets for snaring (Wadley 2010a). Gregarious, fleet-footed creatures like rabbits and hares are also suitable prey for snares, and small carnivores, such as civets and mongooses, tend to become ensnared. The presence of small carnivores in a faunal list can thus point to the use of remote capture technologies such as snares; in effect, snares target a wide range of prey, resulting in taxonomically diverse collections (Lupo and Schmitt 2002; Schmitt and Lupo 2008), cross-cutting age and sex categories (Noss 1998). Consequently, diverse prey species and age profiles that reflect living populations might imply snaring. As with any archaeological interpretation, the use of multi-stranded (albeit circumstantial) evidence is wise (for full discussion see Wadley 2010a; also see Wadley 1998, 2006; Clark and Plug 2008; Clark 2011).

Mortality profiles can provide useful information on hunting techniques providing that potential problems with their use are acknowledged (e.g., Wilkinson 1976; Lupo 2001). For example, they can indicate ways in which animals died, or provide clues to the accumulation of their bones at archaeological sites. The analytical value of the method is, however, dependent on understanding the full range of possible causes for the prototypes (e.g., Stiner 1990; Wadley 2010a). It is critical to understand taxa-specific behavior to make plausible interpretations from mortality profiles. The attritional model produces a profile where very young and/or old individuals exceed other age classes (Klein and Cruz-Uribe 1984). It implies purposeful selection of vulnerable age classes, and is therefore not associated with snaring, which samples animal populations randomly. Rather, attritional mortality yields a relatively high proportion of very young individuals, which are particularly at risk during active hunting (e.g., Klein 1978). The prime-aged adult mortality pattern is rare in nature, but is typical of cervid, bovid and equid remains in some archaeological assemblages. The focus on prime adult prey entails selective and controlled procurement, and only humans regularly produce this pattern (Stiner 1990). The use of weapons such as spears or bow-and-arrow sets might produce a predominantly adult mortality profile, but it seems unlikely that the use of snares would do so (Wadley 2010a).

A catastrophic mortality profile exemplifies the full age structure of a live herd (Klein and Cruz-Uribe 1984). High percentages of juveniles can thus indicate non-selective capture techniques, but, because of their small size and limited body fat, juveniles probably did not rank as high as their adult counterparts (Speth and Clark 2006). The catastrophic mortality profile may suggest mass killing through natural disasters or human intervention. Natural disasters occur sporadically, but archaeological sites with repeated, stratified occurrences of faunal remains exhibiting catastrophic mortality profiles cannot be considered coincidental. Unnatural catastrophic mortality profiles can also be expected where humans drive game over cliffs or into traps, but such strategies do not qualify as a remote capturing. Snares, which are remote-capturing devices, are also likely to produce a catastrophic mortality pattern, although the

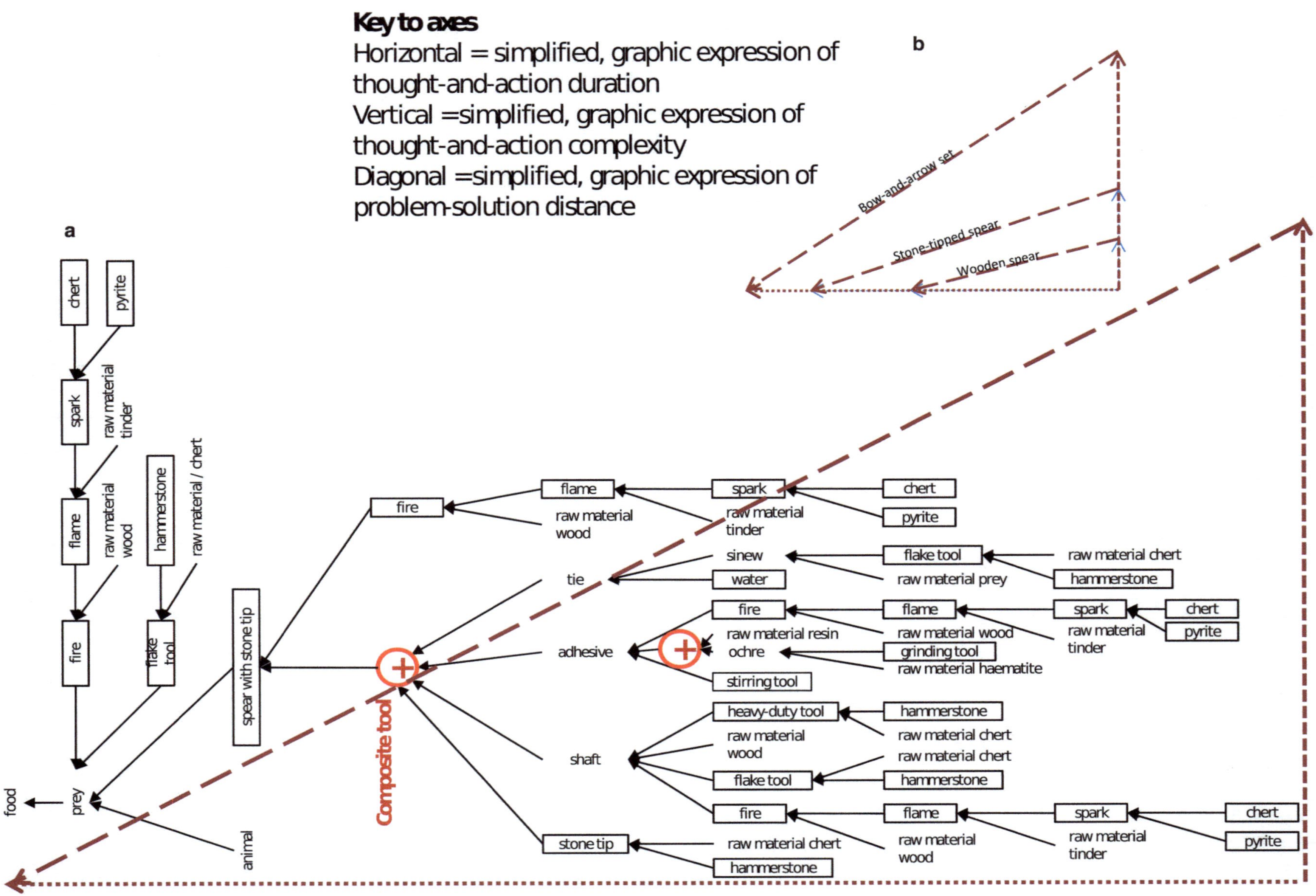

Fig. 19.3 Effective chains of production and use of weapon systems can be compared. **a** That of a stone-tipped spear illustrates the new cognitive element of composition (encircled +) during the production of adhesives and during the final assembly of the composite artifact. **b** Direct comparison of the hypothetical dimensions associated with production and use of wooden spears (single-unit technologies), stone-tipped spears (composite technologies) and a bow-and-arrow set (symbiotic technologies). **c** That of a bow-and-arrow set illustrates the novel cognitive concept of complementary or symbiotic technologies ("{" in illustration). Figure adapted from Lombard 2012, Fig. 2

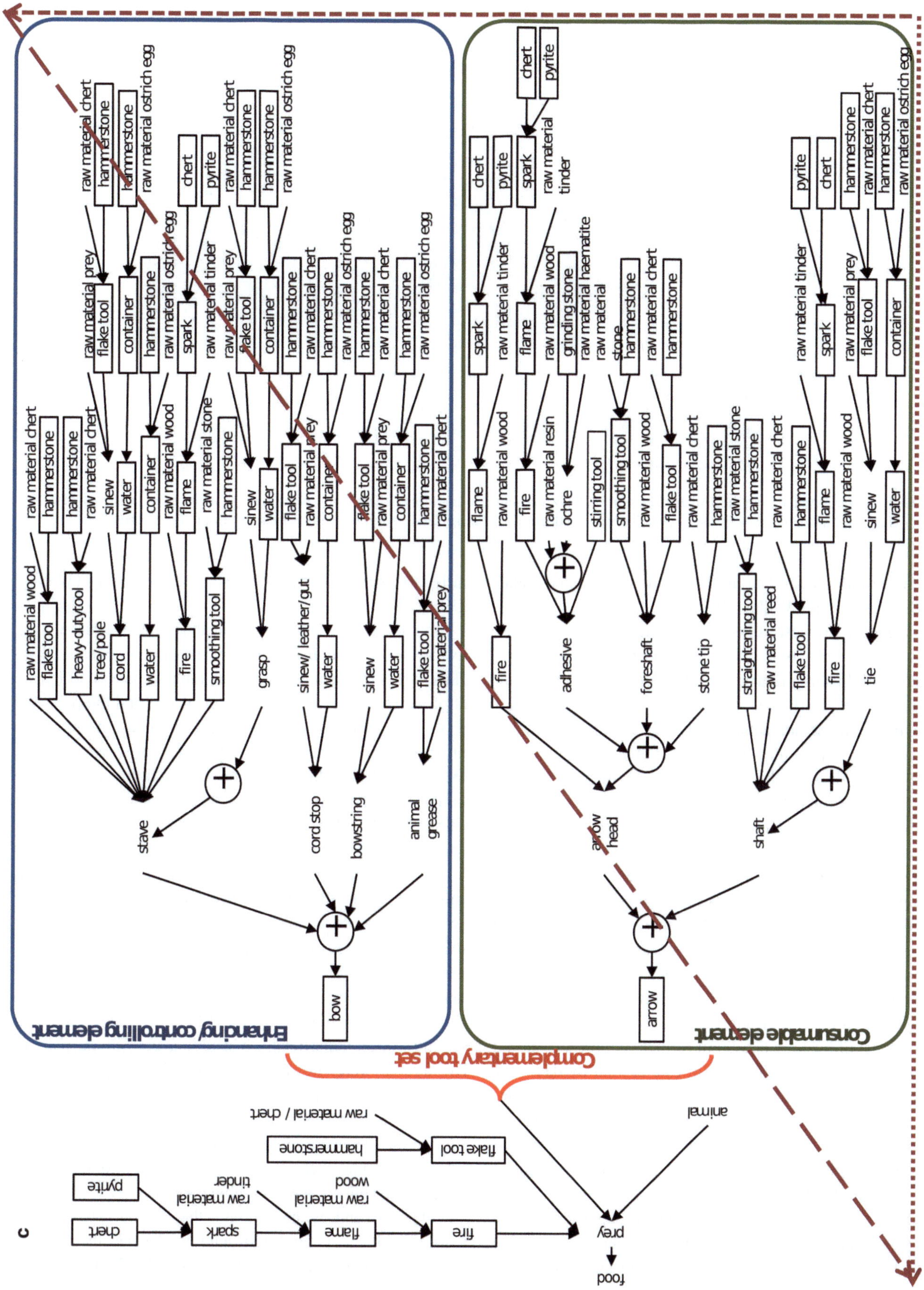

Fig. 19.3 (continued)

Fig. 19.4 A Naro hunter from the Kalahari Desert in Botswana setting a traditional snare using plant cordage and the latent energy of a bent sapling (Lombard and Phillipson 2010; © Ariadne van Zandberg used for academic purposes with permission from AfriPics.com)

pattern would accumulate as the result of repeated rather than single events (Wadley 2010a).

During the Howiesons Poort at Sibudu small bovids (Size Class I), especially blue duiker, are most commonly represented, with bushpig remains the next most common (Clark and Plug 2008; Clark 2011). Most of the blue duiker remains from this context are adult, with only about 11% of the duikers being juveniles (Clark personal communication 2009). Blue duikers that are inclined to move in spaces between forest thickets seem more aptly caught in snares set on their frequented pathways (Apps 2000). Other small antelope represented in the Howiesons Poort (see Clark 2011), such as *Raphicerus campestris* (steenbok), *Sylvicapra grimmia* (grey or common duiker), *Cephalophus natalensis* (red duiker) and bushbuck are also suited to being caught in snares because their home ranges are small and their non-migratory behavior is predictable. It seems likely that only rare chance encounters would have resulted in them being speared or shot (Wadley 2010a).

Remains of bushpig at Sibudu further support the snaring hypothesis. These animals are difficult to hunt by day, they are aggressive and treacherous to deal with when encountered, and flushing them out of cover is unwise because, when cornered, they tend to turn on hunters (Skinner and Chimimba 2005). In southeastern Cameroon (Yasuoka 2006), Central Africa and West Africa (Fa et al. 2005), bushpig are traditionally caught in snares. Capturing them in this way requires rather robust snares because bushpig are much heavier (69–72 kg) than blue duiker (<5 kg), though fairly large pitfalls can also be used to trap the pigs (Wadley 2010a).

Sibudu's Howiesons Poort faunal assemblage has a more diverse taxonomic list than the subsequent phases (e.g., Clark 2011). It includes monkeys, rabbits, hares and hyrax, all of which are prone to capture in snares, and the presence of small carnivores such as felids, viverrids, mongooses, mustelids and canids, also supports the snare scenario (Wadley 2010a), although there seems little evidence for human use of these carnivores (Clark personal communication 2012). What is more, the Sibudu faunal list from ~65–62 ka fulfils Lupo and Schmitt's (2002) conditions for the range of prey likely to be caught in snares. Without careful analysis, though, the age structure of the Sibudu Howiesons Poort sample of blue duiker, with only some juveniles, seems to counter the snaring hypothesis, but as previously mentioned animal behavior impacts on mortality patterns. For example, the live population estimate from Central Africa included only about 11% juveniles (Lupo and Schmitt 2002), superficially resembling the Sibudu pattern. The figures were, however, obtained from capturing blue duiker in nets, which may underestimate the true number of juveniles in a given area. After birth, which can occur throughout the year, offspring are sedentary and they are hidden for some weeks, and tend to lie motionless when danger threatens. Lambs of about eight weeks will walk roughly 55 m and by three months they venture about 73 m, a distance that is upheld until adulthood when they voluntarily leave their parents' territory (Estes 1997). In contrast, adults travel an average minimum distance of 979 m every day, which is more than 13 times that of juveniles. Adult blue duiker range over about 40% of their territory daily (Estes 1997, 1999), habitually using the same paths to move from sleeping places to feeding areas. People who are aware of these habits can thus readily catch them in snares set along their routes (Skinner and Chimimba 2005). In short, juvenile blue duiker have much less chance of running into snares than the wider-ranging adults, and a catastrophic mortality pattern can consequently not be expected for blue duiker caught in snares (Wadley 2010a).

What the Use of Snares Reveals About Human Cognitive Evolution

Constructing snares with readily-obtained components, and setting them in the paths of potential prey is technologically relatively simple, yet the concept of remote capture that enabled the invention of such equipment is complicated (Wadley 2010a). Equipment designed to function, not immediately, but sometime in the future without human presence, provides evidence for the ability to perceive and integrate action across space and through time (Wynn and Coolidge 2003). Snares are a good example of equipment used for remote capture; they function out-of-sight, but not out-of-mind (Wadley 2010a). The use of snares implies delayed gratification, with capture of a prey animal intended to be distant and unseen by the hunter. Being able to envisage action that is removed from human supervision in both space and time engages modern executive functions of the brain; in turn, these brain functions typify enhanced working memory and modern cognition (Wynn and Coolidge 2003, 2007a, b; Coolidge and Wynn 2005).

The central executive is the decision-making component of working memory, and its functions include paying attention to the goals of an immediate task and reducing superfluous thought and action (Wynn and Coolidge 2007a). The use of aspects of the working memory model as bridging theory (in the sense used by Botha 2008), enables us to link the concept of snaring with modern executive functions of the brain (Wadley 2010a). Evidence for the use of snares in the past thus seems to offer an example of complex cognition because remote capturing devices incorporate sophisticated concepts. It could be argued that spiders create webs as snares, but such behavior cannot be compared to human manufacturing, setting and tending snares. Spider web-making is instinctive with simply coded operational sequences, i.e., as shown in cognigrams, in which the problem-solution distance is far smaller than that demonstrated by similar human thought-and-action sequences (Haidle 2011).

The humble snare and its products are not usually valued in the same way as hunting with encounter weapons such as spears or bow-and-arrow sets. This is obvious from behavior amongst extant hunter-gatherers. Yet, the concept of a snare necessitates the sort of mental abilities that we associate with complex cognition. By operating out-of-sight, while not being forgotten, they can be argued to be the outcome of minds with capabilities that overlapped with ours. Snares reduce search costs for prey and they can provide regular food sources that are of special value for children and the elderly. People who made snares in the Middle Stone Age observed animal behavior, planned the positioning of snares accordingly, and then waited for the remote capture that they could visualize from a distance. Snares and traps are such important behavioral indicators that much attention needs to be given to recognizing their use in the past (Wadley 2010a).

Discussion and Conclusion

Encounter hunting is assumed to be an essential part of the package of behavior associated with hunter-gatherers like those still found in some parts of the world. The activity carries with it implications for complex social interaction, co-operation and reciprocity. Spear-hunting, almost certainly carried out in groups (Wadley 1998), appears to carry with it high status as well as high risk and a promise of irregular, but large meat parcels. The manufacture of the spears themselves can vary from simple, single component, wooden weapons to composites, with stone tips and/or barbs hafted to wooden shafts with compound adhesives. The creation of composite weapons can follow a protracted chain of operations involving long-term planning to accumulate the ingredients, and multi-tasking to ensure that correct conditions are maintained through the manufacturing process. In some instances, for example, in the assembling of compound adhesives, there is no set procedure to follow because the natural ingredients are so variable (Wadley et al. 2009). An essential process is the use of pyrotechnology with careful control of temperatures, duration of heat, and proximity of ingredients to the heat. Experiments imply that so much "thinking on one's feet" is required during the manufacture of composite weapons that artisans are unlikely to be able to impart their knowledge to novices without the use of language. Furthermore, the long and complex "string of beads" represented by the sequence of actions needed to make a composite weapon would be difficult to learn by rote without also understanding the principles behind the actions.

Working with cognigrams and effective chains of production and use demonstrates that the levels of complexity observed in the manufacture and use of composite weapons is accommodated by an increased decoupling of satisfaction and basic need. This means that the small operational units (e.g., the search for components and ingredients, or the production/maintenance of fires, adhesives, materials or tools) are self-sufficient so that each action sequence has its own intermediate aim, independent of immediate basic needs, e.g., satisfying hunger (Lombard and Haidle 2012). Such a modular way of solving problems enables almost unrestricted combinations of units, side-by-side or in effective chains. This allows for levels of behavioral complexity that are barely conceivable without modular simplification (Haidle 2010). Composite spears thus represent the mind's ability for advanced conceptual and technological modularization and flexibility that facilitates relatively high levels of behavioral and cognitive complexity.

The use of arrows with bows opens the way for social change in meat acquisition. It is a light-weight, portable system, providing a lone hunter with numerous shots that can be fired in quick succession into a range of prey types from a distance and/or a concealed position. Although it requires considerable skill, it permits an individual to do alone what can only be accomplished in a group or at great risk using hand-delivered weapons. Bow-and-arrow technology is thus not only niche-broadening in terms of prey type and/or landscape, but it also increases the fitness profile of a single person or a small (core family) group. The manufacturing process for bows and composite arrows is complex, but, when the components are considered individually, the procedure does not indicate behavior that is cognitively more complex than that required by making composite artifacts such as stone-tipped spears. However, as soon as a bow-and-arrow set is used simultaneously as an effective unit, a novel cognitive component is revealed in the form of technological symbiosis. In turn this represents amplified conceptual, technological and behavioral modularization that facilitates levels of complexity and flexibility that are not possible with non-symbiotic technologies. Once humans were able fully to decouple tools and satisfaction of basic needs, and assemble ideas, objects and actions in enhanced modules, the scope for innovative and/or creative problem-solving became infinite. It allows a range of cognitive and behavioral complexity and flexibility that is basic to current human behavior (Lombard and Haidle 2012).

In the case of the stone-tipped arrows inferred in the Howiesons Poort, the sequence of operations is particularly long and demanding. In principle, the stone tips of arrows could be miniature versions of the stone tips used for spears; however, this seems not always to have been the case in the Howiesons Poort. The remarkable innovation of the Howiesons Poort technology involves the rotation of a single stone tool type, the backed segment, to create different types of weapon inserts that were glued to their shafts using variable adhesive recipes depending on the requirements of the design (Wadley and Mohapi 2008). The ability mentally to rotate objects is an indicator of complex cognitive abilities (Wadley 2010b), comparable to those of people living today.

The modest snare, with its seemingly simple technology, is perhaps the most remarkable of all the meat-acquisition techniques (Wadley 2010a). Using snares for meat procurement has several economical and social advantages. For example, the devices can be set and tended by males and females of all ages. They are a safe and reliable means of providing protein and they eliminate search costs by bringing meat to the hunter, rather than requiring the hunter to pursue prey (e.g., Wadley 1998). Because they capture remotely, snares also free time for people to engage in other activities that could include social engagements, ritual activities, collection of plant foods or partaking in group hunting. The snares themselves are completely absent from archaeological records and their use can only be inferred from circumstantial evidence. Nonetheless, the cognitive implications of remote capture are considerable. Any action that is conducted out-of-sight of the actor implies mental abilities that are as sophisticated as those of people like us.

With this contribution we show that it is not necessarily the artifacts themselves, the apparent complexity of their production, their effectiveness, or even their value to the people who used them, that indicate cognitive aptitude in the past. Rather, the cognitive components or concepts that hunting technologies represent can be explored by dealing with well-contextualized archaeological evidence theoretically and using new methodology. Understanding these mental components and concepts is key to the study of how and when humans started to think like us.

Acknowledgements We thank the editors for inviting us to contribute to this volume. We also thank friends and colleagues who worked with us on the presented material and/or ideas, as well as those who reviewed the manuscript. The research of both authors was funded by grants received from the South African National Research Foundation (NRF). Opinions expressed are not necessarily shared by the NRF and any errors in the paper are ours.

References

Allott, L. F. (2006). Archaeological charcoal as a window on palaeovegetation and wood use during the Middle Stone Age at Sibudu Cave. *Southern African Humanities, 18*, 173–201.

Ambrose, S. (2010). Coevolution of composite-tool technology, constructive memory, and language. *Current Anthropology, 51*, S135–S147.

Apps, P. (2000). *Smithers' mammals of Southern Africa: A field guide*. Cape Town: Struik.

Backwell, L., d'Errico, F., & Wadley, L. (2008). Middle Stone Age bone tools from the Howiesons Poort layers, Sibudu Cave, South Africa. *Journal of Archaeological Science, 35*, 566–1580.

Barham, L. (2010). A technological fix for 'Dunbar's dilemma'? *Proceedings of the British Academy, 158*, 367–389.

Botha, R. (2008). Prehistoric shell beads as a window on language evolution. *Language & Communication, 28*, 197–212.

Bradfield, J., & Lombard, M. (2011). A macrofracture study of bone points used in experimental hunting with reference to the South African Middle Stone Age. *South African Archaeological Bulletin, 66*, 67–76.

Brooks, A., Nevell, L., Yellen, J., & Hartman, G. (2006). Projectile technologies of the African MSA: Implications for modern human origins. In E. Hovers & S. Kuhn (Eds.), *Transitions before the transition: Evolution and stability in the Middle Palaeolithic and Middle Stone Age* (pp. 233–255). New York: Springer.

Clark, J. L. (2011). The evolution of human culture during the later Pleistocene: Using fauna to test models on the emergence and nature of "modern" human behaviour. *Journal of Anthropological Anthropology, 30*, 273–291.

Clark, J. L., & Plug, I. (2008). Animal exploitation strategies during the South African Middle Stone Age: Howiesons Poort and post-Howiesons Poort fauna from Sibudu Cave. *Journal of Human Evolution, 54*, 886–898.

Cochrane, G. W. G. (2006). An analysis of lithic artefacts from the 60 ka layers of Sibudu Cave. *Southern African Humanities, 18*, 69–88.

Cooledge, F. L., & Wynn, T. (2005). Working memory, its executive functions, and the emergence of modern thinking. *Cambridge Archaeological Journal, 15*, 5–26.

Delagnes, A., Wadley, L., Villa, P., & Lombard, M. (2006). Crystal quartz backed tools from the Howiesons Poort at Sibudu Cave. *Southern African Humanities, 18*, 43–56.

Estes, R. D. (1997). *The behavior guide to African mammals*. Halfway House: Russel-Friedman.

Estes, R. D. (1999). *The safari companion: A guide to watching African mammals* (Revised edition). Halfway House: Russel Friedman.

Fa, J. E., Ryan, S. F., & Bell, D. J. (2005). Hunting vulnerability, ecological characteristics and harvest rates of bushmeat species in afrotropical forests. *Biological Conservation, 121*, 167–176.

Glenny, W. (2006). Report on the micromammal assemblage analysis from Sibudu Cave. *Southern African Humanities, 18*, 279–288.

Haidle, M. N. (2009). How to think a simple spear. In S. A. De Beaune, F. L. Coolidge & T. Wynn (Eds.), *Cognitive archaeology and human evolution* (pp. 57–73). Cambridge: Cambridge University Press.

Haidle, M. N. (2010). Working-memory capacity and the evolution of modern cognitive potential: Implications from animal and early human tool use. *Current Anthropology, 51*, S149–S166.

Haidle, M. N. (2011). How to think tools? A comparison of cognitive aspects in tool behavior of animals and during human evolution. In *Cognitive perspectives in tool behaviour* (Vol. 1). Tübingen: University of Tübingen.

Hall, G., & Woodborne, S. (2010). Ecosystem change during MIS4 and early MIS 3: Evidence from Middle Stone Age sites in South Africa. *Palaeoecology of Africa, 30*, 1–33.

Jacobs, Z., Roberts, R. G., Galbraith, R. F., Deacon, H. J., Grün, R., Mackay, A., et al. (2008). Ages for the Middle Stone Age of Southern Africa: Implications for human behavior and dispersal. *Science, 322*, 733–735.

Klein, R. G. (1978). Stone Age predation on large African bovids. *Journal of Archaeological Science, 5*, 195–217.

Klein, R. G., & Cruz-Uribe, K. (1984). *The analysis of animal bones from archaeological sites*. Chicago: University of Chicago.

Lombard, M. (2008). Finding resolution for the Howiesons Poort through the microscope: Micro-residue analysis of segments from Sibudu Cave, South Africa. *Journal of Archaeological Science, 35*, 26–41.

Lombard, M. (2011). Quartz-tipped arrows older than 60 ka: Further use-trace evidence from Sibudu, KwaZulu-Natal, South Africa. *Journal of Archaeological Science, 38*, 1918–1930.

Lombard, M. (2012). Thinking through the Middle Stone Age of sub-Saharan Africa. *Quaternary International, 270*, 140–155.

Lombard, M., & Haidle, M. N. (2012). Thinking a bow-and-arrow set: Cognitive implications of Middle Stone Age bow and stone-tipped arrow technology. *Cambridge Archaeological Journal, 22*, 237–264.

Lombard, M., & Pargeter, J. (2008). Hunting with Howiesons Poort segments: Pilot experimental study and the functional interpretation of archaeological tools. *Journal of Archaeological Science, 35*, 2523–2531.

Lombard, M., & Parsons, I. (2008). Blade and bladelet function and variability in risk management during the last 2000 years in the Northern Cape. *South African Archaeological Bulletin, 63*, 18–27.

Lombard, M., & Phillipson, L. (2010). Indications of bow and stone-tipped arrow use 64,000 years ago in KwaZulu-Natal, South Africa. *Antiquity, 84*, 635–648.

Lombard, M., Wadley, L., Deacon, J., Wurz, S., Parsons, I., Mohapi, M., et al. (2012). South African and Lesotho Stone Age sequence updated (I). *South African Archaeological Bulletin, 67*, 120–144.

Lupo, K. D. (2001). Archaeological skeletal part profiles and differential transport: An ethnoarchaeological example from Hadza bone assemblages. *Journal of Anthropological Archaeology, 20*, 361–378.

Lupo, K. D., & Schmitt, D. N. (2002). Upper Paleolithic net-hunting, small prey exploitation, and women's work effort: A view from the ethnographic and ethnoarchaeological record of the Congo Basin. *Journal of Archaeological Method Theory, 9*, 147–179.

McBrearty, S., & Tryon, C. (2005). From Acheulean to Middle Stone Age in the Kapthurin Formation, Kenya. In E. Hovers & S. L. Kuhn (Eds.), *Transitions before The Transition: Evolution and stability in the Middle Paleolithic and Middle Stone Age* (pp. 257–277). New York: Springer.

Mercier, N., & Valladas, H. (2003). Reassessment of TL age estimates of burnt flint from the Paleolithic site of Tabun Cave, Israel. *Journal of Human Evolution, 45*, 401–409.

Mohapi, M. (2012). Point morphology and the Middle Stone Age cultural sequence of Sibudu Cave, KwaZulu-Natal. South Africa. *South African Archaeological Bulletin, 67*, 5–15.

Noss, A. J. (1998). The impacts of cable snaring on wildlife populations in the forests of the Central African Republic. *Conservation Biology, 12*, 390–398.

Pageter, J. (2007). Howiesons Poort segments as hunting weapons: Experiments with replicated projectiles. *South African Archaeological Bulletin, 62*, 147–153.

Pétillon, J.-M., Bignon, O., Bodu, P., Cattelain, P., Debaut, G., Langlais, M., et al. (2011). Hard core and cutting edge: Experimental manufacture and use of Magdalenian composite projectile tips. *Journal of Archaeological Science, 38*, 1266–1283.

Pickering, R. (2006). Regional geology, setting and sedimentology of Sibudu Cave. *Southern African Humanities, 18*, 123–129.

Plug, I. (2004). Resource exploitation: Animal use during the Middle Stone Age at Sibudu Cave, KwaZulu-Natal. *South African Journal of Science, 100*, 151–158.

Plug, I. (2006). Aquatic animals and their associates from the Middle Stone Age levels at Sibudu. *Southern African Humanities, 18*, 289–299.

Schiegl, S., & Conard, N. (2006). The Middle Stone Age sediments at Sibudu: Results from FTIR spectroscopy and microscopic analyses. *Southern African Humanities, 18*, 149–172.

Schmitt, D. N., & Lupo, K. D. (2008). Do faunal remains reflect socioeconomic status? An ethnoarchaeological study among Central African farmers in the northern Congo Basin. *Journal of Anthropological Archaeology, 27*, 315–325.

Semenov, S. A. (1964). *Prehistoric technology*. Bath: Adams & Dart.

Shea, J. J. (2006). The origins of lithic projectile point technology: Evidence from Africa, the Levant, and Europe. *Journal of Archaeological Science, 33*, 823–846.

Shea, J. J., & Sisk, M. L. (2010). Complex projectile technology and *Homo sapiens* dispersal into western Eurasia. *PaleoAnthropology, 2010*, 100–122.

Sievers, C. (2006). Seeds from the Middle Stone Age layers at Sibudu Cave. *Southern African Humanities, 18*, 203–222.

Sievers, C. (2011). Sedges from Sibudu, South Africa: Evidence for their use. In A. G. Fahmy, S. Kahlheber & A. C. d'Andrea (Eds.), *Windows on the African past: Current approaches to African archaeobotany. Reports in African Archaeology* (Vol. 3, pp. 9–18). Frankfurt am Main: Africa Magna Verlag.

Sievers, C., & Muasya, A. M. (2011). Identification of the sedge *Cladium mariscus* subsp. *jamaicense* and its possible use in the Middle Stone Age at Sibudu, KwaZulu-Natal. *Southern African Humanities, 23*, 77–86.

Sisk, M. L., & Shea, J. J. (2009). Experimental use and quantitative performance analysis of triangular flakes (Levallois points) used as arrowheads. *Journal of Archaeological Science, 36*, 2039–2047.

Skinner, J. D., & Chimimba, C. T. (2005). *The mammals of the Southern African Sub-region*. Singapore: Cambridge University Press.

Soriano, S., Villa, P., & Wadley, L. (2009). Ochre for the toolmaker: Shaping still bay points at Sibudu (KwaZulu-Natal, South Africa). *Journal of African Archaeology, 7*, 41–54.

Speth, J. D., & Clark, J. L. (2006). Hunting and overhunting in the Levantine Late Middle Paleolithic. *Before Farming, 3*, 1–42.

Stiner, M. C. (1990). The use of mortality patterns in archaeological studies of hominid predatory adaptations. *Journal of Anthropological Archaeology, 9*, 305–351.

Villa, P., & Lenoir, M. (2006). Hunting weapons of the Middle Stone Age and the Middle Paleolithic: Spear points from Sibudu, Rose Cottage and Bouheben. *Southern African Humanities, 18*, 89–122.

Villa, P., & Soriano, S. (2010). Hunting weapons of Neanderthals and early modern humans in South Africa. *Journal of Anthropological Research, 66*, 5–38.

Villa, P., Delagnes, A., & Wadley, L. (2005). A late Middle Stone Age artefact assemblage from Sibudu (KwaZulu-Natal): Comparisons with the European Middle Paleolithic. *Journal of Archaeological Science, 32*, 399–422.

Villa, P., Soriano, S., Teyssandier, N., & Wurz, S. (2010). The Howiesons Poort and MSA III at Klasies River Mouth, Cave 1A. *Journal of Archaeological Science, 37*, 630–655.

Wadley, L. (1998). The invisible meat providers: Women in the Stone Age of South Africa. In S. Kent (Ed.), *Gender in African prehistory* (pp. 69–81). Walnut Creek: AltaMira Press.

Wadley, L. (2004). Vegetation changes between 61 500 and 26 000 years ago: The evidence from seeds in Sibudu Cave, KwaZulu-Natal. *South African Journal of Science, 100*, 167–173.

Wadley, L. (2005). A typological study of the final Middle Stone Age stone tools from Sibudu Cave, KwaZulu-Natal. *South African Archaeological Bulletin, 60*, 1–13.

Wadley, L. (2006). Partners in grime: Results of multi-disciplinary archaeology at Sibudu Cave. *Southern African Humanities, 18*, 315–341.

Wadley, L. (2007). Announcing a Still Bay industry at Sibudu Cave, South Africa. *Journal of Human Evolution, 52*, 681–689.

Wadley, L. (2010a). Were snares used in the Middle Stone age and does it matter? A review and case study from Sibudu, South Africa. *Journal of Human Evolution, 58*, 179–192.

Wadley, L. (2010b). Compound-adhesive manufacture as a behavioral proxy for complex cognition in the Middle Stone Age. *Current Anthropology, 51*, S111–S119.

Wadley, L. (2010c). Cemented ash as a receptacle or work surface for ochre powder production at Sibudu, South Africa, 58,000 years ago. *Journal of Archaeological Science, 37*, 2397–2406.

Wadley, L., & Jacobs, Z. (2006). Sibudu Cave: Background to the excavations, stratigraphy and dating. *Southern African Humanities, 18*, 1–26.

Wadley, L., & Mohapi, M. (2008). A Segment is not a Monolith: Evidence from the Howiesons Poort of Sibudu, South Africa. *Journal of Archaeological Science, 35*, 2594–2605.

Wadley, L., Hodgskiss, T., & Grant, M. (2009). Implications for complex cognition from the hafting of tools with compound adhesives in the Middle Stone Age, South Africa. *Proceedings of the National Academy of Sciences, 106*, 9590–9594.

Wadley, L., Sievers, C., Bamford, M., Goldberg, P., Berna, F., & Miller, C. (2011). Middle Stone Age bedding construction and settlement patterns at Sibudu, South Africa. *Science, 334*, 1388–1391.

Wilkinson, P. F. (1976). Random hunting and the composition of faunal samples from archaeological excavations: A modern example from New Zealand. *Journal of Archaeological Science, 3*, 321–328.

Wynn, T., & Coolidge, F. L. (2003). The role of working memory in the evolution of managed foraging. *Before Farming, 2003*(2), 1–16.

Wynn, T., & Coolidge, F. L. (2007a). Did a small but significant enhancement in working memory capacity power the evolution of modern thinking? In P. Mellars, K. Boyle, O. Bar-Yosef & C. Stringer (Eds.), *Rethinking the human revolution: New behavioural and biological perspectives on the origin and dispersal of modern humans* (pp. 79–90). Cambridge: McDonald Institute Monographs.

Wynn, T., & Coolidge, F. L. (2007b). A Stone-Age meeting of minds. *American Scientist, 96*, 44–51.

Yaroshevich, A., Kaufman, D., Nuzhnyy, D., Bar-Yosef, O., & Weinstein-Evron, M. (2010). Design and performance of microlith implemented projectiles during the Middle and the Late Epipaleolithic of the Levant: Experimental and archaeological evidence. *Journal of Archaeological Science, 37*, 368–388.

Yasuoka, H. (2006). The sustainability of duiker (*Cephalophus* spp.) hunting for the Baka hunter-gatherers in southeastern Cameroon. *African Study Monographs Supplement, 33*, 95–120.

Part VI

Conclusions

Chapter 20
Summary and Conclusions

Radu Iovita and Katsuhiro Sano

Stone Age Weapons in the Context of Major Debates in Human Evolution

That tools, and especially weaponry, as a substitute for sharp teeth, claws, and physical stature, should have played a major, constitutive role in human evolution was clear from the very beginning of the discipline (e.g., Lamarck 1820; Darwin 1871). The idea that early humans "freed" themselves from the constraints of the environment by using tools to hunt and defend themselves can even be traced back to classical times (e.g., Lucretius, cited in Stoczkowski 2002), and has reappeared frequently in explanations of technological evolution. The theoretical postulation of their existence before archaeology even existed as a discipline can perhaps be held responsible for the delay with which providing real, objective proof of it came into the focus of research. In a certain sense, weapons were obvious and could be identified by inspection, with description and classification being the only tasks remaining. This attitude reflects the essentialist approach which characterized the earliest treatments of all stone implements (e.g., Holdaway and Douglass 2011).

In order to place this book in its relevant historical context, we mined the available article databases for research trends and themes that might have driven or informed past bursts in activity related to this topic. Although quantitative data on articles published before 1955 are not available, a quick survey of the literature gives the impression that most accounts of the subject before the middle of the 20th century simply assume or infer the presence of weapons in archaeological contexts from simple, common-sense morphological characteristics that weapons should possess, such as being "pointed". In regions where such formally "obvious" points were common (e.g., the Americas, Upper Paleolithic Europe), most of the effort then went into creating complex classificatory schemes for identifying culture areas a culminating example for North American being Bell and Perino (1958). This trend continued despite early experimental studies on performance and its relationship to morphology, some of which contradicted the mainstream claims (e.g., Browne 1940, who recognized that the size of the projectile point does not necessarily allow conclusions about the delivery mode). From the second half of the 20th century, a survey of two of the most important databases of scholarly articles, Web of Science (Thomson Reuters) and Scopus (Elsevier) (illustrated below in Fig. 20.1) shows several peaks in the occurrence of the terms "weapons" or "projectiles" in journal articles.[1]

The timing of these peaks is for the most part, explicable in the context of the known debates in the fields of anthropology and archaeology.

The first peak appears shortly after the Second World War, in a period when several of the lines of work that would later turn out to be important for the study of Stone Age weapons first entered the fore. The first incontrovertible evidence of Paleolithic weapons came slightly before this time, with the discovery of arrow shafts at the late Upper Paleolithic site of Stellmoor in northern Germany (Rust 1943). However, further discoveries of tipped arrows from Mesolithic water-logged contexts in Denmark (Bröndsted 1957) and Germany (Troels-Smith 1959) followed in the 1950s. At roughly the same time, the first English translation of Semenov's *Prehistoric technology* (1964) brought

R. Iovita (✉)
MONREPOS Archaeological Research Centre and Museum for Human Behavioural Evolution, Römisch-Germanisches Zentralmuseum – Leibniz-Forschungsinstitut für Archäologie, 56567 Neuwied, Germany
e-mail: iovita@rgzm.de

K. Sano
The University Museum, The University of Tokyo, Hongo 7-3-1, Bunkyo-Ku, Tokyo, 113-0033, Japan
e-mail: sano@um.u-tokyo.ac.jp

[1]The list of journals was limited to those which specialize in Stone Age archaeology or general anthropology journals. It is possible that some irrelevant results might have crept in, but the overall trends make sense. Note also that only English-language results were included in the computation.

Radu Iovita and Katsuhiro Sano (eds.), *Multidisciplinary Approaches to the Study of Stone Age Weaponry*, Vertebrate Paleobiology and Paleoanthropology, DOI 10.1007/978-94-017-7602-8_20

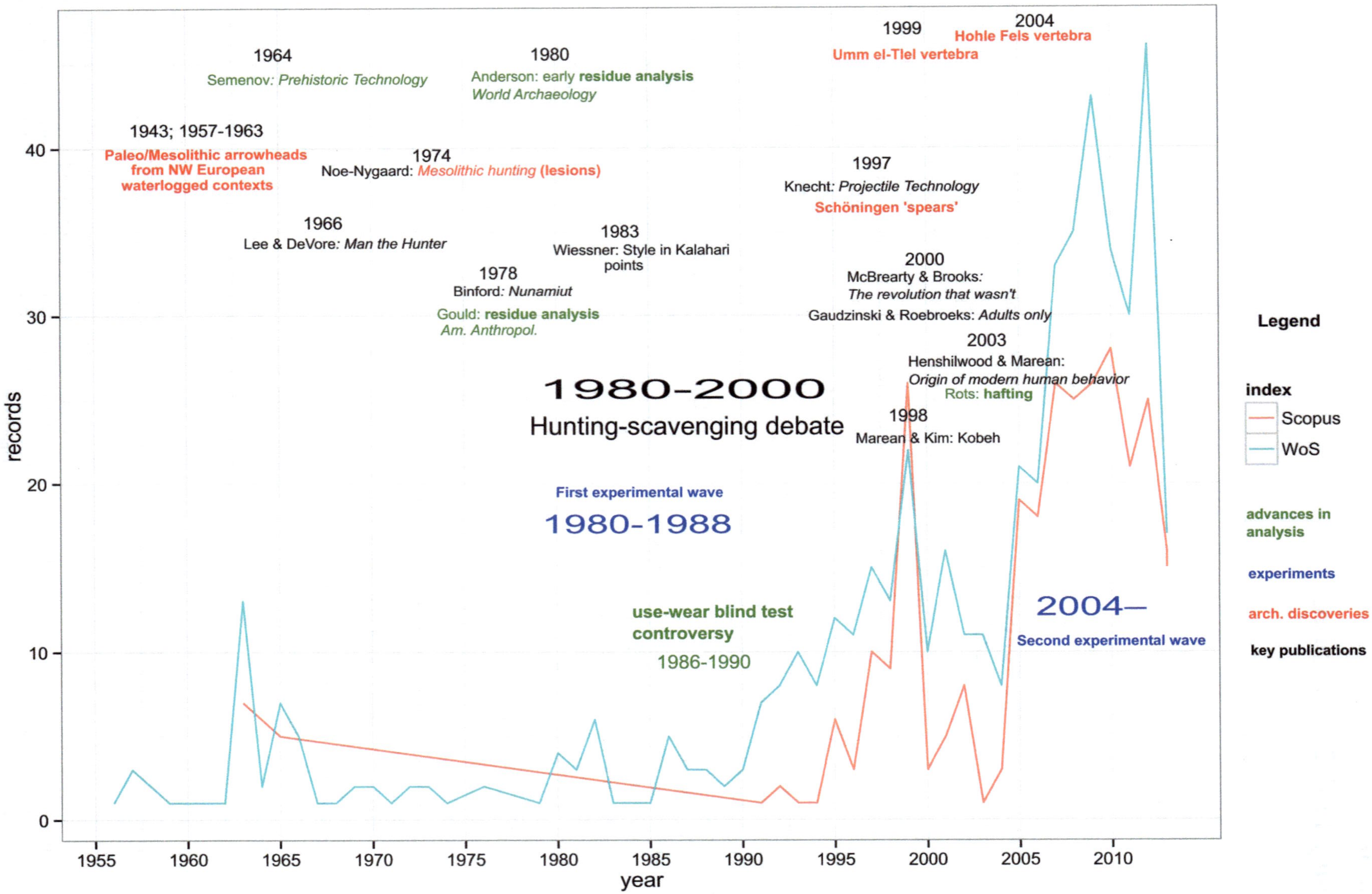

Fig. 20.1 Number of records containing the terms “weapons” or “projectiles” by year in the Scopus (Elsevier, red) and Web of Science (Thomson Reuters, cyan) databases of scientific articles. The trend shows two major peaks (1990s and mid 2000s) and a general increase in the number of articles devoted to the topic

attention to the role of traces that human actions leave on stone tools. Finally, the role of hunting and the hunting-gathering economy and way of life, a topic that had concerned cultural anthropologists ever since the beginnings of the discipline, was entering a crystallization phase at this time, which culminated in the publication of *Man the hunter* (Lee and DeVore 1968).

We are not trying to suggest that this first peak reflects a genuine focus by the entire research community on weapons and the role they might have played in human evolution. Rather, the 1960s were a period when many of the necessary "ingredients" that later gave impetus to the effervescence of the 1980s and 1990s first became available. That they did not immediately result in a synthesis with a lasting effect is evidenced by the fact that many of the early discoveries were only recently incorporated into the body of knowledge about prehistoric weapons. For example, many of the tips of the preserved arrows from Northern Europe were simple, unretouched flakes, and, most importantly, they were not pointed (so-called, "transverse arrowheads"). In that sense, they emphatically contradicted the "common-sense" view of what weapon tips should look like, and it is remarkable that decades were later spent focusing only on "pointed" pieces as possible weapon tips (sadly, this is often still the case, although see Lombard and Pargeter 2008; Yaroshevich et al. 2010 for new studies in various areas of the world).

The most important "missing ingredient" to the systematic study of hunting implements was a theoretical link between the knowledge gained from hunter-gatherer ethnographies and the archaeological record. For this reason, the development of the New Archaeology in the 1960s and 1970s, and especially the establishment of ethnoarchaeology as a discipline of its own contributed enormously to focusing research on recreating real, actual behavior from inanimate objects. The fact that the Bordes-Binford debate (Bordes 1961; Binford and Binford 1966) centered on almost comically very fundamental aspects of the nature of lithic assemblages – whether variation in the composition of the Mousterian was a result of cultural tradition or of functional differences in site use – illustrates just how much of Stone Age archaeology had been carried out on the basis of unchallenged assumptions about the meaning of the archaeological record.

Once exposed, these assumptions began to be tested, ushering in one of the most active periods of interdisciplinary research, which featured advances in experimentation, quantitative analyses, and ethnoarchaeology. This can be said to have led to a real paradigm shift in Stone Age archaeology, which largely determined the direction of research for decades to come. Following the publication of Binford's Nunamiut work (Binford 1981), lithic assemblages began to be seen as reflecting various mobility patterns and/or economic strategies of landscape exploitation, rather than mere units of cultural tradition frozen in time. In North America in particular, where bifacially flaked points dominated the lithic record, putative weapons were quickly integrated in economic models of lithic procurement, use, and maintenance (for a review see Shott 1996). Especially the idea that weapons would have formed an important part of curated technologies also influenced the understanding of point morphology as a result of retooling (Ahler 1971; Frison 1974; Hoffman 1985).

At the same time, archaeologists became aware of the effect that natural processes had on archaeological site formation (e.g., Schiffer 1987), and, in particular, on the preservation of behaviorally-relevant aspects of bone accumulations (e.g., Bunn 1991). Skepticism regarding the timing and role of hunting (and, implicitly, weapon use) in the accumulation of faunal remains in hominin sites eventually led to the famous hunting-scavenging debate focused in large part on the earliest part of the record (Bunn 1981; Blumenschine 1987) but playing an important role in interpreting the more recent, Late Pleistocene sites as well (Stiner 1994; Marean and Kim 1998). The methodological question at the center of the debate, namely deciding whether cut marks overlay or underlay carnivore marks on the bones of prey species, prompted a much-needed emphasis on isolating and interpreting micro and macroscopic traces of behavior on both artifacts and bones. Although most zooarchaeological studies were dedicated to the study of butchery marks, some (e.g., Noe-Nygaard 1974; Bratlund 1991) also examined damage left behind by weapons.

Systematic studies of damage on the weapons themselves appeared relatively late, as part of the wave of use-wear experiments in the 1980s and 1990s. Using principles of fracture mechanics and the "low-power approach" (Tringham et al. 1974; Cotterell and Kamminga 1979; Tsirk 1979; Odell and Odell-Vereecken 1980) allowed for a more systematic description of macrofractures on experimental specimens and their correlation with archaeologically-known patterns of damage (e.g., Barton and Bergman 1982; Moss and Newcomer 1982; Huckell 1982; Bergman and Newcomer 1983; Fischer et al. 1984; Odell and Cowan 1986; Shea 1988). The subsequent projectile experiments in the 1990s (Midoshima 1991, 1996; Geneste and Plisson 1993; Caspar 1996; Kelterborn 1999) confirmed that "diagnostic impact fractures" indeed frequently occur on tips made on a variety of raw materials and having diverse shapes. Unfortunately, the 1980s boom in use-wear studies on lithic materials was rarely in direct dialogue with similar developments in zooarchaeology, and relatively few of the early experimental studies carried out in the 1980s explicitly framed their research within the context of the hunting-scavenging debate. However, looking back at this crucial decade, it makes sense that the convergence of the above-mentioned lines of work gave an impetus to projectile experiments and the systematic investigation of weapons.

It is in this climate of research, concerned with fundamental methodological problems and the challenging of common-sense assumptions that the synthesis present in Knecht's volume (1997) took form. And Knecht's book was not the only attempt at synthesizing the knowledge in the 1990s: books on *The Evolution of Human Hunting* (Nitecki and Nitecki 1987) and on *Hunting and Animal Exploitation in the Later Palaeolithic and Mesolithic of Eurasia* (Peterkin et al. 1993) were published in this time period. Similar edited volumes on hunting were produced in the French-speaking research community (Bellier et al. 2000). The sum total of these articles makes up the second peak in the number of articles, in the mid-late 1990s.

Since the early 2000s, zooarchaeological evidence in favor of early hunting has been mounting, and there is now little doubt that hominins had primary access to herbivore carcasses (Domínguez-Rodrigo 2002; Gaudzinski 2004; Domínguez-Rodrigo et al. 2007; Rabinovich et al. 2011). Certainly Neandertals successfully hunted very large and dangerous animals (Marean and Kim 1998; Gaudzinski and Roebroeks 2000), although the evidence for the use of weapons to gain access to the carcasses is not as convincing as that which we find in the Upper Paleolithic (but see Callow 1986; Shea 1988), begging the question of how they might have done it exactly (see also Gaudzinski-Windheuser 2016).

Oddly, it is not the question of how hunting developed in the first place that has dominated the post-scavenging debate research on weapons (although see Wilkins et al. 2012; Wilkins and Schoville 2016). Instead, it is the search for archaeological criteria to distinguish the first modern humans from a behavioral point of view that has encouraged much of the recent work, especially in sub-Saharan Africa. McBrearty and Brooks' (2000) seminal review of the evidence for a gradual accumulation of "modern" human behavior during the African Middle Stone Age relied in part on the observed regional variation in pointed tools, interpreted as weapon tips. Thereby they opened several research avenues for the future: weapons, and especially long-range projectiles, could be seen as both technological achievement conferring an adaptive advantage, and, at the same time, a sign of cognitive development and even an identity marker (Henshilwood and Marean 2003; Brown et al. 2012), the latter being largely derived from the ethnographic work by Wiessner (1983). The mounting evidence in support for MSA hunting weapons (e.g., Lombard 2005; Lombard and Pargeter 2008; Brown et al. 2012; Wilkins et al. 2012), led to a similarly intensive search for such weapon tips in the European Middle Paleolithic, providing several candidates (Rots 2009, 2013; Villa et al. 2009; Lazuén 2012). Further driven by the debate on the origin and meaning of "complex projectile weaponry" (Shea and Sisk 2010; Sisk and Shea 2011), aided by a few notable archaeological discoveries of possible weapons of great antiquity (Thieme 1997; Boëda et al. 1999; Münzel and Conard 2004), and accompanied by a revitalization of use-wear studies and experiments in the mainstream literature, weapons research can be said to have entered a new and productive phase in the new millennium.

Towards a New Synthesis

The present book exhibits a broad spectrum of issues concerning Stone Age weaponry, from the technicalities of forensic reconstructions to the cognitive implications of an inferred hunting-related behavior. Despite the diversity of subjects and geographical areas of the individual chapters, a few strong themes emerge to form a framework for directing future studies. A particularly important development is the clarification of the roles played by different types of approaches to reconstructing behaviors associated with the use of weapons. Below, we summarize some of the main topics addressed by the papers in this volume.

Identifying weapons in some areas, such as *the forensic study of impact traces*, new controlled experimental protocols (Iovita et al. 2016; Sano et al. 2016) and science-based redefinition of the relevant study parameters (Hutchings 2016) are introduced in order to remove causal ambiguities resulting from the complexity of the object of study and the equifinality inherent in many of the processes involved. Hutchings (2016) opens by discussing logical and methodological pitfalls common to most current approaches to identifying past weapon tips and makes some suggestions for improving standards of scientific rigor. He puts a strong accent on the search for analytical units that are causally linked with the behaviors on which inference is sought. Iovita et al. (2016) present the results of two controlled experiments using copies of Levallois points cast in soda-lime glass, synthetic targets, and dynamically-monitored launching mechanisms in the lab. They attempt to distinguish between spear thrusting and projectiles of various speeds based on macroscopic (fracture morphology) and microscopic (Wallner Lines) criteria. Sano et al. (2016) use a calibrated cross-bow set-up for testing a correlation between impact-related fractures and impact velocities in siliceous hard shale replicas of Japanese Upper Paleolithic trapezoids in order to examine whether or not the trapezoids were mechanically delivered. Both studies aim to improve upon aspects of experiments from the 1980s and establish proxies for recognizing launching mode. Pétillon et al. (2016) review 30 years of experiments using bone projectile points, and outline the successes and challenges facing further work. The section closes with a consideration of our evolving understanding of the physics of atlatl propulsion by Whittaker (2016) and its implications for reconstructing prehistoric weapon use.

Another 1980s topic that has been revived in the last few years is the identification of weapon marks on bones. Sabine Gaudzinski-Windheuser (2016) leads the next section with a timely review of the available zooarchaeological evidence for weapon damage in the Pleistocene, concluding that there are surprisingly few such marks in the period before the Late Upper Paleolithic, given how good the evidence for hunting is, even for the earliest periods. Jayne Wilkins and Ben Schoville (2016) dedicate their paper to a more detailed study of the combined effects of taphonomy and spear-use on the damage patterns in what may be the oldest stone-tipped weapons to date, from Kathu Pan in South Africa. Alla Yaroshevich et al. (2016) continue the archaeological applications with a discussion of Early Middle Paleolithic impact damage on points from the site of Misliya in Israel. The next two papers concern Gravettian implements. Milks et al. (2016) question the projectile function of tanged "Font-Robert" points, whereas Marreiros et al. (2016) present evidence for impact-related wear in microliths previously thought to have been involved in piercing tasks. The last two chapters in this section continue the discussion on the use of composite weapons, with Richard Fullagar (2016) weighing in on the debate about whether Australian microlithic composite tools were primarily used as weapons or for cutting/piercing household tasks, whereas Veerle Rots (2016) examines the role of the identification of hafting arrangements for the identification of armatures as a whole.

Weapon Performance

A particularly interesting avenue of research that has been developed in the last few decades regards the potential of various reconstructed weapons to achieve their purpose successfully. In this vein, Chris Clarkson (2016) evaluates experimentally the use of common morphometric criteria, such as tip-cross-sectional-area (TCSA) for the ballistic efficacy of stone tools as weapon tips. In the next chapter, Paul Salem and Steven Churchill (2016) reevaluate recent results which suggest lithic points penetrate deeper than wooden ones, and constitute a technological improvement over the latter. This research has important implications for assessing differences between Middle Pleistocene hunters in Europe, who are believed to have used wooden implements (such as those from Schöningen) as weapons, and those from Africa (see also Wilkins and Schoville 2016). Joseba Rios-Garaizar (2016) closes the section with an exploratory replicative study of the feasibility of throwing spears tipped with European Middle Paleolithic stone tips.

Curation and Life-History of Weapons

The issue of curation and the life-history of artifacts has long been an important side of discussing the economy and use of weapons in the landscape (e.g., Shott 1996). Langley (2016) presents an overview of curation and maintenance of osseous projectile tips, including its implications for manufacturing skill and performance. In a different take on the same issues, but focusing on lithic points, Shott (2016) compares experimental attrition distributions with those predicted by Gompertz and Gompertz-Makeham models.

Beyond Weapons as Tools

A couple of essays aimed at elucidating the behavioral, cultural, and cognitive implications of the actual use of specific weapons close the volume. Marcelo Cardillo and Judith Charlin (2016) explore the relationship between Patagonian projectile tip shape, quantified with the most modern geometric morphometric methods, and geography. Finally, Marlize Lombard and Lyn Wadley (2016) close with a comparative exploration of the cognitive capacities required for meat acquisition through hunting and those required for the manufacture of composite weapons and snares.

Despite the eclectic nature of this collection of papers, there are several common themes that suggest a crystallization of thought on how to identify and interpret weapons in Stone Age contexts. First comes a desire for more scientific rigor, often accompanied by an increase in quantification and a return to first principles. Second comes a more realistic view of the nature of the archaeological record, with a special regard for taphonomy. Most of the studies presented here acknowledge that a careful consideration of taphonomic factors on use-related wear traces is required, especially when making inferences with great behavioral implications (Hutchings 2016; Wilkins and Schoville 2016). Third, and finally, almost all papers emphasize the need for employing a combination of as many lines of evidence as possible, including macroscopic and microscopic wear traces, along with residue analysis (Fullagar 2016; Rots 2016) when making archaeological identifications of weapons. The latter has serious implications for the amount of data that can be processed on any particular collection and the labor input that is necessary for obtaining the data. Even if progress appears to be slowed down by these strict requirements, it is always preferable to start from a stable basis of first principles.

Final Thoughts and Future Directions

At the close of this volume, what can be said to be the gains in knowledge about the archaeological record in light of these new methodological advances, and how does the medium-term prognosis look? As the papers from this book which have an explicit applied archaeological content show, there are some surprises in store for the future student of Stone Age weapons. Starting with the most basic question, that of identification, it is already clear that some of the tools originally thought to be projectile points may have been involved in other tasks. Moreover, the era of facile guesses of function from inspecting the form looks to be over, as our ability to do this is demonstrably much lower than usually assumed (Fullagar 2016; Marreiros et al. 2016; Milks et al. 2016). Likewise, new discoveries have blurred the easy correspondences between technologies and hominin species, both as far as the difficulty of manufacture and the performance attributes of the respective technologies are concerned (e.g., Yaroshevich et al. 2010; Wilkins and Schoville 2016).

These insights have made it slightly more difficult to use weapons in order to construct a grand narrative about the development of technology during the span of human evolution. While the chronological progression from "simple" (hand-delivered) to "complex" (mechanically-aided) weaponry still stands, that distinction appears less useful than originally thought. There are two main reasons for this: the first, which we have mentioned before, is because simple, unambiguous, easy to measure and at the same time universally applicable proxies for distinguishing launching technologies still do not exist (Clarkson 2016; Hutchings 2016; Iovita et al. 2016; Sano et al. 2016; Rots 2016). The second reason concerns higher-order concepts of technological quality and its implications for biological evolution. More specifically, we must refine our hypotheses about the links between weapons and adaptation to a particular environment, the ability to colonize new realms, as well as the cognition and skills that made these achievements possible.

Regarding the latter topic, it will perhaps be necessary to re-examine some of our most basic assumptions about performance and quality. One such assumption which has been least challenged, is that the spearthrower and, especially, the bow were absolute game-changing technologies, which conferred massive competitive advantages upon the groups which had invented them. However, in most cases, not only the more circumstantial aspects of hunting with a particular gear set [which can be exceedingly complex as shown by Hitchcock and Bleed 1997 where success depends largely on the use of poison (for arrows) or horses (for spears)] remain unknown, but, in the absence of the organic parts themselves, so are even the most basic physical properties of the weapon system itself. In some cases, the advantages and disadvantages of a particular technology, such as the use of stone tips on penetrating weapons, are dependent on the goal (maximum bleeding or maximum penetration Waguespack et al. 2009; Salem and Churchill 2016). The famously low impact energy delivered by San bows (Wannenburgh et al. 1999) constitutes a cautionary tale regarding reading too much into the advantages conferred by use of bows in the past, unless the performance-related physical properties can also be reconstructed. Answers to these questions will no doubt require more experimentation (both controlled and replicative) and a revision of the ethnographic record in search of possible use modalities, which might have been overlooked exactly because they contradict our common-sense understanding of how weapons work. Examples include the use of very large blades as tips for spearthrower darts (Newman and Moore 2013), or the low prey yields for users of bow-shot poisoned arrows among the Hadza (e.g., O'Connell 1988).

The flip side of the performance question is that neither complexity of manufacture nor of use is equivalent to performance superiority. This means that incredibly simple technologies may have persisted despite the development of a brain capable of "doing better". Although some weapons systems, such as composite projectiles and snares do imply a depth of planning consistent with modern human cognitive capacities (Lombard and Wadley 2016), their absence does not imply an absence of those capacities. As the zooarchaeological record shows (Gaudzinski-Windheuser 2016), the lack of stone points embedded in bone and characteristic lesions on faunal material from much of the Lower and Middle Paleolithic suggests that perhaps simple, one-piece wooden spears were used for a very long time. Given that savannah chimpanzees hunt with wooden sticks, some of which are sharpened with their teeth (Pruetz and Bertolani 2007), it is possible that weapons per se were a part of the "package" of complex tool use since the time of the last common ancestor. Interestingly, since chimpanzees are very strong compared with humans, it also contradicts the age-old assumption that it was an inherent physical weakness that somehow forced our ancestors to invent and use tools (Lamarck 1820; Darwin 1871). The difference between what is possible in terms of intelligence and what is manifest in the behavior of a particular species is also clearly present in the case of contemporary and recent hunter-gatherers (*H. sapiens*), which have different levels of technological complexity ranging from very complex to very simple (Oswalt 1976). These examples should serve as a warning for taking the first appearance of a particular technology or technique as evidence for the existence of a cognitive capacity allowing it to develop, and, especially, for taking negative evidence to mean the contrary.

Given all these complexities, it is quickly becoming clear that, despite a strong interest and revival of interest in the topic, we are still a long way from a real evaluation of the evolutionary role of weapons. For this purpose, a much tighter integration of the data is needed, and that will only be possible when entire behavioral contexts become available. The establishment of standards in terminology, as well as a good communication between researchers carrying out experimental studies is crucial, as was unanimously agreed in Mainz in September 2011. As we write this, a flurry of new papers is coming out, probably a sign that it will soon be time for a new meeting. With that in mind, we close this book with the hope for many exciting and lasting discoveries which will help put the use of weapons in the Stone Age in their proper evolutionary context.

References

Ahler, S. (1971). *Projectile point form and function at Rodgers Shelter, Missouri* (p. 8). Columbia: Missouri Archaeological Society Research Series no. 8

Anderson, P. (1980). A testimony of prehistoric tasks: Diagnostic residues on stone tool working edges. *World Archaeology, 12*, 181–194.

Barton, R. N. E., & Bergman, C. A. (1982). Hunters at Hengistbury: Some evidence from experimental archaeology. *World Archaeology, 14*, 237–248.

Bergman, C. A., & Newcomer, M. H. (1983). Flint arrowhead breakage: Examples from Ksar Akil, Lebanon. *Journal of Field Archaeology, 10*, 238–243.

Bell, R. E., & Perino, G. (1958). *Guide to the identification of certain American Indian projectile points*. Oklahoma City: Oklahoma Anthropological Society

Bellier, C., Cattelain, P., & Otte, M. (Eds.). (2000). *La Chasse dans la prehistoire, actes du colloque international de Treignes, 3–7 octobre 1990, Bruxelles*. Liège: société royale belge d'Anthropologie et de Préhistoire.

Binford, L. R. (1981). *Bones: Ancient men and modern myths*. New York: Academic Press.

Binford, L. R., & Binford, S. R. (1966). A preliminary analysis of functional variability in the Mousterian of Levallois facies. *American Anthropologist, 68*, 238–295.

Blumenschine, R. J. (1987). Characteristics of an early hominid scavenging niche. *Current Anthropology, 38*, 383–407.

Boëda, E., Geneste, J. M., Griggo, C., Mercier, N., Muhesen, S., Reyss, J. L., et al. (1999). A Levallois point embedded in the vertebra of a wild ass (*Equus africanus*): Hafting, projectiles and Mousterian hunting weapons. *Antiquity, 73*, 394–402.

Bordes, F. (1961). Mousterian cultures in France: Artifacts from recent excavation dispel some popular misconceptions about Neanderthal man. *Science, 134*, 803.

Bratlund, B. (1991). A study of hunting lesions containing flint fragments on reindeer bones at Stellmoor, Schleswig-Holstein, Germany. *CBA Research Report, 77*, 193–207.

Brown, K. S., Marean, C. W., Jacobs, Z., Schoville, B. J., Oestmo, S., Fisher, E. C., et al. (2012). An early and enduring advanced technology originating 71,000 years ago in South Africa. *Nature, 491*, 590–593.

Browne, J. (1940). Projectile points. *American Antiquity, 5*, 209–213.

Bröndsted, J. (1957). *Danmarks oldtid* (Vol. 1). Copenhagen: Gyldendal.

Bunn, H. T. (1981). Archaeological evidence for meat-eating by Plio-Pleistocene hominids from Koobi Fora and Olduvai Gorge. *Nature, 291*, 574–577.

Bunn, H. T. (1991). A taphonomic perspective on the archaeology of human origins. *Annual Review of Anthropology, 20*, 433–367.

Callow, P. (1986). The flint tools. In P. Callow & J. Conrnford (Eds.), *La Cotte de St Brelade* (pp. 251–314). Norwich, United Kingdom: Geo Books.

Cardillo, M., & Charlin, J. (2016). Morphological diversification of stemmed projectile points of Patagonia (Southernmost South America). Assessing spatial patterns by means of phylogenies and comparative methods. In R. Iovita & K. Sano (Eds.), *Multidisciplinary approaches to the study of Stone Age weaponry* (pp. 261–272). Dordrecht: Springer.

Caspar, J. P. B. M. (1996). Preparing for the hunt in the Late Paleolithic camp at Rekem, Belgium. *Journal of Field Archaeology, 23*, 437–460.

Clarkson, C. (2016). Testing archaeological approaches to determining past projectile delivery systems using ethnographic and experimental data. In R. Iovita & K. Sano (Eds.), *Multidisciplinary approaches to the study of Stone Age weaponry* (pp. 189–201). Dordrecht: Springer.

Cotterell, B., & Kamminga, J. (1979). The mechanics of flaking. In B. Hayden (Ed.), *Lithic use-wear analysis* (pp. 97–112). New York: Academic Press.

Darwin, C. G. (1871). *The Descent of Man*. London: Murray.

Domínguez-Rodrigo, M. (2002). Hunting and scavenging by early humans: The state of the debate. *Journal of World Prehistory, 16*, 1–54.

Domínguez-Rodrigo, M., Egeland, C. P., & Barba, R. (2007). *Deconstructing Olduvai: A taphonomic study of the Bed I sites*. Dordrecht: Springer.

Fischer, A., Hansen, P. V., & Rasmussen, P. (1984). Macro and microwear traces on lithic projectile points. Experimental results and prehistoric examples. *Journal of Danish Archaeology, 3*, 19–46.

Frison, G. C. (1974). Archeology of the Casper Site. In G. C. Frison (Ed.), *The Casper site: A Hell Gap bison kill on the High Plains*. New York: Academic Press.

Fullagar, R. (2016). Uncertain evidence for weapons and craft tools: Functional investigations of Australian microlith. In R. Iovita & K. Sano (Eds.), *Multidisciplinary approaches to the study of Stone Age weaponry* (pp. 159–166). Dordrecht: Springer.

Gaudzinski, S. (2004). Subsistence patterns of early Pleistocene hominids in the Levant–taphonomic evidence from the 'Ubeidiya formation (Israel). *Journal of Archaeological Science, 31*, 65–75.

Gaudzinski-Windheuser, S. (2016). Hunting lesions in Pleistocene and early Holocene European bone assemblages and their implications for our knowledge on the use and timing of lithic projectile technology. In R. Iovita & K. Sano (Eds.), *Multidisciplinary approaches to the study of Stone Age weaponry* (pp. 77–100). Dordrecht: Springer.

Gaudzinski, S., & Roebroeks, W. (2000). Adults only. Reindeer hunting at the Middle Palaeolithic site Salzgitter Lebenstedt, Northern Germany. *Journal of Human Evolution, 38*, 497–521.

Geneste, J. M., & Plisson, H. (1993). Hunting technologies and human behavior: Lithic analysis of Solutrean shouldered points. In H. Knecht, A. Pike-Tay & R. White (Eds.), *Before Lascaux: The complex record of the Early Upper Paleolithic* (pp. 117–135). Boca Raton, Ann Arbor, London, Tokyo: CRC Press.

Henshilwood, C. S., & Marean, C. W. (2003). The origin of modern human behavior. *Current Anthropology, 44*, 627–651.

Hitchcock, R., & Bleed, P. (1997). Each according to need and fashion: Spear and arrow use among San hunters of the Kalahari. In H. Knecht (Ed.), *Projectile technology* (pp. 345–368). New York: Plenum Press.

Hoffman, C. (1985). Projectile point maintenance and typology: Assessment with factor analysis and canonical correlation. In C. Carr (Ed.), *For concordance in archaeological analysis: Bridging data structure, quantitative technique, and theory* (pp. 566–612). Kansas City, MO: Wesport Press.

Holdaway, S., & Douglass, M. (2011). A twenty-first century archaeology of stone artifacts. *Journal of Archaeological Method and Theory, 19*, 101–131.

Huckell, B. B. (1982). The Denver elephant project: A report on experimentation with thrusting spears. *Plains Anthropologist, 27*, 217–224.

Hutchings, K. (2016). When is a point a projectile? Morphology, impact fractures, scientific rigor, and the limits of inference. In R. Iovita & K. Sano (Eds.), *Multidisciplinary approaches to the study of Stone Age weaponry* (pp. 3–12). Dordrecht: Springer.

Iovita, R., Schönekeß, H., Gaudzinski-Windheuser, S., & Jäger, F. (2016). Identifying weapon delivery systems using macrofracture analysis and fracture propagation velocity: A controlled experiment. In R. Iovita & K. Sano (Eds.), *Multidisciplinary approaches to the study of Stone Age weaponry* (pp. 13–27). Dordrecht: Springer.

Kelterborn, P. (1999). Analysen und Experimente zu Herstellung und Gebrauch von horgener Pfeilspitzen. *Jahrbuch der schweizerischen Gesellschaft für Ur- und Frühgeschichte, 83*, 37–64.

Knecht, H. (Ed.). (1997). *Projectile technology*. New York: Plenum Press.

Lamarck, J.-B. (1820). *Système analytique des connaissances positives de l'homme* (pp. 1–371). Paris: CNRS/A Belin.

Langley, M. C. (2016). More to the point: Developing a multi-faceted approach to investigating the curation of Magdalenian osseous projectile points. In R. Iovita & K. Sano (Eds.), *Multidisciplinary approaches to the study of Stone Age weaponry* (pp. 229–244). Dordrecht: Springer.

Lazuén, T. (2012). European Neanderthal stone hunting weapons reveal complex behaviour long before the appearance of modern humans. *Journal of Archaeological Science, 39*, 2304–2311.

Lee, R. B., & DeVore, I. (Eds.). (1968). *Man the hunter*. New York: Aldine de Gruyter.

Lombard, M. (2005). Evidence of hunting and hafting during the Middle Stone Age at Sibudu Cave, KwaZulu-Natal, South Africa: A multianalytical approach. *Journal of Human Evolution, 48*, 279–300.

Lombard, M., & Pargeter, J. (2008). Hunting with Howiesons Poort segments: Pilot experimental study and the functional interpretation of archaeological tools. *Journal of Archaeological Science, 35*, 2523–2531.

Lombard, M., & Wadley, L. (2016). Hunting technologies during the Holwiesons Poort at Sibudu Cave: What they reveal about human cognition in KwaZulu-Natal, South Africa, between 65 and 62 ka. In R. Iovita & K. Sano (Eds.), *Multidisciplinary approaches to the study of Stone Age weaponry* (pp. 273–286). Dordrecht: Springer.

Marean, C. W., & Kim, S. Y. (1998). Mousterian large-mammal remains from Kobeh Cave: Behavioral implications for Neanderthals and early modern humans. *Current Anthropology, 39* (supplement), 79–113.

Marreiros, J., Bicho, N., Gibaja, J., Cascalheira, J., & Pereira, T. (2016). Early Gravettian projectile technology in Southwestern Iberian Peninsula: The double backed and bipointed bladelets of Vale Boi (Portugal). In R. Iovita & K. Sano (Eds.), *Multidisciplinary approaches to the study of Stone Age weaponry* (pp. 147–158). Dordrecht: Springer.

McBrearty, S., & Brooks, A. S. (2000). The revolution that wasn't: A new interpretation of the origin of modern human behavior. *Journal of Human Evolution, 39*, 453–563.

Midoshima, T. (1991). Collisional flaking of the stone arrowhead and tanged point. *Kodai, 92*, 79–97.

Midoshima, T. (1996). Projectile experiments with backed points. *Kanagawa Koko, 32*, 79–97.

Milks, A., Dinnis, R., & Pope, M. (2016). Morphometric variability of Early Gravettian tanged "Font-Robert" points, and functional implications. In R. Iovita & K. Sano (Eds.), *Multidisciplinary approaches to the study of Stone Age weaponry* (pp. 135–146). Dordrecht: Springer.

Moss, E. H., & Newcomer, M. H. (1982). Reconstruction of tool use at Pincevent: Microwear and experiments in Tailler! Pourquoi faire: Préhistoire et technologie lithique II, Recent progress in microwear studies. *Studia Praehistorica Belgica, 2*, 289–312.

Münzel, S. C., & Conard, N. J. (2004). Cave bear hunting in the Hohle Fels, a cave site in the Ach Valley, Swabian Jura. *Revue de Paléobiologie, Genève, 23*, 877–885.

Newman, K., & Moore, M. W. (2013). Ballistically anomalous stone projectile points in Australia. *Journal of Archaeological Science, 40*, 2614–2620.

Nitecki, M. H., & Nitecki, D. V. (1987). *The evolution of human hunting*. Springer.

Noe-Nygaard, N. (1974). Mesolithic hunting in Denmark illustrated by bone injuries caused by human weapons. *Journal of Archaeological Science, 1*, 217–248.

O'Connell, J. F. (1988). Hadza scavenging: Implications for Plio/Pleistocene hominid subsistence. *Current Anthropology, 29*, 356–363.

Odell, G. H., & Cowan, F. (1986). Experiments with spears and arrows on animal targets. *Journal of Field Archaeology, 13*, 195–212.

Odell, G. H., & Odell-Vereecken, F. (1980). Verifying the relationships of lithic use wear assessments by "blind tests": The low-power approach. *Journal of Field Archaeology, 7*, 87–120.

Oswalt, W. H. (1976). *An anthropological analysis of food-getting technology*. New York: John Wiley & Sons.

Peterkin, G. L., Bricker, H. M., & Mellars, P. (1993). Hunting and animal exploitation in the later Palaeolithic and Mesolithic of Eurasia. *Archaeological papers of the American Anthropological Association*.

Pétillon, J.-M., Plisson, H., & Cattelain, P. (2016). Thirty years of experimental research on the breakage patterns of Stone Age osseous points. Overview, methodological problems and current perspectives. In R. Iovita & K. Sano (Eds.), *Multidisciplinary approaches to the study of Stone Age weaponry* (pp. 47–63). Dordrecht: Springer.

Pruetz, J. D., & Bertolani, P. (2007). Savannah chimpanzees, Pan troglodytes verus, hunt with tools. *Current Biology, 17*, 412–417.

Rabinovich, R., Gaudzinski-Windheuser, S., Kindler, L., & Goren-Inbar, N. (2011). *The Acheulian site of Gesher Benot Ya'aqov* (Vol. III). Berlin: Springer.

Rios-Garaizar, J. (2016). Experimental and archeological observations of Northern Iberian Peninsula Middle Paleolithic Mousterian point assemblages. Testing the potential use of throwing spears among Neanderthals. In R. Iovita & K. Sano (Eds.), *Multidisciplinary approaches to the study of Stone Age weaponry* (pp. 213–225). Dordrecht: Springer.

Rots, V. (2009). The functional analysis of the Mousterian and Micoquian assemblages of Sesselfelsgrotte, Germany: Aspects of tool use and hafting in the European late Middle Palaeolithic. *Quartär, 56*, 37–66.

Rots, V. (2013). Insights into early Middle Palaeolithic tool use and hafting in western Europe. The functional analysis of level IIa of the early Middle Palaeolithic site of Biache Saint-Vaast (France). *Journal of Archaeological Science, 40*, 497–506.

Rots, V. (2016). Projectiles and hafting technology. In R. Iovita & K. Sano (Eds.), *Multidisciplinary approaches to the study of Stone Age weaponry* (pp. 167–185). Dordrecht: Springer.

Rust, A. (1943). *Die alt- und mittelsteinzeitlichen Funde von Stellmoor*. Neumünster: Karl Wachholtz Verlag.

Sano, K., Denda, Y., & Oba, M. (2016). Experiments in fracture patterns and impact velocity with replica hunting weapons from Japan. In R. Iovita & K. Sano (Eds.), *Multidisciplinary approaches to the study of Stone Age weaponry* (pp. 29–46). Dordrecht: Springer.

Salem, P. E., & Churchill, S. E. (2016). Penetration, tissue damage, and lethality of wood- *versus* lithic-tipped projectiles. In R. Iovita & K. Sano (Eds.), *Multidisciplinary approaches to the study of Stone Age weaponry* (pp. 203–212). Dordrecht: Springer.

Schiffer, M. B. (1987). *Formation processes of the archaeological record*. Albuquerque: University of New Mexico Press.

Semenov, S. A. (1964). *Prehistoric technology: An experimental study of the oldest tools and artefacts from traces of manufacture and wear*. London: Cory, Adams, & Mackey.

Shea, J. J. (1988). Spear points from the Middle Paleolithic of the Levant. *Journal of Field Archaeology, 15*, 441–450.

Shea, J. J., & Sisk, M. L. (2010). Complex projectile technology and *Homo sapiens* dispersal into western Eurasia. *PaleoAnthropology, 2010*, 100–122.

Shott, M. J. (1996). An exegesis of the curation concept. *Journal of Anthropological Research, 52*, 259–280.

Shott, M. J. (2016). Survivorship distribution in experimental spear points: Implications for tool design and assemblage formation. In R. Iovita & K. Sano (Eds.), *Multidisciplinary approaches to the study of Stone Age weaponry* (pp. 245–258). Dordrecht: Springer.

Sisk, M. L., & Shea, J. J. (2011). The African origin of complex projectile technology: An analysis using tip cross-sectional area and perimeter. *International Journal of Evolutionary Biology, 2011*, 1–8. doi:10.4061/2011/968012.

Stiner, M. C. (1994). *Honor among thieves: A zooarchaeological study of neandertal ecology*. Princeton, NJ: Princeton University Press.

Stoczkowski, W. (2002). *Explaining human origins*. Newcastle upon Tyne: Cambridge University Press.

Thieme, H. (1997). Lower Palaeolithic hunting spears from Germany. *Nature, 385*, 807–810.

Tringham, R., Cooper, G., Odell, G. H., Voytek, B., & Whitman, A. (1974). Experimentation in the formation of edge damage: A new approach to lithic analysis. *Journal of Field Archaeology, 1*, 171–196.

Troels-Smith, J. (1959). En elmetreaes bue fra aamosen og andre traesager fra tidlig-neolitisk tid. *Aarboger for nordisk*, 122–146.

Tsirk, A. (1979). Regarding fracture initiations. In B. Hayden (Ed.), *Lithic use-wear analysis* (pp. 83–96). New York: Academic Press.

Villa, P., Boscato, P., Ranaldo, F., & Ronchitelli, A. (2009). Stone tools for the hunt: Points with impact scars from a Middle Paleolithic site in southern Italy. *Journal of Archaeological Science, 36*, 850–859.

Waguespack, N. M., Surovell, T. A., Denoyer, A., Dallow, A., Savage, A., Hyneman, J., et al. (2009). Making a point: Wood versus stone-tipped projectiles. *Antiquity, 83*, 786–800.

Wannenburgh, A., Johnson, P., & Bannister, A. (1999). *The Bushmen*. Cape Town: Struik.

Whittaker, J. C. (2016). Levers, not springs: How a spearthrower works and why it matters. In R. Iovita & K. Sano (Eds.), *Multidisciplinary approaches to the study of Stone Age weaponry* (pp. 65–74). Dordrecht: Springer.

Wiessner, P. (1983). Style and social information in Kalahari San projectile points. *American Antiquity, 48*, 253–276.

Wilkins, J., & Schoville, B. J. (2016). Edge damage on 500-thousand-year-old spear tips from Kathu Pan 1, South Africa: The combined effects of spear use and taphonomic processes. In R. Iovita & K. Sano (Eds.), *Multidisciplinary approaches to the study of Stone Age weaponry* (pp. 101–117). Dordrecht: Springer.

Wilkins, J., Schoville, B. J., Brown, K. S., & Chazan, M. (2012). Evidence for early hafted hunting technology. *Science, 338*, 942–946.

Yaroshevich, A., Zaidner, Y., & Weinstein-Evron, M. (2016). Projectile damage and point morphometry at the Early Middle Paleolithic Misliya Cave, Mount Carmel (Isral): Preliminary results and interpretations. In R. Iovita & K. Sano (Eds.), *Multidisciplinary approaches to the study of Stone Age weaponry* (pp. 119–134). Dordrecht: Springer.

Yaroshevich, A., Kaufman, D., Nuzhnyy, D., Bar-Yosef, O., & Weinstein-Evron, M. (2010). Design and performance of microlith implemented projectiles during the Middle and the Late Epipaleolithic of the Levant: Experimental and archaeological evidence. *Journal of Archaeological Science, 37*, 368–388.

Index

Note Page numbers followed by *f*, *t* and *n* indicate figures, tables and footnotes respectively

Radu Iovita and Katsuhiro Sano (eds.), *Multidisciplinary Approaches to the Study of Stone Age Weaponry*, Vertebrate Paleobiology and Paleoanthropology, DOI 10.1007/978-94-017-7602-8

MIX
Papier aus verantwortungsvollen Quellen
Paper from responsible sources
FSC® C105338

If you have any concerns about our products,
you can contact us on
ProductSafety@springernature.com

In case Publisher is established outside the EU,
the EU authorized representative is:
Springer Nature Customer Service Center GmbH
Europaplatz 3, 69115 Heidelberg, Germany

Printed by Libri Plureos GmbH
in Hamburg, Germany